Advances in Hydrogen Production, Storage, and Utilization

Advances in Hydrogen Production, Storage, and Utilization

Guest Editor

Guo-Ming Weng

MDPI

Basel • Beijing • Wuhan • Barcelona • Belgrade • Novi Sad • Cluj • Manchester

Guest Editor
Guo-Ming Weng
School of Materials Science
and Engineering
Shanghai Jiao Tong University
Shanghai
China

Editorial Office
MDPI AG
Grosspeteranlage 5
4052 Basel, Switzerland

This is a reprint of the Special Issue, published open access by the journal *Hydrogen* (ISSN 2673-4141), freely accessible at: https://www.mdpi.com/journal/hydrogen/special_issues/4TZNDI5WXW.

For citation purposes, cite each article independently as indicated on the article page online and as indicated below:

Lastname, A.A.; Lastname, B.B. Article Title. *Journal Name* **Year**, *Volume Number*, Page Range.

ISBN 978-3-7258-6584-0 (Hbk)
ISBN 978-3-7258-6585-7 (PDF)
https://doi.org/10.3390/books978-3-7258-6585-7

Contents

About the Editor

Guo-Ming Weng

Guo-Ming Weng is an Associate Professor at the Center of Hydrogen Science, School of Materials Science and Engineering, Shanghai Jiao Tong University. He earned his Ph.D. in Chemistry from The University of Hong Kong and has completed postdoctoral research at prestigious institutions, including The Chinese University of Hong Kong, Yale University, and New York University. His primary research interests lie in hydrogen science, recycling, and new energy technologies, with a focus on advancing both fundamental science and practical applications in these fields. Dr. Weng has authored or co-authored about 80 high-impact research papers published in top-tier international journals, such as *Energy & Environmental Science, Angewandte Chemie,* and *Advanced Functional Materials.* He has also contributed two book chapters and holds multiple granted invention patents in both the United States and China. Throughout his career, Dr. Weng has successfully led and participated in numerous pioneering international collaborative projects.

Preface

The global shift toward sustainable and decarbonized energy systems has highlighted hydrogen as a key player in reducing carbon emissions while enabling a diverse range of applications across industries, transportation, and power generation. Hydrogen's role as a clean, efficient, and versatile energy carrier is integral to advancing the energy transition and securing a sustainable future.

The Reprint of the Special Issue "Advances in Hydrogen Production, Storage, and Utilization" serves as a comprehensive collection of cutting-edge research that explores significant advancements in the hydrogen value chain. This Special Issue brings together insights into novel hydrogen production pathways, such as electrolysis, photochemical, thermochemical, and biological processes, along with innovations in hydrogen storage technologies. Additionally, it highlights the emerging applications of hydrogen in fuel cells, industrial processes, and integrated energy systems.

The motivation for this Reprint stems from the growing need for interdisciplinary research that bridges materials science, engineering, chemistry, and policy. Hydrogen technologies are not only at the heart of technical innovation but are also essential in addressing critical economic, environmental, and social challenges. This collection aims to foster collaboration and inspire the continued development and deployment of hydrogen technologies at scale.

This Reprint is addressed to researchers, engineers, industry professionals, and policymakers involved in hydrogen-related fields. It seeks to provide a platform for sharing knowledge, exchanging ideas, and accelerating the adoption of hydrogen solutions that are crucial to achieving global climate and energy goals.

We hope this Reprint will contribute meaningfully to the scientific community, encouraging further exploration and advancement in hydrogen technologies and their applications.

Guo-Ming Weng
Guest Editor

hydrogen

MDPI

Article

Theoretically Estimated Experimental Adiabatic Electron Affinities of Hydrogen and Cyclohexane Molecules

Igor Y. Shchapin [1,2,*] and Andrey I. Nekhaev [3]

1 Scientific Research Institute of Space Device Engineering, Aviamotornaya Street 53, 111024 Moscow, Russia
2 Department of High Energy Chemistry and Radioecology, D.I. Mendeleev University of Chemical Technology of Russia, Miusskaya Ploshchad 9, 125047 Moscow, Russia
3 A.V. Topchiev Institute of Petrochemical Synthesis, Russian Academy of Sciences, Leninsky Prospect 29, 119991 Moscow, Russia; nekhaev.andrey@yandex.ru
* Correspondence: shchapin@yandex.ru

Abstract

Molecular hydrogen is the basis of hydrogen energy. It is formed and used in many fields of industry, physics, and chemistry. Molecular hydrogen is the main product formed during the gamma radiolysis of liquid cyclohexane. When studying the mechanism of molecular hydrogen formation during the gamma radiolysis of liquid cyclohexane, we found that the values of adiabatic electron affinity, one of the fundamental characteristics of atoms and molecules, had not yet been experimentally determined for hydrogen and cyclohexane molecules. Theoretical estimates of the adiabatic electron affinity of the hydrogen molecule made by other authors varied widely ([−0.3; −5.771] eV) and could not be compared with experimental values due to the absence of such data. Using DFT calculations at the PBE0/TZVPP level of theory, and a constructed correlation with experimental values of the adiabatic first ionization potential and electron affinity for a number of molecules, neutral radicals, and atoms, we estimated, for the first time, the experimental adiabatic electron affinities of hydrogen (−3.08 eV) and cyclohexane (−2.13 eV) molecules in the gas phase. When an electron is attached to a cyclohexane molecule, a cyclohexane radical anion is formed, a new, highly reactive species that has not been studied before. A new perspective on molecular hydrogen formation during the gamma radiolysis of liquid cyclohexane was introduced and discussed.

Keywords: molecular hydrogen; cyclohexane; gamma-rays; DFT calculations; ionization potential; electron affinity

1. Introduction

The importance of hydrogen in the structure of the universe, our solar system, and, in particular, the Earth, all living creatures on it, and the organization of organic and inorganic matter cannot be overestimated [1–4]. Hydrogen is the most common element in the universe, accounting for about 90% of all atoms. It is the main component of stars and interstellar gas. The role of hydrogen on Earth is determined by its abundance: hydrogen atoms make up about 17% of all atoms (second only to oxygen). Therefore, hydrogen plays a crucial role in the chemical processes occurring on Earth. It is a component of almost all organic substances and is present in all living cells, where it accounts for nearly 63% of the total number of atoms. Hydrogen also participates in geochemical processes: it is found in volcanic gases, flows along faults in rifting zones, and is released in some coal deposits.

Most hydrogen is used to produce ammonia, methanol, and hydrogen chloride. In the oil refining industry, hydrogen is used in hydroprocessing to remove sulfur from fuels. High-purity hydrogen is used in semiconductor manufacturing for cleaning wafers, etching, alloying, and growing semiconductor films. In metallurgy, hydrogen is used to extract metals from their ores and as a shielding atmosphere during heat treatment.

In the food industry, hydrogen is used for the hydrogenation of fats and oils; to improve the appearance (glazing) of confectionery products; and as a protective environment for food packaging, which helps prevent oxidation and spoilage, thereby extending shelf life. The antioxidant and anti-inflammatory properties of hydrogen allow it to be used in medicine to reduce organ and tissue damage in diseases and conditions associated with oxidative stress.

Liquid hydrogen is used as a fuel in rocket and space technology and also serves as a refrigerant, while gaseous hydrogen acts as a carrier gas in gas chromatography.

The above list shows that, today, humanity primarily focuses on the numerous practical applications of hydrogen.

In many natural reactions and practical applications of hydrogen, single electron transfer (SET) processes occur. Two characteristics of the hydrogen molecule are fundamental to describing the energy physics of such processes: ionization potential (IP) and electron affinity (EA).

The first characteristic has been reliably established through various experimental and theoretical methods, while the second has not yet been experimentally determined, and the available theoretical data are inconsistent.

In this paper, this gap is addressed through a theoretical study aimed at determining the EA of a cyclohexane molecule.

Molecular hydrogen is the main end product of the gamma radiolysis of liquid cyclohexane [5–9]. The EA values of hydrogen and cyclohexane molecules were theoretically determined in order to clarify the energy physics and mechanism of the early stages of gamma radiolysis of liquid cyclohexane, which belong to the SET class of processes.

In this paper, we used data on gamma-^{60}Co radiolysis of liquid cyclohexane [5–9]. Gamma rays from ^{60}Co were represented by equal amounts of rays with energies of 1.1732 and 1.3325 MeV, with an average ray energy of $E_{av} \approx 1.25$ MeV [10].

Upon gamma irradiation of liquid cyclohexane (**1**), molecular hydrogen (H_2) (**2**), cyclohexene ($c\text{-}C_6H_{10}$) (**2′**), and dicyclohexyl ($(c\text{-}C_6H_{11})_2$) (**2″**) are formed as the main final products (FP) (reactions (1)–(9)) [5–9].

$$c\text{-}C_6H_{12} + \gamma\text{-rays} \rightarrow k\{c\text{-}C_6H_{12}{}^{\bullet+} + e^-{}_{(E \approx 20\text{–}100\ eV)}\} + l\{c\text{-}C_6H_{12}{}^{*(S\ and\ T)}\} \quad (1)$$

$$e^-{}_{(E \approx 20\text{–}100\ eV)} + m(c\text{-}C_6H_{12}) \rightarrow m(c\text{-}C_6H_{12}{}^{\bullet+}) + (m+1)e^-{}_{(E < IP \approx 10\ eV)} \quad (2)$$

$$e^-{}_{(E < IP \approx 10\ eV)} \rightarrow e^-{}_{(E(kT) \approx 0.025\ eV)} \quad (3)$$

$$c\text{-}C_6H_{12}{}^{\bullet+} + e^-{}_{(E(kT) \approx 0.025\ eV)} \rightarrow c\text{-}C_6H_{12}{}^{*(S\ and\ T)} \quad (4)$$

$$c\text{-}C_6H_{12}{}^{*(S)} \rightarrow H_2 + c\text{-}C_6H_{10} \quad (5)$$

$$c\text{-}C_6H_{12}{}^{*(T)} \rightarrow H^\bullet + c\text{-}C_6H_{11}{}^\bullet \quad (6)$$

$$H^\bullet + c\text{-}C_6H_{12} \rightarrow H_2 + c\text{-}C_6H_{11}{}^\bullet \quad (7)$$

$$2\ c\text{-}C_6H_{11}{}^\bullet \rightarrow c\text{-}C_6H_{10} + c\text{-}C_6H_{12} \quad (8)$$

$$2\ c\text{-}C_6H_{11}{}^\bullet \rightarrow (c\text{-}C_6H_{11})_2 \quad (9)$$

Reactions (1)–(9) are a simplified general scheme of gamma radiolysis **1**, which shows how gamma ray energy is absorbed (reaction (1)), distributed (reactions (2) and (3)), and transformed into final products **2** (reactions (4)–(7)), **2′**, and **2″** (reactions (6)–(9)) over characteristic times $t \approx 10^{-15}$–10^{-14} s (the physical stage), $t \approx 10^{-12}$ s (the physical and chemical stage), and $t \approx 10^{-10}$–10^{-6} s (the chemical stage), respectively.

The absorption of gamma ray energy occurs in portions, which form a set of cyclohexane radical cations (RCs) ($c\text{-}C_6H_{12}{}^{\bullet+}$) and high-energy free electrons (FEs) ($e^-{}_{(E \approx 20\text{–}100\ eV)}$). It is believed that after the energy is spent on the ionization of cyclohexane molecules (reaction (2)) and the electrons reach thermal energy ($e^-{}_{(E(kT) \approx 0.025\ eV)}$) (reaction (3)), charge recombination occurs (reaction (4)), which forms the main number of or all (then l = 0 in Equation (1)) electron-excited molecules (EEMs) ($c\text{-}C_6H_{12}{}^*$ (S and T are singlet and triplet EEM, respectively))—precursors of molecular hydrogen (reactions (5)–(7)) [5–9].

The vertical and adiabatic IPs (VIP and AIP) of cyclohexane are 10.3 ± 0.2 eV [11] and 9.88 ± 0.02 eV [11,12], respectively. The same values (10.3 and 9.88 eV) were reported in [13,14], and a more accurate value ($79{,}720 \pm 10$ cm^{-1} (9.8840 ± 0.0012 eV)) is reported in [15].

Therefore, in reactions (2) and (3), using the condition $E < IP \approx 10$ eV, emphasis is placed on the fact that when the kinetic energy of an FE decreases to less than ≈ 10 eV, ionization of cyclohexane molecules becomes impossible, and other mechanisms of energy absorption must be considered.

More than 85 years ago, a theory was proposed in which, after the formation of an ion pair, the kinetic energy of an FE is continuously reduced to thermal energy ($E(kT) \approx 0.025$ eV) due to Coulomb interaction with a complementary cation (radical cation) (reaction (3)) [16]. For illustrative purposes, this theoretical model is still used today [8,17].

According to modern data, instead of continuous energy absorption (reaction (3)), two new channels of partial FE energy absorption should be considered.

First, there is reaction (3′), in which electronically excited molecules $c\text{-}C_6H_{12}{}^{*(S\ and\ T)}$ are formed. In liquid cyclohexane, electronically excited molecules in the ground electronic states are formed upon the absorption of energy $E \approx 7$ eV, since it is known that the wavelength of the absorption edge $\lambda_{\ (absorption\ edge)}$ = 177.5 nm (energy 6.99 eV) [18]. This value of $\lambda_{\ (absorption\ edge)}$ increases to 180 nm (energy 6.89 eV) [19] with an increase in the amount of impurities [20].

Second, this is reaction (3″), in which cyclohexane radical anions (RAs) ($c\text{-}C_6H_{12}{}^{\bullet-}$) are formed, since it is known that cyclohexane molecules, like many other hydrocarbon molecules, have a negative EA [5–8].

A negative EA means that, in order for an FE to attach to a cyclohexane molecule, energy must be expended (i.e., has a positive value), numerically equal to the EA of the cyclohexane molecule (which has a negative value), taken with the opposite sign (see Section 2 and the graphical abstract). This energy can be provided by an FE with a kinetic energy of $E \approx 10 - 7 = 3$ eV (reaction (3″)).

If reactions (3′) and (3″) occur, they replace reaction (3), which ceases to be the main reaction in the general scheme of cyclohexane gamma radiolysis.

This means that in gamma-irradiated cyclohexane, charge recombination (charge neutralization) should occur not according to reaction (4), but according to reaction (4′), in which RCs and RAs of cyclohexane participate.

$$e^-{}_{(E < IP \approx 10\ eV)} + c\text{-}C_6H_{12} \rightarrow c\text{-}C_6H_{12}{}^{*(S\ and\ T)}{}_{(E \approx 7\ eV)} + e^-{}_{(E \approx 3\ eV)} \quad (3')$$

$$e^-{}_{(E \approx 3\ eV)} + c\text{-}C_6H_{12} \rightarrow c\text{-}C_6H_{12}{}^{\bullet-} + E \approx 3\ eV - (\text{-}EA) \quad (3'')$$

$$c\text{-}C_6H_{12}{}^{\bullet+} + c\text{-}C_6H_{12}{}^{\bullet-} \rightarrow c\text{-}C_6H_{12}{}^{*(S\ and\ T)} + c\text{-}C_6H_{12} \quad (4')$$

For reaction (3″) to occur, the following condition must be met: E = 3 eV − (-EA) ≥ 0, i.e., EA(**1**) ≥ −3 eV. Only in this case will the kinetic energy of the electron E = 3 eV be sufficient for the FE to attach itself to the cyclohexane molecule.

If we were to rely on the single experimental value EA(**1**) = −4.11 eV [21], since −4.11 eV < −3 eV, we would conclude that reaction (3″) cannot occur.

For this reason, it was very important to estimate the experimental EA of the cyclohexane molecule in the gas phase as accurately and reliably as possible and to compare it with the experimental value of −4.11 eV [21]. Theoretically and experimentally, this value (−4.11 eV [21]) has not been reverified.

To this end, we performed DFT calculations for a number of chemical structures, correlated them with experimental data (see Section 2), obtained a reliable estimate of the experimental value EA(**1**) = −2.13 eV > −3 eV, and showed that the value −4.11 eV [21] is an outlier in the established correlation (see Section 3).

Thus, the findings of this paper support the conclusion that reaction (3″) can and should occur during the gamma radiolysis of liquid cyclohexane.

2. Materials and Methods

Orca, an ab initio, DFT, and the semiempirical SCF-MO package, version 3.0.1 was used for all DFT calculations at the PBE0/TZVPP level of theory [22,23]. The TD-DFT method was used for calculations of the first single state of excited molecules [18,19]. The libint2 library for the computation of 2-el integrals [24], the basis Ahlrichs-TZV [25], and the Ahlrichs (2df, 2pd) polarization functions from the TurboMole basis set library [26] were utilized for all our calculations.

Since the IP and EA of molecules, radicals, and atoms represent the energetic cost of transferring one electron from or to the initial structure, respectively, the IP and EA values were calculated based on the energy difference (ΔE) between the formation energies of the initial and final structures, taking into account the sign change of ΔE depending on the direction of electron transfer.

The RHF method was used to calculate structures with closed electron shells (even-electron systems: molecules and ions), while the UHF method was applied to structures with open electron shells (odd-electron systems: radical ions and radicals).

To calculate the adiabatic IP (AIP) and adiabatic EA (AEA) values, full geometry optimization was performed for both the initial and final structures. For these structures, vibrational frequencies were calculated to confirm that the global minimum on the potential energy surface had been found.

To determine the vertical IP (VIP) and vertical EA (VEA) values, the formation energies of the final structures were calculated at the fixed geometries of the optimized initial structures (single-point calculations).

For a theoretical evaluation of the experimental AEA values of hydrogen and cyclohexane molecules, independent of the chosen computational method, a correlation was constructed between the calculated and experimental AIP and AEA values of specially selected molecules, radicals, and atoms.

Since the goal was to calculate the lowest possible values of EA(**1**) and EA(**2**), the correlation was based on reference and calculated AIP and AEA values, rather than the typically slightly higher VIP and VEA values.

The ChemCraft 1.7 program was used to create input files and visualize and design the calculation results [27]. All the calculations were carried out on personal computers: I (Intel(R) Core(TM) i7 CPU X 980 Processor, 6 cores to run 12 threads, 3.33 GHz, 24 GB RAM (16 GB available), x64 processor) and II (AMD Ryzen Threadripper 2990WX Processor, 32-cores to run 64 threads, 3.00 GHz, 128 GB RAM, x64 processor).

The calculation results are given in the text of the article and in the "Supplementary Materials" file.

3. Results and Discussion

3.1. Estimation by DFT of Experimental Values of the Adiabatic Electron Affinity of Hydrogen and Cyclohexane Molecules

The hydrogen molecule sets the range of experimental characteristics under consideration: AIP(H_2) = 15.43 eV and AEA(H_2) = −2.66 eV represent the highest and lowest values, respectively (Figure 1, Tables 1 and 2 [28–50]). As an experimental value in our work, we used the value AEA(H_2) = −2.66 eV, determined by us from the data presented in Figure 1 of the theoretical study [38]. This choice requires additional explanation.

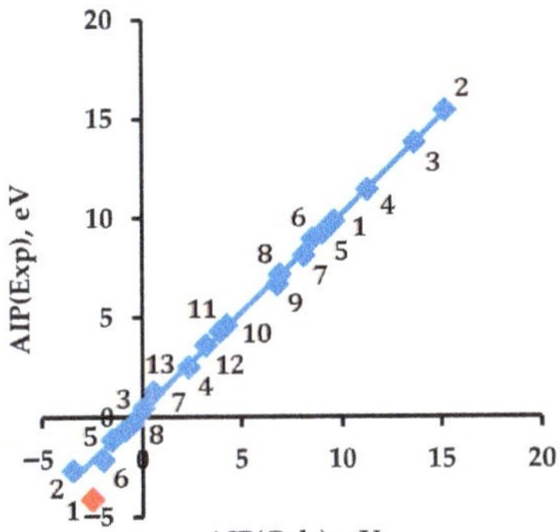

Figure 1. Correlation between DFT-calculated and experimental values of AIP(Mol or Rad) and AIP(RA or An) = AEA(Mol or Rad): AIP(Exp) = 0.9874·AIP(DFT) + 0.2702, R^2 = 0.9984. Different points with the same number refer to different physical characteristics (AIP and AEA) of one substance (see Tables 1 and 2). The red-marked point (−2.43; −4.11), which refers to RA **1**, is an outlier from the correlation.

Table 1. Experimental and DFT-calculated adiabatic ionization potential (AIP, eV) and DFT calculated energies (E, E_h) of molecules (Mol) or radicals and radical cations (RC) or cations *.

Formula	No.	AIP, eV			E (Mol or Radical)	E (RC or Cation)
		Exp.	DFT	DFT′		
c-C_6H_{12}	1	9.88 ± 0.02 [14]	9.66	9.81	−235.671105366798	−235.316052419430
H_2	2	15.43 ** [28,29]	15.22	15.30	−1.168486584480	−0.609131622748
CO_2	3	13.778 ± 0.002 [30]	13.73	13.83	−188.466575767139	−187.962068024449
Cl_2	4	11.481 ± 0.003 [31]	11.34	11.47	−920.085323334828	−919.668654767299
C_6H_6	5	9.24 *** [32,33]	9.13	9.29	−232.047538110431	−231.712153897878
c-C_6H_{10}	6	8.94 ± 0.01 [14,34]	8.6	8.76	−234.442623563656	−234.126538720488
$CH_2CHCH_2^•$	7	8.13 ± 0.02 [35]	8.13	8.30	−117.153855236725	−116.855265113306
c-$C_6H_{11}^•$	8	7.15 ± 0.04 [36]	6.98	7.16	−235.008340307973	−234.751829087788
$C_9H_{18}NO^•$	9	6.73 [37]	6.88	7.06	−483.328660873514	−483.075698854896

* DFT-calculated AIP = (E(RC or Cation) − E(Mol or Radical))·27.2113834. DFT′ = A·DFT + B. The coefficients A = 0.9874 and B = 0.2702 were used. ** 15.425942 ± 0.00001 [28] and 15.425932 ± 0.000002 [29]. *** 9.24372 ± 0.00005 [32] and 9.24384 ± 0.00006 [33].

Table 2. Experimental and DFT calculated adiabatic electron affinities (AEA, eV) and DFT-calculated energies (E, E_h) of molecules (Mol) or radicals and radical anions (RA) or anions *.

Formula	No.	AEA, eV			E (Mol or Radical)	E (RA or Anion)
		Exp.	DFT	DFT′		
c-C_6H_{12}	1	−4.11 [21]	−2.43	−2.13	−235.671105366798	−235.581799233619
H_2	2	−2.66 ** [38]	−3.39	−3.08	−1.168486584480	−1.043961074553
CO_2	3	−0.60 *** [41]	−0.81	−0.53	−188.466575767139	−188.436770199596
Cl_2	4	2.50 ± 0.20 [42]	2.43	2.67	−920.085323334828	−920.174678186930
C_6H_6	5	−1.12 ± 0.03 [43–45]	−1.48	−1.19	−232.047538110431	−231.993234723623
c-C_6H_{10}	6	−2.13 [21]	−1.86	−1.57	−234.442623563656	−234.374272467846
$CH_2CHCH_2^{\bullet}$	7	0.481 ± 0.008 [46]	0.16	0.43	−117.153855236725	−117.159654973015
c-$C_6H_{11}^{\bullet}$	8	−0.24 ± 0.11 [47]	−0.42	−0.15	−235.008340307973	−234.992920354572
Cu_2Cl_4	10	4.6 ± 0.1 [48]	4.32	4.54	−5120.975980754558	−5121.134650902112
$CuCl_2$	11	4.3 ± 0.1 [48]	4.00	4.22	−2560.458515677480	−2560.605611152270
Cl	12	3.6127 ± 0.0001 [49]	3.27	3.50	−459.996618462930	−460.116771209037
Cu	13	1.235 ± 0.005 [50]	0.67	0.93	−1640.236451203338	−1640.261078219124

* DFT-calculated AEA = (E(Mol or Radical) − E(RA or Anion))·27.2113834. DFT′ = A·DFT + B. The coefficients A = 0.9874 and B = 0.2702 were used. ** Our estimate made from the data of the theoretical work [38] was used as the experimental value of AEA(H_2). An explanation based on the works [39,40] is given in the text. *** −0.599986 eV [41].

To date, the experimental value of AEA(H_2) (and VEA(H_2) too) has not been determined. Such data are absent in reviews [39,40]. Many studies have noted that the direct experimental determination of AEA(H_2) is complicated by the population of excited states of both the H_2 molecule and the negative ion $H_2^{-\bullet}$ [51,52] in combination with the very short lifetime of $H_2^{-\bullet}$ [38,53,54].

Given the importance of estimating the experimental value of AEA(H_2), we analyzed the available theoretical data. The theoretical values of AEA(H_2), either calculated and reported by the authors or estimated by us from their published potential curves of the ground states of the H_2 molecule and the RA $H_2^{-\bullet}$, have a large scatter.

The highest value (−0.3 eV) was reported in [55]. The lowest value (−5.771 eV) was reported in [56]. Other studies reported intermediate values (eV): −0.78, −1.04, −1.82 [57], −1.9 [58], −2.0 [59,60], −2.5 [55], −2.66 [38], −2.79, −3.32, −3.53 [39], −3.26 [61], −3.6 [62]. As a fairly reliable theoretical estimate of the experimental value, we used AEA(H_2) = −2.66 eV [38]. Since this value is close to the middle of the range of the above-listed values, this work was published relatively recently, and Figure 1 from that study, which we used for our estimation, has been reproduced in more modern theoretical publications [63,64].

The experimental value of AEA(H_2) = −3.08 eV (Table 2) determined by us using the constructed correlation (Figure 1) is closest to the theoretical value of AEA(H_2) = −3.26 eV [61] and 0.42 eV less than the theoretical value of −2.66 eV [38], which we initially selected as the theoretical reference for building the correlation.

The values of VEA(H_2) calculated by the DFT method are −3.83 eV (DFT) and −3.51 eV (DFT′), respectively, without and taking into account the correlation with experimental data (E($H_2^{\bullet-}$) = −1.02788541467 Eh (energy of a single point $H_2^{\bullet-}$ on the geometry of the H_2 molecule), DFT′ = A·DFT+B, A = 0.9874 and B = 0.2702).

The following characteristics were calculated for cyclohexane by the DFT method: VEA(c-C_6H_{12}) = −2.11 eV (E(c-$C_6H_{12}^{\bullet-}$) = −235.579469844684 Eh) and AEA(c-C_6H_{12}) = −2.43 eV (Table 2). Using the constructed correlation (DFT′ = A·DFT + B, A = 0.9874 and B = 0.2702), we obtain theoretical estimates of experimental values, which are, respectively, equal to VEA(c-C_6H_{12}) = −1.81 eV and AEA(c-C_6H_{12}) = −2.13 eV (Table 2).

Our correlation (Figure 1) shows that the experimental value of AEA(c-C_6H_{12}) = −4.11 eV [21] is an outlier (marked in red) and has about two times less value than our estimate of the experimental value of AEA(c-C_6H_{12}) = −2.13 eV.

It is this most reliable and accurate value of AEA(c-C_6H_{12}) = −2.13 eV > −3 eV that indicates the possibility of implementing a sequence of reactions (3′), (3″) and (4′) related to the initial stages of gamma radiolysis of liquid cyclohexane.

3.2. The Boundary Orbitals and Chemical Stability of the Cyclohexane Molecule, RA and RC

The composition and energies of the frontier orbitals of the cyclohexane molecule determine the composition and energies of the frontier orbitals, as well as the reactivity, of its radical ions, RC and RA. Due to the presence of a third-order symmetry axis in the cyclohexane molecule (overall D3d symmetry), the highest occupied molecular orbital (HOMO) exhibits twofold degeneracy [65]. According to Koopmans' theorem, for the cyclohexane molecule, the experimental value of VIP = 10.3 ± 0.2 eV [11], taken with a negative sign, determines the HOMO energy, that is, E(HOMO) = -VIP.

The lowest unoccupied molecular orbital (LUMO) of the cyclohexane molecule is the Rydberg (3s-C) molecular orbital [66–70]. The LUMO energy determines the VEA value of the cyclohexane molecule (our estimated value is −1.81 eV). Thus, the experimental distance ΔE = ∣E(HOMO) − E(LUMO)∣ = 8.49 eV determined by VIP = 10.3 eV and VEA = −1.81 eV for the gas phase is 1 eV lower than the value of 9.5 eV calculated using the DFT method for the cyclohexane molecule.

The experimental distance ΔE = ∣E(HOMO) – E(LUMO)∣ = 7.75 eV determined by AIP = 9.88 eV and AEA = −2.13 eV for the gas phase is only 0.76 eV higher than the value of 6.99 eV [18] determined for liquid cyclohexane. If spectral line broadening is taken into account, then in the gas phase, the experimental distance ΔE = ∣E(HOMO) – E(LUMO)∣ will decrease further, bringing it closer to the value determined for liquid cyclohexane.

The structures of the cyclohexane molecule, RC and RA, calculated by the DFT method, correspond to the energy minima, since all vibration frequencies have real positive values (provided in the "Supplementary Materials" file).

The cyclohexane RA is formed by attaching one unpaired electron to a cyclohexane molecule, which occupies its LUMO. Since this Rydberg (3s-C) type MO (LUMO of molecule, Figure 2A) is non-degenerate and has high symmetry, when it is populated with one unpaired electron (the single occupied molecular orbital (SOMO) of RA, Figure 2B), the symmetry (D_{3d}) of the original molecule and the sequence of energy arrangement of its occupied and vacant MO are preserved, and all geometric changes are distributed evenly over all one type of C-C bonds and two types of C-H bonds (Figure 2C).

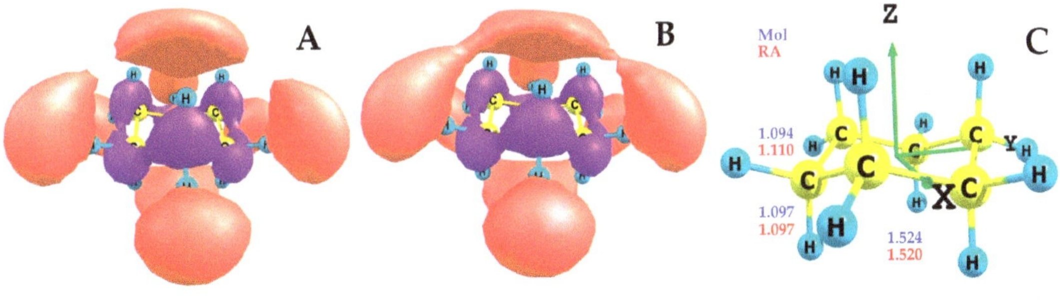

Figure 2. (**A**) LUMO of the c-C_6H_{12} molecule; (**B**) SOMO of c-$C_6H_{12}^{\bullet-}$ RA; (**C**) Axes and DFT-calculated bond lengths of the cyclohexane molecule and RA.

When the cyclohexane RA is formed, the lengths (1.097 Å) of axial C-H bonds remain unchanged, the lengths of C-C bonds are slightly reduced (by 0.004 Å), and the lengths of equatorial C-H bonds increase significantly (by 0.016 Å) (Figure 2C). This indicates that, in accordance with the type of MO, during the transition from the LUMO of the molecule to the SOMO of the RA of cyclohexane, it is the equatorial C-H bonds that are activated. These bonds are weakened and become more susceptible to further chemical transformations associated with the rupture of C-H bonds.

It is convenient to compare the chemical stability of the cyclohexane RA and its reactivity with respect to the rupture of the activated C-H bond (reactions (10) and (11)) with the same characteristics of the homolytic rupture of the C-H bond of the original cyclohexane molecule (reaction (12)) and the hydrogen molecule (reaction (13)).

$$c\text{-}C_6H_{12}^{\bullet -} \rightarrow c\text{-}C_6H_{11}^{\bullet} + H^- \quad (10)$$

$$c\text{-}C_6H_{12}^{\bullet -} \rightarrow c\text{-}C_6H_{11}^{-} + H^{\bullet} \quad (11)$$

$$c\text{-}C_6H_{12} \rightarrow c\text{-}C_6H_{11}^{\bullet} + H^{\bullet} \quad (12)$$

$$H_2 \rightarrow H^{\bullet} + H^{\bullet} \quad (13)$$

According to experimental data, the energies of homolytic rupture of the C-H bond of cyclohexane and the H-H bond of the hydrogen molecule are 4.31 and 4.46 eV, respectively [69,70] (Figure 3, Table 3). According to our experimental data, the formation of a cyclohexane RA from a cyclohexane molecule requires an energy input of 2.13 eV (reaction (3″), Figure 3, Tables 2 and 3).

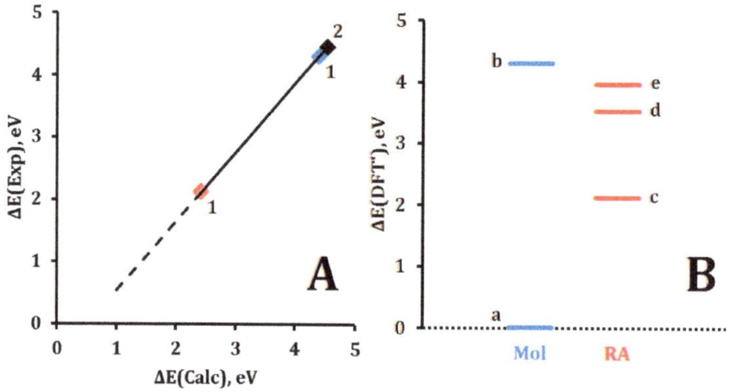

Figure 3. (**A**) Correlation between DFT-calculated and experimental values of ΔE parameters of **1** (Blue), **2** (Black) molecules and RA **1** (Red): ΔE(Exp) = 1.167·ΔE(Calc) − 0.8258, R^2 = 1; (**B**) DFT-calculated with correction values of ΔE parameters: (a) ΔE(DFT′) = 0 corresponds to E(c-C_6H_{12}) = −235.671105366798 Eh, (b) dissociation energy D(C-H) of c-C_6H_{12}, (c) relative energy ΔE(c-$C_6H_{12}^{\bullet -}$) of RA **1** formation, (d) and (e) dissociation energy D(C-H) of c-$C_6H_{12}^{\bullet -}$ plus ΔE(c-$C_6H_{12}^{\bullet -}$) for reactions (10) and (11), correspondently (see Table 3).

As described earlier, this energy corresponds to the negative value AEA(**1**) = −2.13 eV which makes it possible to plot both theoretically calculated and experimental values of AEA and AIP on a single correlation curve (Figures 1 and 3A). At the same time, to compare the dissociation energies of the molecule and the RA of cyclohexane taking into account AEA(**1**), it is necessary to use the opposite sign value ΔE(c-$C_6H_{12}^{\bullet -}$) = −AEA(**1**). This reflects a shift from an energy accounting scheme based on the direction of SET to a unified scheme used the formation energy of molecule **1** as a single initial structure (Figure 3B).

This adjustment is also necessary because, by definition, positive values of EA and IP correspond to opposite energy effects: energy release and energy absorption, respectively.

Table 3. Experimental and DFT-calculated ΔE parameters without and with corrections, eV *.

Reaction No.	ΔE Parameter	Exp.	DFT	DFT′ = A·DFT + B
(3″)	$\Delta E(c\text{-}C_6H_{12}^{\bullet -}) = -AEA(\mathbf{1})$	2.13 [**]	2.43	2.13
(10)	D(C-H) of $c\text{-}C_6H_{12}^{\bullet -}$	–	2.02	1.68
(11)	D(C-H) of $c\text{-}C_6H_{12}^{\bullet -}$	–	2.39	2.09
(12)	D(C-H) of $c\text{-}C_6H_{12}$	4.31 ± 0.05 [69]	4.40	4.31
(13)	D(H-H) of H_2	4.46 *** [70]	4.53	4.46

* DFT-calculated ΔE = (E(Products) − E(Initial substances))·27.2113834. $E(H^{\bullet}) = -0.501036722759$ Eh, $E(H^{-}) = -0.499283610882$ Eh, $E(H_2)$, $E(c\text{-}C_6H_{12})$, $E(c\text{-}C_6H_{11}^{\bullet})$, $E(c\text{-}C_6H_{11}^{-})$ and $E(c\text{-}C_6H_{12}^{\bullet -})$, see Table 2. The coefficients A = 1.1082 and B = −0.5632 (Figure 3) were used. ** This work. *** 35,999.582834 (11) cm^{-1} [70].

The dissociation energy values D(C-H) of $c\text{-}C_6H_{12}^{\bullet -}$ for reactions (10) and (11) have not been determined experimentally. Using DFT calculations and a correlation with experimental data, we made theoretical estimates (DFT′) of these values (Figure 3, Table 3).

The smallest value (1.68 eV) was obtained for D(C-H) $c\text{-}C_6H_{12}^{\bullet -}$ in reaction (10). This value is 2.57 times lower than D(C-H) $c\text{-}C_6H_{12}$ in reaction (12). If the values AEA = 2.13 eV and D(C-H) = 1.68 eV are used for RA **1**, then their total energy E = AEA + D(C-H) = 3.81 eV sets the threshold energy range for FEs: when the kinetic energy of an FE falls within this range, RA 1 can form and remain chemically stable. Specifically, for the formation and stability of RA **1**, the FE energy must satisfy the condition: 2.13 eV < E(e−) < 3.81 eV (estimated gas-phase values were used).

This inequality also means that, in the case of reaction (3′), the remaining portion of energy E(e−) = 10 − 7 = 3 eV is sufficient to form RA **1**, but not enough to implement reaction (10). In this sense, during the implementation of reaction (3′), we can speak of the chemical stability of RA **1**, since it does not possess sufficient excess energy to break even one of its C-H bonds.

If we were interested in the chemical stability of RA **1** in a condensed medium, then instead of considering the monomolecular reaction (10), which pertains to the gas phase, we would calculate the energy effect of a similar but bimolecular reaction (10′).

$$c\text{-}C_6H_{12}^{\bullet -} + c\text{-}C_6H_{12} \rightarrow c\text{-}C_6H_{11}^{\bullet} + c\text{-}C_6H_{13}^{-} \quad (10')$$

However, this is not necessary, since the main reaction RA **1** should be the neutralization of charges (reaction (4′)), rather than a 10′ type reaction. This is because the rates of SET reactions in which a lighter particle, a single electron, is transferred, are high and do not require physical contact between the reacting particles. Since reaction (4′) is also the main reaction for consumption of RC **1**, in a condensed medium, the lifetimes of RC **1** and RA **1** should be equal or very close.

For energetic reasons, it can be assumed that in this ion pair, the negative ion RA **1** is chemically more stable and has a longer lifetime than the positive ion RC **1**. This is because the formation of RC **1** requires five times more energy (E ≈ 10 eV) than the formation of RA **1** (E ≈ 2 eV), meaning the chemical bonds in RC **1** are more strongly activated than in RA **1**. Thus, in our opinion, the previously unexplored primary particles of RA **1** exhibit sufficient chemical stability to be included in the general scheme of the gamma radiolysis of cyclohexane.

It is worth recalling that, due to the presence of a third-order symmetry axis in the cyclohexane molecule (general symmetry D_{3d}), the HOMO of the molecule is twofold degenerate. Using the ESR method in combination with the results of theoretical calculations, it was established that with single ionization, the double degeneracy of the HOMO of the cyclohexane molecule was removed and RC **1** was formed in the ground state ^2A_g (the D_{3d} symmetry of the original molecule was reduced to C_{2h}) [71].

The type of SOMO and LUMO of RC **1** calculated by us using the DFT method in the ^2A_g ground state and the geometric changes occurring during single ionization are shown in Figure 4. The SOMO of RC **1** is mainly concentrated on two C-C bonds, which are greatly elongated (by 0.089 Å), and on two codirectional equatorial C-H bonds, which are also greatly elongated (by 0.027Å), become less strong and more reactive (Figure 4A,C). The other four C–C bonds are shortened (by 0.030 Å) and strengthened (Figure 4A,C).

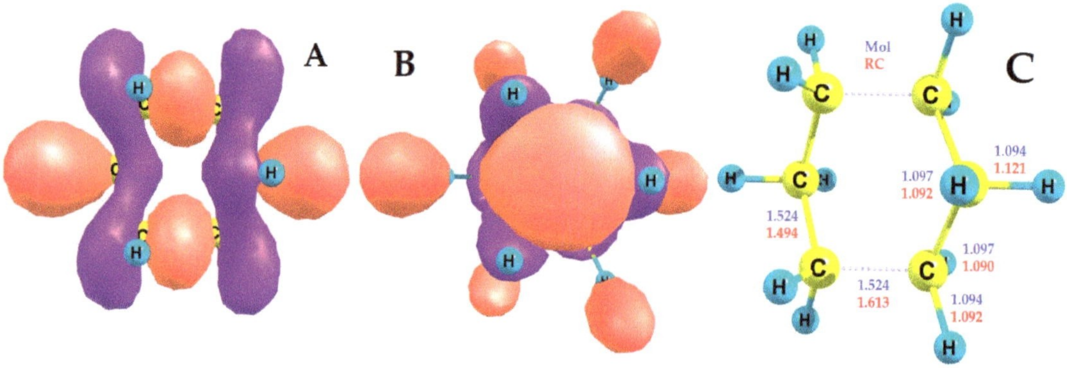

Figure 4. (**A**) SOMO of the $c\text{-}C_6H_{12}{}^{\bullet+}$ RC; (**B**) LUMO of the $c\text{-}C_6H_{12}{}^{\bullet+}$ RC; (**C**) DFT-calculated bond lengths of the cyclohexane molecule and RC.

With such strong geometric changes (Figure 4C), the same type of LUMO remains in RC **1** (Figure 4B) as in the parent molecule (Figure 2A). Figure 4 shows the characteristics of a cyclohexane RC and molecule when viewed along the "-z" axis (see Figure 2C).

A strong elongation of the C-H bond in RC **1** leads to its significant weakening, which in liquid cyclohexane promotes the ion-molecular reaction (14). One of the products of this reaction is the protonated form of the molecule ($c\text{-}C_6H_{13}{}^+$) [72].

$$c\text{-}C_6H_{12}{}^{\bullet+} + c\text{-}C_6H_{12} \rightarrow c\text{-}C_6H_{11}{}^{\bullet} + c\text{-}C_6H_{13}{}^{+} \quad (14)$$

However, it should be noted that reaction (14) is a very fast, assumed (but not experimentally confirmed) reaction. Further studies are needed to assess the contributions of reactions (14) and (10′) to the general mechanism of cyclohexane radiolysis. Depending on the specific research objectives and the accepted model simplifications, these reactions may or may not need to be considered.

In our opinion, the main reaction of cyclohexane radical ions (RC and RA) is the charge neutralization reaction (4′) (electron transfer), which forms EEM. Cyclohexane RC and RA can also participate in reactions (14) and (10′). But in these two reactions, heavier particles are transferred: H^+ and H^-, respectively. Therefore, reactions (14) and (10′) should be slower compared to reaction (4′). The concentration factor acts in the opposite direction: the concentration of molecules **1** is much higher than the concentrations of RC **1** and RA **1**. The definition and comparison of the rates of reactions (4′), (10′) and (14), taking into account the contributions of various factors, is the subject of a separate study.

3.3. The Initial Stage of Gamma Radiolysis of Liquid Cyclohexane, Taking into Account the Formation of RC, EEM, and RA

In the introduction, reactions (1)–(9) were presented, which described the formation of three main end products (**2**, **2′**, **2″**) of gamma radiolysis of liquid cyclohexane, all associated with the breaking of C-H bonds. In the introduction, we did not mention the formation of minor products of gamma radiolysis of liquid cyclohexane associated with C-C bond breaks [5–8]. We also did not describe the photophysical processes responsible for the fluorescence of irradiated systems (cyclohexane without and with additives of special substances) [72–74]. Against the background of chemical transformations, such photophysical processes have a very low intensity and manifest themselves at the final stage of radiation-stimulated processes. The quantum yield of fluorescence is characterized by a wavelength λ (maximum) = 201 nm (energy 6.17 eV) [75] and a quantum yield of $\varphi \approx 10^{-3}$–10^{-2} [74,75].

At the same time, it is known that during gamma radiolysis of liquid cyclohexane, a negligible number of thermalized FEs (0.11 e−/100 eV) is detected [76]. Consequently, thermalized FEs are minor products of gamma radiolysis of liquid cyclohexane, though they are traditionally included in the main reaction scheme (reactions (1)–(9)), particularly in reaction (4).

This approach is based on the long-standing assumption in radiation chemistry that, due to its negative electron affinity (EA), the cyclohexane molecule cannot attach an FE and therefore cannot form RA **1**. As a result, the thermalized FE has traditionally been considered the only negative charge participating in the charge neutralization reaction (reaction (4)).

In our study, we demonstrated for the first time that this widely accepted view is incorrect. RA **1** is a chemically active species that can be formed during gamma radiolysis of liquid cyclohexane and has sufficient chemical stability to be included in the general mechanism.

Moreover, considering that the initial number of thermalized FEs is extremely low and that the concentration of cyclohexane molecules is many orders of magnitude greater than the total concentration of all primary, secondary, and final radiolysis products, the formation of RA **1** via reaction (2′) should occur much more rapidly than the neutralization reaction (4).

Therefore, when describing the mechanism of gamma radiolysis of liquid cyclohexane, reaction (3′), rather than reaction (4), should be considered the primary pathway. Furthermore, the energy E ≈ 3 eV used in reactions (3′) and (3″), more precisely the -EA(**1**), should be defined with greater accuracy. It is worth noting that E ≈ 3 eV is also the arithmetic mean of the previously established range 2.13 eV < E(e−) < 3.81 eV (estimated for the gas-phase), which reflects the energy window required for both the formation and chemical stability of RA **1**. Thus, 3 eV represents the center of this energy range.

If we consider the sequence of gamma ray energy absorption reactions—firstly, reaction (1), in which k = 1 and l = 0 (only ion pairs are formed), and, secondary, reactions (2), (3′), and (3″) (the kinetic energy of FE is consumed)—then the initial stage of the gamma radiolysis of liquid cyclohexane can be written as reaction (1′).

$$c\text{-}C_6H_{12} + \gamma\text{-rays} \rightarrow c\text{-}C_6H_{12}^{\bullet+} + c\text{-}C_6H_{12}^{*(\text{S and T})} + c\text{-}C_6H_{12}^{\bullet-} \quad (1')$$

In reaction (1′), charge separation occurs, resulting in the formation of pairs of cyclohexane ions (RC and RA). This means that upon gamma-ray energy absorption, one electron is transferred from one cyclohexane molecule to another.

During gamma radiolysis, the two cyclohexane molecules involved may be adjacent or separated by some distance. In some RC–RA ion pairs, spin correlation may be preserved; in others, it may not. Cyclohexane EEM is formed from a third cyclohexane molecule.

In reaction (1′), the active particles RC:EEM:RA of cyclohexane are formed in a ratio of 1:1:1. To realize the reaction (1′) in liquid cyclohexane, energy expenditure is required: E(1′, liquid) = IP(liquid) + ΔE(liquid) + (-EA(liquid)), where IP(liquid) = 8.43 ± 0.05 eV [77–79] (IP(solid) = 8.2 ± 0.1 eV [80]), ΔE(liquid) = | E(HOMO) − E(LUMO) | = 6.99 eV [18].

The experimental value of EA(liquid) for cyclohexane is unknown, as no direct measurements have been performed. However, it can be estimated in two ways.

Before we proceed to their practical discussion, it should be clarified that the experimental papers [77–79] reported two close values of IP(liquid **1**) without explanation: 8.43 ± 0.05 eV and 8.75 ± 0.1 eV. Earlier, for illustrative purposes, we used the first value, since it is smaller. Here, we will make numerical estimates; therefore, we will use both experimental IP values (liquid **1**) for comparison [77–79].

The first method is based entirely on experimental data related to reaction (1′).

When 100 eV of gamma ray energy is absorbed by liquid cyclohexane, $\Sigma(H_2)$ = 5.6 hydrogen molecules are formed and the total number of Σ(FP) = 6.4 molecules of all final products (FP) associated with the breaking of C-H bonds (Σ(C-H) = $\Sigma(H_2)$ = 5.6 molecules) and C-C bonds (Σ(C-C) = 0.8 molecules) [1–5].

Thus, the experimental value of E(1′, liquid **1**) = 100/5.6 = 17.86 eV. Then the experimental (adiabatic) value EA(liquid **1**) = − (E(1′, liquid **1**) − ((IP(liquid **1**) + ΔE(liquid **1**))) = − (17.86 − (8.43 + 6.99)) = −2.44 eV, if IP(liquid **1**) = 8.43 eV, and EA(liquid **1**) = − (17.86 − (8.75 + 6.99)) = −2.12 eV, if IP(liquid **1**) = 8.75 eV.

According to the second method, we assume that EA(liquid **1**) = AEA(gas **1**) = −2.13 eV (see Table 2).

In the case IP(liquid **1**) = 8.43 eV, E(1′, liquid **1**) = 8.43 + 6.99 + (−(−2.13)) = 17.58 eV. Therefore, per 100 eV of absorbed gamma ray energy, Σ(PAP) = 100/17.58 = 5.69 PAP is formed, where Σ(PAP) is the total number of primary active particles (PAP) (Σ(PAP) = RC + EEM + RA), which lead to the formation of hydrogen molecules.

In the case IP(liquid **1**) = 8.75 eV, E(1′, liquid **1**) = 8.75 + 6.99 + (−(−2.13)) = 17.87 eV. Therefore, per 100 eV of absorbed gamma ray energy, Σ(PAP) = 100/17.87 = 5.60 PAP is formed, which coincides with the experimental value $\Sigma(H_2)$ = 5.6 molecules/100 eV.

Thus, both values of IP(liquid **1**) yield similar results, but in the case of IP(liquid **1**) = 8.75 eV, both methods of estimating EA(liquid **1**) give virtually identical results. Therefore, it can be assumed that during gamma radiolysis of liquid cyclohexane, an ionization channel corresponding to molecules with IP(liquid **1**) = 8.75 eV is realized.

In this case, all the experimental data related to the 1′ (liquid) reaction agree well with each other: $\Sigma(H_2)$ = 5.6 molecules/100 eV, the ratio RC:EEM:RA = 1:1:1, IP(liquid **1**) = 8.75 eV, ΔE(liquid **1**) = 6.99 eV, EA(liquid **1**) = −2.13 eV.

The result also implies that during the transition from the gas to the liquid phase, IP(**1**) decreases from IP(gas **1**) ≈ 10 eV to IP(liquid **1**) ≈ 9 eV, while the value of EA(**1**) remains unchanged: EA(gas **1**) = AEA(liquid **1**) ≈ −2 eV. This finding is new and requires further interpretation, clarification, and investigation into the reasons behind it.

During the transition from the gas to the liquid and solid phases, the IP of cyclohexane decreases: 9.88 eV [12], 8.43–8.75 eV [77–79], and 8.2 eV [80], respectively. This effect is usually associated with the structuring of liquids and solids within the framework of band theory and ideas about the polarization of molecules in condensed media [77–80]. Within the framework of this approach, attention could be drawn to the fact that during the transition from gaseous to condensed media, the FE does not go to the zero-vacuum level $E(e^-)$ = 0 eV ≈ 0.025 eV (E(kT)), but remains in an ionized condensed medium

(E(e^-) ≈ 2 eV (-AEA of gas and liquid **1**, with work)). This implies the accession of FE to LUMO **1**. This important point has not been previously addressed in the literature.

The main FP of gamma radiolysis of cyclohexane were determined for the gas, liquid, and solid phases [5–9]. Even at a gas density of 0.0047 g/cm^3, the absorption of 100 eV of gamma ray energy leads to the formation of 5.3 H_2 molecules [6,81]. This value, defined for a gas (5.3 H_2 molecules), is close to the value defined for a liquid (5.6 H_2 molecules [5,6,82]). In other experiments, the formation of identical amounts of molecular hydrogen was measured for the gas and solid phases: 4.7 molecules [6,83] and 4.73 molecules, respectively, [6,84]. These facts indicate that the total reaction (1′) of the formation of PAP has a more universal meaning and applies to all three states of aggregation of cyclohexane: gas, liquid and solid.

3.4. Relative Energy Location of the Boundary Orbitals of the Cyclohexane Molecule and Its Primary Active Particles of Gamma Radiolysis

Figure 5 shows the relative energy location of the boundary MOs of the cyclohexane molecule: the doubly degenerate HOMO (MOs **a** and **b**) and the Rydberg (3s-C) type LUMO (MO **c**), as well as all the PAP discussed in the article: RC, RA and EEM in electronic states S and T. The distances between MOs were chosen not from the results of our DFT calculations, but based on experimental data, known general trends, and the convenience of conveying the main ideas in graphical form.

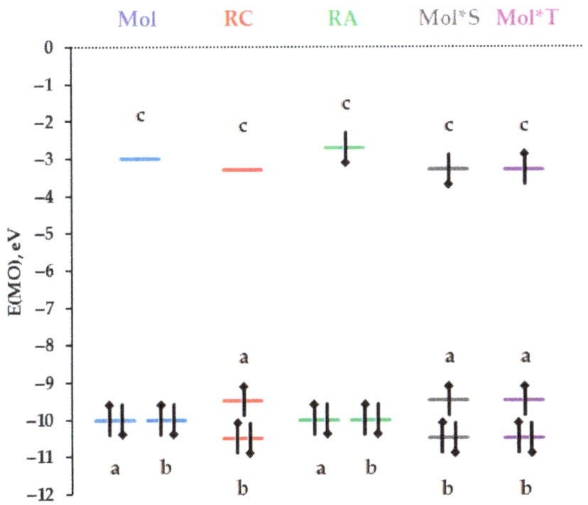

Figure 5. The energy arrangement of the boundary molecular orbitals **a**, **b** (HOMO and SOMO) and **c** (LUMO and SOMO) of the ground electronic states of the molecule (Mol), radical cation (RC) and radical anion (RA) of cyclohexane, as well as the first singlet (Mol*S) and triplet (Mol*T) electronically excited states of the same molecule.

The HOMO energy of a cyclohexane molecule is equal to the first vertical potential, taken with the opposite sign, E(HOMO) = −VIP. In Figure 5, the value E(HOMO) = −10 eV is used instead of the experimental value VIP = 10.3 ± 0.2 eV [11]. The distance between the HOMO and LUMO of the cyclohexane molecule is chosen to be 7 eV [18]—the minimum energy of light quanta at which a cyclohexane molecule begins to absorb ultraviolet light in the liquid phase. For the convenience of conveying the main ideas in graphical form, another simplification has been made in Figure 5, namely, MOs **a** and **b** of RC and EEM are separated by a distance of 1 eV, which is actually unknown.

When one electron is removed from the HOMO Mol **1**, RC **1** is formed, in which, due to the removal of double degeneracy of MOs **a** and **b** and large structural changes, the distance between the boundary MOs—**a** (HOMO RC **1**) and **c** (LUMO RC **1**)—decreases due to reducing binding in MO **a** (MO energy increases) and reducing loosening in MO **c** (MO energy decreases).

In the case of RC **1**, geometric changes (see Figure 4C) are specified by a change in the population of MO **a** from double (in the molecule) to single (in the RC). If we trace how the geometric changes that occurred during single ionization of the cyclohexane molecule affect the effects of binding and loosening in MO **c**, it becomes clear that binding in two C-C bonds decreases, and in four C-C bonds increases. In this case, the loosening will decrease in two activated C-H bonds, and in other C-H bonds, it will remain approximately the same as in the original molecule. Therefore, we assume that MO **c** in RC **1** will decrease in energy somewhat relative to its position in the initial molecule (see Figure 5).

In the case of RA **1**, single electron occupancy of MO **c** produces the expected geometric changes: a slight shortening and strengthening of all C-C bonds and a significantly greater lengthening and weakening of all equatorial C-H bonds (see Figure 2C). The effect of increased loosening in C-H bonds is stronger than the effect of increasing binding in C-C bonds. Therefore, we assume that MO **c** in RA **1** increases in energy somewhat relative to its position in the original molecule (see Figure 5).

In this regard, it should be noted that, although MO **c** is a non-binding Rydberg (3s-C) type MO, it has one nodal surface, which is located between carbon atoms and hydrogen atoms (see Figure 2A (Mol), 4B (RC) and 2B (RA)). According to the type of MO **c** in it, the binding between all carbon atoms is compensated by loosening in all C-H bonds.

It is important that MO **c** occupied by one electron in RA **1** is located somewhat higher in energy than the free (vacant) MO **c** in RC **1**. This is an additional favorable energy factor that promotes the occurrence of reaction (4′)—the transfer of one electron from RA **1** to RC **1** with the participation of the same type and symmetry of MO **c**.

Figure 5 shows a spin-correlated pair of RC **1** and RA **1**, from which, according to reaction (4′), the cyclohexane EEM is formed in the singlet (S) state (Mol*S = $c\text{-}C_6H_{12}{}^{*(S)}$) (see Figure 5 and reactions (1),(4),(5),(1′),(3′),(4′)). If the spin of electron, occupied MO **c** in RA **1**, is changed to the opposite spin, then a pair of RC **1** and RA **1** will be obtained, which, by reaction (4′), will lead to the formation of EEM in the triplet (T) state (Mol*T = $c\text{-}C_6H_{12}{}^{*(T)}$) (see Figure 5 and reactions (1),(4),(6),(1′),(3′),(4′)).

4. Conclusions

1. Using DFT calculations and the constructed correlation with experimental data, it was found that the experimental values of the adiabatic electron affinity (AEA) of cyclohexane (**1**) and hydrogen (**2**) molecules in the gas phase are −2.13 eV (AEA(gas **1**)) and −3.08 eV (AEA(gas **2**)), respectively.
2. Using independent experimental data related to the liquid phase, it was found that the experimental value of the adiabatic electron affinity of the cyclohexane molecule in the liquid phase is −2.12 eV (EA(liquid **1**)).
3. Thus, for the first time, it was found that EA(liquid **1**) = AEA(gas **1**) ≈ -2 eV.
4. Upon gamma irradiation of liquid cyclohexane, all molecular hydrogen is formed at the final stage of the transformation of the energy of ^{60}Co gamma rays (E $\approx$ 1.25 MeV) into the energy of three primary particles—a radical cation (E $\approx$ 9 eV (liquid)), an electronically excited molecule (E $\approx$ 7 eV (liquid)) and a radical anion (E = -AE(liquid) $\approx$ 2 eV)—at a ratio of 1:1:1, with total energy ΣE $\approx$ 18 eV (liquid).
5. Using DFT calculations and the literature to date, it is shown that the cyclohexane radical anion has sufficient chemical stability to be included in the general scheme of gamma radiolysis of gaseous, liquid and solid cyclohexane.

Supplementary Materials: The following supporting information can be downloaded at https://www.mdpi.com/article/10.3390/hydrogen6040115/s1, Table S1: The Charge, Spin state, Energy and XYZ coordinates of all atoms of all structures optimized by the DFT method.

Author Contributions: Conceptualization, I.Y.S.; methodology, I.Y.S.; software, I.Y.S.; validation, A.I.N. and I.Y.S.; formal analysis, A.I.N. and I.Y.S.; investigation (DFT calculations), I.Y.S.; resources, A.I.N. and I.Y.S.; data curation, A.I.N. and I.Y.S.; writing, original draft preparation, I.Y.S.; writing, review and editing, A.I.N. and I.Y.S.; visualization, I.Y.S.; supervision, A.I.N. and I.Y.S.; project administration, A.I.N. and I.Y.S.; funding acquisition, A.I.N. and I.Y.S. All authors have read and agreed to the published version of the manuscript.

Funding: This research received no external funding.

Institutional Review Board Statement: Not applicable.

Informed Consent Statement: Not applicable.

Data Availability Statement: The original contributions presented in this study are included in the article/Supplementary Material. Further inquiries can be directed to the corresponding author.

Acknowledgments: The authors express their sincere gratitude to F. Neese and members of his team for creating the free package of quantum-chemical programs "Orca,—an ab initio, DFT, and the semiempirical SCF-MO package—version 3.0.1" and detailed instructions for its use [22,23], thanks to which this work and the previous one [85] became possible.

Conflicts of Interest: The authors declare no conflicts of interest.

Abbreviations

AIP	Adiabatic Ionization Potential
AEA	Adiabatic Electron Affinity
An	Anion ($^-$)
Cat	Cation ($^+$)
C-H	Carbon-Hydrogen bond
C-C	Carbon- Carbon bond
c-C_6H_{10}	Cyclohexene molecule (**2′**)
c-$C_6H_{11}^{\bullet}$	Cyclohexyl radical
c-C_6H_{12}	Cyclohexane molecule (**1**)
$(c$-$C_6H_{11})_2$	Dicyclohexyl (**2″**)
D	Dissociation energy
DFT	Density Functional Theory
E	Energy
EA	Electron Affinity
EEM	Electronically Excited Molecule (*)
FE	Free Electron (e^-)
FP	Final Product
GGA	Generalized Gradient Approximation
$H^{\bullet}$	Hydrogen radical (atom)
H_2	Hydrogen Molecule (**2**)
HOMO	Highest Occupied Molecular Orbital
IP	Ionization Potential
LUMO	Lowest Unoccupied Molecular Orbital
MO	Molecular Orbital
Mol	Molecule
NFR	Neutral Free Radical
PAP	Primary Active Particle
PBE0	One-parameter hybrid version of PBE

PBE	Perdew-Burke-Erzerhoff GGA functional
PP	Three sets of first polarization functions on all atoms
RA	Radical Anion ($^{\bullet -}$)
Rad	Radical ($^{\bullet}$)
RC	Radical Cation ($^{\bullet +}$)
RHF	Restricted Hartree-Fock
S	Singlet
SCF-MO	Self-Consistent Field-Molecular Orbital
SET	Single Electron Transfer
SOMO	Single Occupied Molecular Orbital
T	Triplet
TZV	Ahlrichs Triple-Zeta Valence basis set
TZVPP	TZV + PP
UHF	Unrestricted Hartree-Fock
VEA	Vertical Electron Affinity
VIP	Vertical Ionization Potentials

References

1. Ishaq, H.; Dincer, I.; Crawford, C. A review on hydrogen production and utilization: Challenges and opportunities. *Int. J. Hydrogen Energy* **2022**, *47*, 26238–26264. [CrossRef]
2. Zainal, B.S.; Ker, P.J.; Mohamed, H.; Ong, H.C.; Fattah, I.M.R.; Rahman, S.M.A.; Nghiem, L.D.; Mahlia, T.M.I. Recent advancement and assessment of green hydrogen production technologies. *Renew. Sustain. Energy Rev.* **2024**, *189 Pt A*, 113941. [CrossRef]
3. Mose, M.P.; Kannaiyan, S.; Huang, S.-J. Hydrogen carriers for hydrogen transport and storage (hydrogen Storage): A review. *Mater. Chem. Phys.* **2025**, *345*, 131252. [CrossRef]
4. Serik, A.; Kuspanov, Z.; Daulbayev, C. Cost-effective strategies and technologies for green hydrogen production. *Renew. Sustain. Energy Rev.* **2026**, *226 Pt A*, 116242. [CrossRef]
5. Ho, S.K.; Freeman, G.R. Radiolysis of Cyclohexane. V. Purified Liquid Cyclohexane and Solutions of Additives. *J. Phys. Chem.* **1964**, *68*, 2189–2197. [CrossRef]
6. Földiak, G. (Ed.) *Radiation Chemistry of Hydrocarbons*; Akadémiai Kiadó: Budapest, Hungary, 1981.
7. LaVerne, J.A.; Pimblott, S.M.; Wojnarovits, L. Diffusion–kinetic modeling of the γ-radiolysis of liquid cycloalkanes. *J. Phys. Chem. A* **1997**, *101*, 1628–1634. [CrossRef]
8. Wojnárovits, L. Radiation Chemistry. In *Handbook of Nuclear Chemistry 3*; Vértes, A., Nagy, S., Klencsár, Z., Lovas, R.G., Rösch, F., Eds.; Chemical Applications of Nuclear Reactions and Radiations; Springer Science + Business Media B.V: Berlin/Heidelberg, Germany, 2011; Chapter 23; pp. 1263–1331. [CrossRef]
9. Shchapin, I.Y.; Makhnach, O.V.; Klochikhin, V.L.; Nekhaev, A.I. Radiolysis products of the cyclohexane–bicyclic diene binary system. *Pet. Chem.* **2017**, *57*, 897–903. [CrossRef]
10. Pikaev, A.K. *Modern Radiation Chemistry: Main Regularities, Experimental Technique and Methods*; Nauka: Moscow, Russia, 1985; p. 375. (In Russian)
11. Ikuta, S.; Yoshihara, K.; Shiokawa, T.; Jinno, M.; Yokoyama, Y.; Ikeda, S. Photoelectron spectroscopy of cyclohexane, cyclopentane, and related compounds. *Chem. Lett.* **1973**, *2*, 1237–1240. [CrossRef]
12. Watanabe, K. Ionization potentials of some molecules. *J. Chem. Phys.* **1957**, *26*, 542–547. [CrossRef]
13. Bischof, P.; Hashmall, J.A.; Heilbronner, E.; Hornung, V. Photoelektronspektroskopische Bestimmung der Wechselwirkung zwischen nicht-konjugierten Doppelbindungen [1]. Vorläufige Mitteilung. *Helv. Chim. Acta* **1969**, *52*, 1745–1749. [CrossRef]
14. Bieri, G.; Burger, F.; Heilbronner, E.; Maier, J.P. Valence ionization energies of hydrocarbons. *Helv. Chim. Acta* **1977**, *60*, 2213–2233. [CrossRef]
15. Raymonda, J.W. Rydberg states in cyclic alkanes. *J. Chem. Phys.* **1972**, *56*, 3912–3920. [CrossRef]
16. Onsager, L. Initial Recombination of Ions. *Phys. Rev.* **1938**, *54*, 554. [CrossRef]
17. Klein, G. Charge carrier generation and recombination in cyclohexane solutions excited by VUV light. *Chem. Phys. Lett.* **1986**, *124*, 147–151. [CrossRef]
18. Ostafin, A.E.; Lipsky, S. The fluorescence action spectra of some saturated hydrocarbon liquids for excitation energies above and below their ionization thresholds. *J. Chem. Phys.* **1993**, *98*, 5408–5418. [CrossRef]
19. Costner, E.A.; Long, B.K.; Navar, C.; Jockusch, S.; Lei, X.; Zimmerman, P.; Campion, A.; Turro, N.J.; Willson, C.G. Fundamental optical properties of linear and cyclic alkanes: VUV absorbance and index of refraction. *J. Phys. Chem. A* **2009**, *113*, 9337–9347. [CrossRef] [PubMed]

20. Choi, H.T.; Askew, D.; Lipsky, S. A note on the G value for the production of the lowest excited singlet state of cyclohexane. *Radiat. Phys. Chem. (1977)* **1982**, *19*, 373–375. [CrossRef]
21. Howard, A.E.; Staley, S.W. Negative Ion States of Three- and Four-Membered Ring Hydrocarbons: Studied by Electron Transmission Spectroscopy. In *Resonances*; ACS Symposium Series; American Chemical Society: Washington, DC, USA, 1984; Volume 263, pp. 183–192. [CrossRef]
22. Neese, F. The ORCA program system. *WIREs Comput. Mol. Sci.* **2012**, *2*, 73–78. [CrossRef]
23. Neese, F. *ORCA–An Ab Initio, DFT and Semiempirical SCF-MO Package*, version 3.0.1; Max-Plank-Institute for Chemical Energy Conversion: Ruhr, Germany, 2013. Available online: https://orcaforum.kofo.mpg.de/ (accessed on 26 November 2022).
24. Valeev, E. Libint: High-Performance Library for Computing Gaussian Integrals in Quantum Mechanics. Available online: http://libint.valeyev.net (accessed on 19 December 2023).
25. Schaefer, A.; Horn, H.; Ahlrichs, R. Fully optimized contracted Gaussian basis sets for atoms Li to Kr. *J. Chem. Phys.* **1992**, *97*, 2571–2577. [CrossRef]
26. The Turbomole Basis Set Library. Available online: http://ftp.chemie.uni-karlsruhe.de/pub/bases (accessed on 27 January 2024).
27. Chemcraft—Graphical Software for Visualization of Quantum Chemistry Computations. Available online: https://www.chemcraftprog.com (accessed on 27 January 2024).
28. Glab, W.L.; Hessler, J.P. Multiphoton excitation of high singlet *np* Rydberg states of molecular hydrogen: Spectroscopy and dynamics. *Phys. Rev. A* **1987**, *35*, 2102–2110. [CrossRef]
29. McCormack, E.; Gilligan, J.M.; Cornaggia, C.; Eyler, E.E. Measurement of high Rydberg states and the ionization potential of H_2. *Phys. Rev. A.* **1989**, *39*, 2260(R). [CrossRef] [PubMed]
30. Wang, L.; Reutt, J.E.; Lee, Y.T.; Shirley, D.A. High resolution UV photoelectron spectroscopy of $CO^{+}{}_2$, COS^+, and $CS^{+}{}_2$ using supersonic molecular beams. *J. Electr. Spectrosc. Relat. Phenom.* **1988**, *47*, 167–186. [CrossRef]
31. Yencha, A.J.; Hopkirk, A.; Hiraya, A.; Donovan, R.J.; Goode, J.G.; Maier, R.R.J.; King, G.C.; Kvaran, A. Threshold photoelectron spectroscopy of Cl_2 and Br_2 up to 35 eV. *J. Phys. Chem.* **1995**, *99*, 7231–7241. [CrossRef]
32. Chewter, L.A.; Sander, M.; Müller-Dethlefs, K.; Schlag, E.W. High resolution zero kinetic energy photoelectron spectroscopy of benzene and determination of the ionization potential. *J. Chem. Phys.* **1987**, *86*, 4737–4744. [CrossRef]
33. Nemeth, G.I.; Selzle, H.L.; Schlag, E.W. Magnetic ZEKE experiments with mass analysis. *Chem. Phys. Lett.* **1993**, *215*, 151–155. [CrossRef]
34. Kimura, K.; Katsumata, S.; Achiba, Y.; Yamazaki, T.; Iwata, S. Ionization Energies, Ab Initio Assignments, and Valence Electronic Structure for 200 Molecules. In *Handbook of HeI Photoelectron Spectra of Fundamental Organic Compounds*; Japan Scientific Soc. Press: Tokyo, Japan; Halsted Press: New York, NY, USA, 1981; Volume 28.
35. Houle, F.A.; Beauchamp, J.L. Detection and investigation of allyl and benzyl radicals by photoelectron spectroscopy. *J. Am. Chem. Soc.* **1978**, *100*, 3290–3294. [CrossRef]
36. Houle, F.A.; Beauchamp, J.L. Thermal decomposition pathways of alkyl radicals by photoelectron spectroscopy. Application to cyclopentyl and cyclohexyl radicals. *J. Phys. Chem.* **1981**, *85*, 3456–3461. [CrossRef]
37. Morishima, I.; Yoshikawa, K.; Yonezawa, T.; Matsumoto, H. Photoelectron spectral studies of organic free radicals. The nitroxide radical. *Chem. Phys. Lett.* **1972**, *16*, 336–339. [CrossRef]
38. Cízek, M.; Horácek, J.; Domcke, W. Nuclear dynamics of the $H_2{}^-$ collision complex beyond the local approximation: Associative detachment and dissociative attachment to rotationally and vibrationally excited molecules. *J. Phys. B At. Mol. Opt. Phys.* **1998**, *31*, 2571–2583. [CrossRef]
39. Schulz, G.J. Resonances in electron impact on diatomic molecules. *Rev. Mod. Phys.* **1973**, *45*, 423–486. [CrossRef]
40. Rienstra-Kiracofe, J.C.; Tschumper, G.S.; Schaefer, H.F., III; Nandi, S.; Barney Ellison, G. Atomic and molecular electron affinities: Photoelectron experiments and theoretical computations. *Chem. Rev.* **2002**, *102*, 231–282. [CrossRef]
41. Knapp, A.; Echt, O.; Kreisle, D.; Mark, T.D.; Recknagel, E. Formation of long-lived $CO_2{}^-$, N_2O^- and their dimer anions, by electron attachment to van der Waals clusters. *Chem. Phys. Lett.* **1986**, *126*, 225–231. [CrossRef]
42. Bowen, K.H.; Liesegang, G.W.; Sanders, R.A.; Herschbach, D.W. Electron attachment to molecular clusters by collisional charge transfer. *J. Phys. Chem.* **1983**, *87*, 557–565. [CrossRef]
43. Sanche, L.; Schulz, G.J. Electron transmission spectroscopy: Resonances in triatomic molecules and hydrocarbons. *J. Chem. Phys.* **1973**, *58*, 479–493. [CrossRef]
44. Modelli, A.; Jones, D.; Distefano, G. ETS study of the negative ion states of t-butyl and trimethylsilyl derivatives of ethylene and benzene. *Chem. Phys. Lett.* **1982**, *86*, 434–437. [CrossRef]
45. Burrow, P.D.; Michejda, J.A.; Jordan, K.D. Electron transmission study of the temporary negative ion states of selected benzenoid and conjugated aromatic hydrocarbons. *J. Chem. Phys.* **1987**, *86*, 9–24. [CrossRef]
46. Wenthold, P.G.; Polak, M.L.; Lineberger, W.C. Photoelectron spectroscopy of the allyl and 2-methylallyl anions. *J. Phys. Chem.* **1996**, *100*, 6920–6926. [CrossRef]

47. Peerboom, R.A.L.; Rademaker, G.J.; de Koning, L.J.; Nibbering, N.M.M. Stabilization of cycloalkyl carbanions in the gas phase. *Rapid Commun. Mass Spectrom.* **1992**, *6*, 394–399. [CrossRef]
48. Ko, Y.J.; Wang, H.; Pradhan, K.; Koirala, P.; Kandalam, A.K.; Bowen, K.H.; Jena, P. Superhalogen properties of Cu_mCl_n clusters: Theory and experiment. *J. Chem. Phys.* **2011**, *135*, 244312. [CrossRef]
49. Trainham, R.; Fletcher, G.D.; Larson, D.J. One- and two-photon detachment of the negative chlorine ion. *J. Phys. B* **1987**, *20*, L777. [CrossRef]
50. Leopold, D.G.; Ho, J.; Lineberger, W.C. Photoelectron spectroscopy of mass–selected metal cluster anions. I. Cu^-_n, n = 1–10. *J. Chem. Phys.* **1987**, *86*, 1715–1726. [CrossRef]
51. Jordon-Thaden, B.; Kreckel, H.; Golser, R.; Schwalm, D.; Berg, M.H.; Buhr, H.; Gnaser, H.; Grieser, M.; Heber, O.; Lange, M.; et al. Structure and stability of the negative hydrogen molecular ion. *Phys. Rev. Lett.* **2011**, *107*, 193003. [CrossRef]
52. Kreckel, H.; Herwig, P.; Schwalm, D.; Čížek, M.; Golser, R.; Heber, O.; Jordon-Thaden, B.; Wolf, A. Metastable states of diatomic hydrogen anions. *J. Phys. Conf. Ser.* **2014**, *488*, 012034. [CrossRef]
53. Golser, R.; Gnaser, H.; Kutschera, W.; Priller, A.; Steier, P.; Wallner, A.; Čížek, M.; Horáček, J.; Domcke, W. Experimental and theoretical evidence for long-lived molecular hydrogen anions H_2^- and D_2^-. *Phys. Rev. Lett.* **2005**, *94*, 223003. [CrossRef]
54. Heber, O.; Golser, R.; Gnaser, H.; Berkovits, D.; Toker, Y.; Eritt, M.; Rappaport, M.L.; Zajfman, D. Lifetimes of the negative molecular hydrogen ions: H_2^-, D_2^-, and HD^-. *Phys. Rev. A* **2006**, *73*, 060501(R). [CrossRef]
55. Fischer-Hjalmars, I. Theoretical investigation of the negative hydrogen molecule ion. *J. Chem. Phys.* **1959**, *30*, 1099–1100. [CrossRef]
56. Zhan, C.-G.; Nichols, J.A.; Dixon, D.A. Ionization potential, electron affinity, electronegativity, hardness, and electron excitation energy: Molecular properties from density functional theory orbital energies. *J. Phys. Chem. A* **2003**, *107*, 4184–4195. [CrossRef]
57. Meunier, M.; Quirke, N.; Binesti, D. The calculation of the electron affinity of atoms and molecules. *Mol. Simul.* **1999**, *23*, 109–125. [CrossRef]
58. McWeeny, R. The electron affinity of H_2: A valence bond study. *J. Mol. Struct. THEOCHEM* **1992**, *261*, 403–413. [CrossRef]
59. Sharp, T.E. Potential-energy curves for molecular hydrogen and its ions. *At. Data Nucl. Data Tables* **1971**, *2*, 119–169. [CrossRef]
60. Linder, F.; Schmidt, H. Rotational and vibrational excitation of H_2 by slow electron impact. *Z. Naturforsch. A* **1971**, *26*, 1603–1617. [CrossRef]
61. Bruna, P.J.; Lushington, G.H.; Grein, F. Electron-spin g-factors of H_2^-. An ab initio study. *Chem. Phys. Lett.* **1996**, *258*, 427–430. [CrossRef]
62. Moiseiwitsch, B.L. Electron Affinities of Atoms and Molecules. *Adv. At. Mol. Phys.* **1965**, *1*, 61–83. [CrossRef]
63. Nieto, P.; Pijper, E.; Barredo, D.; Laurent, G.; Olsen, R.A.; Baerends, E.-J.; Kroes, G.-J.; Farías, D. Reactive and nonreactive scattering of H_2 from a metal surface is electronically adiabatic. *Science* **2006**, *312*, 86–89. [CrossRef]
64. Kroes, G.-J.; Díaz, C. Quantum and classical dynamics of reactive scattering of H_2 from metal surfaces. *Chem. Soc. Rev.* **2015**, *45*, 3658–3700. [CrossRef]
65. Deleuze, M.S.; Delhalle, J. Outer-valence green's function study of cycloalkane and cycloalkyl−alkane compounds. *J. Phys. Chem. A* **2001**, *105*, 6695–6702. [CrossRef]
66. Jung, J.M. On the photoionisation of liquid cyclohexane, 2,2,4 trimethylpentane and tetramethylsilane. *Chem. Phys. Lett.* **2003**, *380*, 190–195. [CrossRef]
67. Jung, J.M.; Gress, H. Single-photon absorption of liquid cyclohexane, 2,2,4 trimethylpentane and tetramethylsilane in the vacuum ultraviolet. *Chem. Phys. Lett.* **2003**, *377*, 495–500. [CrossRef]
68. Shang, Q.Y.; Bernstein, E.R. (σ3s) Rydberg states of cyclohexane, bicyclo[2.2.2]octane, and adamantane. *J. Chem. Phys.* **1994**, *100*, 8625–8632. [CrossRef]
69. Tian, Z.; Fattahi, A.; Lis, L.; Kass, S.R. Cycloalkane and Cycloalkene C-H Bond Dissociation Energies. *J. Am. Chem. Soc.* **2006**, *128*, 17087–17092. [CrossRef]
70. Hölsch, N.; Beyer, M.; Salumbides, E.J.; Eikema, K.S.E.; Ubachs, W.; Jungen, C.; Merkt, F. Benchmarking theory with an improved measurement of the ionization and dissociation energies of H_2. *Phys. Rev. Lett.* **2019**, *122*, 103002. [CrossRef]
71. Iwasaki, M.; Toriyama, K.; Nunome, K. Electron spin resonance studies of structures and reactions of radical cations of a series of cycloalkanes in low-temperature matrices. *Faraday Disc. Chem. Soc.* **1984**, *78*, 19–33. [CrossRef]
72. Wojnárovits, L. Energy transfer from excited cyclohexane aromatic solutes. *J. Photochem.* **1984**, *24*, 341–353. [CrossRef]
73. Wojnarovits, L. Photochemistry and Radiation Chemistry of Liquid Alkanes: Formation and Decay of Low-Energy Excited States. In *Charged Particle and Photon Interactions with Matter: Chemical, Physicochemical, and Biological Consequences with Applications*; Mozumder, A., Hatano, Y., Eds.; Marcel Dekker: New York, NY, USA, 2004; Chapter 13; pp. 365–402; ISBN 0-8247-4623-6.
74. Ausloos, P.; Rebbert, R.E.; Schwarz, F.P.; Lias, S.G. Pulse- and gamma ray-radiolysis of cyclohexane: Ion recombination mechanisms. *Radiat. Phys. Chem. (1977)* **1983**, *21*, 27–43. [CrossRef]
75. Hirayama, F.; Lipsky, S. Fluorescence of saturated hydrocarbons. *J. Chem. Phys.* **1969**, *51*, 3616–3617. [CrossRef]

76. Tewari, P.H.; Freeman, G.R. Dependence of radiation-induced conductance of liquid hydrocarbons on molecular structure. *J. Chem. Phys.* **1968**, *49*, 4394–4399. [CrossRef]
77. Schmidt, W.F. Electrons in nonpolar dielectric liquids. *IEEE Transact. Electr. Insul.* **1991**, *26*, 560–567. [CrossRef]
78. Schmidt, W.F.; Illenberger, E. Low energy electrons in non-polar liquids. *Nukleonika* **2003**, *48*, 75–82. Available online: https://scholar.google.com/scholar?hl=ru&as_sdt=0%2C5&q=78.%09Schmidt%2C+W.F.%3B+Illenberger%2C+E.+Low+energy+electrons+in+non-polar+liquids.+Nukleonika+2003%2C+48%2C+75%E2%88%9282.&btnG= (accessed on 27 January 2024).
79. Casanovas, J.; Grob, R.; Delacroix, D.; Guelfucci, J.P.; Blanc, D. Photoconductivity studies in some nonpolar liquids. *J. Chem. Phys.* **1981**, *75*, 4661–4668. [CrossRef]
80. Seki, K.; Inokuchi, H. The ultraviolet photoelectron spectroscopy of aliphatic hydrocarbons and tetramethylsilane in the solid state. *Bull. Chem. Soc. Jpn.* **1983**, *56*, 2212–2219. [CrossRef]
81. Blachford, J.; Dyne, P.J. Vapor-phase radiolysis of cyclohexane and mixtures of benzene and cyclohexane. *Can. J. Chem.* **1964**, *42*, 1165–1173. [CrossRef]
82. Wojnárovits, L.; Földiak, G. Influence of cyclic structure on the radiolysis of hydrocarbons. II. Radiolysis of alkylcyclopentanes and alkylcyclohexanes. *Acta Chim. Acad. Sci. Hung.* **1974**, *82*, 285–303. Available online: https://inis.iaea.org/records/1yrck-cjb36 (accessed on 27 January 2024).
83. Theard, L.M. Effects of Additives on the Radiolysis of Cyclohexane Vapor at 100°. *J. Phys. Chem.* **1965**, *69*, 3292–3298. [CrossRef]
84. Bouillot, M.S. Sur la radiolyse du cyclohexane en phase solide. *Int. J. Radiat. Phys. Chem.* **1970**, *2*, 117–118. [CrossRef]
85. Shchapin, I.Y.; Nekhaev, A.I. The boundary between two modes of gas evolution: Oscillatory (H_2 and O_2) and conventional redox (O_2 only), in the hydrocarbon/H_2O_2/Cu(II)/CH_3CN system. *Hydrogen* **2023**, *4*, 74–102. [CrossRef]

hydrogen

MDPI

Article

Influence of Catalytic Support on Hydrogen Production from Glycerol Steam Reforming

Jorge Feijoo [1,*], Rocío Maceiras [1], Victor Alfonsín [1], Nevin Aly [2] and Alejandro de la Fuente [3]

[1] Defense University Center at Spanish Naval Academy, University of Vigo, Plaza de España 2, Marín, 36920 Pontevedra, Spain; rmaceiras@cud.uvigo.es (R.M.); valfonsin@cud.uvigo.es (V.A.)
[2] Geological Engineering and Geophysics Department, Faculty of Petroleum and Mining Engineering, Suez University, Suez 43721, Egypt; nevin.aly@suezuni.edu.eg
[3] Spanish Naval Academy, Plaza de España 2, Marín, 36920 Pontevedra, Spain
* Correspondence: jfeijoo@cud.uvigo.es

Abstract

The use of hydrogen as an energy carrier represents a promising alternative for mitigating climate change. However, its practical application requires achieving a high degree of purity throughout the production process. In this study, the influence of the type of catalytic support on H_2 production via steam glycerol reforming was evaluated, with the objective of obtaining syngas with the highest possible H_2 concentration. Three types of support were analyzed: two natural materials (zeolite and dolomite) and one metal oxide, alumina. Alumina and dolomite were coated with Ni at different loadings, while zeolite was only evaluated without Ni. Reforming experiments were carried out at a constant temperature of 850 °C, with continuous monitoring of H_2, CO_2, CO, and CH_4 concentrations. The results showed that zeolite yielded the lowest H_2 concentration (51%), mainly due to amorphization at high temperatures and the limited effectiveness of physical adsorption processes. In contrast, alumina and dolomite achieved H_2 purities of around 70%, which increased with Ni loading. The improvement was particularly significant in dolomite, owing to its higher porosity and the recarbonation processes of CaO, enabling H_2 purities of up to 90%.

Keywords: catalytic reforming; alumina; dolomite; zeolite; Ni; H_2 purity

1. Introduction

Historically, fossil fuels have dominated the global energy matrix. Although significant advances have been made in the development of renewable energies, hydrocarbons continue to be the main source of primary energy. Currently, according to data provided by the Energy Institute [1], more than 60% of global electricity is generated from fossil fuels, with coal being the main contributor, with a share of 35%. Obtaining energy from these fuels generates significant pollution due to emissions of nitrogen oxides, sulfur dioxide, greenhouse gases, and particulate matter. These pollutants contribute to climate change, ocean acidification, and air quality degradation, with serious economic and environmental consequences.

In this context, the United Nations Framework Convention on Climate Change (UNFCCC), held in Glasgow (COP26), marked a milestone in the global effort to decarbonize and limit global warming. To this end, renewable energies such as hydroelectric, wind, and solar power are presented as crucial alternatives for a sustainable energy future, as their inexhaustible nature and absence of polluting emissions make them key elements of

https://doi.org/10.3390/hydrogen6040088

the energy transition. However, the variability in its production requires coexistence with conventional energy sources to guarantee supply. That is why H_2 is emerging as an energy carrier with great potential to contribute to reducing the carbon footprint by providing secure, competitive, and CO_2-free energy.

The versatility of hydrogen allows it to be used in both internal combustion engines and fuel cells, generating only water vapor. Despite the promising potential, large-scale implementation of H_2 faces challenges such as high production costs, the need for specific infrastructure, and the requirement for high purity. Currently, H_2 can be produced from various renewable and non-renewable sources—such as water, natural gas, methane, ethanol, and glycerol—using different processes, including electrolysis, gasification, partial oxidation, autothermal reforming, dark fermentation, and steam reforming [2,3]. Other routes have also been explored, including photocatalysis, which harnesses solar energy to produce H_2 with excellent purity [4,5].

With the growing global emphasis on sustainability and the circular economy, the use of inexhaustible resources such as seawater in electrolysis processes [6,7] or by-products from other industries, such as glycerin for use in steam reforming processes [8,9], are becoming promising options for hydrogen production. However, it is necessary to ensure the quality of the H_2 produced, particularly its purity. The presence of contaminants such as carbon monoxide (CO), carbon dioxide (CO_2), methane (CH_4), and sulfur-containing compounds, often generated in many of the current hydrogen production processes, such as steam reforming, the most widely used among the production techniques [3,10], can seriously affect the subsequent use of this energy carrier. This underscores the critical role of purification in the hydrogen production chain to meet the high purity standards required for various hydrogen applications. In this context, hydrogen carriers have recently gained attention as an alternative route, since they allow the selective release of hydrogen streams with negligible CO content [11–13].

There are different ways to improve the purity of hydrogen before its final application. One of them is based on increasing the purity of the gas during the production process using different types of catalysts [14,15]. The choice of catalyst plays a decisive role in determining the effectiveness of the steam reforming process, since it should have the capacity to promote the cleavage of C–C, O–H, and C–H bonds and inhibit the formation of the C–O bonds. Furthermore, it should promote the shift of CO to CO_2, as well as promote dehydrogenation and hydrogenation processes.

A wide range of studies have explored the application of various catalysts, spanning from noble metals such as Pt [16], Pd [17], Ir [18], Rh [19], and Ru [20] to non-noble metals like Ni [21] and Co [22]. Pt-based catalysts have been reported to reach H_2 selectivities above 90% in glycerol reforming at moderate temperatures [23], while Ru/Al_2O_3 can achieve H_2 yields of 95% with high stability for the steam reforming of ethanol, with methane being the only byproduct [24]. Noble metals, such as Au and Ag, have occasionally been tested, but their application has been mostly limited to photocatalytic systems and remains scarce in thermal catalytic reforming [25,26]. While noble metal catalysts (e.g., Pt, Ru, Rh) typically exhibit higher activity and stronger resistance to coke formation compared to their non-noble counterparts [27] (an important parameter that causes the deactivation of the catalyst), their high cost limits their practicality for large-scale industrial use. Among the non-noble metals, Ni-based systems typically produce H_2 yields in the range of 60–75% with lower purity due to the formation of CO and CH_4 [28], and Cu-based catalysts show moderate activity with faster deactivation than Ni due to sintering and coke deposition [29–31]. Recent studies have also investigated other non-noble metal catalysts as cost-effective alternatives, reporting promising activity and selectivity in reforming reactions like zinc and indium [32,33].

Unlike the noble metal catalysts, the use of Ni as a catalyst improves the reforming process at an affordable production cost [34], allowing it to be subsequently extrapolated to industrial applications. For this reason, Ni is the most widely used commercial catalyst, especially on Al_2O_3 catalytic supports [9,35]. However, as previously reported in [21,35], Ni-based catalysts are prone to carbon deposition and metal particle sintering, phenomena that inevitably result in their gradual deactivation [36]. To improve the stability of Ni catalysts, their performance has been investigated on different types of supports, particularly various metal oxides (e.g., Al_2O_3, SiO_2, La_2O_3, CeO_2, ZrO_2) [23,37,38], which may even be doped with transition metals (e.g., Fe, Co, Sn) [39,40], noble metals (e.g., Ag, Pt, Pd, Ir) [41,42], lanthanide metals (e.g., La, Ce, Pr) [39,43], and alkaline earth metals (e.g., Mg, Ca, Ba) [39,43].

The support material plays an important role in the overall performance of a catalytic system since some physicochemical properties of the support—such as thermal stability at elevated temperatures, surface area (which influences catalyst dispersion), pore size distribution, and the basicity of active sites— strongly affect process efficiency [21]. Soares et al. [44] assessed the gas-phase conversion of glycerol into synthesis gas over supported Pt catalysts at 623 K. Five different supports were evaluated, and they found that Pt supported on Al_2O_3, ZrO_2, CeO_2/ZrO_2, and MgO/ZrO_2 exhibited deactivation during the operating time, while Pt supported on carbon showed stable conversion of glycerol to synthesis gas for at least 30 h. Pompeo et al. [16] evaluated the effect of different supports (specifically SiO_2, ZrO_2, γ-Al_2O_3, and α-Al_2O_3 modified with Ce and Zr) in steam reforming of glycerol using Pt catalysts at temperatures lower than 450 °C. This study found that neutral supports (SiO_2) are better than acidic ones (ZrO_2 and γ-Al_2O_3) for promoting selective hydrogen production and catalyst stability. This is because, as pointed out in [37], greater basicity of the support also leads to CO_2 activation, which helps to oxidize deposited coke to CO and also reduces coke formation from methane decomposition.

Although Ni-based catalysts are widely used in steam reforming, the influence of the support remains a key challenge. Alumina, while common, is prone to deactivation through coke deposition and Ni sintering; dolomite has been less explored despite its favorable porosity and CO_2 capture capacity; zeolites often suffer structural instability at high temperatures. These knowledge gaps highlight the importance of evaluating different supports under identical conditions. Accordingly, the aim of this study is to assess the optimization of the hydrogen production process through a reforming process, comparing the yields obtained by three different supports and a Ni catalyst to increase the purity of the outlet gas mixture before it is introduced into a subsequent purification process. This study will contribute to deeper knowledge of the influence of different catalytic supports, both with and without active catalysts, which were selected considering the cost of the process for a possible scalability of the study to industry.

2. Glycerol Reforming Process

Glycerol steam reforming (GSR) is considered a promising alternative for H_2 production [45], as it closely resembles current industrial processes, and as a by-product of other industries, glycerol is readily available and economically viable.

The feasibility of this process has been extensively examined, as glycerol's molecular structure provides a favorable hydrogen-to-carbon ratio, enabling high H_2 yields. According to the overall reaction (Equation (1)), one mole of $C_3H_8O_3$ can theoretically generate up to seven moles of $H_{2\,(g)}$.

$$\text{Glycerol steam reforming } C_3H_8O_3 + 3H_2O \leftrightarrow 3CO_{2\,(g)} + 7H_{2\,(g)} \quad (1)$$

Thus, it is the combination of the following two reactions:

$$\text{Glycerol decomposition } C_3H_8O_{3\,(g)} \leftrightarrow 3CO_{\,(g)} + 4H_{2\,(g)} \tag{2}$$

$$\text{Water–gas shift } CO_{(g)} + H_2O_{\,(g)} \leftrightarrow CO_{2\,(g)} + H_{2\,(g)} \tag{3}$$

The reaction pathway is highly complex, with several competing processes potentially occurring. In particular, CO_2 and CO hydrogenation can result in methane formation, while secondary reactions such as dehydration, dehydrogenation, cyclization, and polymerization may take place, ultimately contributing to coke deposition [46].

$$\text{Methanation reaction } CO_2 + 4H_{2\,(g)} \leftrightarrow CH_4 + 2H_2O \tag{4}$$

$$\text{Methanation reaction } CO_{(g)} + 3H_{2(g)} \leftrightarrow CH_4 + H_2O \tag{5}$$

$$\text{Coke formation } H_2 + CO \leftrightarrow C + H_2O \tag{6}$$

$$\text{Coke formation } 2CO \leftrightarrow CO_2 + C \tag{7}$$

$$\text{Coke formation } CH_4 \leftrightarrow 2H_2 + C \tag{8}$$

The quantities of H_2, CO_2, CO, and CH_4 produced, as well as the C formation, depend on reaction conditions such as concentration of glycerol, supply flow rate, temperature, and the pressure of the reaction.

3. Materials and Methods

Glycerol steam reforming tests were carried out in a stainless-steel fixed bed reactor with a total length of 700 mm and an inner diameter of 48 mm, located in a vertical electrical tubular furnace (TR4, Hobersal) capable of heating to 1200 °C and connected with a gas analyzer (Mamos, Madur), which allows a continuous measurement of the concentrations of H_2, CO_2, CO, and CH_4 (Figure 1, similar setup to that used in [8]).

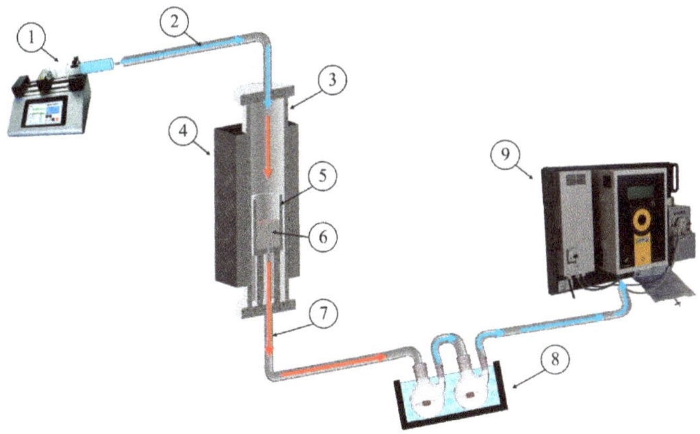

Figure 1. Scheme of the glycerol steam reforming system: syringe pump (1), glycerol flow (2), cylindrical reactor (3), electrical tubular furnace (4), perforated cylinder bucket (5), catalyst support as fixed bed (6), syngas (7), condensate tank (8), and gas analyzer (9).

Glycerol was supplied to the top of the reactor using a syringe pump with a feed rate set at 0.855 mL/min, which is the optimal value using this setup, according to [8], to maximize the production and purity of the hydrogen generated. This parameter was kept constant in order to compare the influence of different supports under identical conditions, rather than to re-optimize the operating point for each catalyst.

The tubular reactor houses a perforated bucket where the support is placed up to a fixed bed height (in this case, 4 cm), so that the glycerol must flow through the bed until it reaches the lower outlet of the reactor. After leaving the reactor, the gaseous products were condensed in a thermostatic bath, and the resulting dry gas was sent to the gas analyzer, where both its composition and flow rate were measured.

Prior to initiating the glycerol reforming process, a catalyst support conditioning step was performed. This procedure involved the activation of the Ni catalyst by heating the reactor to 700 °C at a rate of 10 °C/min. Once the target temperature of 700 °C was reached, it was maintained for 30 min while a gas mixture of H_2 and N_2 (50:50 vol.%) was introduced. After this period, only N_2 was supplied, which was maintained until all residual gases were purged from the reactor and the test temperature of 850 °C was achieved. This process was carried out both in the tests where the supports were impregnated with Ni and in those where only the support was used without impregnation.

In this study, three types of supports were assessed: one an oxide metallic (porous γ-Al_2O_3) usually used in steam reforming processes, and two natural porous materials (dolomite and zeolite).

The alumina used was γ-alumina, which is suitable for high-temperature applications. This alumina had a diameter of approximately 0.5 cm, an apparent porosity of around 7%, and a water absorption capacity of around 3%. As reported in several studies, this support is characterized by a high specific surface area, which promotes effective metal dispersion [47]. However, at elevated temperatures, it undergoes phase transformations, sintering, and a consequent loss of surface area [34]. In all catalytic tests with alumina as support, a total of 43 g of material was used.

The zeolite employed was clinoptilolite $(Ca,Na,K)_6(Si_{30}Al_6)O_{72}\cdot 20H_2O)$, supplied by Zeopol, with an average grain size of 0.6 cm. Zeolites are crystalline aluminosilicates characterized by a three-dimensional framework of SiO_4 and AlO_4 tetrahedra. Owing to their high specific surface area and the selective capture of impurities such as CO_2, they are commonly used in subsequent hydrogen purification processes by physisorption [48–50], thereby yielding purified hydrogen. However, several studies also report their application as catalytic supports during high-temperature hydrogen production. Yao et al. [51] evaluated different reforming temperatures (650, 750, and 850 °C) using Ni with zeolite as the catalyst for hydrogen production. They found that the optimal operation temperature was 850 °C. In the zeolite tests, 47 g of support was used in each experiment.

Dolomite is a double carbonate of Ca and Mg ($CaMg(CO_3)_2$) that decomposes into its respective oxides as temperature increases (in this case, after calcination, 57% CaO and 37% MgO). MgO is formed at around 500 °C, while CaO is generated at higher temperatures (800–900 °C), so a partial decomposition of CaO and complete decomposition of MgO was expected under the operation conditions of the test (i.e., 850 °C) [52,53]. This material has been widely applied for H_2 purification through the chemisorption of impurities, primarily carbon dioxide (CO_2). However, several studies have also explored its use as a catalytic support during H_2 production. Zhang et al. [54] reported that NiO-impregnated dolomite, when employed as a reactor bed in high-temperature steam gasification of biomass, enhanced H_2 yield. Similarly, Gallucci et al. [55] demonstrated the effectiveness of dolomite as a CO_2 sorbent in steam gasification processes, attributing its performance to CaO recarbonation at approximately 800 °C, a temperature comparable to that used in the present study. In the experiments with dolomite as support, 42 g of material were employed.

Each of the three supports exhibited a different degree of basicity. Alumina is amphoteric, containing both acidic and basic sites, which is why it is generally considered a neutral or slightly acidic support. The acidity of γ-alumina is lower than that of natural zeolites, which, like alumina, possess both Brønsted and Lewis acid sites [56]. Finally, dolomite is the most basic of the three supports, since upon thermal decomposition it generates alkaline earth oxides (CaO and MgO), both of which are strongly basic.

The impregnation of the support with the Ni catalyst was carried out by the wet impregnation method, because it is a simple, reproducible, and scalable technique that allows for adequate dispersion of Ni species on porous supports. The process involves the following steps: the support was first calcined at 650 °C (heating rate of 10 °C/min) in an air atmosphere for 3 h to remove residual impurities, and in the case of the dolomite, to create active MgO and part of CaO. Once cooled to room temperature, the support was immersed, with continuous stirring, in a nickel-nitrate solution of predetermined concentration for 20 min (two solutions with different concentrations were used, one at 10 wt% and the other at 12 wt%, to obtain two different amounts of Ni deposition). After impregnation, the material was dried in an oven at 110 °C for 2 h, followed by cooling to room temperature. This procedure was repeated by immersing the sample in the nickel-nitrate solution once more to promote the formation of active phases such as nickel oxide (NiO) and to enhance the stability and dispersion of nickel species.

The amount of nickel dry matter that remained on the corresponding support was measured by the gravimetric method according to Equation (9), which is often applied when a consolidant or water repellent is applied to a stone substrate [57–59].

$$\text{Ni dry matter (\%)} = 100 \times (W_f - W_d)/W_d \quad (9)$$

where W_d corresponds with the initial dry weight of the support before the impregnation process, and W_f is the weight of the support after 7 days of drying.

Each support was evaluated in duplicate, both in the absence and presence of the catalyst, with two different Ni loadings. This approach enabled the comparison of the performance of the bare support with that of the Ni-loaded supports, as well as the assessment of the influence of Ni content on the H_2 production process. In all cases, the production temperature was maintained at 850 °C for 40 min, the bed height at 4 cm, and the particle diameter of all supports between 5 and 6 mm.

Mineral phase characterization was performed by X-ray diffraction using a Philips X'Pert Bragg Brentano X-ray diffractometer with Cu Kα radiation (λ = 1.5406 Å) operating at 40 kV (generator voltage), 30 mA (tube current) with 0.02° scan step size, 200 s/step of counting time, and a data angle range (2θ) of 5–80°.

The hydrogen production efficiency of the reforming process was determined using the following equation, which compares the average H_2 concentration obtained in the reforming tests with the catalytic support (CS) with those obtained with the nickel-impregnated support (CNi/S).

$$\text{Efficiency (\%)} = 100 \times (\text{CNi/S} - \text{CS})/\text{CS} \quad (10)$$

4. Results and Discussion

4.1. Supports Without Ni

Figure 2 and Table 1 present the concentrations of the main components of the synthesis gas at the reactor outlet for the different supports evaluated—alumina, dolomite, and zeolite—in the absence of Ni impregnation.

The data obtained during the reforming process indicate that dolomite is the support that provides the highest H_2 concentrations, both in terms of maximum and average values throughout the steam reforming process. In addition, this support leads to a significant reduction in CO_2 and CO concentrations compared to the other two supports evaluated. This behavior can be attributed, as reported by Gallucci et al. [55], to the ability of CaO, released at temperatures above 800 °C, to undergo recarbonation to $CaCO_3$ in the presence of CO_2, which would reduce the concentration of this gas in the environment. This direct removal of CO_2 by carbonation with CaO promotes the thermodynamic equilibrium of the water–gas shift reaction to the hydrogen production side [10,54]. Moreover, the enhanced CO_2 capture also leads to a reduction in CO content, as this process can shift the CO equilibrium toward higher CO_2 conversions (Equation (7)). This effect may promote greater coke formation, which would explain the increased CH_4 production observed when using this support (Equation (8)).

Table 1. Maximum and average content (in %) of the main compounds contained in the syngas composition using different supports.

Alumina	H_2	CO_2	CO	CH_4
Maximum (%)	70.8	20.8	7.3	7.4
Average (%)	62.0	16.8	5.8	5.7
Dolomite	H_2	CO_2	CO	CH_4
Maximum (%)	71.6	16.6	5.9	9.2
Average (%)	65.1	13.8	4.8	7.7
Zeolite	H_2	CO_2	CO	CH_4
Maximum (%)	54.2	20.9	9.6	8.7
Average (%)	51.0	18.2	7.5	7.4

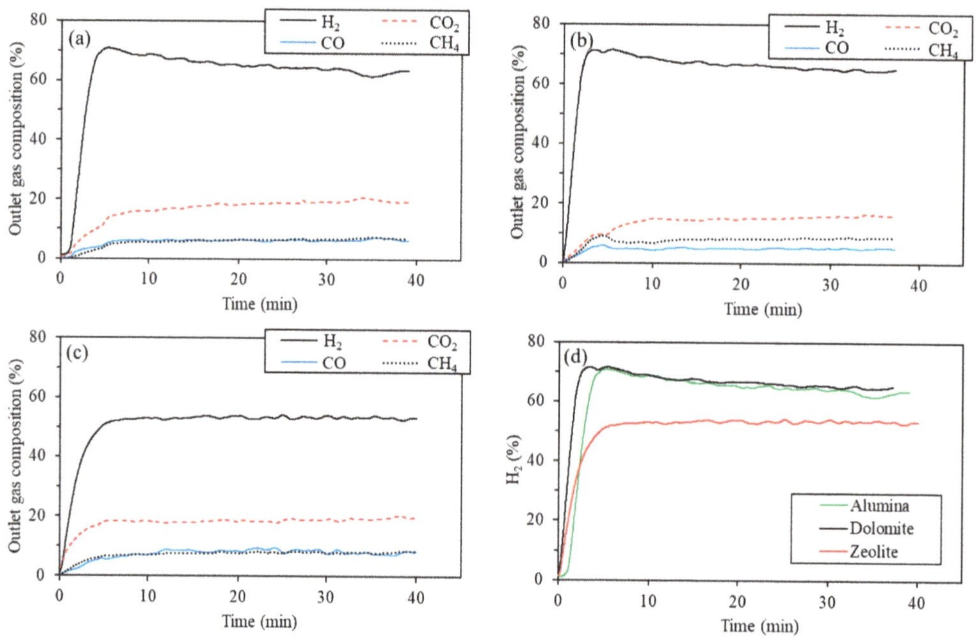

Figure 2. Average concentration of H_2, CO_2, CO, and CH_4 in syngas using different supports (n = 2): alumina (**a**), dolomite (**b**), zeolite (**c**). A comparison between the supports with respect to H_2 concentration (in %) is also shown (**d**).

The catalytic support that showed the worst performance was zeolite, both in terms of H_2 (with a final concentration of around 50%), CO_2, and CO concentration. This behavior could be attributed to the following: (1) The fact that the primary adsorption mechanism of zeolite is based on physical adsorption processes, whose effectiveness decreases at temperatures above 100 °C as thermal energy disrupts weak physical bonds [60]. The higher acidity of this support means that CO_2 adsorption occurs through these weak interactions, such as electrostatic forces or hydrogen bonds. At elevated temperatures, chemisorption mechanisms become more dominant, involving the formation of stronger chemical bonds and thus offering better capture performance [60]. (2) The possible collapse of the structure due to the calcination process such as occurred in previous studies [50]. XRD pattern performed on a zeolite sample previously calcined at 800 °C, a temperature 50 °C lower than the operating temperature of the oven, shows a large reduction in peak intensity and an increase in peak broadening compared to the natural zeolite pattern, suggesting a significant loss of crystallinity and amorphization of the samples due to thermal decomposition (Figure 3). This loss of crystallinity and the possible generation of sintering processes, which usually occurs at temperatures higher than 630 °C [61,62], influence the adsorption capacity of the zeolite since they reduce the specific surface area of the material.

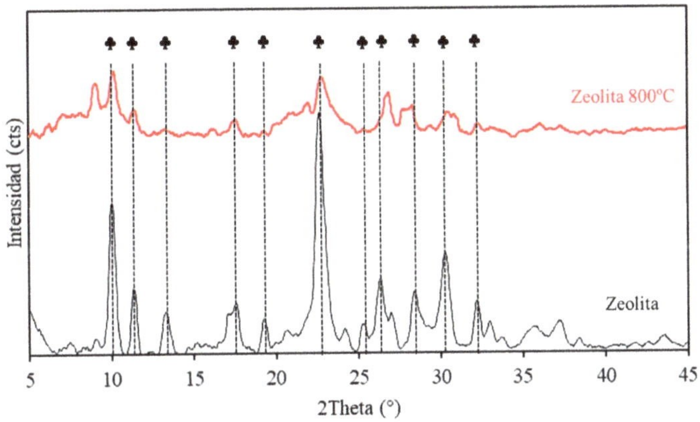

Figure 3. XRD pattern of the natural zeolite (black line) and that calcined at 800 °C (red line). ♣: clinoptilolite (JCPDS card 83-1260).

Due to its low adsorption capacity at elevated temperatures, this adsorbent was excluded from the subsequent stages of the study.

4.2. Supports with Different Concentrations of Ni

Table 2 shows the dry matter percentages for both alumina and dolomite achieved after the wet impregnation process with the two nickel-nitrate solutions at 10 and 12 wt%. It can be observed that, after the impregnation process, dolomite is able to retain a greater amount of nickel. This behavior is attributed to the fact that, as an ornamental rock, dolomite exhibits higher accessible porosity than porous alumina, along with a broader distribution within the mesopore and macropore ranges; both aspects facilitate greater metal deposition and good dispersion throughout the particle. This behavior is consistent with prior reports showing that dolomite supports, due to their basicity and porous structure, facilitate more uniform dispersion and stronger anchoring of Ni species [63]. It is also evident that increasing the concentration of the nitrate solution leads to a greater amount of nickel deposited on the support surface.

Table 2. Dry matter of NiO in each support (in %) obtained with different nickel-nitrate concentrations (10 and 12 wt%).

$Ni(NO_3)_2$ Concentration	10 wt%	12 wt%
Alumina	9%	12%
Dolomite	31%	37%

The results obtained from the reforming test (Figure 4 and Table 3) indicate that the addition of Ni as a catalyst enhances the efficiency of the reforming process, leading to a higher H_2 purity. This effect becomes more pronounced with increasing Ni loading on the alumina support. Furthermore, a higher NiO content results in a lower amount of CH_4 produced. This behavior is associated with the reduced concentrations of CO_2 and CO generated, which otherwise may promote methanation reactions, as described in Section 2 (Equations (4) and (5)).

A notable observation in Figure 4d is that, with lower NiO loads on the Al_2O_3 support, specifically 9 wt%, gradual catalyst deactivation occurs. This deactivation causes a decrease in H_2 purity, which approaches values similar to those obtained with the uncoated support. This observation agrees with previous reports showing that Ni/Al_2O_3 catalysts are prone to deactivation under reforming conditions. The main causes are coke deposition, sintering of Ni particles, and blockage of active sites, which progressively reduce catalyst activity and H_2 purity during time-on-stream [64]. Similar behavior has been reported in both glycerol and methane reforming systems, where carbon accumulation and structural changes significantly compromise the long-term stability of Ni/Al_2O_3 catalysts [36].

Regarding the dolomite support (Figure 5), it is again observed that the addition of Ni enhances the reforming efficiency and hydrogen concentration compared with the non-impregnated support. However, beyond a certain Ni loading point (approximately 31 wt%), no significant improvement is observed, indicating a saturation effect in the distribution of active sites. Specifically, at 31 wt% Ni, higher loadings do not justify the additional cost, as the gain in efficiency is negligible.

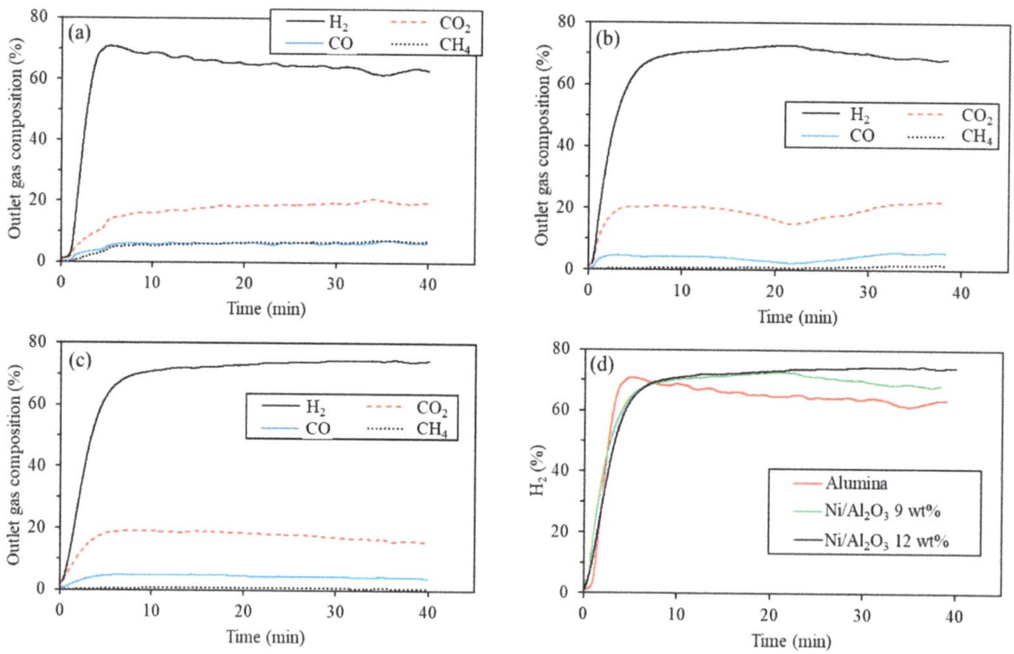

Figure 4. Average concentration of H_2, CO_2, CO, and CH_4 in syngas using different alumina supports (n = 2): alumina without Ni (**a**), Ni/Al_2O_3 9 wt% (**b**), and Ni/Al_2O_3 12 wt% (**c**). A comparison between the three supports with respect to H_2 concentration (in %) is also shown (**d**).

Table 3. Maximum and average content (in %) of the main compounds contained in the syngas composition using different supports, with and without a catalyst.

Alumina	**H_2**	**CO_2**	**CO**	**CH_4**
Maximum (%) Average (%)	70.8 62.0	20.8 16.8	7.3 5.8	7.4 5.7
Ni/Al_2O_3 9 wt%	**H_2**	**CO_2**	**CO**	**CH_4**
Maximum (%) Average (%)	72.8 66.2	22.3 18.7	5.9 4.1	1.8 0.9
Ni/Al_2O_3 12 wt%	**H_2**	**CO_2**	**CO**	**CH_4**
Maximum (%) Average (%)	76.0 70.1	19.2 16.4	5.0 4.2	0.8 0.6
Dolomite	**H_2**	**CO_2**	**CO**	**CH_4**
Maximum (%) Average (%)	71.6 65.1	16.6 13.8	5.9 4.8	9.2 7.7
Ni/D 31 wt%	**H_2**	**CO_2**	**CO**	**CH_4**
Maximum (%) Average (%)	90.3 81.0	16.7 8.0	3.3 2.0	1.2 0.7
Ni/D 37 wt%	**H_2**	**CO_2**	**CO**	**CH_4**
Maximum (%) Average (%)	87.5 81.3	13.6 5.9	3.6 2.4	1.1 0.7

The same thing happens when comparing the other gas concentrations, i.e., CO_2, CO, and CH_4. In all cases, insignificant differences are also obtained.

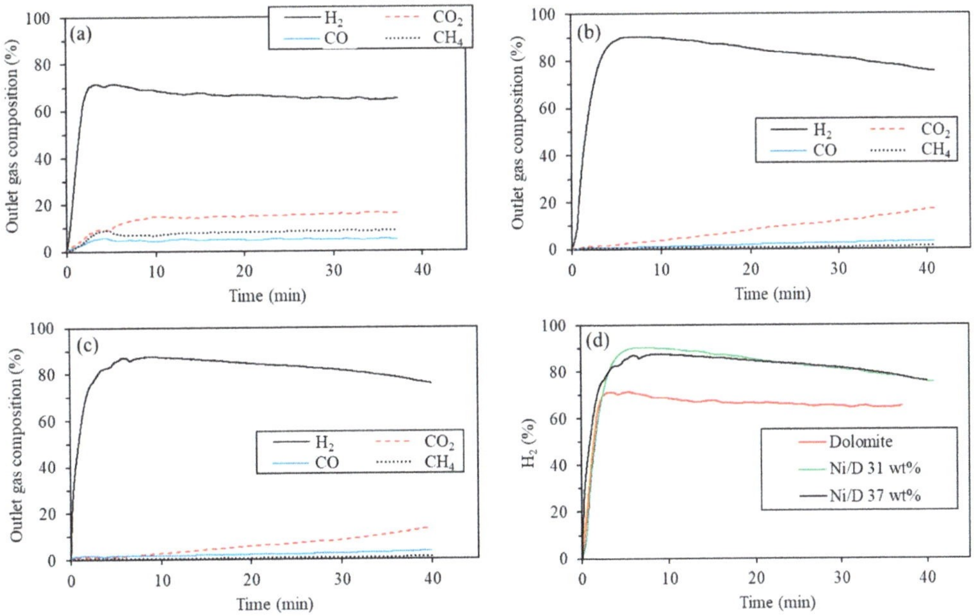

Figure 5. Average concentration of H_2, CO_2, CO, and CH_4 in syngas using different alumina supports (n = 2): dolomite without Ni (**a**), Ni/D 31 wt% (**b**), Ni/D 37 wt% (**c**). A comparison between the three supports with respect to H_2 concentration (in %) is also shown (**d**).

Comparatively, among the supports evaluated, the highest performance is achieved using dolomite as the support with Ni as the catalyst, reaching efficiencies of approximately 25%, corresponding to an average hydrogen concentration exceeding 80% and a maximum higher than 90% (Table 4). This could be related to (1) the better dispersion of Ni in this support than in Al_2O_3 support due to the presence of both MgO [65] and CaO [66]; (2) the higher content of Ni adsorbed by the support increasing the number of active sites on the catalyst surface, as occurs in [67]; (3) the presence of basic oxides, such as MgO and CaO, able to promote the water–gas shift reaction (Equation (3)), increasing H_2 production [68]; and (4) the different porous structure, since alumina is a support characterized by a larger specific surface area than dolomite due to its smaller pores. The presence of these smaller pores inhibits intra-particle diffusion of reactants and products [23]. Moreover, like what was observed with the nickel-free supports, the concentrations of CO_2, CO, and CH_4 are lower than those obtained with the alumina support, partly due to the carbonation processes occurring in dolomite and to the higher basicity of this support compared to alumina.

Table 4. Maximum and average hydrogen concentrations, in %, obtained from the reforming processes using alumina and dolomite, both with and without Ni. The efficiency achieved with each support when is coated with Ni is also shown.

Support	Maximum (%)	Average (%)	Efficiency (%)
Alumina	70.8	62.0	-
Ni/Al_2O_3 9 wt%	72.8	66.2	6.8
Ni/Al_2O_3 12 wt%	76.0	70.1	13.1
Dolomite	71.6	65.1	-
Ni/D 31 wt%	90.3	81.0	24.4
Ni/D 37 wt%	87.5	81.3	24.9

5. Conclusions

This study demonstrates that the incorporation of Ni significantly improves the reforming efficiency and hydrogen concentration compared with non-impregnated supports. Among the tested materials, zeolite showed the lowest performance and poor thermal stability at high temperature, while alumina exhibited moderate activity with partial deactivation at lower Ni contents. Dolomite, on the other hand, showed the most promising performance due to its higher porosity, greater nickel dispersion, the presence of basic oxides (CaO and MgO) that promote the water–gas shift reaction, and its capacity to capture CO_2 via recarbonation, which shifts the equilibrium toward hydrogen production. However, beyond a certain Ni loading of approximately 31 wt%, no further improvement in efficiency was observed, indicating a saturation effect in the distribution of active sites.

Future work should focus on correlating catalyst structure and performance, optimizing key operating parameters, and exploring Ni-based catalyst modifications to improve stability and coking resistance. Additionally, scale-up tests and preliminary techno-economic assessments under industrially relevant conditions would allow a broader evaluation of dolomite- and alumina-supported catalysts.

Author Contributions: Conceptualization, J.F. and R.M.; methodology, J.F., A.d.l.F. and V.A.; validation, A.d.l.F., J.F. and V.A.; formal analysis, J.F., A.d.l.F. and N.A.; investigation, A.d.l.F. and J.F.; resources, J.F. and R.M.; data curation, J.F. and A.d.l.F.; writing—original draft preparation, J.F., R.M. and V.A.; writing—review and editing, J.F., R.M. and N.A.; supervision, J.F.; project administration, J.F.; funding acquisition, J.F. and R.M. All authors have read and agreed to the published version of the manuscript.

Funding: This research was funded by the Defense University Center at the Spanish Naval Academy (CUD-ENM) under projects PICUD-2021-07 and PICUD-2024-01.

Data Availability Statement: The original contributions presented in the study are included in the article; further inquiries can be directed to the corresponding author.

Acknowledgments: The authors would like to acknowledge the use of scientific and technical services of Servicios de Seguridad Alimentaria y Desarrollo Sostenible from Centro de Apoio Científico e Tecnolóxico á Investigación (CACTI-Universidade de Vigo).

Conflicts of Interest: The authors declare no conflicts of interest.

References

1. Energy Institute. *Statistical Review of World Energy 2025*, 74th ed.; Energy Institute: London, UK, 2025.
2. Martino, M.; Ruocco, C.; Meloni, E.; Pullumbi, P.; Palma, V. Main Hydrogen Production Processes: An Overview. *Catalysts* **2021**, *11*, 547. [CrossRef]
3. Schwengber, C.A.; Alves, H.J.; Schaffner, R.A.; Da Silva, F.A.; Sequinel, R.; Bach, V.R.; Ferracin, R.J. Overview of Glycerol Reforming for Hydrogen Production. *Renew. Sustain. Energy Rev.* **2016**, *58*, 259–266. [CrossRef]
4. Tao, X.; Zhao, Y.; Wang, S.; Li, C.; Li, R. Recent Advances and Perspectives for Solar-Driven Water Splitting Using Particulate Photocatalysts. *Chem. Soc. Rev.* **2022**, *51*, 3561–3608. [CrossRef]
5. Song, H.; Luo, S.; Huang, H.; Deng, B.; Ye, J. Solar-Driven Hydrogen Production: Recent Advances, Challenges, and Future Perspectives. *ACS Energy Lett.* **2022**, *7*, 1043–1065. [CrossRef]
6. Yu, L.; Ning, M.; Wang, Y.; Yuan, C.; Ren, Z. Direct Seawater Electrolysis for Hydrogen Production. *Nat. Rev. Mater.* **2025**. [CrossRef]
7. Mishra, A.; Park, H.; El-Mellouhi, F.; Suk Han, D. Seawater Electrolysis for Hydrogen Production: Technological Advancements and Future Perspectives. *Fuel* **2024**, *361*, 130636. [CrossRef]
8. Maceiras, R.; Feijoo, J.; Alfonsín, V.; Pérez, L.; Álvarez-Feijoo, M.A.; Falcón, P.; Vallejo, J.P. Influence of Alumina Fixed-Bed in Steam Reforming of Glycerol for Hydrogen Production. *Energy Rep.* **2023**, *9*, 309–315. [CrossRef]
9. Gujar, J.P.; verma, A.; Modhera, B. Optimizing Glycerol Conversion to Hydrogen: A Critical Review of Catalytic Reforming Processes and Catalyst Design Strategies. *Int. J. Hydrogen Energy* **2025**, *109*, 823–850. [CrossRef]
10. Florin, N.H.; Harris, A.T. Enhanced Hydrogen Production from Biomass with in Situ Carbon Dioxide Capture Using Calcium Oxide Sorbents. *Chem. Eng. Sci.* **2008**, *63*, 287–316. [CrossRef]
11. Diglio, M.; Contento, I.; Impemba, S.; Berretti, E.; Della Sala, P.; Oliva, G.; Naddeo, V.; Caporali, S.; Primo, A.; Talotta, C.; et al. Hydrogen Production from Formic Acid Decomposition Promoted by Gold Nanoparticles Supported on a Porous Polymer Matrix. *Energy Fuels* **2025**, *39*, 14320–14329. [CrossRef]
12. Erdemir, D.; Dincer, I. A Quicker Route to Hydrogen Economy with Ammonia. *Int. J. Hydrogen Energy* **2024**, *82*, 1230–1237. [CrossRef]
13. Coşkuner Filiz, B.; Civelek Yörüklü, H.; Açıkalın, K.; Demirci, U.B.; Kantürk Figen, A. Boron-Based Hydrogen Storage Materials towards Power-to-X Technology on the Path to Carbon Neutrality. *Int. J. Hydrogen Energy* **2023**, *48*, 39389–39407. [CrossRef]
14. Goren, A.Y.; Temiz, M.; Erdemir, D.; Dincer, I. The Role of Effective Catalysts for Hydrogen Production: A Performance Evaluation. *Energy* **2025**, *315*, 134257. [CrossRef]
15. Ghaffari Saeidabad, N.; Noh, Y.S.; Alizadeh Eslami, A.; Song, H.T.; Kim, H.D.; Fazeli, A.; Moon, D.J. A Review on Catalysts Development for Steam Reforming of Biodiesel Derived Glycerol; Promoters and Supports. *Catalysts* **2020**, *10*, 910. [CrossRef]
16. Pompeo, F.; Santori, G.; Nichio, N.N. Hydrogen and/or Syngas from Steam Reforming of Glycerol. Study of Platinum Catalysts. *Int. J. Hydrogen Energy* **2010**, *35*, 8912–8920. [CrossRef]
17. Lin, K.H.; Chang, A.C.C.; Lin, W.H.; Chen, S.H.; Chang, C.Y.; Chang, H.F. Autothermal Steam Reforming of Glycerol for Hydrogen Production over Packed-Bed and Pd/Ag Alloy Membrane Reactors. *Int. J. Hydrogen Energy* **2013**, *38*, 12946–12952. [CrossRef]
18. Lai, H.J.; Liu, Y.C.; Nachimuthu, S.; Lin, S.D.; Jiang, J.C. Small Iridium Clusters Supported on TiO2 as Catalysts for Intensifying Low-Temperature Methane Activation and Reforming. *Chem. Eng. J.* **2024**, *492*, 152352. [CrossRef]
19. Kuznetsov, V.V.; Vitovsky, O.V.; Gasenko, O.A. Methane Steam Reforming in an Annular Microchannel with Rh/Al_2O_3 Catalyst. *J. Eng. Thermophys.* **2009**, *18*, 187–196. [CrossRef]
20. Namioka, T.; Saito, A.; Inoue, Y.; Park, Y.; Min, T.J.; Roh, S.A.; Yoshikawa, K. Hydrogen-Rich Gas Production from Waste Plastics by Pyrolysis and Low-Temperature Steam Reforming over a Ruthenium Catalyst. *Appl. Energy* **2011**, *88*, 2019–2026. [CrossRef]
21. Charisiou, N.D.; Douvartzides, S.L.; Siakavelas, G.I.; Tzounis, L.; Polychronopoulou, K.; Goula, M.A. The Relationship Between Reaction Temperature and Carbon Deposition on Nickel Catalysts Based on. *Catalysts* **2019**, *9*, 676. [CrossRef]
22. Cheng, C.K.; Foo, S.Y.; Adesina, A.A. H_2-Rich Synthesis Gas Production over Co/Al_2O_3 Catalyst via Glycerol Steam Reforming. *Catal. Commun.* **2010**, *12*, 292–298. [CrossRef]

23. Charisiou, N.D.; Siakavelas, G.I.; Papageridis, K.N.; Motta, D.; Dimitratos, N.; Sebastian, V.; Polychronopoulou, K.; Goula, M.A. The Effect of Noble Metal (M: Ir, Pt, Pd) on M/Ce_2O_3-γ-Al_2O_3 Catalysts for Hydrogen Production via the Steam Reforming of Glycerol. *Catalysts* **2020**, *10*, 790. [CrossRef]
24. Liguras, D.K.; Kondarides, D.I.; Verykios, X.E. Production of Hydrogen for Fuel Cells by Steam Reforming of Ethanol over Supported Noble Metal Catalysts. *Appl. Catal. B Environ.* **2003**, *43*, 345–354. [CrossRef]
25. López-Tenllado, F.J.; Estévez, R.; Hidalgo-Carrillo, J.; López-Fernández, S.; Urbano, F.J.; Marinas, A. Hydrogen Photo-Production from Glycerol on Platinum, Gold and Silver-Modified TiO_2-USY62 Catalysts. *Catal. Today* **2022**, *390–391*, 92–98. [CrossRef]
26. Marques, T.M.F.; Morais, R.N.; Nobre, F.X.; Rocha, J.M.; Ghosh, A.; Soares, T.A.S.; Viana, B.C.; Machado, G.; Costa, J.C.S.; De Matos, J.M.E. Hydrogen Production from Aqueous Glycerol Using Titanate Nanotubes Decorated with Au Nanoparticles as Photocatalysts. *An. Acad. Bras. Cienc.* **2019**, *91*, 1–15. [CrossRef]
27. Pompeo, F.; Santori, G.F.; Nichio, N.N. Hydrogen Production by Glycerol Steam Reforming with Pt/SiO_2 and Ni/SiO_2 Catalysts. *Catal. Today* **2011**, *172*, 183–188. [CrossRef]
28. Fakeeha, A.H.; Arafat, Y.; Ibrahim, A.A.; Shaikh, H.; Atia, H.; Abasaeed, A.E.; Armbruster, U.; Al-Fatesh, A.S. Highly Selective Syngas/H_2 Production via Partial Oxidation of CH_4 Using (Ni, Co and Ni-Co)/ZrO_2 -Al_2O_3 Catalysts: Influence of Calcination Temperature. *Processes* **2019**, *7*, 141. [CrossRef]
29. Thyssen, V.V.; Maia, T.A.; Assaf, E.M. Cu and Ni Catalysts Supported on γ-Al_2O_3 and SiO_2 Assessed in Glycerol Steam Reforming Reaction. *J. Braz. Chem. Soc.* **2015**, *26*, 22–31. [CrossRef]
30. Charisiou, N.D.; Papageridis, K.N.; Siakavelas, G.; Tzounis, L.; Goula, M.A. Effect of Active Metal Supported on SiO_2 for Selective Hydrogen Production from the Glycerol Steam Reforming Reaction. *BioResources* **2016**, *11*, 10173–10189. [CrossRef]
31. Abiso, A.M.; Fasanya, O.O.; Suleiman, M.Y.; Atta, A.Y.; Dutta, J.; Jibril, B.E.-Y. Advances in Copper-Based Catalysts for Sustainable Hydrogen Production via Methanol Steam Reforming. *Chem. Eng. J. Adv.* **2024**, *19*, 100625. [CrossRef]
32. Impemba, S.; Provinciali, G.; Filippi, J.; Salvatici, C.; Berretti, E.; Caporali, S.; Banchelli, M.; Caporali, M. Engineering the Heterojunction between TiO_2 and In_2O_3 for Improving the Solar-Driven Hydrogen Production. *Int. J. Hydrogen Energy* **2024**, *63*, 896–904. [CrossRef]
33. Zheng, X.; Song, Y.; Liu, Y.; Yang, Y.; Wu, D.; Yang, Y.; Feng, S.; Li, J.; Liu, W.; Shen, Y.; et al. ZnIn2S4-Based Photocatalysts for Photocatalytic Hydrogen Evolution via Water Splitting. *Coord. Chem. Rev.* **2023**, *475*, 214898. [CrossRef]
34. Iriondo, A.; Cambra, J.F.; Güemez, M.B.; Barrio, V.L.; Requies, J.; Sánchez-Sánchez, M.C.; Navarro, R.M. Effect of ZrO_2 Addition on Ni/Al_2O_3 Catalyst to Produce H_2 from Glycerol. *Int. J. Hydrogen Energy* **2012**, *37*, 7084–7093. [CrossRef]
35. Li, G.; Hu, L.; Hill, J.M. Comparison of Reducibility and Stability of Alumina-Supported Ni Catalysts Prepared by Impregnation and Co-Precipitation. *Appl. Catal. A Gen.* **2006**, *301*, 16–24. [CrossRef]
36. El Doukkali, M.; Iriondo, A.; Cambra, J.F.; Gandarias, I.; Jalowiecki-Duhamel, L.; Dumeignil, F.; Arias, P.L. Deactivation Study of the Pt and/or Ni-Based γ-Al_2O_3 Catalysts Used in the Aqueous Phase Reforming of Glycerol for H_2 Production. *Appl. Catal. A Gen.* **2014**, *472*, 80–91. [CrossRef]
37. Charisiou, N.D.; Siakavelas, G.; Papageridis, K.N.; Baklavaridis, A.; Tzounis, L.; Goula, G.; Yentekakis, I.V.; Polychronopoulou, K.; Goula, M.A. The Effect of WO_3 Modification of ZrO_2 Support on the Ni-Catalyzed Dry Reforming of Biogas Reaction for Syngas Production. *Front. Environ. Sci.* **2017**, *5*, 66. [CrossRef]
38. Shao, S.; Shi, A.-W.; Liu, C.-L.; Yang, R.-Z.; Dong, W.-S. Hydrogen Production from Steam Reforming of Glycerol over Ni/CeZrO Catalysts. *Fuel Process. Technol.* **2014**, *125*, 1–7. [CrossRef]
39. Alessio, H.J.; Pestana, G.L.; Comelli, R.A.; Grau, J.M. Hydrogen Production via Aqueous Phase Reforming of Glycerol over Ni-Co/γ-Al_2O_3 Catalysts: Effect of Support Modification with Lanthanides and Alkaline Earth Metals. *Fuel* **2025**, *404*, 136217. [CrossRef]
40. Luo, N.; Ouyang, K.; Cao, F.; Xiao, T. Hydrogen Generation from Liquid Reforming of Glycerin over Ni–Co Bimetallic Catalyst. *Biomass Bioenergy* **2010**, *34*, 489–495. [CrossRef]
41. Profeti, L.P.R.; Dias, J.A.C.; Assaf, J.M.; Assaf, E.M. Hydrogen Production by Steam Reforming of Ethanol over Ni-Based Catalysts Promoted with Noble Metals. *J. Power Sources* **2009**, *190*, 525–533. [CrossRef]
42. Karimi, S.; Bibak, F.; Meshkani, F.; Rastegarpanah, A.; Deng, J.; Liu, Y.; Dai, H. Promotional Roles of Second Metals in Catalyzing Methane Decomposition over the Ni-Based Catalysts for Hydrogen Production: A Critical Review. *Int. J. Hydrogen Energy* **2021**, *46*, 20435–20480. [CrossRef]
43. Iriondo, A.; Barrio, V.L.; Cambra, J.F.; Arias, P.L.; Güemez, M.B.; Navarro, R.M.; Sánchez-Sánchez, M.C.; Fierro, J.L.G. Hydrogen Production from Glycerol Over Nickel Catalysts Supported on Al_2O_3 Modified by Mg, Zr, Ce or La. *Top. Catal.* **2008**, *49*, 46–58. [CrossRef]
44. Soares, R.R.; Simonetti, D.A.; Dumesic, J.A. Glycerol as a Source for Fuels and Chemicals by Low-Temperature Catalytic Processing. *Angew. Chemie Int. Ed.* **2006**, *45*, 3982–3985. [CrossRef]
45. Bepari, S.; Kuila, D. Steam Reforming of Methanol, Ethanol and Glycerol over Nickel-Based Catalysts-A Review. *Int. J. Hydrogen Energy* **2020**, *45*, 18090–18113. [CrossRef]

46. Luo, N.; Zhao, X.; Cao, F.; Xiao, T.; Fang, D. Thermodynamic Study on Hydrogen Generation from Different Glycerol Reforming Processes. *Energy Fuels* **2007**, *21*, 3505–3512. [CrossRef]
47. Alberton, A.L.; Souza, M.M.V.M.; Schmal, M. Carbon Formation and Its Influence on Ethanol Steam Reforming over Ni/Al_2O_3 Catalysts. *Catal. Today* **2007**, *123*, 257–264. [CrossRef]
48. Maceiras, R.; Feijoo, J.; Perez-Rial, L.; Alfonsin, V.; Falcon, P. Study of Natural Zeolites for Hydrogen Purification: CO_2 Adsorption Capacity and Kinetic Mechanism. *Fuel* **2024**, *376*, 132732. [CrossRef]
49. van Heijst, J.; Martin-Calvo, A.; Calero, S. Hydrogen Recovery from Steam Methane Reforming Using the ITQ-12 Zeolite. *Sep. Purif. Technol.* **2024**, *350*, 127895. [CrossRef]
50. Maceiras, R.; Feijoo, J.; Alfonsin, V.; Alvarez-Feijoo, M.A.; Azofra, L. Enhanced CO_2 Adsorption for Hydrogen Purification Using Calcined Zeolite: Process Analysis and Kinetic Study. *Microporous Mesoporous Mater.* **2025**, *399*, 113869. [CrossRef]
51. Yao, D.; Yang, H.; Chen, H.; Williams, P.T. Investigation of Nickel-Impregnated Zeolite Catalysts for Hydrogen/Syngas Production from the Catalytic Reforming of Waste Polyethylene. *Appl. Catal. B Environ.* **2018**, *227*, 477–487. [CrossRef]
52. Grasa, G.S.; Abanades, J.C. CO2 Capture Capacity of CaO in Long Series of Carbonation/Calcination Cycles. *Ind. Eng. Chem. Res.* **2006**, *45*, 8846–8851. [CrossRef]
53. Sun, P.; Grace, J.R.; Lim, C.J.; Anthony, E.J. The Effect of CaO Sintering on Cyclic CO_2 Capture in Energy Systems. *AIChE J.* **2007**, *53*, 2432–2442. [CrossRef]
54. Zhang, B.; Zhang, L.; Yang, Z.; He, Z. An Experiment Study of Biomass Steam Gasification over NiO/Dolomite for Hydrogen-Rich Gas Production. *Int. J. Hydrogen Energy* **2017**, *42*, 76–85. [CrossRef]
55. Gallucci, K.; Stendardo, S.; Foscolo, P.U. CO_2 Capture by Means of Dolomite in Hydrogen Production from Syn Gas. *Int. J. Hydrogen Energy* **2008**, *33*, 3049–3055. [CrossRef]
56. Weitkamp, J.; Hunger, M. Chapter 22—Acid and Base Catalysis on Zeolites. In *Introduction to Zeolite Science and Practice*; Čejka, J., van Bekkum, H., Corma, A., Schüth, F., Eds.; Studies in Surface Science and Catalysis; Elsevier: Amsterdam, The Netherlands, 2007; Volume 168, pp. 787–835. [CrossRef]
57. Feijoo, J.; Gomez-Villalba, L.S.; de los Ríos, A.; Fort, R. Electroprecipitation of Inorganic Borates, with Different Solubility, within Monumental Stones to Avoid Fungal Colonization. *Constr. Build. Mater.* **2023**, *368*, 130435. [CrossRef]
58. de Rosario, I.; Rivas, T.; Buceta, G.; Feijoo, J.; Mosquera, M.J. Surfactant-Synthesized Consolidants Applied To A Granitic Medieval Necropolis In NW Spain. Laboratory And In Situ Effectiveness Evaluation. *Int. J. Archit. Herit.* **2017**, *11*, 1166–1176.
59. Feijoo, J.; de Rosario, I.; Rivas, T.; Mosquera, M.J.; Benavides, R. Influence of a Pre-Consolidation Treatment on the Desalination Effectiveness of a Highly Deteriorated Granite Façade of Medieval Age. *Int. J. Archit. Herit.* **2023**, *17*, 1965–1983. [CrossRef]
60. Mambetova, M.; Dossumov, K.; Baikhamurova, M.; Yergaziyeva, G. Sorbents Based on Natural Zeolites for Carbon Dioxide Capture and Removal of Heavy Metals from Wastewater: Current Progress and Future Opportunities. *Processes* **2024**, *12*, 2071. [CrossRef]
61. Yaşyerli, S.; Ar, İ.; Doğu, G.; Doğu, T. Removal of Hydrogen Sulfide by Clinoptilolite in a Fixed Bed Adsorber. *Chem. Eng. Process. Process Intensif.* **2002**, *41*, 785–792. [CrossRef]
62. Kuldeyev, E.; Seitzhanova, M.; Tanirbergenova, S.; Tazhu, K.; Doszhanov, E.; Mansurov, Z.; Azat, S.; Nurlybaev, R.; Berndtsson, R. Modifying Natural Zeolites to Improve Heavy Metal Adsorption. *Water* **2023**, *15*, 2215. [CrossRef]
63. Shamsuddin, M.R.; Asikin-Mijan, N.; Marliza, T.S.; Miyamoto, M.; Uemiya, S.; Yarmo, M.A.; Taufiq-Yap, Y.H. Promoting Dry Reforming of Methane via Bifunctional NiO/Dolomite Catalysts for Production of Hydrogen-Rich Syngas. *RSC Adv.* **2021**, *11*, 6667–6681. [CrossRef] [PubMed]
64. Zulqarnain; Kim, S.; Hyuk Chun, D.; Yoo, J.; Kim, S.; Im, H.; Choi, H.; Lim, J. Improving the Activity and Long-Term Durability of the Ni/Al_2O_3 Catalyst in the Low-Temperature Steam Reforming of Highly Concentrated Volatile Organic Compounds. *Energy Convers. Manag. X* **2024**, *24*, 100819. [CrossRef]
65. Bastan, F.; Kazemeini, M.; Larimi, A.; Maleki, H. Production of Renewable Hydrogen through Aqueous-Phase Reforming of Glycerol over Ni/Al_2O_3MgO Nano-Catalyst. *Int. J. Hydrogen Energy* **2018**, *43*, 614–621. [CrossRef]
66. Md Radzi, M.R.; Azizan, M.T.; Daud, N.F.D.M.; Topek, M.A.; Abidin, S.Z. Aqueous Phase Reforming of Sorbitol over Ca Doped Ni/Al_2O_3 for Value-Added Chemicals Production. *Mater. Today Proc.* **2018**, *5*, 21728–21736. [CrossRef]
67. Kiren, B.; Ayas, N. Nickel Modified Dolomite in the Hydrogen Generation from Sodium Borohydride Hydrolysis. *Int. J. Hydrogen Energy* **2022**, *47*, 19702–19717. [CrossRef]
68. Guo, Y.; Azmat, M.U.; Liu, X.; Wang, Y.; Lu, G. Effect of Support's Basic Properties on Hydrogen Production in Aqueous-Phase Reforming of Glycerol and Correlation between WGS and APR. *Appl. Energy* **2012**, *92*, 218–223. [CrossRef]

hydrogen

MDPI

Article

Assessing the Effect of Mineralogy and Reaction Pathways on Geological Hydrogen (H_2) Generation in Ultramafic and Mafic (Basaltic) Rocks

Abubakar Isah *, Hamidreza Samouei and Esuru Rita Okoroafor

Harold Vance Department of Petroleum Engineering, Texas A&M University, College Station, TX 77843-3116, USA; samouei@tamu.edu (H.S.); ritaok@tamu.edu (E.R.O.)
* Correspondence: abubakar.isah@tamu.edu

Abstract

This study evaluates the impact of mineralogy, elemental composition, and reaction pathways on hydrogen (H_2) generation in seven ultramafic and mafic (basaltic) rocks. Experiments were conducted under typical low-temperature hydrothermal conditions (150 °C) and captured early and evolving stages of fluid–rock interaction. Pre- and post-interactions, the solid phase was analyzed using X-ray Diffraction (XRD) and X-ray Photoelectron Spectroscopy (XPS), while Inductively Coupled Plasma Mass Spectrometry (ICP-MS) was used to determine the composition of the aqueous fluids. Results show that not all geologic H_2-generating reactions involving ultramafic and mafic rocks result in the formation of serpentine, brucite, or magnetite. Our observations suggest that while mineral transformation is significant and may be the predominant mechanism, there is also the contribution of surface-mediated electron transfer and redox cycling processes. The outcome suggests continuous H_2 production beyond mineral phase changes, indicating active reaction pathways. Particularly, in addition to transition metal sites, some ultramafic rock minerals may promote redox reactions, thereby facilitating ongoing H_2 production beyond their direct hydration. Fluid–rock interactions also regenerate reactive surfaces, such as clinochlore, zeolite, and augite, enabling sustained H_2 production, even without serpentine formation. Variation in reaction rates depends on mineralogy and reaction kinetics rather than being solely controlled by Fe oxidation states. These findings suggest that ultramafic and mafic rocks may serve as dynamic, self-sustaining systems for generating H_2. The potential involvement of transition metal sites (e.g., Ni, Mo, Mn, Cr, Cu) within the rock matrix may accelerate H_2 production, requiring further investigation. This perspective shifts the focus from serpentine formation as the primary driver of H_2 production to a more complex mechanism where mineral surfaces play a significant role. Understanding these processes will be valuable for refining experimental approaches, improving kinetic models of H_2 generation, and informing the site selection and design of engineered H_2 generation systems in ultramafic and mafic formations.

Keywords: geologic hydrogen; serpentinization; redox cycling; surface-mediated reactions; mafic and ultramafic rocks

1. Introduction

The geological process known as serpentinization has garnered significant attention in recent years due to its potential as a method for generating low-carbon hydrogen (H_2). This process involves the reaction of water with ultramafic rocks, primarily composed of olivine and pyroxene, leading to the oxidation of ferrous iron (Fe^{2+}) and the production of hydrogen gas as a byproduct. Hydrogen generation through fluid–rock interactions has been extensively studied, with a focus on ultramafic and mafic rocks due to their rich Fe-bearing mineral compositions. An extreme case is the hydrous alteration of ultramafic rocks, known as serpentinization, where most of the original rock minerals are replaced by serpentine. The alteration of mafic rocks, such as basalts, also produces H_2, but in smaller amounts compared to those generated in ultramafic environments [1–4]. Experimental studies suggest that the alteration of felsic rocks under hydrothermal conditions can also lead to significant H_2 production [5].

Early studies, such as those by [6], primarily investigated serpentinization kinetics, focusing on the hydration rates of forsterite and enstatite, which led to the formation of serpentine, brucite, and talc. However, these studies did not assess H_2 production, leaving a gap in understanding the direct role of mineral transformation in hydrogen generation. Kita et al. [7] explored H_2 generation in granite and quartz systems at temperatures ranging from 25 to 270 °C, attributing H_2 production to reactions involving Si- and Si-O-based transformations. These studies laid the foundation for understanding geochemical transformations, but did not fully address the iron (Fe) oxidation pathways as a major driver of H_2 evolution.

Recent discoveries underscore the growing importance of natural hydrogen systems in diverse geological contexts. For example, studies in the Songliao Basin [8] and Bohai Bay Basin [9] in China demonstrate that large-scale H_2 generation is actively occurring in mafic and ultramafic lithologies, highlighting their potential as sustainable subsurface energy resources. Similarly, well-characterized sites in Oman, the Philippines, and Turkey [10] confirm that serpentinization-driven H_2 production is a global phenomenon with significant implications for future energy portfolios. Despite these advances, fundamental mechanisms governing H_2 fluxes remain poorly constrained, particularly the interplay of mineralogy, geochemical conditions, and fluid–rock interactions. Such knowledge is critical for assessing the viability of natural hydrogen as a sustainable energy resource and for developing strategies to optimize its production, utilization, and storage in subsurface systems.

More recent research has demonstrated that H_2 generation is strongly linked to the oxidation of Fe^{2+} to Fe^{3+} in ultramafic systems. Marcaillou et al. [11] investigated peridotite at 300 °C and 300 bar, demonstrating that Fe^{2+} oxidation in olivine and pyroxene, resulting in Fe^{3+} in magnetite and serpentine, contributed to H_2 production. Mayhew et al. [12] expanded upon these findings by examining a broader range of ultramafic rocks and minerals, including fayalite, pyroxene, olivine, and magnetite, under ambient pressure and at temperatures of 55 °C and 100 °C over a period of 100 days. Their results indicated a direct correlation between H_2 production and metal oxide minerals with the general formula $[M^{2+}M_2{}^{3+}]O_4$, though their experiments did not reach equilibrium conditions. Similarly, [13] conducted computational simulations on various compositions of forsterite-fayalite, establishing a direct correlation between the Fe content of olivine and the extent of H_2 generation across a temperature range of 25 °C to 400 °C at 50 MPa.

Experimental studies by [14,15] have further refined our understanding of H_2 production mechanisms in rocks. Using synthetic samples at 230 °C and 35 MPa over 146 days, the authors of [14] found that H_2 generation increased significantly with pH and that the reaction of olivine and orthopyroxene proceeded at a slower rate initially. Their later study [15] focused on high-Mg, low-Fe minerals, demonstrating that Fe-rich olivine produced more H_2 per mole than Fe-poor olivine. These results reinforce the importance of Fe oxidation states in controlling H_2 generation and suggest that mineral composition plays an important role in determining reaction kinetics and efficiency. However, these studies examined fewer rock types, which limits their broader applicability.

Some previous studies used only major elements data, including MgO, SiO_2, FeO, CaO, Na_2O, and K_2O, to estimate H_2 generation potential. Although these estimates of past ultramafic rock distribution rely on assumptions based on a single bulk rock composition [16,17], natural ultramafic rocks exhibit compositional variability, leading to differences in H_2 production potential. Leong et al. [18] addressed this variability by incorporating rocks with a range of compositions, from Mg-rich to Mg-depleted. However, variations in Fe and other trace minerals were not considered, and the simulations were conducted only under ambient temperature conditions.

Leong et al. [18] utilized thermodynamic simulations of water–rock interactions combined with mass-transport calculations to estimate H_2 production from the alteration of Fe-bearing igneous rocks under low-temperature conditions (25 °C). Their simulations predicted H_2-generation potential for the hydrous alteration of Fe-rich igneous rocks, ranging from ultramafic rocks with high MgO content, such as peridotites, to those with lower MgO content, like basalts. The authors observed a decrease in H_2-generation potential between rocks with MgO content above and below 35 wt.%, even when FeO concentrations were similar. Additionally, their model results indicated a gradual reduction in H_2 generation as rock compositions became less Mg-rich, approaching 20 wt.% MgO. They do not promote serpentine formation during rock alteration. Instead, the alteration of these Mg-deficient but Si- and Al-rich rocks favors the stabilization of minerals like chlorite, talc, and clay minerals.

Recent research has also emphasized the impact of environmental conditions on serpentinization and H_2 evolution. Leong et al. [19] investigated H_2 outgassing from the Samail ophiolite in Oman under ambient temperature and pressure conditions (25 °C, 1 bar). Their study reported serpentinization rates and H_2 yields of approximately 8×10^{-14} s^{-1} and 0.3 mol H_2 kg^{-1} of ultramafic rock, respectively, emphasizing the potential for natural H_2 reservoirs in ultramafic formations. The most prominent reactions being reported in the literature involve the reaction of olivine and water to form serpentine [14,15] or serpentine, magnetite, brucite, and hydrogen [11,14,20–22], while others reported the formation of magnetite, silica, and hydrogen generation, as shown in reactions (1) through (4).

Olivine (Fosterite) + Water → Serpentine

$$3Mg_2SiO_4 + SiO_2 + 4H_2O \rightarrow 2Mg_3Si_2O_5(OH)_4 \quad (1)$$

Olivine (Fosterite) + Water → Serpentine + Brucite + Hydrogen

$$2Mg_2SiO_4 + 3H_2O \rightarrow Mg_3Si_2O_5(OH)_4 + Mg(OH)_2 \quad (2)$$

$$2Mg_2SiO_4 + 4H_2O + H^+ \rightarrow Mg_3Si_2O_5(OH)_4 + Mg(OH)_2 + H_2 + OH^- \quad (3)$$

Olivine (Fayalite) + Water → Magnetite + Silica + Hydrogen

$$3Fe_2SiO_4 + 2H_2O \rightarrow 2Fe_3O_4 + 3SiO_2 + 2H_2 \quad (4)$$

The amount of H_2 generated through Fe oxidation, as seen in reaction (4), is influenced far more by the bulk composition of the rock [18,23]. Furthermore, the thermodynamic modeling by [23,24] suggests that other reactions may be contributing to H_2 production. These authors indicate that iron partitioning and oxidation state are highly dependent on reaction conditions, while the availability of an external silica source influences magnetite formation.

The body of literature demonstrates that Fe oxidation and partitioning, rock mineral composition, and the impact of pH and temperature are important factors in determining H_2 yields from ultramafic and mafic rocks. Rock mineralogy is a fundamental factor influencing H_2 generation during serpentinization and hydro-geochemical reactions in mafic and ultramafic rocks [18,23,25]. Accordingly, rock minerals may not only participate as reactants but could also act as goecatalysts, influencing reaction rates and pathways without necessarily being consumed or transformed [26–32]. The potential for catalysis in geochemical reactions due to the rock minerals is likely to have significant implications for their progression.

The hypothesis of this study is that H_2 generation from ultramafic and mafic rocks is not solely controlled by mineral hydration or serpentinization, but is also influenced by Fe oxidation–reduction cycles, and the presence of trace reactive minerals in ultramafic and mafic rocks. The objective of this study is to test this hypothesis by investigating what mineralogy, elemental composition, and reaction mechanisms influence H_2 generation in ultramafic and basaltic mafic rocks. It aims to identify key mineral phases, and geochemical interactions that control H_2 production during water–rock interactions in these natural rock systems.

The current literature primarily focuses on either a limited selection of rock samples or relies heavily on simulations to study hydrogen generation processes. The novelty of this work lies in its comprehensive experimental investigation of a diverse range of rock types, up to seven, including ultramafic and mafic rocks sourced from different geological regions. This broader scope allows for a more in-depth understanding of H_2 generation beyond the widely studied serpentinization process, revealing additional reaction pathways and complexities that may not have been extensively explored. Employing X-ray Photoelectron Spectroscopy (XPS) for surface elemental analysis provides detailed information on the elemental composition and oxidation states of these rocks. This involves assessing natural ultramafic and mafic (basaltic) rocks obtained from various locations, comparing H_2 generation potential, mineral phase modifications, reaction rates, and changes in the oxidation states of some major elements via spectroscopic methods. Understanding the role of minerals in accelerating reactions or acting as buffers could lead to new insights into natural hydrogen generation, mineral transformations, and broader geochemical processes to improve hydrogen generation yield from mafic and ultramafic rocks.

2. Methodology

2.1. Rock Samples

This study utilized a total of seven rock samples, consisting of five ultramafic and two mafic (basaltic) rock types (see Figure 1), to investigate their potential for H_2 generation under controlled experimental conditions. The ultramafic rocks consisted of two samples sourced from the Trinity Ophiolite, at Eunice Bluff, California, designated as Samples A and E. Additionally, the ultramafic category included three peridotite samples from different sources: Peridotite Sample B, obtained via Ward's Science from Balsam, North Carolina; Olivine Sample C obtained via Northern Geological Supplies Ltd., Bolton, UK, from the Gusdal Olivine Pit in Åheim, Norway; and Sample D, acquired from the Twin Sisters Peridotite in Washington. The mafic (basaltic) group included two samples: Sample F obtained via Northern Geological Supplies Ltd., Bolton, UK, from Finland, and Sample G, sourced through a supplier, EISCO, on Amazon.com, processed in Victor, New York, whose exact source location could not be determined. These samples represent a diverse selection of specimens collected from both within and outside the United States.

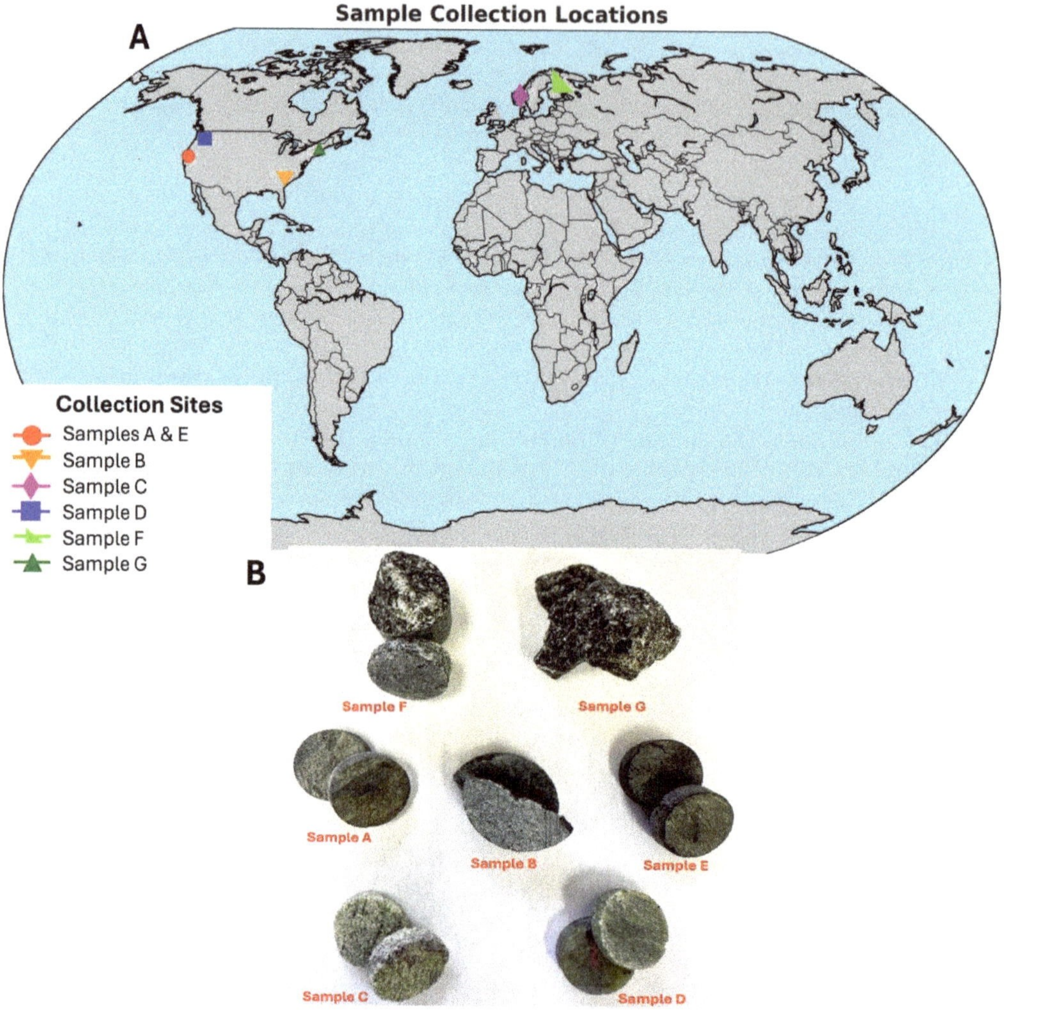

Figure 1. Geographical map showing the collection locations (**A**) and photographs (**B**) of the seven rock samples investigated, including five ultramafic (Dunite ophiolite (Samples A and E), Sample B, Sample C, and Sample D) and two mafic rock samples (Samples F and G).

The rock samples were initially crushed using a porcelain mortar and pestle to avoid any potential metal contamination that could influence subsequent geochemical reactions. The crushed material was then dry-sieved to obtain a uniform particle size distribution. To remove adhering fines and dust that might artificially enhance reaction kinetics, the sieved fractions were carefully rinsed with de-ionized (DI) water and subjected to ultrasonic agitation (sonication). This cleaning step ensured that only well-defined grains were retained for experimentation. A target particle size range of 150–300 μm was selected, as this size is commonly used in laboratory-scale fluid–rock interaction studies to balance sufficient surface reactivity with representative bulk rock properties. The prepared samples were characterized to confirm a consistent specific surface area of approximately 12 m^2/kg of rock, which was maintained across all experiments to ensure comparability of results.

2.2. Experimental Description

The experiments were conducted in a 150 mL Hastelloy mini-autoclave using a water–rock ratio of approximately 2. The water–rock mixture was placed in the Hastelloy autoclave reactor. The system was pressurized to 320 psi using high-purity nitrogen (99.99%). Before pressurization, nitrogen gas was used to purge the free space, ensuring the removal of air from the system. The autoclave was then heated in an oven set to 150 °C at a heating rate of 10 °C per minute, with a precision of ±1 °C. As the temperature reached 150 °C, the reaction reactor pressure increased to 500 psi. The rock samples were subjected to temperature and pressure hydrothermal conditions reported in some serpentinized systems [12,22,33], simulating natural subsurface environments where natural hydrogen is produced through rock–water interactions. The hydrothermal rock–fluid interactions were carried out for 3 days in the first batch and 7 days in the second batch of the experiment. The experiments and measurements were conducted at 3 and 7 days because the samples were crushed to provide high-surface-area powders, which accelerate serpentinization. Under these conditions, diagnostic changes such as reactive minerals dissolution and H_2 release can be detected, effectively capturing the early stages of reaction rates. Extended experiments were also performed for Sample A, with reaction durations of 14 and 28 days. Low-temperature serpentinization, mineral alterations, and H_2 production over these time durations have been reported elsewhere [14,34–36].

2.2.1. Fluids Analysis

At the end of the experiment, the system was allowed to cool at an ambient temperature for one hour, and the H_2 generated was measured. Hydrogen gas detection was performed using an electrochemical gas analyzer with a measurement accuracy of ±2 ppm. The H_2 gas was carefully collected in a transparent, sealed plastic bag to ensure reliable and contamination-free measurements. Inside the sealed plastic bag, the H_2 detector was placed to measure the gas concentration in a controlled environment. A known volume of gas was introduced, ensuring that the concentration remained within the detection range of the analyzer. Once the gas was stabilized, the H_2 concentration was recorded.

The supernatant aqueous fluid was collected in separate vials and analyzed using Inductively Coupled Plasma Mass Spectrometry (ICP-MS). This technique analyzes the chemical composition by measuring the levels of various elements. This was used to detect elements in the liquids, including those present in trace amounts. ICP-MS to assess the elemental compositions necessary for understanding the precipitation/dissolution processes during the rock–fluid interactions. In addition, the total dissolved solids (TDS), conductivity, and pH of the reaction fluids were measured using a pH/TDS/conductivity meter to monitor changes in the geochemical properties of the system over time. These parameters serve as indicators of fluid chemistry evolution and potential mineral dissolution or precipitation processes. Measuring TDS provides the solution's overall concentration of dissolved ions, while conductivity reflects the ionic strength and mobility of charged species. Meanwhile, pH variations can indicate acid–base reactions and shifts in equilibrium conditions. These measurements were conducted to evaluate the precipitation and dissolution kinetics of rock–fluid interactions during hydrogen production.

2.2.2. Mineral Characterization Using X-Ray Diffraction (XRD)

X-ray Diffraction (XRD) analysis was performed on the unreacted rock samples to establish a baseline mineralogical composition, which was then compared to the mineralogical changes observed in the hydrothermally reacted samples. This enables the identification of any phase transformations, mineral dissolution, or secondary mineral formations that result from hydrothermal reactions. At the end of the experiment, the solids were retrieved and dehydrated in an oven at 100 °C for 2 h. Alteration of mineral phases was evaluated through XRD analysis using a Malvern Empyrean PANalytical diffractometer equipped with Ni-filtered Cu Kα radiation. The dried bulk solids were homogenized into fine powder using a porcelain mortar and pestle and analyzed under operating conditions of 45 kV and 40 mA, over a 2θ range of 4.5–70°, with a scan time of 8.7 s. The resulting data were analyzed using the HighScore Plus (v. 4.9) search-match module, integrated with the ICDD PDF-4-2025 mineralogical library.

2.2.3. X-Ray Photoelectron Spectroscopy (XPS)

The X-ray Photoelectron Spectroscopy (XPS) spectra of the Fe 2p, Si, Mg, Al 2p, and O 1s photoelectron lines were recorded, and data were acquired using an Omicron ESCA+ system equipped with a magnesium (Mg) X-ray source. The emission current was set to 20 mA, while the applied voltage was maintained at 15 kV to ensure optimal excitation of core electrons. The pass energy was set at 20 eV, providing high-resolution spectra with enhanced peak definition and minimal background noise. The system was calibrated before analysis, and data acquisition was conducted under ultra-high-vacuum (UHV) conditions to minimize contamination and ensure reliable results. The collected spectra were used to determine the elemental composition, oxidation states, and chemical bonding environments of the rock samples before and after the reaction. Deconvolution and interpretation of the components' peaks were performed using CasaXPS software, version 2.3.26PR1.0. The determination of the oxidic Al 2p core level binding energies and the measured spectra were resolved into oxidic and metallic components in their respective binding energy regions. This procedure is discussed in detail in the works by [37–39].

3. Results

3.1. Hydrogen Generation

Hydro-geochemical reactions, often broadly referred to as serpentinization, are commonly associated with H_2 generation across various rock types. In the ultramafic and mafic rocks, both high and low levels of H_2 production have been observed, as shown in Figure 2. Among ultramafic rocks, Sample C generated as little as 1.2 mmol/kg of rock, while Sample B produced up to 9.6 mmol/kg of rock. This variation shows that H_2 generation can vary significantly within the same rock class. Despite both being ultramafic, there is an apparent disparity in their H_2 output.

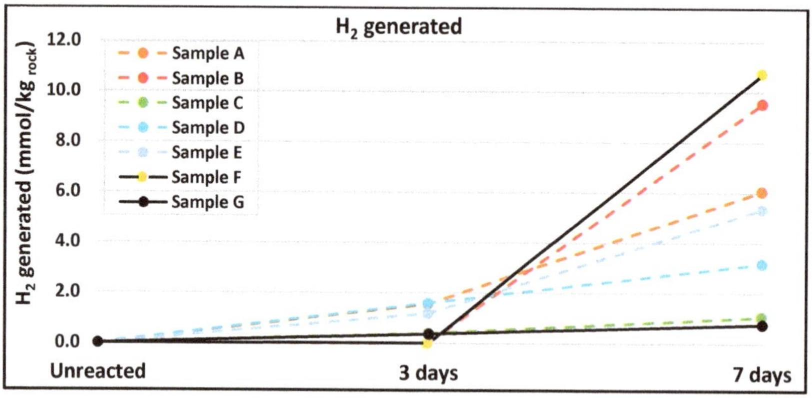

Figure 2. Hydrogen generation recorded for different ultramafic (dashed lines; Samples A, B, C, D, and E) and mafic (solid lines; Samples F and G) rocks over time.

Similarly, for mafic rocks, Sample F produced up to 10.8 mmol/kg of rock, whereas Sample G yielded only 0.76 mmol/kg of rock. These results indicate that H_2 generation is not uniform across ultramafic and mafic rocks, suggesting the influence of mineralogical and geochemical differences on H_2 production. The mineralogical composition of rocks plays a significant role in governing H_2 generation during serpentinization and hydrogeochemical reactions in mafic and ultramafic formations. The primary difference between mafic and ultramafic rocks lies in their silica and magnesium content [18]; however, this is likely not the only factor influencing H_2 production in these rocks. Replicate experiments, particularly the 3-day batch tests, were carried out on selected rock types under identical conditions. The results showed strong consistency in hydrogen production patterns and associated mineralogical changes. The subsequent sections examine how the mineralogy, elemental composition, and reaction pathways influence H_2 production in these rock types. The key mineral phases and geochemical processes that regulate H_2 release during water–rock interactions in ultramafic and mafic environments are further analyzed.

3.2. *Mineralogical Characterization*

3.2.1. Elemental Composition

The elemental composition of the rock samples using X-Ray Fluorescence (XRF) analysis is presented in Figure 3. Although both ultramafic and mafic rocks exhibit comparable concentrations of silicon (Si), iron (Fe), manganese (Mn), and zinc (Zn), they display notable differences in the composition of other major cations. Specifically, the mafic rocks (Sample F and G) contain substantially higher levels of aluminum (Al), calcium (Ca), sodium (Na), and potassium (K), while exhibiting significantly lower concentrations of magnesium (Mg) in comparison to the ultramafic rocks (Sample A to E). These variations in elemental composition likely reflect differences in the mineralogical makeup and geochemical evolution of the rocks, which could, in turn, result in varying hydrogen H_2 generation potentials across the different rock types.

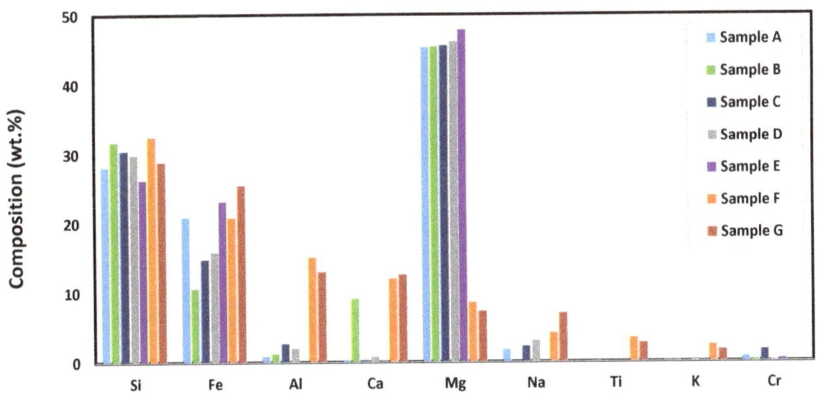

Figure 3. Elemental composition of the unreacted seven rock samples using X-Ray fluorescence (XRF) analysis: Five ultramafic (Sample A to E) and two mafic rock samples (Sample F and G).

3.2.2. Ultramafic Rocks

There are significant variations in the mineral composition of the assessed rocks, with each rock exhibiting unique mineralogical alterations over different treatment durations. For ultramafic rocks, notable mineralogical changes occur during the generation of hydrogen. For example, significant mineral alteration was observed in Sample A (Figure 4). The olivine content, initially 59.5 wt.%, decreased to 57.3 wt.% after 3 days and decreased to 55.3 wt.% after 7 days of interaction, and no brucite formation was detected in this sample over this reaction period. In Sample B (Figure 5), a substantial increase in the lizardite phase was observed after 3 days (64.3 wt.%) and 7 days (68.5 wt.%) of hydrothermal exposure, compared to its initial composition (38.4 wt.%). This aligns with the commonly reported serpentinization reaction responsible for H_2 production in hydrothermal systems. Additionally, Fe-rich minerals such as olivine, hematite, and clinopyroxene showed a decrease, indicating their transformation into lizardite, which correlates with H_2 generation. Furthermore, birnessite, a manganese hydroxide ($Mn(OH)_4$), was also observed to substantially decrease after 7 days of exposure, suggesting additional geochemical interactions during the reaction process.

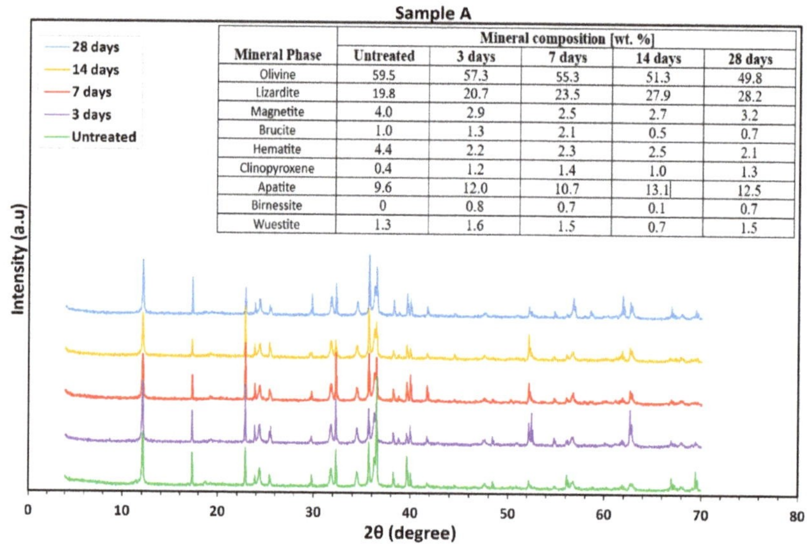

Mineral Phase	Mineral composition [wt. %]				
	Untreated	3 days	7 days	14 days	28 days
Olivine	59.5	57.3	55.3	51.3	49.8
Lizardite	19.8	20.7	23.5	27.9	28.2
Magnetite	4.0	2.9	2.5	2.7	3.2
Brucite	1.0	1.3	2.1	0.5	0.7
Hematite	4.4	2.2	2.3	2.5	2.1
Clinopyroxene	0.4	1.2	1.4	1.0	1.3
Apatite	9.6	12.0	10.7	13.1	12.5
Birnessite	0	0.8	0.7	0.1	0.7
Wuestite	1.3	1.6	1.5	0.7	1.5

Figure 4. XRD spectra of the progressive mineralogical transformations occurring in Dunite ophiolite Sample A throughout the duration of hydrothermal reaction experiments for H_2 generation.

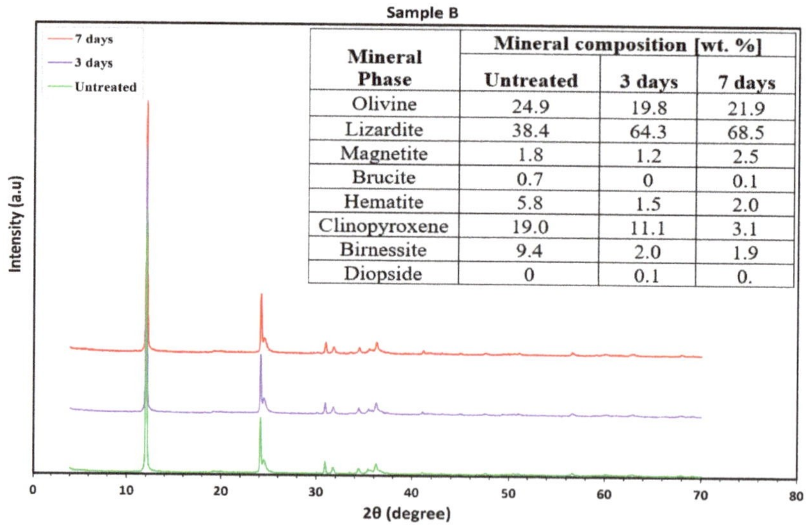

Mineral Phase	Mineral composition [wt. %]		
	Untreated	3 days	7 days
Olivine	24.9	19.8	21.9
Lizardite	38.4	64.3	68.5
Magnetite	1.8	1.2	2.5
Brucite	0.7	0	0.1
Hematite	5.8	1.5	2.0
Clinopyroxene	19.0	11.1	3.1
Birnessite	9.4	2.0	1.9
Diopside	0	0.1	0.

Figure 5. XRD spectra of the progressive mineralogical transformations in peridotite-Sample B for the duration of hydrothermal reaction experiments for H_2 generation.

Similar mineral alterations were observed for Sample C (Figure 6), with a decrease in olivine content from the initial 82.4 wt.% to 77.6 wt.% and 74.6 wt.% after 3 and 7 days of reaction, respectively. Correspondingly, lizardite increased from 4.8 wt.% to 7.3 wt.% and 8.2 wt.%. Additionally, this sample showed an increase in brucite content from 1.8 wt.% to 2.6 wt.% after 7 days. Contrary to Sample A, which showed a decreasing trend in hematite, Sample C exhibited an increase in hematite content from 1.6 wt.% to 3.9 wt.% and 5.7 wt.% after 3 and 7 days of hydrothermal interaction, respectively. No magnetite was observed in this sample. XRD analysis revealed changes in the composition of iron-rich minerals such as olivine, hematite, and magnetite, indicating their role as major reactants in the transformation process.

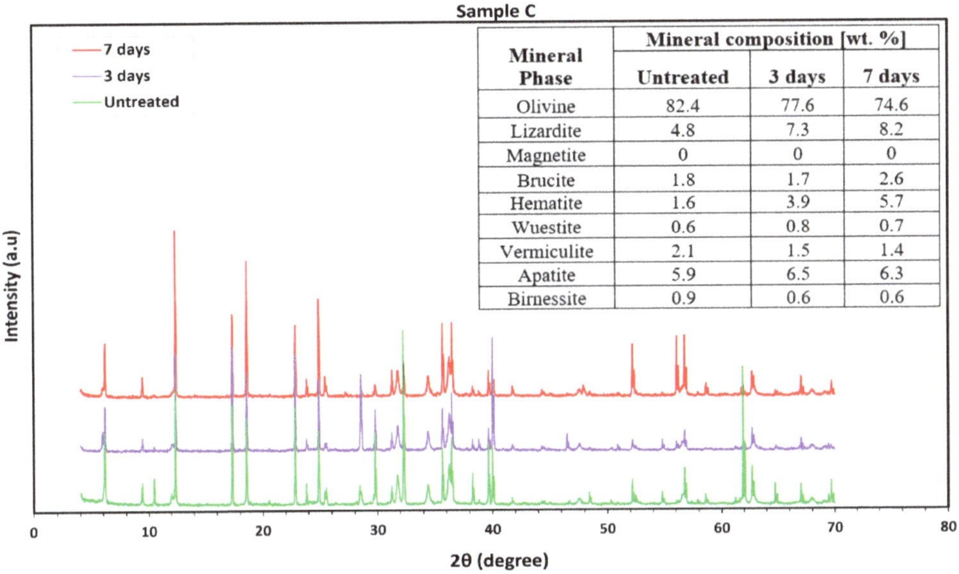

Mineral Phase	Mineral composition [wt. %]		
	Untreated	3 days	7 days
Olivine	82.4	77.6	74.6
Lizardite	4.8	7.3	8.2
Magnetite	0	0	0
Brucite	1.8	1.7	2.6
Hematite	1.6	3.9	5.7
Wuestite	0.6	0.8	0.7
Vermiculite	2.1	1.5	1.4
Apatite	5.9	6.5	6.3
Birnessite	0.9	0.6	0.6

Figure 6. XRD spectra of the progressive mineralogical transformations in Sample C throughout the duration of hydrothermal reaction experiments for H_2 generation.

For Sample D (Figure 7) and Sample E (Figure 8), although a decent amount of H_2 was generated by these samples (3.2 and 5.4 mmol/kg of rock, respectively), there was no notable magnetite presence as shown in the XRD. Considering Sample D (Figure 4), olivine content decreased from 70.7 wt.% to 66.8 wt.% and 62.8 wt.% after 3 and 7 days, respectively. Meanwhile, lizardite increased from 15.2 wt.% in the untreated rock to 23.9 wt.% after 7 days of interaction, accompanied by the formation of brucite, which reached 1.4 wt.% after 7 days of exposure. This is the well-established pathway for H_2 generation during serpentinization. It involves serpentine (lizardite) formation through the hydration of olivine, along with brucite precipitation, where Mg-rich forsterite reacts with water to produce $Mg(OH)_2$. Additionally, the oxidation of Fe-rich wuestite (FeO) facilitates H_2 release as Fe^{2+} undergoes oxidation.

For Sample E (Figure 8), where larger H_2 was produced (compared to Sample D), there was considerable mineral phase transformation. Specifically, the lizardite content increased from 19.6 wt.% in the untreated rock to 36.7 wt.% after 7 days of hydrothermal exposure. No brucite or magnetite was formed. This suggests that not all geological H_2-generating reactions in ultramafic rocks result in the formation of brucite or magnetite.

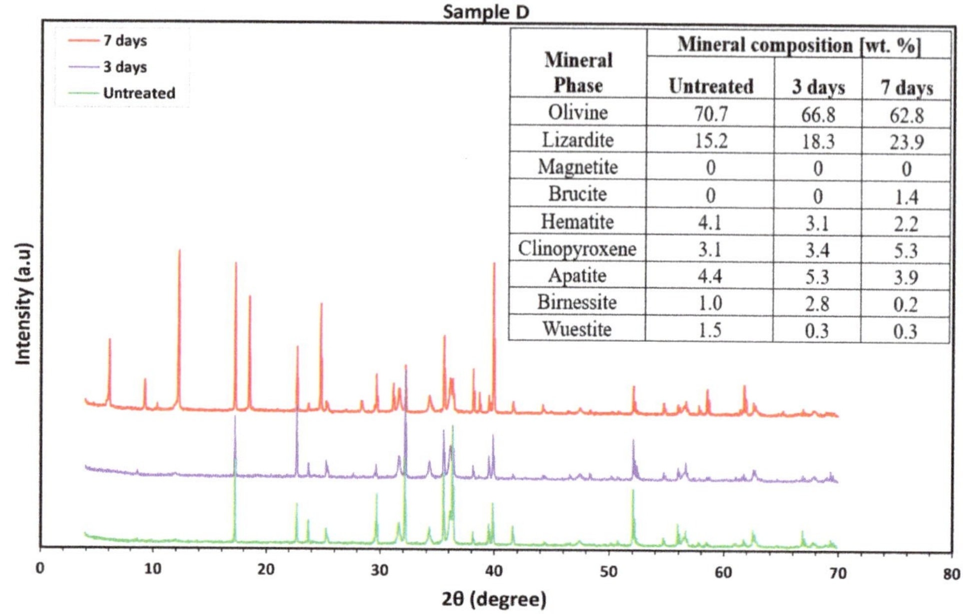

Mineral Phase	Mineral composition [wt. %]		
	Untreated	3 days	7 days
Olivine	70.7	66.8	62.8
Lizardite	15.2	18.3	23.9
Magnetite	0	0	0
Brucite	0	0	1.4
Hematite	4.1	3.1	2.2
Clinopyroxene	3.1	3.4	5.3
Apatite	4.4	5.3	3.9
Birnessite	1.0	2.8	0.2
Wuestite	1.5	0.3	0.3

Figure 7. XRD spectra of the progressive mineralogical transformations occurring in peridotite Sample D throughout the duration of hydrothermal reaction experiments for H_2 generation.

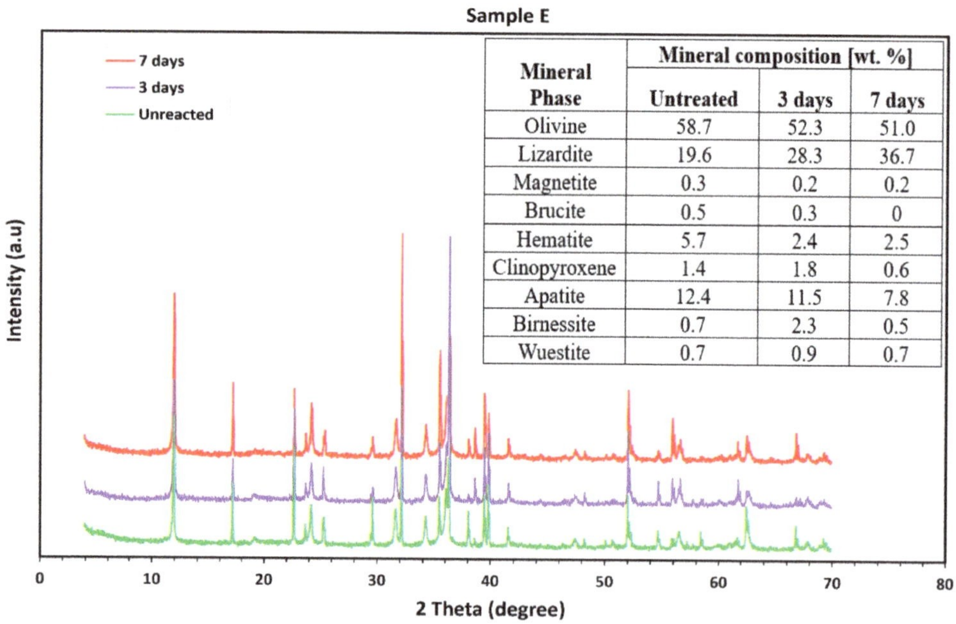

Mineral Phase	Mineral composition [wt. %]		
	Untreated	3 days	7 days
Olivine	58.7	52.3	51.0
Lizardite	19.6	28.3	36.7
Magnetite	0.3	0.2	0.2
Brucite	0.5	0.3	0
Hematite	5.7	2.4	2.5
Clinopyroxene	1.4	1.8	0.6
Apatite	12.4	11.5	7.8
Birnessite	0.7	2.3	0.5
Wuestite	0.7	0.9	0.7

Figure 8. XRD spectra of the progressive mineralogical transformations in Dunite ophiolite Sample E for the duration of hydrothermal reaction experiments for H_2 generation.

3.2.3. Mafic (Basaltic) Rocks

The mafic rocks investigated primarily comprise plagioclase minerals, with calcic labradorite and anorthite being the dominant phases. In addition to these, olivine and trace amounts of silicate minerals, including biotite, orthopyroxene (enstatite), clinopyroxene (diopside), and zeolite, were also identified. Despite the general similarity in mineralogical composition between the two mafic rock samples examined, a significant difference in H_2 generation was observed. Sample F showed a remarkably high H_2 production, reaching up to 10.8 mmol/kg of rock, whereas Sample G yielded only 0.76 mmol/kg of rock. This stark contrast suggests that factors beyond mere bulk mineral composition influence H_2 production, with the reaction pathway and mineral transformation mechanisms likely playing an important role.

Before hydrothermal treatment, both rock types contained similar primary mineral phases, including labradorite, andesine, biotite, olivine, and enstatite. However, after 7 days of reaction, Sample F exhibited a transformation of plagioclase labradorite into andesine, suggesting a gain of Na or a loss of Ca from the solid matrix (Figure 9). In contrast, sample G transformed into anorthite, indicating Na depletion or Ca enrichment within the crystalline phase (Figure 10). These variations in element redistribution suggest distinct geochemical processes that may have influenced the extent of H_2 production in each rock type.

Interestingly, Fe-rich mineral phases, including clinopyroxene (enstatite) and maghemite, were observed in Sample F, while olivine decreased after 7 days of reaction. This transformation was also accompanied by the consumption of cristobalite (SiO_2), a silica polymorph that may indicate enhanced reaction kinetics. The presence of these Fe-rich phases is significant, as iron oxidation and redox cycling are known to play a critical role in H_2 generation [11,14,15,40,41]. This suggests that the higher H_2 production observed in Sample F could be attributed to active Fe-driven redox reactions and the formation of reactive mineral surfaces, including maghemite and biotite. This could be the primary reason for the high H_2 generation observed in this mafic rock type, surpassing even the H_2 produced by all the ultramafic rocks studied. In contrast, Sample G exhibited the lowest H_2 production among all the rocks investigated, which could be linked to its lower total content of Fe-rich minerals in the unreacted state compared to the other analyzed rocks, as shown in Figure 11.

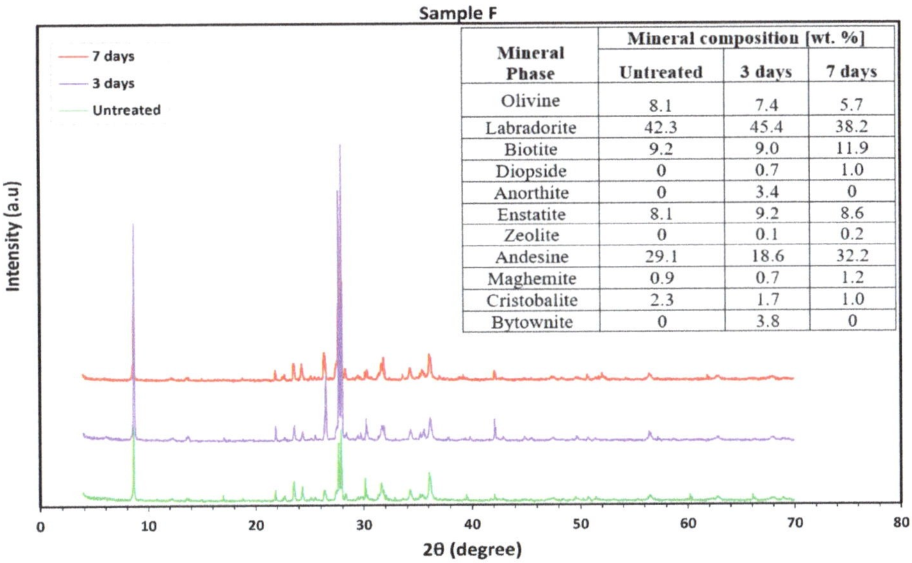

Mineral Phase	Mineral composition [wt. %]		
	Untreated	3 days	7 days
Olivine	8.1	7.4	5.7
Labradorite	42.3	45.4	38.2
Biotite	9.2	9.0	11.9
Diopside	0	0.7	1.0
Anorthite	0	3.4	0
Enstatite	8.1	9.2	8.6
Zeolite	0	0.1	0.2
Andesine	29.1	18.6	32.2
Maghemite	0.9	0.7	1.2
Cristobalite	2.3	1.7	1.0
Bytownite	0	3.8	0

Figure 9. Mineral phase transformations by XRD for the unreacted and reacted mafic rocks (Sample F) for H_2 generation hydrothermal reactions.

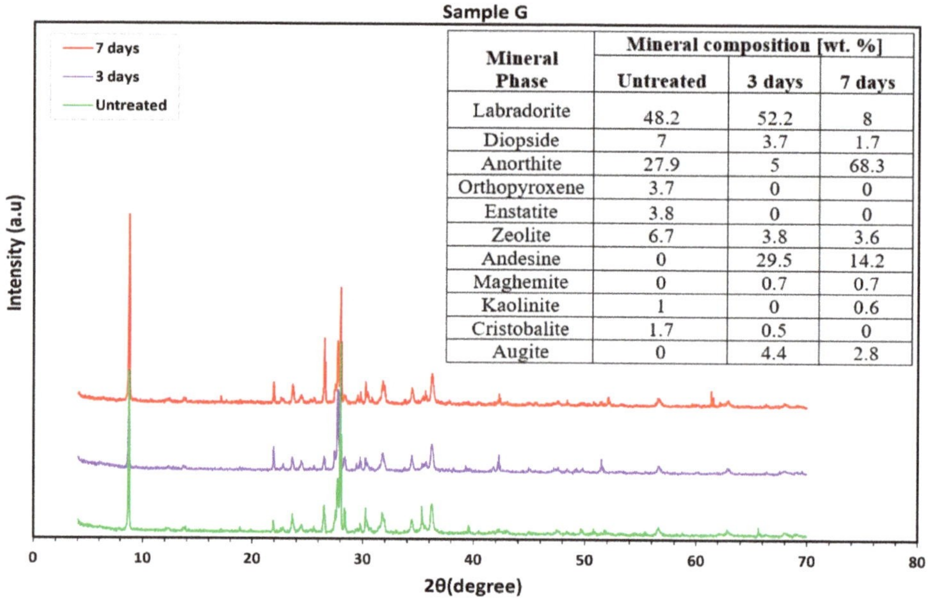

Figure 10. Mineral phase transformations by XRD for the unreacted and reacted mafic rocks (Sample G) for H_2 generation hydrothermal reactions.

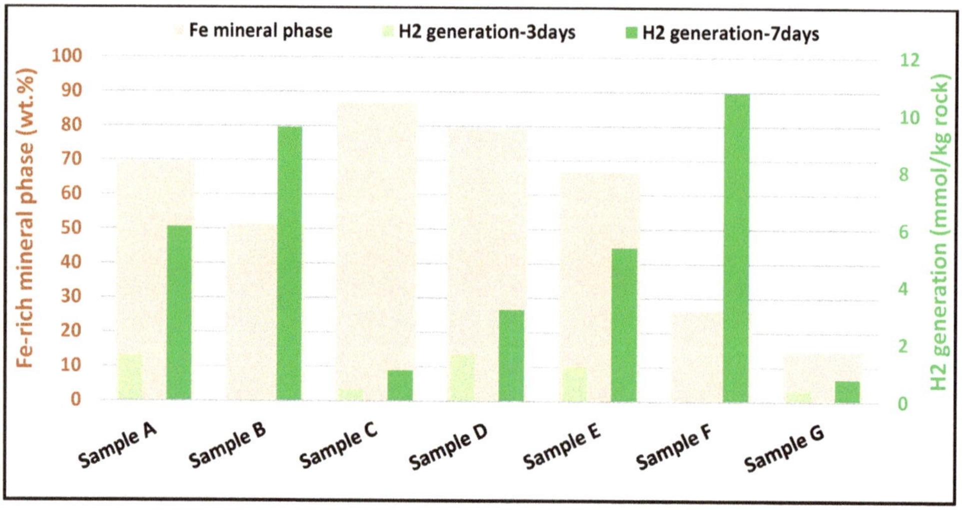

Figure 11. Iron-rich mineral phase against H_2 generation for the ultramafic and basaltic rocks over 3- and 7-day reaction durations.

3.3. *Elemental Analysis and Iron (Fe) Oxidation State*

3.3.1. XPS Elemental Analysis

X-ray Photoelectron Spectroscopy (XPS) was performed to determine the elemental composition and chemical state of the rock samples before and after reaction by measuring the binding energies. The XPS atomic ratios of the major elements in the ultramafic and mafic rocks are analyzed. The mafic rock samples, Sample F and Sample G, exhibited a higher silicon (Si) content compared to the ultramafic rocks. At the same time, their magnesium (Mg) composition is relatively lower in the bulk surface. Basalts originate from smaller extents of mantle melting, which usually results in rocks with lower Mg and higher Si, Fe, and Al contents [18,42]. This compositional variation is significant because Mg-rich minerals, such as olivine and pyroxenes, are commonly associated with serpentinization reactions in ultramafic rocks. However, the higher Fe and Al content in these mafic rocks, compared to their ultramafic counterparts, suggests that factors beyond just Mg content play an important role in determining the reactivity of these rocks in H_2 generation.

The increasing Fe content with treatment duration observed in the bulk surface of these mafic rocks is believed to be a key factor influencing their H_2 generation potential, specifically in Sample F. Iron is a key component in redox reactions, particularly in converting Fe^{2+} to Fe^{3+}, a process closely linked to H_2 production. The availability of Fe-bearing minerals, such as clinopyroxene, maghemite, and biotite, may facilitate surface-mediated electron transfer, potentially enhancing H_2 evolution in this basaltic rock. Additionally, aluminum (Al) in the crystalline structure can influence mineral stability and reaction pathways by affecting cation exchange and fluid–mineral interactions. The compositional differences between mafic and ultramafic rocks suggest that while Mg content is traditionally associated with H_2 production through serpentinization, the presence of Fe-rich phases in mafic rocks could provide an alternative mechanism for sustained hydrogen generation.

3.3.2. Iron (Fe) Oxidation States and H_2 Generation

Iron (Fe) can exist in multiple oxidation states, with Fe^{2+} and Fe^{3+} being the most common in geochemical systems. Different iron oxides, such as magnetite (Fe_3O_4), hematite (Fe_2O_3), and wuestite (FeO), exhibit distinct binding energy (B.E) peaks in X-ray Photoelectron Spectroscopy (XPS). However, due to the overlap of peak positions among these oxides, accurately determining the Fe oxidation state requires careful analysis. In this study, the Fe oxidation state was identified using XPS by referencing literature-established peak, ensuring precise differentiation between Fe^{2+} and Fe^{3+} species. The input parameters for the deconvolution of the XPS spectra [37–39] are detailed in Table 1.

Table 1. Peak model input parameters for deconvolution of XPS spectra using CasaXPS software [37–39].

S/#	Name	Pos. Constraint (eV)	FWHM Constraint (eV)	Area Constraint	Position (eV)	FWHM (eV)
A	Fe	729.108, 704.398	0.9, 0.95	0.0, 10,000,000.0	706.70	0.95
B	Fe^{2+} 1	708.75, 708.25	1.35, 1.45	0.0, 10,000,000.0	708.48	1.45
C	Fe^{2+} 2	710.05, 708.258	1.55, 1.75	0.0, 12,400,000.0	710.05	1.75
D	Fe^{2+} 3	711.25, 708.25	1.55, 1.75	0.0, 5,990,000.0	708.25	1.75
E	Fe^{2+} 4	712.45, 708.25	2.85, 3.05	0.0, 10,580,000.0	709.42	3.05
F	Fe^{2+} 5	715.75, 708.25	2.45, 2.55	0.0, 2,310,000.0	714.36	2.5
G	Fe^{3+} 1	710.35, 709.75	1.15, 1.35	0.0, 10,000,000.0	710.35	1.15
H	Fe^{3+} 2	711.35, 709.75	1.25, 1.35	0.0, 9,520,000.0	711.13	1.25
I	Fe^{3+} 3	712.25, 709.75	1.35, 1.55	0.0, 7,320,000.0	712.25	1.55
J	Fe^{3+} 4	713.35, 709.75	1.35, 1.55	0.0, 3,980,000.0	709.75	1.55
K	Fe^{3+} 5	714.45, 709.75	1.85, 1.95	0.0, 3,010,000.0	713.65	1.95
L	Fe^{3+} 6	719.85, 709.75	2.65, 2.75	0.0, 3,270,000.0	719.85	2.75

This approach helps understand the redox transformations and their role in H_2 generation and mineral reactivity. The peak models provide constraints for the peak position, area, and the full width at half maximum (FWHM) for each of the Fe components input into the model. With known peak position and FWHM, the components are quantified as area percentages. The XPS spectra reveal variations in Fe oxidation states, with each rock type displaying distinct oxidation behavior.

The resolved components in the Fe core level peak, obtained through curve-fitting of the various iron oxides, are identified. The XPS spectra of the ultramafic rocks with different Fe-oxide components are provided in Figures 12–14. Distinct variations in Fe oxidation were observed across different rock types and at various interaction times. This suggests that the hydro-geochemical reactions responsible for H_2 production differ among rock types, indicating that multiple reaction pathways are involved. These differences arise because each rock type exhibits unique mineralogical and geochemical properties, as well as crystallinity and surface reactivity, which influence how iron undergoes Fe oxidation–reduction cycles under identical reaction conditions.

In addition, the presence of trace minerals, such as biotite and zeolite, as identified by XRD, can further modulate reaction kinetics by acting as mediators, buffers, or inhibitors, which influence reactivity and the potential for H_2 generation. Even minor variations in mineral composition can introduce changes, alter reaction kinetics, and influence Fe^{2+}/Fe^{3+} cycling. Consequently, the observed differences in Fe oxidation states across rock types reflect the complex interplay of bulk mineralogy, trace element chemistry, and reaction dynamics, ultimately leading to distinct pathways for hydrogen production.

The Fe^{2+}/Fe^{3+} ratios obtained through deconvolution of the XPS spectra were plotted against the amount of H_2 generated by the various rocks, as presented in Figure 15. This is to investigate potential correlations between iron oxidation states and hydrogen production. Sample A showed a direct correlation between its Fe^{2+}/Fe^{3+} ratio and hydrogen production. Although Sample B produced the highest amount of H_2 among the ultramafic rocks analyzed, no clear correlation was observed between its Fe^{2+}/Fe^{3+} ratio and the hydrogen generated. This lack of correlation suggests that Fe oxidation alone is unlikely to be the primary controlling factor for H_2 generation in this rock type within the investigated time frame and reaction conditions. Among the other ultramafic samples, only Samples A and C exhibited an observable relationship between Fe^{2+}/Fe^{3+} ratios and H_2 production, which will be discussed in detail in subsequent sections. However, no direct correlation was observed for most ultramafic rocks analyzed, indicating that additional geochemical factors, such as trace mineral composition, or secondary reaction pathways, may play a role in H_2 production.

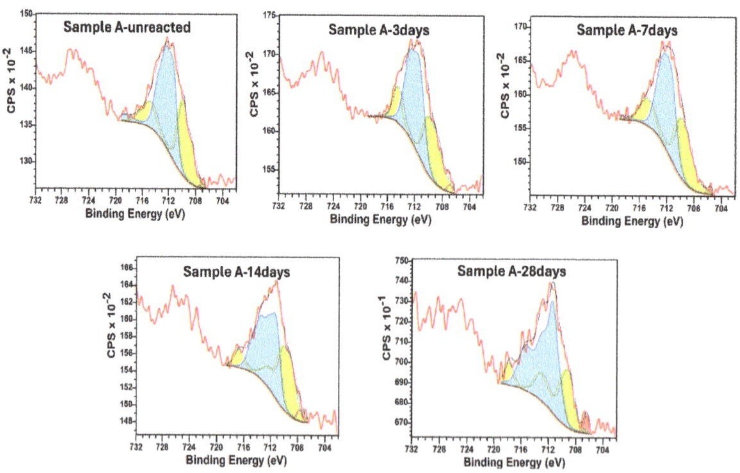

Figure 12. Deconvoluted XPS spectra of changes in iron oxidation (Fe metal—Red region; Fe^{2+}—Green region; Fe^{3+}—Blue region) for the extended reaction durations in ultramafic Sample A.

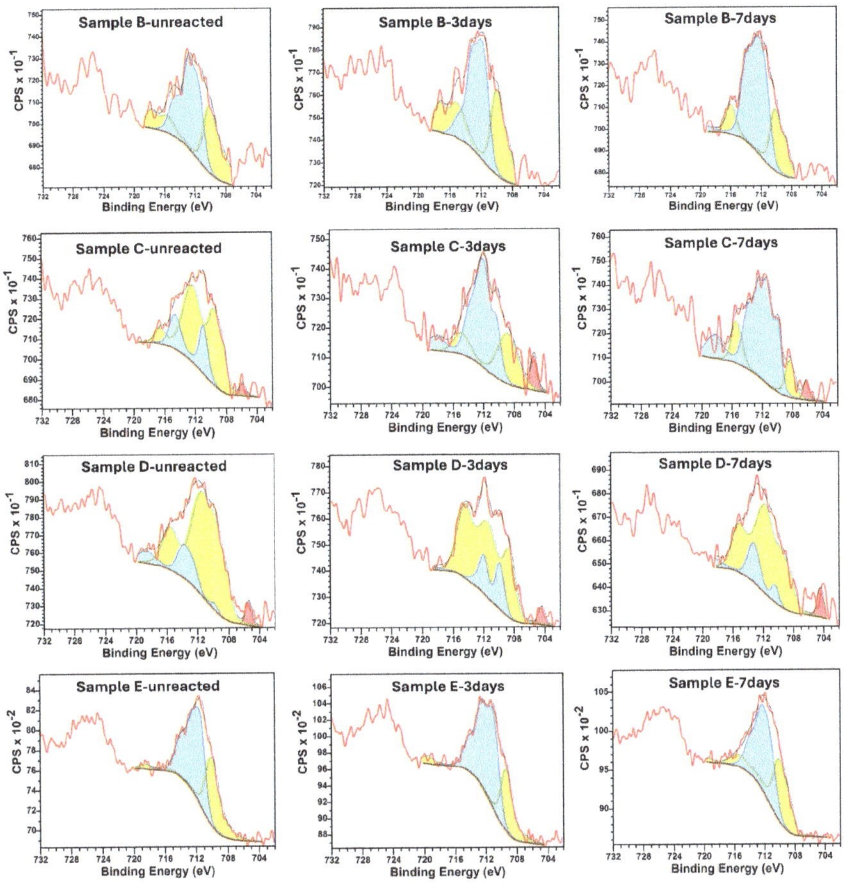

Figure 13. Deconvoluted XPS spectra showing changes in iron oxidation states (Fe metal—Red region; Fe^{2+}—Green region; Fe^{3+}—Blue region) at different reaction durations in the ultramafic rocks.

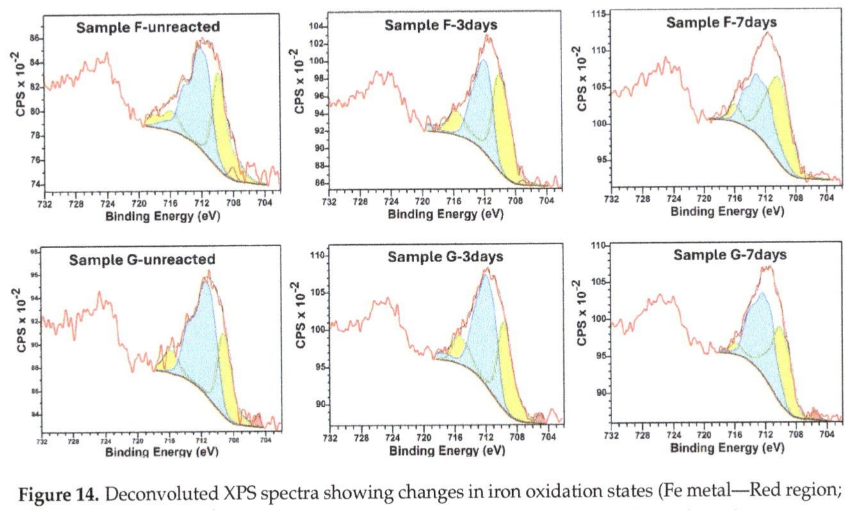

Figure 14. Deconvoluted XPS spectra showing changes in iron oxidation states (Fe metal—Red region; Fe^{2+}—Green region; Fe^{3+}—Blue region) at different reaction durations in the mafic rocks.

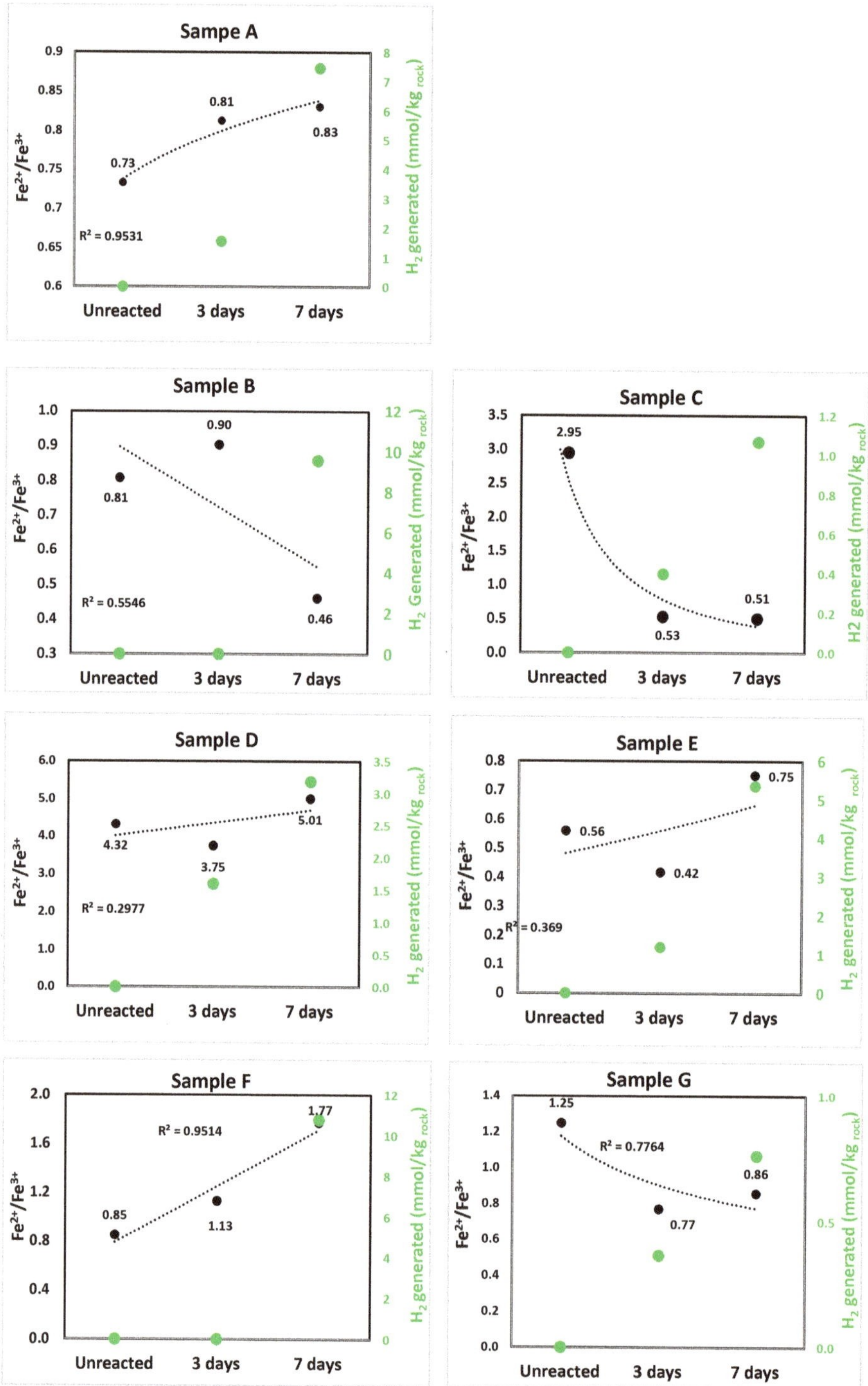

Figure 15. The Fe^{2+}/Fe^{3+} ratios (black dots with black dash trendline) obtained through deconvolution of the XPS spectra against the amount of H_2 generated (green dots) by the ultramafic and mafic rocks.

For the mafic (basaltic) rocks (Figure 15 (Samples E and F)), Sample F showed an increasing trend between the Fe^{2+}/Fe^{3+} ratio and the amount of H_2 generated. A positive relationship suggests that a lower proportion of Fe^{3+} was detected at the surface after reaction, as observed in the XPS analysis. This finding indicates that the oxidation of Fe^{2+} to Fe^{3+} was not the dominant mechanism driving mineral phase transformations and H_2 production in this sample over the investigated period (0 to 7 days). In contrast, Sample G did not show such a correlation, emphasizing the notion that variations in bulk and surface Fe oxidation states do not uniformly control H_2 generation across all rock types. Instead, other geochemical and mineralogical factors likely influence the reaction pathways involved in serpentinization and H_2 production.

3.4. *Aqueous Fluids Analysis*

3.4.1. Dissolution/Precipitation and Components Partitioning

The composition of the major elements in the supernatant, including Al, Mg, K, Ca, Zn, and Na, was analyzed to assess the geochemical evolution of the aqueous fluids. ICP-MS analysis of the aqueous fluids revealed that both dissolution and precipitation processes occurred in most of the samples, including both ultramafic and mafic rocks. However, each rock type exhibited distinct dissolution/precipitation kinetics, indicating that different reaction pathways contribute to H_2 generation.

For the ultramafic rocks, Sample A (Figure 16) showed an increase in Mg and Ca concentrations in the aqueous solution, accompanied by a slight decrease in Zn content. In contrast, Sample B showed minimal modification of the aqueous fluid composition after reaction. A reduction in Mg and Zn concentrations suggests that these elements were removed from the solution through precipitation, potentially forming secondary mineral phases. Despite their low concentrations, this implies that Mg- and Zn-bearing minerals in the bulk solids are more likely to precipitate than dissolve. Importantly, no significant dissolution was observed in this rock, indicating limited release of elements into the fluid phase. Sample C displayed the opposite trend, with a substantial increase in Mg concentration in the supernatant, suggesting that dissolution of Mg-bearing minerals occurred, releasing Mg into the fluid. The most notable change for Sample D (Figure 16) was an increase in K concentration in the reacted water compared to the initial unreacted water, indicating the dissolution of K-bearing minerals into solution, possibly due to phyllosilicate dissolution. For Sample E, a significant increase in Ca concentration after 7 days of hydrothermal reaction and a slight increase in Mg concentration were observed. This suggests that Ca-rich phases, such as clinopyroxenes or apatite, underwent dissolution, enriching the fluid with Ca.

Regarding the mafic rocks, notable geochemical changes were observed in the aqueous fluid composition, particularly in Sample G, where significant increases in K, Ca, and Zn concentrations were detected. This suggests that the dissolution of K-, Ca-, and Zn-bearing minerals occurred, releasing these elements into the fluid phase. The increase in Ca concentration is likely due to the breakdown of calcic silicates, resulting in the transformation of labradorite to andesine or the dissolution of clinopyroxene (diopside). The presence of high K levels points to the leaching of K-rich feldspars. In contrast, Sample F exhibited a more selective geochemical response, characterized by an increase in K concentration. This suggests that K-bearing minerals, such as phyllosilicate or mica, specifically biotite ($K(Mg,Fe)_3AlSi_3O_{10}(OH,F)_2$), were more susceptible to dissolution in this rock type. It has been argued that an elevated Si concentration in the reacting fluid can suppress H_2 production [25]. Aside from other solutes in the reacting fluid, the initial dissolved Si concentrations [25] in the reacting fluids can also influence the overall process of rock alteration [18,25]. Notably, no Si was detected in the aqueous phase for any of the rock–water systems investigated herein.

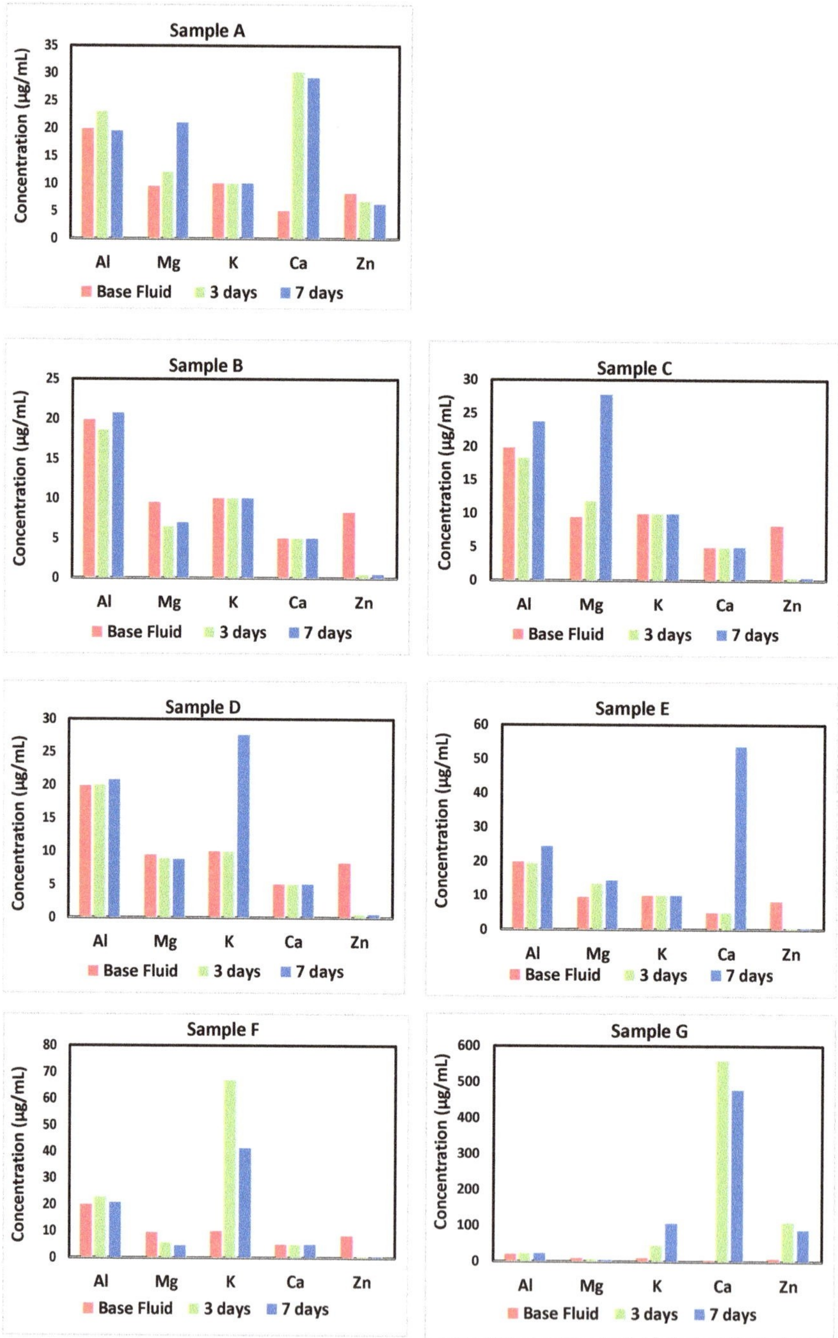

Figure 16. The composition of the major elements in the aqueous solution, including Al, Mg, K, Ca, and Zn, was analyzed to assess the geochemical evolution of the aqueous fluids.

The variation in elemental release between these mafic and ultramafic rocks shows differences in mineral composition, solubility, and fluid–rock interactions, which influence the pathways of element mobilization and potential contributions to hydrogen generation. The differences in reaction kinetics emphasize that H_2 production is influenced not only by Fe oxidation but also by mineralogy, which impacts the solubility and mobility of key elements within the rock–fluid system.

3.4.2. Other Trace Elements

Traces of additional elements, including Ni, Cu, and Mo, were detected in the supernatant, indicating their mobilization during fluid–rock interactions, as presented in Figure 17. These elements are transition metals, whose oxides are commonly known for their geocatalytic properties in redox and hydrothermal reactions [29–31]. Their presence in the reaction fluid suggests potential geochemical transformations involving metal dissolution and reprecipitation.

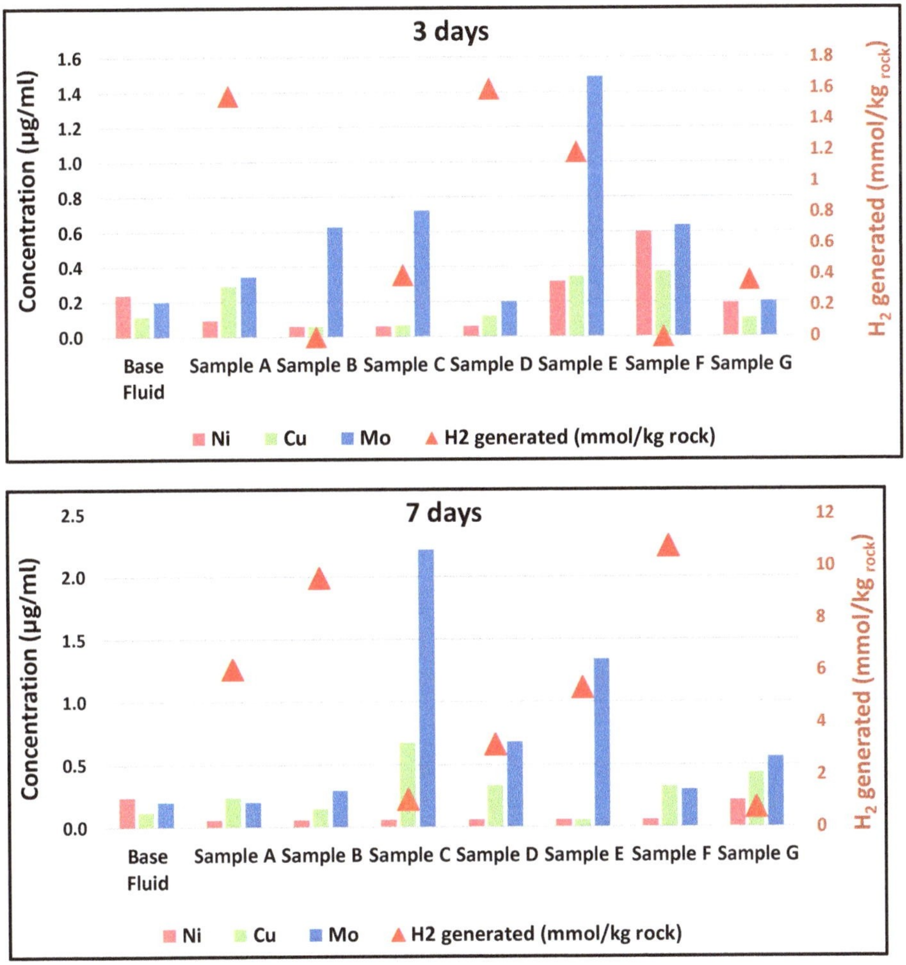

Figure 17. Trace transition metal elements detected in the aqueous fluids using ICP-MS analysis and the H_2 generated after 3- and 7-day hydrothermal interactions.

Interestingly, apart from Samples D and F, all other rocks exhibited increased concentrations of one or more of these trace elements, implying that their presence was more prominent in certain rock types. The varied release of these metals could be linked to differences in mineral composition and changes in oxidation state. Given that Ni, Cu, and Mo are often associated with oxides or silicate minerals, their mobilization may result from the breakdown of specific mineral phases under hydrothermal conditions. We speculate that the presence of these transition metals could contribute to geocatalytic mechanisms in hydrogeochemical H_2 generation. Transition metals are known to facilitate electron transfer reactions, potentially accelerating the oxidation of Fe^{2+} to Fe^{3+} and enhancing H_2 production [29,30]. Their role in surface-mediated redox reactions and mineral–fluid interactions may influence reaction kinetics, affecting the overall efficiency of hydrogen generation in ultramafic and mafic rock systems.

While the leaching of trace elements such as Ni, Cu, and Mo coincided with enhanced H_2 yields in the same samples, these observations should be regarded as correlative rather than conclusive. Leaching data alone does not establish direct catalytic involvement, and other mineralogical or geochemical processes may contribute to the observed variability. Therefore, we interpret the role of trace elements as a potential but unconfirmed factor in H_2 production. Further investigation into the speciation, coordination environment, and reactivity of these metals could provide a deeper understanding of their influence on natural hydrogen production.

3.4.3. Physicochemical Properties

Figure 18 presents the total dissolved solids (TDS), conductivity, and pH of the reaction fluids. The TDS represents the total concentration of dissolved ions in the aqueous system, which originates from the dissolution of minerals during the rock–fluid interaction process. Electrical conductivity is directly related to the ionic strength of the fluid, as the presence of more dissolved ions enhances the conductivity of water. Conventionally, as TDS increases, conductivity also increases, and vice versa. This is because the breakdown of minerals releases ions such as Mg^{2+}, Ca^{2+}, K^+, and Na^+, as observed in their elemental form, which contribute to higher ionic concentrations and thereby increase conductivity. Conversely, a decrease in TDS and conductivity suggests that precipitation processes are occurring, where dissolved ions are removed from the solution and incorporated into newly formed mineral phases.

The pH of the supernatant water is another important parameter that reflects the nature of dissolution and precipitation reactions. For example, an increase in pH may indicate the precipitation of hydroxides or the consumption of protons in the reactions. Among the ultramafic rocks, the most pronounced pH increase was observed in Sample A and Sample C, while among the mafic rocks, Sample F exhibited a similar trend. However, using pH, TDS, and conductivity alone to predict the H_2 generation potential of rocks is challenging. Although both Sample A and Sample C showed an increase in pH, Sample A generated significantly more H_2, whereas Sample C produced the lowest amount of H_2 among the ultramafic rocks. Similarly, Sample G, a mafic rock, exhibited increased TDS, conductivity, and water pH, yet it produced less H_2 than Sample F, another mafic rock. These observations suggest that multiple geochemical factors, beyond just fluid pH and ion concentration, influence the efficiency of H_2 production in hydrothermal systems.

The observed changes in TDS, conductivity, and pH across different rock types suggest that hydrothermal dissolution and precipitation processes vary, influencing reaction kinetics and hydrogen generation efficiency. For instance, ultramafic rocks rich in olivine and pyroxene may show high Mg^{2+} and Fe^{2+} release, driving H_2 production via Fe oxidation, whereas mafic rocks with plagioclase and amphiboles may exhibit different dissolution patterns. Correlating TDS, conductivity, and pH trends with mineralogical changes, as observed in the aqueous fluids' composition, is important in fluid-driven geochemical reactions that control hydrogen production in natural rock systems.

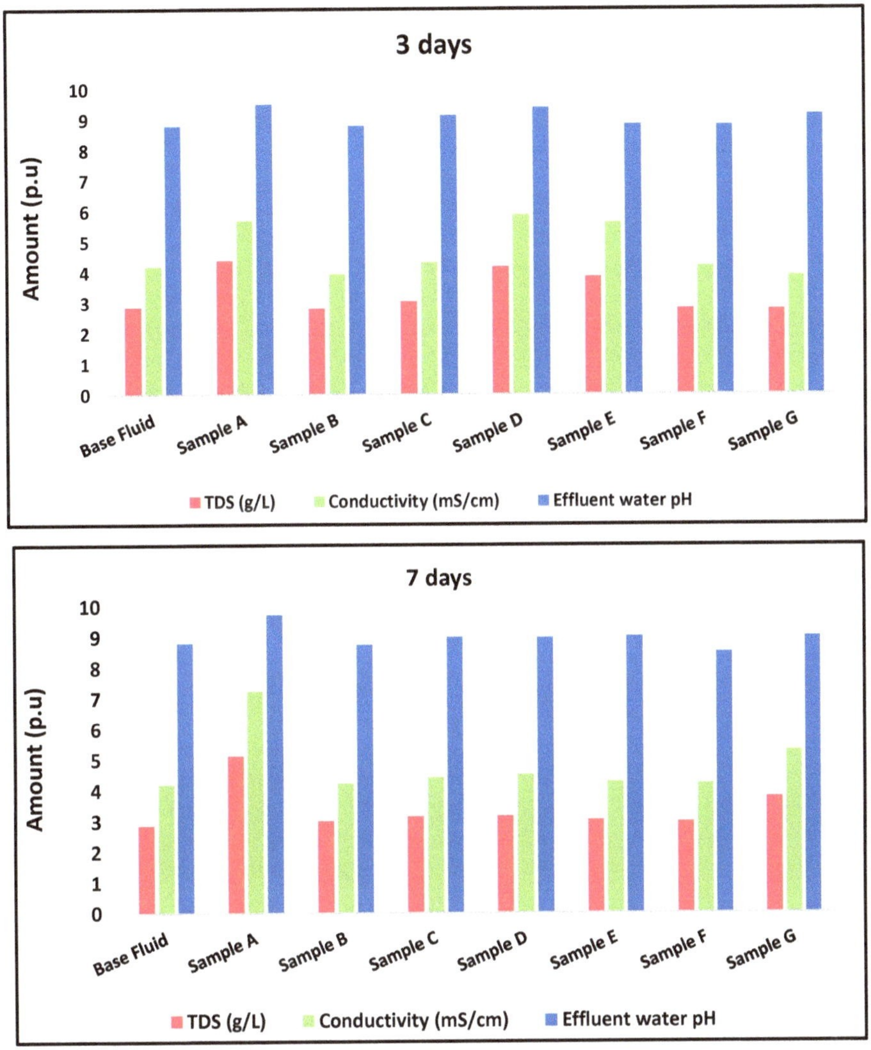

Figure 18. The total dissolved solids (TDS), conductivity, and pH of the reaction fluids.

4. Discussion

4.1. H_2 Generation in Ultramafic and Mafic (Basaltic) Rocks

The hydrothermal reactions of different rock types exhibit distinct reactivity patterns. Even trace mineral components in the bulk rock composition are likely to significantly influence overall reaction kinetics and H_2 production. The study revealed that both bulk mineralogy and elemental-scale transformations contribute to the variability in H_2 generation pathways across different rock types. Although H_2 generation is linked to the oxidation of ferrous iron in rocks, the hydrous alteration of igneous rocks with similar ferrous iron content does not always result in equivalent H_2 amounts. For instance, basaltic and ultramafic rocks contain similar amounts of ferrous iron (with XPS atomic ratio 1 to 4% Fe), and one might expect their hydrous alteration to generate comparable amounts of H_2 through the oxidation of ferrous to ferric iron. However, both high and low H_2 yields were observed in ultramafic and mafic rocks, with ultramafic rocks producing as little as 1 mmol and as much as 9.6 mmol of H_2 per kilogram of rock. In contrast, the mafic rocks exhibited a broader range, from 0.36 mmol to 10.8 mmol, the latter being larger than most hydrogen generation rates of the ultramafic rocks in this study. This difference is primarily attributed to distinct reaction pathways, influenced by the presence of trace minerals that affect hydrogen generation mechanisms.

4.1.1. Reaction Pathways in Ultramafic Rocks

Among the ultramafic rocks, variations in H_2 generation rates and reaction mechanisms were apparent. Our results suggest that rocks with a higher abundance of Fe-rich minerals tend to exhibit greater H_2 generation, confirming that iron availability plays a vital role in hydrothermal hydrogen production. This trend was consistent across both ultramafic and basaltic rock types, reinforcing the significance of Fe-bearing minerals in driving redox reactions that lead to the release of H_2. However, despite this broad trend, individual rock types display variations in reaction pathways. For example, Sample B of ultramafic rock did not generate any H_2 during the short-term hydrothermal exposure (3 days). The noticeable mineral transformation was the conversion of clinopyroxene ($Ca(Mg,Fe)Si_2O_6$) and hematite (Fe_2O_3) with some olivine into increased amounts of lizardite ($Mg_3Si_2O_5(OH)_4$), as indicated in the XRD, with this reaction defined by reaction 5. Significant mineral alterations and H_2 generation (the largest among the ultramafic rocks) were observed after 7 days of interaction (reaction 6). Most Fe-rich minerals, including olivine, hematite, and clinopyroxene (which decreased from 24.9 to 21.9 wt.%, 5.8 to 2.0 wt.%, and 19 to 3.1 wt.%, respectively), were depleted as shown in the XRD analysis. The primary newly formed minerals were serpentine (lizardite), which increased from 38.4 to 68.5 wt.%, and magnetite, which rose from 1.8 to 2.57 wt.%. Notably, no brucite was formed. Detailed information about the XRD mineral transformations is provided in Figures 4–10 above. The mineral names and nominal stoichiometry are assigned based on a restricted compositional range within a "solid solution series," characterized by the free isomorphic substitution of two elements (e.g., Fe and Mg) within the crystal lattice (see [43]). The reaction equations are approximated based on the nominal stoichiometry of each mineral identified through XRD; the elements in parentheses may freely substitute for one another.

Sample B 3 days

Olivine + Lizardite + Birnessite + Hematite + Water → Lizardite + Olivine + Birnessite + Hematite

$$\begin{aligned}&(Mg,Fe)_2SiO_4 + Mg_3Si_2O_5(OH)_4 + Mn(OH)_4 + Fe_2O_3 + H_2O\\&\rightarrow Mg_3Si_2O_5(OH)_4 + (Mn,Fe)_2SiO_4 + Mn(OH)_4 + Fe_2O_3\end{aligned} \quad (5)$$

Sample B 7 days

Lizardite + Birnessite + Olinive + Clinopyroxene + Magnetite + Hematite + Water → Lizardite + Birnessite + Olivine + Clinopyroxene + magnetite + Hematite + Hydrogen

$$\begin{aligned}&Mg_3Si_2O_5(OH)_4 + Mn(OH)_4 + (Mg,Fe)_2SiO_4 + Ca(Mg,Fe)Si_2O_6 + Fe_3O_4 + Fe_2O_3 + H_2O + H^+\\&\rightarrow Mg_3Si_2O_5(OH)_4 + Mn(OH)_4 + (Mg,Fe)_2SiO_4 + Ca(Mg,Fe)Si_2O_6 + Fe_3O_4 + Fe_2O_3 + H_2 + OH^-\end{aligned} \quad (6)$$

For Sample C, this rock type produced the lowest H_2 yield among the ultramafic rocks, generating only 0.76 mmol of H_2 per kilogram of rock. After 7 days of reaction, the most notable mineral transformations included the decrease in olivine and vermiculite

($(Mg,Fe,Al)_6(Si,Al)_4O_{10}(OH)_8$), accompanied by the formation of hematite and brucite through the alteration of Fe-Mg silicates like olivine. These mineral modifications contributed to the changes in Mg concentration in the aqueous solution, as shown in Figure 16 (Sample B). The approximate chemical reaction pathways for these transformations can be represented by Reaction 7. Although serpentine formation was not the sole mineral transformation, H_2 generation in these rocks appears to be primarily driven by the presence of Fe-containing minerals.

Sample C 3 to 7 days interaction

Olivine + Lizardite + hematite + brucite + water → Olivine + Lizardite + hematite + brucite + Hydrogen

$$\begin{aligned}&(Fe, Mg)_2SiO_4 + Mg_3Si_2O_5(OH)_4 + Fe_2O_3 + Mg(OH)_2 + H_2O + H^+ \\ &\rightarrow (Fe, Mg)_2SiO_4 + Mg_3Si_2O_5(OH)_4 + Fe_2O_3 + Mg(OH)_2 + H_2 + OH^-\end{aligned} \quad (7)$$

Although serpentine formation was not the only mineral transformation observed, hematite was also formed, as evidenced by its increased XRD composition in Sample C. It is important to note that this is the only rock sample that exhibited an increase in hematite content, whereas all other analyzed samples showed a decreasing trend compared to the untreated rocks. In these rocks, H_2 generation is primarily attributed to the presence of Fe-containing minerals based on their mineralogical composition. The ultramafic rocks did not exhibit a consistent trend between H_2 production and the formation of brucite or magnetite, as shown in Figures 4–10. Specifically, magnetite decreased (or was consumed) in Sample A, increased in Sample B, was not detected in Samples C and D, and showed no significant change in Sample E. For brucite, precipitation occurred in Samples A, C, and D after hydrothermal reaction, while Samples B and E showed no change over the 3- and 7-day interaction periods. Particularly, the reaction pathway for Sample A is similar to that of Sample E, with the exception of greater serpentine formation in Sample E over the 3- and 7-day reaction durations. Similarly, the reaction pathway or mechanism responsible for H_2 generation in Sample C resembles that of Sample D. The key difference lies in the transformation of hematite, which increased in Sample C but decreased in Sample D from the untreated state through the 3- and 7-day hydrothermal reactions. This unique feature in the mineralogical transformation of Sample C, among all the examined ultramafic rocks, could be the primary reason for the lowest H_2 generation observed compared to the other ultramafic samples.

Thermodynamic simulations yielded similar results, indicating that magnetite is unlikely to form during the serpentinization of orthopyroxene, as all Fe initially present in orthopyroxene is instead incorporated into serpentine [13]. Meanwhile, ferroan brucite oxidation has been proposed as a major contributor to H_2 production in ophiolites at temperatures below 423 K [44,45]. This aligns with petrographic observations of natural samples, which indicate that brucite undergoes oxidation to form magnetite under open-system conditions [46,47].

The thermodynamic simulation work of [18] demonstrated that H_2 production is strongly influenced by rock composition. In particular, the serpentinization process in lithologies with larger magnesium content results in significantly greater H_2 generation, attributing higher H_2 generation to increased MgO composition. Specifically, rocks with MgO content exceeding 35 wt.% have the highest potential for H_2 generation. However, the amounts vary significantly due to differences in their ability to stabilize various Fe(III)-bearing phases during the process of hydrous alteration. Even among rocks with similar MgO content, variations in SiO_2 levels influence the precipitation of Fe-bearing secondary minerals. Higher SiO_2 concentrations promote the formation of greenalite over cronstedtite, resulting in lower H_2 production, as Fe(II) from primary minerals is incorporated into greenalite without oxidation [24,25]. In contrast, rocks with MgO content below 20 wt.% do not produce significant H_2 at any water-to-rock ratio in their simulations [18].

It is important to mention that the thermodynamic computations by [13] reported a somewhat contrasting finding: the serpentinization of Fe-rich olivine within the forsterite–fayalite (Fo-Fa) solid solution led to the formation of a free H_2 gas phase, with H_2 generation increasing from Fo_{10}-Fa to Fo_{70}-Fa. The complete diadochy between Mg^{2+} and Fe^{2+} results in a broad range of olivine compositions, ranging from forsterite to fayalite [48]. The chemical makeup of olivine significantly affects its stability when exposed to water [13]. Consistent with our experimental findings, mineral alteration and H_2 generation are highly dependent on the specific starting minerals.

It was documented that in the early stages of serpentinization, Fe-bearing serpentine minerals commonly form, incorporating ferric iron within their structure [18]. Interestingly, not all H_2-generating hydrothermal reactions resulted in serpentine ($(Mg_{2.70}Fe_{0.18}Al_{0.11})(Si_{1.8}1Al_{0.19}O_5)(OH)_4$) formation. This challenges the conventional assumption that H_2 generation in Fe-rich rocks is always linked to serpentinization. Instead, in some cases, birnessite (a Mn-rich layered hydroxide mineral, ($H_{4.424}MnO_{4.212}$)), and hematite (Sample C), could form as a secondary mineral in hydrothermal environments, particularly from the alteration of manganoan olivine and pyroxene, was identified as a major reaction products. This distinction shows that alternative mineralogical pathways contribute to H_2 generation beyond traditional serpentinization reactions.

4.1.2. Reaction Pathways in Mafic (Basalts) Rocks

For the mafic rocks (Samples F and G), there is no evidence of serpentine formation. H_2 production in these systems follows a different mineralogical transformation path. Instead, the secondary minerals formed were andesine and anorthite, both of which belong to the plagioclase feldspar group. This aligns with recent findings that indicate H_2 can be generated from basalts [49]. Basaltic rocks generate H_2 through mechanisms that differ from those governing Fe-rich ultramafic systems, involving feldspar alteration and Fe-rich mineral transformations rather than typical ultramafic serpentinization pathways.

The reactive mineral assemblage in basalts is dominated by Mg- and Fe-bearing phases, with olivine present in lower modal abundance relative to ultramafic counterparts. These compositional differences significantly influence mineral–fluid reactivity and the efficiency of subsequent H_2 generation during water–rock interaction. These minerals include orthopyroxene ($(\mathrm{Mg},\mathrm{Fe})_2)Si_2O_6$), enstatite ($Mg_2Si_2O_6$), diopside ($CaMgSi_2O_6$), and biotite ($K(\mathrm{Mg},\mathrm{Fe})_3AlSi_3O_{10}(OH)_2$). The increased concentrations of K and Ca observed in the aqueous solution, as shown in Figure 16 (Samples F and G), provide clear evidence of the dissolution of biotite and diopside, respectively. Biotite, a K-rich phyllosilicate mineral, releases potassium ions (K^+) into the solution as it undergoes chemical breakdown. Similarly, diopside, a Ca-Mg pyroxene, contributes to the increased Ca^{2+} concentrations through its dissolution. This facilitates geochemical reactions that drive the generation of H_2. The breakdown of Fe- and Mg-bearing silicates promotes redox reactions where ferrous iron (Fe^{2+}) within the mineral structure can be oxidized, concurrently reducing water to produce molecular H_2 as presented in Reactions 8 and 9. This process is particularly significant in mafic rock systems, where Fe^{2+}-rich phases readily participate in redox-driven H_2 generation.

Sample F 3 days

Labradorite + Andesine + Enstatite + Water → Labradorite + Andesine + Enstatite

$$\begin{aligned} &2Al_{0.81}Ca_{0.33}Na_{0.16}Si_{1.19}O_4 + Al_{0.74}Ca_{0.24}Na_{0.26}Si_{1.27}O_4 + Mg_2Si_2O_6 + H_2O \\ &\rightarrow Al_{0.81}Ca_{0.33}Na_{0.16}Si_{1.19}O_4 + Al_{0.74}Ca_{0.24}Na_{0.26}Si_{1.27}O_4 + Mg_2Si_2O_6 + H_2O \end{aligned} \tag{8}$$

Sample F 7 days

Labradorite + Biotite + Enstatite + Olivine + Water → Labradorite + Andesine + Biotite(Mica) + Enstatite + Olivine + Hydrogen

$$\begin{aligned}&2Al_{0.81}Ca_{0.33}Na_{0.16}Si_{1.19}O_4 + K(Mg,Fe)_3(AlSi_3O_{10})(OH)_2 + Mg_2Si_2O_6 + (Mg,Fe)_2SiO_4 + H_2O + H^+\\ &\rightarrow Al_{0.81}Ca_{0.33}Na_{0.16}Si_{1.19}O_4 + Al_{0.74}Ca_{0.24}Na_{0.26}Si_{1.27}O_4 + K(Mg,Fe)_3(AlSi_3O_{10})(OH)_2 + (Mg,Fe)_2SiO_4 + Mg_2Si_2O_6 + OH^- + H_2\end{aligned} \quad (9)$$

Thus, the increasing K and Ca concentrations not only confirm the reactivity of biotite and diopside in the rock–fluid system but also serve as indirect indicators of mineral–fluid interactions that contribute to enhanced H_2 production. As primary sources of Mg and Fe, these minerals are essential for H_2 generation in basalts. Notably, Sample E contains a higher quantitative abundance of these reactive Mg- and Fe-rich phases compared to Sample G (refer to Figures 9 and 10). This greater mineral availability enhances H_2 generation in Sample E. The difference in H_2 yield between the two mafic samples can thus be attributed to variations in mineral composition and reactivity, emphasizing the significance of Fe-Mg silicates in influencing H_2 production rates during rock–fluid interactions. The major minerals' transformation pathway for Sample G is illustrated in Reaction 10.

Sample G 3 to 7 days

Labradorite + Anorthite + Augite(Enstatite + Diopsite) + water → Labradorite + Andesine + Anorthite + Augite + Hydrogen

$$\begin{aligned}&2Al_{0.81}Ca_{0.33}Na_{0.16}Si_{1.19}O_4 + CaAl_2Si_2O_8 + Mg_2Si_2O_6 + CaMgSi_2O_6 + H_2O\\ &\rightarrow Al_{0.81}Ca_{0.33}Na_{0.16}Si_{1.19}O_4 + Al_{0.74}Ca_{0.24}Na_{0.26}Si_{1.27}O_4 + CaAl_2Si_2O_8 + (Ca,Na)(Mg,Fe,Al)(Si,Al)_2O_6 + H_2\end{aligned} \quad (10)$$

To determine the reactive components in basalts responsible for H_2 production, ref. [49] conducted experiments on typical basaltic mineral phases, including olivine, pyroxene, and feldspar. Specifically, the authors investigated basalt minerals, including labradorite, augite, and olivine (fayalite and forsterite). Their findings indicated that minor ferrous minerals present in basalts, such as magnetite (Fe_3O_4), ilmenite ($FeTiO_3$), hematite (Fe_2O_3, a ferric oxide), pyrite (FeS_2), and pyrrhotite (FeS), did not generate detectable levels of H_2. However, olivine end-members—fayalite and forsterite—and augite generated H_2. In addition, ferrous iron-bearing silicate minerals, including biotite, produced only trace amounts of H_2. The study concluded that ferrous silicate minerals, particularly pyroxene and olivine, were the primary contributors to H_2 generation in basalts, suggesting that Fe^{2+} was a key reactant in the H_2-generating reaction.

Considering the elemental and atomic-scale processes, the results revealed that Fe^{2+} oxidation to Fe^{3+} correlated with H_2 generation in only certain rock types or at a later stage of the reaction progression. This suggests that while Fe oxidation is an important mechanism, it is likely not the sole driver of H_2 production; additional geochemical factors, such as the availability of trace elements and mineral dissolution/precipitation rates, also contribute to H_2 generation in certain rocks.

4.2. Extended Reactions of Sample A and Its Impact on H_2 Generation

Sample A (dunite, ultramafic rock) was subjected to extended reaction durations to better understand mineral phase transformations and H_2 generation dynamics over time. This approach aimed to capture both early-stage reactivity and intermediate to long-term trends in mineral-fluid interactions. A comparison of the Fe oxidation states using the Fe^{2+}/Fe^{3+} ratio, mineral transformations, and H_2 generation is presented in Figure 19a,b.

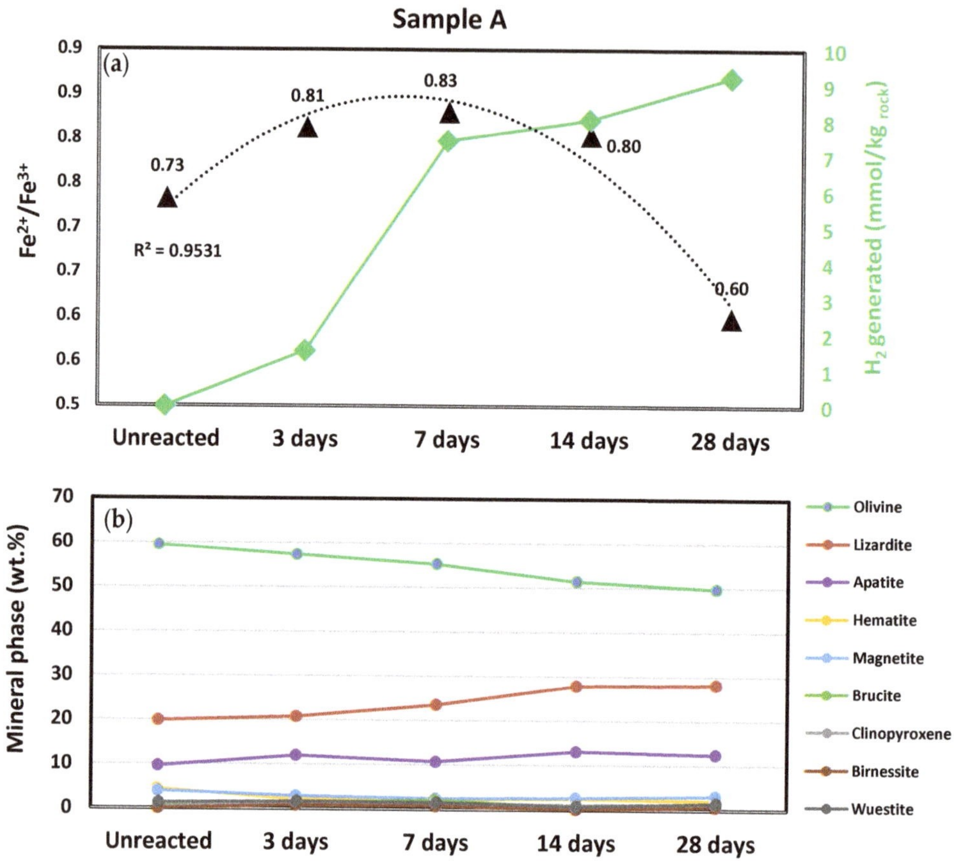

Figure 19. A comparison of Fe oxidation states considering the Fe^{2+}/Fe^{3+} ratio and H_2 generation (**a**), and mineral phase transformations over over 3-, 7-, 14-, and 28 days reaction periods (**b**).

4.2.1. Early-Stage H_2 Generation (0–7 Days)

During the early hydrothermal exposure (0–7 days), H_2 generation was rapid, reaching up to 8 mmol/kg of rock (see Figure 19a). The high rate of hydrogen generation in this early stage suggests that multiple geochemical mechanisms are actively contributing to H_2 release. First, the oxidation of Fe^{2+} to Fe^{3+}: The conversion of Fe-bearing minerals, including olivine and lizardite, to Fe^{3+}-rich phases (including magnetite, Fe_3O_4) releases electrons, which can facilitate water reduction to H_2. The approximate reaction at this stage is represented by Reaction 11.

Sample A 0–7 days

Olivine + Lizardite + Magnetite + Hematite + water → Olivine + Lizardite + Magnetite + Hematite + Wuestite + Hydrogen

$$\begin{aligned}&(Mg,Fe)_2SiO_4 + 2Mg_3Si_2O_5(OH)_4 + 2Fe_3O_4 + 4Fe_2O_3 + H_2O + 3H^+\\&\rightarrow (Mg,Fe)_2SiO_4 + 2Mg_3Si_2O_5(OH)_4 + 3Fe_3O_4 + Fe_2O_3 + 3FeO + H_2 + 3OH^-\end{aligned} \quad (11)$$

Additionally, the presence of preformed reactive intermediates, such as metastable Fe-bearing phases or partially oxidized minerals, enhances H_2 generation by promoting electron transfer reactions. These intermediates include wuestite (FeO), clinopyroxenes, and brucite, which were precipitated and then dissolved as the interaction proceeded (see Figure 4). Their presence suggests that these Fe-rich minerals exist in a reactive, transient state, allowing them to rapidly participate in redox reactions with water. As these metastable phases undergo further oxidation and Fe partitioning, along with olivine conversion (decreasing from 59.5 wt.% in the unreacted rock to 49.8 wt.% after 7 days of interaction) into lizardite, they contribute to an initial burst of H_2 production, increasing H_2 output in the early stages of the hydrothermal interactions. However, as these reactive intermediates are depleted or converted into more stable mineral phases, such as lizardite and birnessite, the rate of H_2 generation slows down. This highlights the significance of mineral composition and oxidation state variability in controlling the temporal dynamics of hydrothermal hydrogen production.

This stage is characterized by an increase in the Fe^{2+}/Fe^{3+} ratio (see Figure 19a), suggesting that the oxidation of ferrous iron (Fe^{2+}) to ferric iron (Fe^{3+}) was not the dominant reaction pathway during this period of interaction. Instead, other geochemical processes, such as mineral dissolution, solid-state Fe redistribution, Fe^{2+} incorporation into newly forming phases, may have played a more significant role in controlling the Fe redox balance and resulting H_2 generation. The persistence of Fe^{2+} suggests that either oxidation was kinetically limited or competing reduction processes were occurring, maintaining a higher Fe^{2+}/Fe^{3+} ratio.

Additionally, the presence of metastable Fe-bearing minerals, such as wuestite and apatite, alongside increasing serpentinization (evidenced by the rise in lizardite formation, as shown in Figure 19b), may have contributed to buffering Fe oxidation. This, in turn, could have slowed or altered the typical redox progression associated with Fe-driven H_2 generation.

4.2.2. Intermediate-Stage H_2 Generation (7–28 Days)

Between 7 and 28 days of the hydrothermal interaction, a decline in the Fe^{2+}/Fe^{3+} ratio was observed, coinciding with a decrease in the rate of hydrogen generation. This shift suggests that the early-stage reactive intermediates were depleted, leaving Fe^{2+} oxidation to Fe^{3+} as the dominant mechanism driving H_2 production in the intermediate phase. The reduction in Fe^{2+} availability may have limited further electron transfer reactions, leading to a gradual decrease in H_2 output. The formation of a stable Fe^{3+}-rich phase (such as serpentine) may have reduced the reactivity of iron minerals, further slowing the rate of hydrogen generation. Precipitation of secondary mineral phases, including lizardite and birnessite, may have altered the surface reactivity of the rock, modifying the kinetics of Fe oxidation and H_2 release, as well as the approximate hydrogeothermal reaction for this stage. In rock systems, this precipitation could lead to surface passivation.

Hydrogen generation in ultramafic rock systems is a complex process, indicating that different reaction pathways dominate at various stages of hydrothermal alteration. Serpentinization and Fe oxidation contribute to H_2 generation through a time-dependent, multi-step reaction process. In the early phase, both Fe oxidation and reactive intermediates contribute to the release of H_2. In the intermediate phase, Fe^{2+} oxidation becomes the primary mechanism, but reaction rates slow as Fe^{2+} is gradually depleted. While the early stage appears to involve a combination of Fe^{2+} oxidation and additional reactive intermediates, the long-term reaction dynamics are primarily controlled by iron oxidation alone. As a result, less H_2 is produced since iron oxidation occurs to a lesser extent compared to scenarios where serpentine forms extensively. Recent studies have highlighted that non-redox-sensitive elements in rocks, such as Mg, Si, Al, and Ca, play a significant role in controlling the distribution and transformation of Fe during secondary mineralization, thereby influencing the redox processes that generate reduced volatiles during fluid–rock interactions. Most of the iron released from primary minerals is integrated into secondary minerals with minimal oxidation, as certain minerals preferentially incorporate ferrous iron into their crystal structures [18].

The availability of Fe^{2+}-bearing minerals and the reaction pathway are important in sustaining H_2 production over extended periods. The mineralogical transformations, effects of trace metals, and changes in fluid composition could provide a deeper understanding and optimize conditions for prolonged and efficient H_2 generation from ultramafic and basaltic rock systems.

4.3. Potential Geocatalysis During H_2 Generation

The presence of transition metals such as Ni, Mo, and Cu within ultramafic and mafic rocks suggests a potential geocatalytic effect [29–31] that enhances H_2 generation during fluid–rock interactions (see Section 3.4.2). These metals can facilitate electron transfer reactions, accelerating Fe^{2+} oxidation and promoting sustained H_2 production beyond typical serpentinization pathways. Additionally, the structural properties of secondary minerals, such as birnessite, vermiculite, and zeolite (as observed by XRD in the basalts), may further support geocatalytic pathways by preserving reactive surfaces, thereby enabling continuous hydrothermal hydrogen evolution over extended periods [27,50]. However, additional research is needed to fully assess the extent of such reactive mediators in geologic H_2 generation.

Recent studies indicate that transition-metal sites within ultramafic and mafic rocks can act as true geocatalysts that enhance abiotic H_2 production beyond what is expected from simple Fe(II) oxidation and serpentinization alone [51,52]. Nickel and Ni–Fe alloys (e.g., awaruite and native Fe–Ni phases) promote H_2 evolution by providing surface sites for electron transfer and by lowering activation barriers for H-forming redox reactions. Experimental and theoretical studies show that Ni^{2+} and Ni–Fe nanoparticles markedly increase H_2 generation rates under low-temperature aqueous conditions. The introduction of only 1% Ni^{2+} boosted the H_2 generation rate by nearly two orders of magnitude at 90 °C [53,54]. Similarly, Fe-rich phases and Fe–Ni alloys formed during alteration can catalyze hydrogen-producing reactions and facilitate redox cycling between reduced metal sites and aqueous species, sustaining H_2 production even after initial mineral transformations. These catalytic pathways are complementary to, and sometimes kinetically dominant over, mineral transformation-controlled H_2 release, meaning that rock compositions rich in catalytic transition metals (Ni, Fe, and trace PGE (platinum-group elements)) may produce sustained H_2 fluxes through surface-mediated reactions and alloy catalysis [51–54].

Meanwhile, aluminum, when present in the reactants, generally acts to stabilize secondary mineral formation and buffers silica activity in olivine-rich serpentinization process. It serves as a viable means of accelerating serpentinization reactions toward an economically feasible timescale for industrial hydrogen production, including geologic means [35,55]. In comparison with other technologies (like steam reforming) for H_2 production, serpentinization also introduces energy minimization through its exothermic nature, and the presence of aluminum can further ensure its occurrence even at lower temperatures. X-ray photoelectron spectroscopy, an effective method for differentiating between Al-hydroxide and Al-oxide phases, utilizes the ratio of oxidic-bound Al to O to analyze Al oxidation and hydration behavior [56]. The investigation of Al-oxide and Al-hydroxide transformations revealed more pronounced alterations in basalts compared to ultramafic rocks. The O/Al atomic ratios of aluminum oxide layers were determined based on O 1s and Al 2p photoelectron intensities, showing an increase from 5.87 to 6.07 in the unreacted and reacted ultramafic Sample E, respectively, and from 3.74 to 4.62 in the unreacted and reacted basalt Sample E, respectively.

In an experimental study, [55] demonstrated a correlation between the onset of serpentinization and the dissolution of aluminum oxide (Al_2O_3) microspheres introduced into the reacting fluid. Importantly, the very first signs of serpentine minerals appeared simultaneously on both the olivine grains and the surface of these dissolving Al_2O_3 microspheres. This suggested that the presence of dissolved aluminum in the fluid significantly increased the solubility of olivine. Additionally, dissolved Al played a significant role in the formation (nucleation) and subsequent growth of serpentine minerals, particularly those rich in aluminum.

Pens et al. [35] also illustrated the contrasting effect of aluminum in both olivine-rich and orthopyroxene-rich serpentinization processes. Their results indicated that in the presence of aluminum (Al), olivine rapidly converts to lizardite with a half-life of only 7 h. In contrast, orthopyroxene shows minimal reaction (only 11% conversion) even after 6 days under the same conditions. Interestingly, orthopyroxene reacts faster without the presence of aluminum (48% conversion in 6 days). Olivine experiments resulted in the formation of lizardite and magnetite, while Orthopyroxene experiments exclusively produced proto-serpentine. The explanation for these contrasting effects lies in how aluminum interacts differently with the surfaces of olivine and orthopyroxene. Aluminum in solutions primarily exists as negatively charged complexes, such as $Al(OH)_4$-. The positively charged surface of olivine attracts and adsorbs these negatively charged Al complexes. This adsorption process likely enhances the rate at which olivine dissolves and also helps in the nucleation and growth of lizardite, which is enriched in aluminum. On the other hand, the neutral surface of orthopyroxene does not significantly interact with the negatively charged Al complexes. This lack of interaction explains the slower reaction rate of orthopyroxene in the presence of aluminum.

5. Conclusions

This study examined the potential for hydrothermal H_2 generation in seven ultramafic and mafic rocks with diverse mineralogical compositions. The study aimed to identify key reaction pathways and geochemical factors influencing H_2 generation efficiency. The results showed how differences in rock composition, Fe oxidation states, and secondary mineral formation influence hydrogen yield and reaction kinetics.

Our findings demonstrate that not all geologic H_2-generating reactions involving ultramafic and mafic rocks result in the formation of serpentine, brucite, or magnetite. Instead, the results suggest that H_2 generation in these systems is a strong function of mineral transformation. Surface-mediated electron transfer and redox cycling play an important role in sustaining hydrogen production beyond the initial mineral phase changes.

The diversity in reaction pathways, especially the formation of birnessite and/or clinochlore instead of serpentine in some instances, suggests that not all Fe-driven hydrogen production processes should be classified strictly as serpentinization. These findings also suggest that categorizing all H_2-producing hydrothermal reactions as "serpentinization" can be misleading, as alternative mineral assemblages such as birnessite, feldspars (andesine and anorthite), and other phyllosilicates (clinochlore) may be more representative of the dominant reaction mechanisms in certain rock types.

Magnetite experienced minimal transformation and, in some cases, was actually formed as a secondary mineral from precursor olivine during hydrothermal exposure. This suggests that the rock mineral may act both as a product and a reactant, indicating that its transformation was not the primary mechanism for H_2 generation. Further investigations into the kinetics of Fe oxidation and the role of trace mineral interactions are needed to fully understand how different geological settings influence natural H_2 production potential in both ultramafic and mafic rock systems.

Major Fe-rich ultramafic minerals, such as olivine and pyroxenes, seem to support sustained H_2 production through continuous solid–solution exchange and phase transformation, while aluminosilicates and transition metal sites (e.g., Ni, Mo, Mn, Cu) likely enhance H_2 yields. Additionally, fluid–rock interactions regenerate reactive surfaces, supporting prolonged H_2 generation even in the absence of serpentine formation. These results suggest that reaction kinetics and mineralogical composition, rather than Fe oxidation states alone, dictate the rates and efficiency of H_2 production.

Ultimately, these insights shift the focus from serpentine-driven mechanisms to a broader mineralogical framework, where transition metal sites and mineral surface properties may play a central role. Understanding these processes can enhance subsurface hydrogen exploration and facilitate the development of engineered H_2 production systems that utilize minerals known to occur within ultramafic and mafic rock formations.

6. Future Works

Future work should investigate longer reaction durations to assess the sustained potential of H_2 generation and mineral transformation over extended time scales. Reactions carried out beyond four-week periods could reveal slower geochemical processes and the stability of hydrogen-producing pathways. Additionally, experiments conducted at a broader range of temperatures and pressures will help determine the thermal thresholds that optimize specific reaction mechanisms such as serpentinization, brucite precipitation, and iron oxidation.

Further investigation is also needed to evaluate the influence of varying fluid chemistries, particularly salinity and pH, on H_2 production efficiency and mineral alteration. These parameters are important in replicating realistic subsurface environments and understanding their role in modulating reaction rates. In addition to serpentinization, this work suggests that surface redox reactions and geocatalytic processes may play a significant role in H_2 production. However, the catalytic contributions of trace minerals and naturally occurring transition metals presence in ultramafic and mafic rocks remain uncertain and warrant further investigation in future studies. Future investigations should focus on explicitly resolving the mechanistic role of trace elements in H_2 production. This could include experiments with isotopically labeled water to track redox processes, selective mineral doping to isolate catalytic effects, and advanced surface-sensitive spectroscopic analyses (e.g., XPS, XAS) to monitor the oxidation states of Ni, Cu, and Mo during reaction. Such approaches will help decouple trace element effects from broader mineralogical pathways and establish whether these elements act as true catalysts or simply correlate with other controlling factors.

Complementing experimental efforts with geochemical modeling will enable the prediction of reaction mechanisms, fluid–mineral equilibria, and long-term system evolution under various conditions, ultimately aiding in the design and optimization of field-scale hydrogen production strategies.

Author Contributions: Conceptualization, A.I. and E.R.O.; methodology, A.I. and E.R.O.; software, A.I. and H.S.; validation, A.I., H.S. and E.R.O. formal analysis, A.I. and E.R.O.; investigation, A.I.; validation, A.I., H.S. and E.R.O.; resources, H.S. and E.R.O.; data curation, A.I.; writing—original draft preparation, A.I.; writing—review and editing, A.I. and E.R.O.; visualization, A.I., H.S. and E.R.O.; supervision, E.R.O.; project administration, E.R.O.; funding acquisition, E.R.O. All authors have read and agreed to the published version of the manuscript.

Funding: The information, data, or work presented herein was funded in part by the Advanced Research Projects Agency-Energy (ARPA-E), U.S. Department of Energy, under Award Number DE-AR0001866. The views and opinions of authors expressed herein do not necessarily state or reflect those of the United States Government or any agency thereof.

Data Availability Statement: The original data presented in the study are openly available in Mendeley Data at https://data.mendeley.com/datasets/vxyzcn98kr/1, accessed on 1 September 2025.

Acknowledgments: The authors acknowledge the support of the Harold Vance Department of Petroleum Engineering at Texas A&M University (TAMU) and the members of the TAMU Subsurface Engineering for Sustainable Energy (SESE) lab for providing the necessary laboratory measurements to support this study. The SESE lab, with Principal Investigator Esuru Rita Okoroafor, was supported by the American Association of University Women (AAUW) Research Publication Grant in Engineering, Medicine, and Science (2024–2025) for experimental equipment. The authors also express their gratitude to Sam Nowak from Optirock Group and Gregory N. Boitnott from New England Research, Inc (NER) for supplying some of the rock samples used in this study.

Conflicts of Interest: The authors declare no conflict of interest.

References

1. Keir, R.S. A note on the fluxes of abiogenic methane and hydrogen from mid-ocean ridges. *Geophys. Res. Lett.* **2010**, *37*, 24609. [CrossRef]
2. Maiga, O.; Deville, E.; Laval, J.; Prinzhofer, A.; Diallo, A.B. Characterization of the spontaneously recharging natural hydrogen reservoirs of Bourakebougou in Mali. *Sci. Rep.* **2023**, *13*, 11876. [CrossRef]
3. Wetzel, L.R.; Shock, E.L. Distinguishing ultramafic-from basalt-hosted submarine hydrothermal systems by comparing calculated vent fluid compositions. *J. Geophys. Res. Solid Earth* **2000**, *105*, 8319–8340. [CrossRef]
4. Zhang, M.; Perkovich, N.; Li, Y.; Weihermann, J.; Crummett, R.N. A geophysical investigation of the fairy circles in Nebraska for geologic hydrogen exploration. *Sci. Rep.* **2025**, *15*, 22344. [CrossRef]
5. Truche, L.; Bourdelle, F.; Salvi, S.; Lefeuvre, N.; Zug, A.; Lloret, E. Hydrogen generation during hydrothermal alteration of peralkaline granite. *Geochim. Cosmochim. Acta* **2021**, *308*, 42–59. [CrossRef]
6. Martin, B.; Fyfe, W.S. Some experimental and theoretical observations on the kinetics of hydration reactions with particular reference to serpentinization. *Chem. Geol.* **1970**, *6*, 185–202. [CrossRef]
7. Kita, I.; Marsuo, S.; Wakita, H. H_2 generation by reaction between H_2O and crushed rock: An experimental study on H_2 degassing from the active fault zone. *J. Geophys. Res. Solid Earth* **1982**, *87*, 10789–10795. [CrossRef]
8. Liu, Q.; Wei, Y.; Li, P.; Huang, X.; Meng, Q.; Wu, X.; Zhu, D.; Xu, H.; Fu, Y.; Zhu, D.; et al. Natural hydrogen in the volcanic-bearing sedimentary basin: Origin, conversion, and production rates. *Sci. Adv.* **2025**, *11*, eadr6771. [CrossRef]
9. Wei, Y.; Liu, Q.; Zhu, D.; Meng, Q.; Xu, H.; Zhang, W.; Wu, X.; Li, P.; Huang, X.; Mou, Y.; et al. Helium and natural hydrogen in the Bohai Bay Basin, China: Occurrence, resources, and exploration prospects. *Appl. Energy* **2025**, *383*, 125398. [CrossRef]
10. Vacquand, C.; Deville, E.; Beaumont, V.; Guyot, F.; Sissmann, O.; Pillot, D.; Arcilla, C.; Prinzhofer, A. Reduced gas seepages in ophiolitic complexes: Evidences for multiple origins of the H_2-CH_4-N_2 gas mixtures. *Geochim. Cosmochim. Acta* **2018**, *223*, 437–461. [CrossRef]
11. Marcaillou, C.; Muñoz, M.; Vidal, O.; Parra, T.; Harfouche, M. Mineralogical evidence for H_2 degassing during serpentinization at 300 °C/300 bar. *Earth Planet. Sci. Lett.* **2011**, *303*, 281–290. [CrossRef]
12. Mayhew, L.E.; Ellison, E.T.; McCollom, T.M.; Trainor, T.P.; Templeton, A.S. Hydrogen generation from low-temperature water–rock reactions. *Nat. Geosci.* **2013**, *6*, 478–484. [CrossRef]
13. Klein, F.; Bach, W.; McCollom, T.M. Compositional controls on hydrogen generation during serpentinization of ultramafic rocks. *Lithos* **2013**, *178*, 55–69. [CrossRef]
14. McCollom, T.M.; Klein, F.; Moskowitz, B.; Berquó, T.S.; Bach, W.; Templeton, A.S. Hydrogen generation and iron partitioning during experimental serpentinization of an olivine–pyroxene mixture. *Geochim. Cosmochim. Acta* **2020**, *282*, 55–75. [CrossRef]
15. McCollom, T.M.; Klein, F.; Moskowitz, B.; Solheid, P. Experimental serpentinization of iron-rich olivine (hortonolite): Implications for hydrogen generation and secondary mineralization on Mars and icy moons. *Geochim. Cosmochim. Acta* **2022**, *335*, 98–110. [CrossRef]
16. Greber, N.D.; Dauphas, N.; Bekker, A.; Ptáček, M.P.; Bindeman, I.N.; Hofmann, A. Titanium isotopic evidence for felsic crust and plate tectonics 3.5 billion years ago. *Science* **2017**, *357*, 1271–1274. [CrossRef]
17. Tang, M.; Chen, K.; Rudnick, R.L. Archean upper crust transition from mafic to felsic marks the onset of plate tectonics. *Science* **2016**, *351*, 372–375. [CrossRef]
18. Leong, J.A.M.; Ely, T.; Shock, E.L. Decreasing extents of Archean serpentinization contributed to the rise of an oxidized atmosphere. *Nat. Commun.* **2021**, *12*, 7341. [CrossRef]

19. Leong, J.A.; Nielsen, M.; McQueen, N.; Karolytė, R.; Hillegonds, D.J.; Ballentine, C.; Darrah, T.; McGillis, W.; Kelemen, P. H_2 and CH_4 outgassing rates in the Samail ophiolite, Oman: Implications for low-temperature, continental serpentinization rates. *Geochim. Cosmochim. Acta* **2023**, *347*, 1–15. [CrossRef]
20. Rezaee, R. Assessment of natural hydrogen systems in Western Australia. *Int. J. Hydrogen Energy* **2021**, *46*, 33068–33077. [CrossRef]
21. Saidy, K.; Fawad, M.; Whattam, S.A.; Al-Shuhail, A.A.; Al-Shuhail, A.A.; Campos, M.; Sulistyohariyanto, F.A. Unlocking the H_2 potential in Saudi Arabia: Exploring serpentinites as a source of H_2 production. *Int. J. Hydrogen Energy* **2024**, *89*, 1482–1491. [CrossRef]
22. Templeton, A.S.; Ellison, E.T.; Kelemen, P.B.; Leong, J.; Boyd, E.S.; Colman, D.R.; Matter, J.M. Low-temperature hydrogen production and consumption in partially-hydrated peridotites in Oman: Implications for stimulated geological hydrogen production. *Front. Geochem.* **2024**, *2*, 1366268. [CrossRef]
23. McCollom, T.M.; Bach, W. Thermodynamic constraints on hydrogen generation during serpentinization of ultramafic rocks. *Geochim. Cosmochim. Acta* **2009**, *73*, 856–875. [CrossRef]
24. Klein, F.; Bach, W.; Jöns, N.; McCollom, T.; Moskowitz, B.; Berquó, T. Iron partitioning and hydrogen generation during serpentinization of abyssal peridotites from 15°N on the Mid-Atlantic Ridge. *Geochim. Cosmochim. Acta* **2009**, *73*, 6868–6893. [CrossRef]
25. Tutolo, B.M.; Seyfried, W.E.; Tosca, N.J. A seawater throttle on H_2 production in Precambrian serpentinizing systems. *Proc. Natl. Acad. Sci. USA* **2020**, *117*, 14756–14763. [CrossRef] [PubMed]
26. Foustoukos, D.I.; Seyfried, W.E. Hydrocarbons in Hydrothermal Vent Fluids: The Role of Chromium-Bearing Catalysts. *Science* **2004**, *304*, 1002–1005. [CrossRef]
27. Geymond, U.; Briolet, T.; Combaudon, V.; Sissmann, O.; Martinez, I.; Duttine, M.; Moretti, I. Reassessing the role of magnetite during natural hydrogen generation. *Front. Earth Sci.* **2023**, *11*, 1169356. [CrossRef]
28. Goldstein, T.P. Geocatalytic Reactions in Formation and Maturation of Petroleum. *AAPG Bull.* **1983**, *67*, 152–159. [CrossRef]
29. Liu, J.; Yang, T.; Peng, Q.; Yang, Y.; Li, Y.W.; Wen, X.D. Theoretical exploration of the interaction between hydrogen and pyrite-type FeS_2 surfaces. *Appl. Surf. Sci.* **2021**, *537*, 147900. [CrossRef]
30. Neubeck, A.; Duc, N.T.; Bastviken, D.; Crill, P.; Holm, N.G. Formation of H_2 and CH_4 by weathering of olivine at temperatures between 30 and 70 °C. *Geochem. Trans.* **2011**, *12*, 6. [CrossRef]
31. Schoonen, M.A.A.; Xu, Y.; Strongin, D.R. An introduction to geocatalysis. *J. Geochem. Explor.* **1998**, *62*, 201–215. [CrossRef]
32. Truche, L.; Berger, G.; Albrecht, A.; Domergue, L. Engineered materials as potential geocatalysts in deep geological nuclear waste repositories: A case study of the stainless steel catalytic effect on nitrate reduction by hydrogen. *Appl. Geochem.* **2013**, *35*, 279–288. [CrossRef]
33. Corre, M.; Brunet, F.; Schwartz, S.; Gautheron, C.; Agranier, A.; Lesimple, S. Quaternary low-temperature serpentinization and carbonation in the New Caledonia ophiolite. *Sci. Rep.* **2023**, *13*, 19413. [CrossRef] [PubMed]
34. Huang, R.; Sun, W.; Ding, X.; Zhao, Y.; Song, M. Effect of pressure on the kinetics of peridotite serpentinization. *Phys. Chem. Miner.* **2020**, *47*, 33. [CrossRef]
35. Pens, M.; Andreani, M.; Daniel, I.; Perrillat, J.P.; Cardon, H. Contrasted effect of aluminum on the serpentinization rate of olivine and orthopyroxene under hydrothermal conditions. *Chem. Geol.* **2016**, *441*, 256–264. [CrossRef]
36. Malvoisin, B.; Brunet, F.; Carlut, J.; Rouméjon, S.; Cannat, M. Serpentinization of oceanic peridotites: 2. Kinetics and processes of San Carlos olivine hydrothermal alteration. *J. Geophys. Res.* **2012**, *117*, 1–13. [CrossRef]
37. Biesinger, M.C.; Payne, B.P.; Grosvenor, A.P.; Lau, L.W.M.; Gerson, A.R.; Smart, R.S.C. Resolving surface chemical states in XPS analysis of first row transition metals, oxides and hydroxides: Cr, Mn, Fe, Co and Ni. *Appl. Surf. Sci.* **2011**, *257*, 2717–2730. [CrossRef]
38. Grosvenor, A.P.; Kobe, B.A.; Biesinger, M.C.; McIntyre, N.S. Investigation of multiplet splitting of Fe 2p XPS spectra and bonding in iron compounds. *Surf. Interface Anal.* **2004**, *36*, 1564–1574. [CrossRef]
39. Molchan, I.S.; Thompson, G.E.; Skeldon, P.; Lindsay, R.; Walton, J.; Kouvelos, E.; Romanos, G.E.; Falaras, P.; Kontos, A.G.; Arfanis, M.; et al. Microscopic study of the corrosion behaviour of mild steel in ionic liquids for CO_2 capture applications. *RSC Adv.* **2015**, *5*, 35181–35194. [CrossRef]
40. Ali, M.; Isah, A.; Yekeen, N.; Hassanpouryouzband, A.; Sarmadivaleh, M.; Okoroafor, R.; Al Kobaisi, M.; Mahmoud, M.; Vahrenkamp, V.; Hoteit, H. Recent Progress in Underground Hydrogen Storage. *Energy Environ. Sci.* **2025**, *18*, 5740–5810. [CrossRef]
41. Mohammed, I.; Isah, A.; Bello, A.; Mahmoud, M.; Al Shehri, D.; Raza, A.; Alafnan, S. Iron-Rich Rocks for Hydrogen Geo-Storage: Induced Storage Capacity by Active Sorption. *Energy Fuels* **2024**, *38*, 16795–16808. [CrossRef]
42. Herzberg, C.; Asimow, P.D.; Arndt, N.; Niu, Y.; Lesher, C.M.; Fitton, J.G.; Cheadle, M.J.; Saunders, A.D. Temperatures in ambient mantle and plumes: Constraints from basalts, picrites, and komatiites. *Geochem. Geophys. Geosyst.* **2007**, *8*, Q02006. [CrossRef]
43. Stevens, T.O.; McKinley, J.P. Abiotic controls on H_2 production from Basalt—Water reactions and implications for aquifer biogeochemistry. *Environ. Sci. Technol.* **2000**, *34*, 826–831. [CrossRef]

44. Carlin, W.; Malvoisin, B.; Brunet, F.; Lanson, B.; Findling, N.; Lanson, M.; Fargetton, T.; Jeannin, L.; Lhote, O. Kinetics of low-temperature H_2 production in ultramafic rocks by ferroan brucite oxidation. *Geochem. Perspect. Lett.* **2024**, *29*, 27–32. [CrossRef]
45. Templeton, A.S.; Ellison, E.T. Formation and loss of metastable brucite: Does Fe(II)-bearing brucite support microbial activity in serpentinizing ecosystems? *Philos. Trans. R. Soc. A* **2020**, *378*, 20180423. [CrossRef]
46. Bach, W.; Paulick, H.; Garrido, C.J.; Ildefonse, B.; Meurer, W.P.; Humphris, S.E. Unraveling the sequence of serpentinization reactions: Petrography, mineral chemistry, and petrophysics of serpentinites from MAR 15°N (ODP Leg 209, Site 1274). *Geophys. Res. Lett.* **2006**, *33*, 13306. [CrossRef]
47. Jöns, N.; Kahl, W.A.; Bach, W. Reaction-induced porosity and onset of low-temperature carbonation in abyssal peridotites: Insights from 3D high-resolution microtomography. *Lithos* **2017**, *268–271*, 274–284. [CrossRef]
48. Deer, W.A.; Howie, R.A.; Zussman, J. *Ortho- and Ring Silicates*; Longmans: London, UK, 1962.
49. Al-Yaseri, A.; Desouky, M.; Aljawad, M.S.; Hassanpouryouzband, A. Evidence of hydrogen release during CO_2 sequestration in basalt. *Int. J. Hydrogen Energy* **2025**, *101*, 1183–1190. [CrossRef]
50. Otake, T.; Wesolowski, D.J.; Anovitz, L.M.; Allard, L.F.; Ohmoto, H. Experimental evidence for non-redox transformations between magnetite and hematite under H_2-rich hydrothermal conditions. *Earth Planet. Sci. Lett.* **2007**, *257*, 60–70. [CrossRef]
51. Preiner, M.; Xavier, J.C.; Sousa, F.L.; Zimorski, V.; Neubeck, A.; Lang, S.Q.; Greenwell, H.C.; Kleinermanns, K.; Tüysüz, H.; McCollom, T.M.; et al. Serpentinization: Connecting geochemistry, ancient metabolism and industrial hydrogenation. *Life* **2018**, *8*, 41. [CrossRef]
52. Schwander, L.; Brabender, M.; Mrnjavac, N.; Wimmer, J.L.; Preiner, M.; Martin, W.F. Serpentinization as the source of energy, electrons, organics, catalysts, nutrients and pH gradients for the origin of LUCA and life. *Front. Microbiol.* **2023**, *14*, 1257597. [CrossRef]
53. Song, H.; Ou, X.; Han, B.; Deng, H.; Zhang, W.; Tian, C.; Chai, C.; Lu, A.; Lin, Z.; Chai, L. An overlooked natural hydrogen evolution pathway: Ni^{2+} boosting H_2O reduction by $Fe(OH)_2$ oxidation during low-temperature serpentinization. *Angew. Chem. Int. Ed.* **2021**, *60*, 24054–24058. [CrossRef]
54. Beyazay, T.; Ochoa-Hernández, C.; Song, Y.; Belthle, K.S.; Martin, W.F.; Tüysüz, H. Influence of composition of nickel-iron nanoparticles for abiotic CO_2 conversion to early prebiotic organics. *Angew. Chem.* **2023**, *135*, e202218189. [CrossRef]
55. Andreani, M.; Daniel, I.; Pollet-Villard, M. Aluminum speeds up the hydrothermal alteration of olivine. *Am. Mineral.* **2013**, *98*, 1738–1744. [CrossRef]
56. Zähr, J.; Oswald, S.; Türpe, M.; Ullrich, H.J.; Füssel, U. Characterisation of oxide and hydroxide layers on technical aluminum materials using XPS. *Vacuum* **2012**, *86*, 1216–1219. [CrossRef]

hydrogen

MDPI

Article

Investigation of Mechano-Electrochemical Effects on Hydrogen Distribution at Corrosion Defects

Zhixiang Dai [1], Jiamin Tang [2], Sijia Zheng [1], Feng Wang [1], Qin Bie [1], Pengcheng Kang [1,*], Xinyi Wang [1], Shiwen Guo [3,*] and Lin Chen [2]

1 Gathering and Transportation Engineering Technology Research Institute, PetroChina Southwest Oil & Gasfield Company, Chengdu 610041, China
2 School of Safety and Ocean Engineering, China University of Petroleum (Beijing), Beijing 102249, China
3 National Institute of Clean-and-Low-Carbon Energy, Beijing 102211, China
* Correspondence: kangpengc@petrochina.com.cn (P.K.); 20101367@ceic.com (S.G.)

Abstract

This study employed tensile test, hydrogen permeation measurements, and potentiodynamic polarization testing to investigate the mechanical properties, hydrogen diffusion coefficients, and electrochemical behavior of X80 steel. A multifield coupled finite element (FE) model was developed that incorporated the mechano-electrochemical (M-E) effect to analyze the stress–strain distribution, anodic equilibrium potential, cathodic exchange current density, and hydrogen distribution characteristics at pipeline corrosion defects under varying tensile strains. The results indicated that tensile strain significantly modulated the anodic equilibrium potential and cathodic exchange current density, leading to localized hydrogen accumulation at corrosion defects. The stress concentration and plastic deformation at the defect site intensified as the tensile strain increased, further promoting hydrogen enrichment. The study concluded that the M-E effect exacerbated hydrogen enrichment at the defect sites, increasing the risk of hydrogen-induced cracking. The simulation results showed that the hydrogen distribution state aligned with the stress–hydrogen diffusion coupling model when considering the M-E effect. However, the M-E effect slightly increased the hydrogen concentration at the defect. These findings provide critical insights for enhancing the safety and durability of hydrogen transmission pipelines.

Keywords: hydrogen diffusion; X80 steel pipeline; finite element simulation; mechano-electrochemical effect

1. Introduction

Fossil fuels release large amounts of greenhouse gases that lead to abnormal climate change and the frequent extreme weather events [1]. Therefore, replacing existing fossil fuels with environmentally friendly and sustainable new energy has become an important issue during the development of current international energy strategies. Hydrogen energy shows promise as a vital future energy source due to its high efficiency, cleanliness, and low carbon emissions [2]. Although transportation via preexisting pipelines can increase hydrogen gas utilization efficiency [3], these pipelines are built for large-scale, long-distance natural gas transportation and are highly sensitive to hydrogen embrittlement (HE) due to their strength [4]. According to the current research findings, the higher the strength of steel, the more sensitive it becomes to hydrogen embrittlement. Hydrogen transportation also increases the risk of corrosion failure on the outer pipeline wall. Therefore, to prevent soil corrosion, a protective coating is applied to the external surfaces of buried pipelines [5], which can be damaged by long-term service and aggravate the degree of corrosion [6]. Therefore, the pipeline transportation of hydrogen presents various challenges that require further investigation.

Although various studies have explored these two typical failure mechanisms, the synergistic effect between them remains largely ignored. First, corrosion defects can form on the steel pipeline surface exposed to soil due to a damaged protective coating [7], which, in turn, can enhance the local stress concentration, resulting in local hydrogen accumulation [8]. Moreover, local stress at the tips of corrosion defects caused by internal pressure can accelerate hydrogen accumulation [9,10], which can aggravate cracking propagation [11]. Therefore, the synergistic effect of the load and electrochemical reaction will exacerbate the corrosion failure of pipeline steel, intensify hydrogen damage, and ultimately pose safety risks to future hydrogen energy transportation.

This study determines the mechanical properties, hydrogen diffusion coefficient, and potential polarization characteristics of X80 steel via tensile test, hydrogen permeation, and potential polarization tests. A finite element (FE) model was established of the mechano-electrochemical (M-E) effect on hydrogen distribution to analyze the Von Mises stress, hydrostatic stress, and equivalent plastic strain distribution at the corrosion defects of a X80 pipeline under different degrees of tensile strain. The anode equilibrium potential, cathode exchange current density, and hydrogen distribution were also analyzed. Finally, the hydrogen diffusion and enrichment at the corrosion defects in the pipeline were revealed.

2. Experimental Procedures

2.1. Materials and Tensile Tests

X80 steel was selected for testing, and the chemical compositions are shown in Table 1.

Table 1. The chemical composition of the X80 steel (wt. $\times 10^{-2}$).

C	Si	Mn	P	S	Cr	Mo	Ni	Nb	V	Cu	B	Al
0.076	0.21	1.89	0.010	0.0022	0.27	0.28	0.055	0.082	0.029	0.059	0.0002	0.038

Figure 1 shows the flat dog-bone-type specimens, which were machined in the rolling direction. The uniaxial tensile tests were performed at a strain rate of 1×10^{-4} s^{-1} at room temperature using an MTS Landmark tensile testing machine.

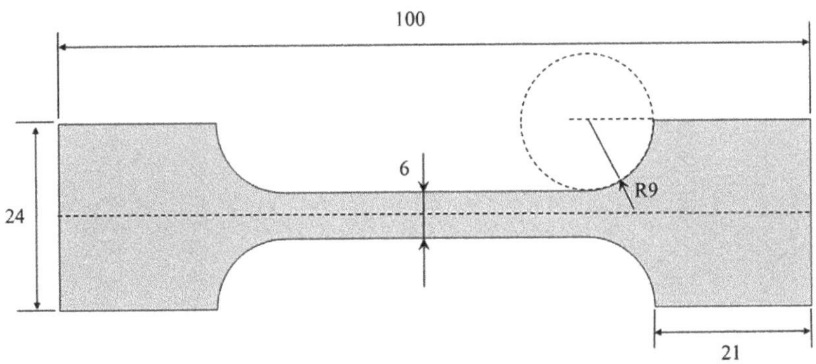

Figure 1. The sizes and shapes of the flat dog-bone-type specimens (mm), with a thickness of 1.5 mm.

2.2. Electrochemical Measurements

An electrochemical workstation (Ivium-n-Stat, Ivium Technologies BV, Eindhoven, The Netherland) was used for the electrochemical measurements. A platinum wire and saturated calomel electrode (SCE) served as the counter and reference electrodes, respectively. The samples were mechanically polished before the electrochemical tests, which were conducted at approximately 298 K. A solution was used for nitrogen purging throughout the experiment to remove oxygen.

2.2.1. Hydrogen Permeation Test

The hydrogen diffusion was characterized using the modified "Devanathan–Stachurski cell" [12] (Figure 2). The exit side of the sample was coated with a 100 nm Ni film to achieve a sufficiently low anodic current. The solution-exposed surface was 4.91 cm^2 (a circle with a diameter of 25 mm). The exit side was immersed in a sodium hydroxide solution (0.1 M NaOH) to determine the anodic overpotential (0.200 V). A mixed charging solution (0.5 M H_2SO_4, 0.2 g/L CH_4N_2S) was used for the entrance while −1100 mV was used as the overpotential. The effective diffusion coefficient (D_{eff}) was calculated using the "time-lag" method [13]:

$$t_{0.63} = \frac{L^2}{6D_{eff}} \tag{1}$$

$$J = \frac{i_\infty}{F} \tag{2}$$

$$C_{app} = JL/D_{eff} \tag{3}$$

where L is the membrane thickness, $t_{0.63}$ is the time where $(i - i_0)/(i_\infty - i_0) = 0.63$, i_0 is the initial current density of the hydrogen permeation curve, i_∞ is the steady-state current density of the hydrogen permeation curve, J is the hydrogen diffusion flux, and C_{app} is the subsurface hydrogen concentration.

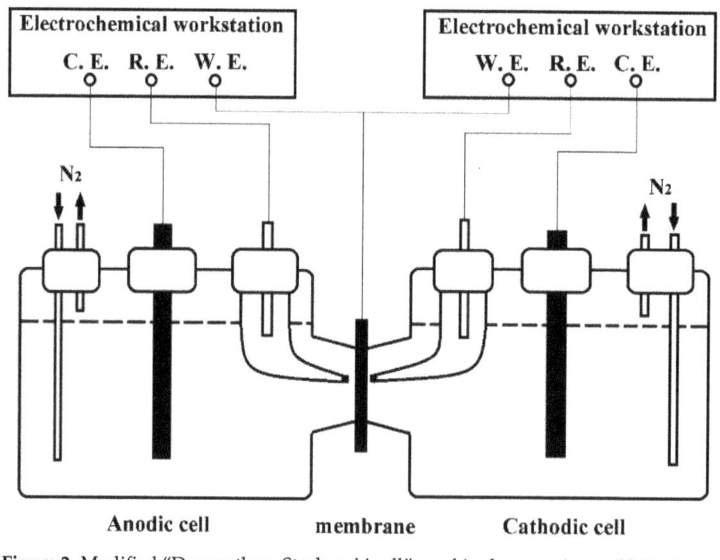

Figure 2. Modified "Devanathan–Stachurski cell" used in the experiment (C.E.: Counter Electrode, R.E.: Reference Electrode, W.E.: Specimens).

2.2.2. Potentiodynamic Polarization Tests

The potentiodynamic polarization tests were performed in an NS4 solution containing 483 mg/L $NaHCO_3$, 122 mg/L KCl, 181 mg/L $CaCl_2 \cdot 2H_2O$, and 131 mg/L $MgSO_4 \cdot 7H_2O$. The potentiodynamic polarization curves were obtained at 0.5 mV/s.

The electrochemical anodic and cathodic reactions of the X80 steel in the deoxygenated, near-neutral pH NS4 solution represented steel oxidation and hydrogen evolution, respectively [14].

Anodic reaction:

$$Fe \rightarrow Fe^{2+} + 2e^- \tag{4}$$

Cathodic reaction:

$$H_2O + e^- \rightarrow H + OH^- \tag{5}$$

It was assumed that the dominant anodic reaction was the dissolution of the iron in the steel, and that the pipeline steel was in an active dissolution state in the test solution. Although the cathodic reaction was described as the single electron production of atomic hydrogen, the destination of the produced atomic hydrogen was not specified [15]. The electrode kinetics of the steel for the anodic and cathodic reactions were activation-controlled and described as follows:

$$i_a = i_{0,a}\exp\left(\frac{\eta_a}{b}\right) \tag{6}$$

$$i_c = i_{0,c}exp\left(\frac{\eta_c}{b}\right) \tag{7}$$

$$\eta = \phi - \phi_{eq} \tag{8}$$

where i and i_0 represent the charge-transfer and exchange current densities of the electrochemical reactions, respectively. Φ and Φ_{eq} denote the electrode and equilibrium electrode potential, respectively, while η is the activation overpotential, and b is the Tafel slope.

Nernst equations were used to obtain the equilibrium potentials of the steel oxidation and hydrogen evolution:

$$\phi^0_{a,eq} = \phi^{0,s}_{a,eq} + \frac{0.0592}{2}log\left[Fe^{2+}\right] \tag{9}$$

$$\phi^0_{c,eq} = \phi^{0,s}_{c,eq} + 0.0592log\left[H^+\right] = -0.0592pH \tag{10}$$

where $\phi^{0,s}_{a,eq}$ and $\phi^{0,s}_{c,eq}$ are the standard equilibrium potentials for the anodic and cathodic reactions, which are calculated as −0.859 V_{SCE} and −0.644 V_{SCE} [15].

2.3. *FE Simulation*

2.3.1. Initial and Boundary Conditions

Commercial COMSOL Multiphysics 6.0 software was used for multifield coupling simulations of the mechanical, electrochemical, and hydrogen diffusion. Figure 3 shows the geometrical model of the X80 steel pipe containing an external corrosion defect. The thickness and length of the pipe were 18.4 mm and 1 m, respectively. The corrosion defect was elliptical, with a length and depth of 40 mm and 7.36 mm, respectively. The depth of the defects changed during the corrosion process.

The contact boundary between the solution and the pipeline was set as the free boundary, while the other boundary conditions of the solution were electrically isolated. During the deformation process, the left side of the pipeline was fixed, which was subjected to various tensile strains (0, 0.1%, 0.2%, 0.3%, and 0.4%). The inner wall of the pipeline was set as the hydrogen source during hydrogen fusion. The initial hydrogen concentration

(mol/m^3) was obtained from the hydrogen permeation test. The unit of mol/m^3 could be converted to ppm by 1 mol/m^3 = 0.13 ppm [16].

A triangular mesh type was used, which consisted of 5503 domain cells and 439 boundary cells. The maximum and minimum element sizes of the pipeline were 6 mm and 1 mm, respectively, with a maximum element growth rate of 1.05. The curvature factor and the narrow area resolution were 0.6 and 1, respectively. Contrarily, the maximum and minimum element sizes of the soil were 134 mm and 8 mm, respectively, with a maximum element growth rate of 1.2. The curvature factor and the narrow area resolution were 0.6 and 1, respectively. The inner wall of the pipeline was set as a hydrogen source, and the initial concentration was determined using a hydrogen permeation test.

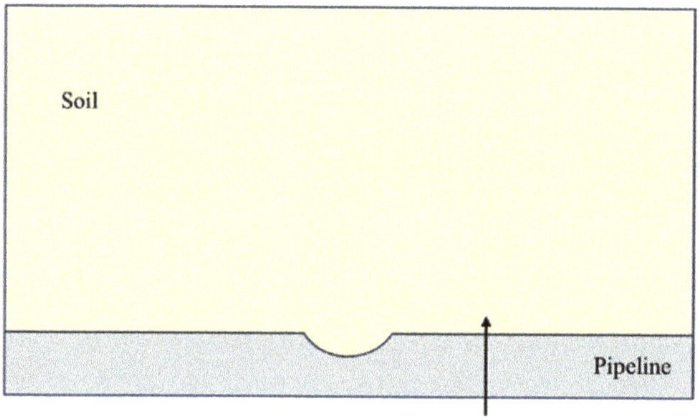

Figure 3. Geometrical model of the X80 steel pipe containing a corrosion defect.

2.3.2. Hydrogen Atom Distribution Simulation

The hydrogen atoms in metals move from higher to lower chemical potential positions, causing hydrogen diffusion. The driving force behind the diffusion may include one or more potential gradients. When the interaction between the diffusing substances is not taken into account, the diffusion rate of hydrogen atoms depends on their mobility [17]. If the metal displays an ideal lattice, the diffusion flux J of hydrogen atoms can be expressed using Equation (11) [18]:

$$J = -MC\nabla\mu \tag{11}$$

where M represents the hydrogen atom mobility in the lattice, M = D/RT, D is the diffusion coefficient of the hydrogen atoms, R is the general gas constant, T is the absolute temperature, C is the hydrogen concentration in the lattice, and μ is the chemical potential of the hydrogen in the lattice.

If stress is not considered, the chemical potential of the hydrogen atoms is expressed using Equation (12):

$$\mu = \mu_0 + RTlnC \tag{12}$$

where μ_0 represents the chemical potential at the reference temperature and pressure.

When the hydrogen atoms occupy tetrahedral or octahedral interstitial positions in the α-iron lattice, the volumes of both interstitial positions expand. If there is external stress, the volume at the gap position is larger, and the chemical potential decreases to a lower level [19]. At this point, the chemical potential μ is expressed using Equation (13):

$$\mu = \mu_0 + RTlnC + \mu_\sigma \tag{13}$$

where μ_σ represents the chemical potential related to external stress, which is expressed using Equation (14) [20]:

$$\mu_\sigma = -\sigma_h V_h \tag{14}$$

where σ_h is the hydrostatic stress, $\sigma_h = \frac{1}{3}\sum_{i=1}^{3} \sigma_{ii}$, σ_{ii} is the principal stress, and V_h is the partial molar volume of the hydrogen.

The relationship between hydrogen diffusion flux and hydrogen concentration is obtained by substituting Equations (13) and (14) into Equation (9), as shown in Equation (15):

$$J = -D\nabla C + \frac{D}{RT} C V_h \nabla \sigma_h \tag{15}$$

When considering a random volume surrounded by surface S, the law of mass conservation requires that the change rate of the total hydrogen volume V is equal to the flux through the surface S [16]. Therefore, Equation (16) is used express mass conservation:

$$\frac{\partial}{\partial t}\int_V CdV + \int_S JndS = 0 \tag{16}$$

where $\partial/\partial t$ is the partial derivative regarding time, and n is the outward unit normal vector.

The mass conservation equations for the hydrogen concentration C and σh hydrostatic stress m can be obtained by replacing the hydrogen diffusion flux J in Equation (16) with Equation (15), as shown in Equation (17):

$$\frac{\partial}{\partial t}\int_V CdV + \int_S \left(-D\nabla C + \frac{D}{RT} C V_h \nabla \sigma_h \right) ndS = 0 \tag{17}$$

Using the divergence theorem, Equation (17) is expressed as Equation (18):

$$\frac{\partial C}{\partial t} + \nabla(-D\nabla C) + \nabla\left(\frac{DCV_h}{RT}\right)\nabla\sigma_h = 0 \tag{18}$$

2.3.3. M-E Effect and Hydrogen Distribution Simulation

Stress Field Simulation

An isotropic hardening model was adopted to simulate the elastoplastic stress–strain on pipelines while the hardening parameter of X80 steel was obtained from the true stress–strain curve.

Electrochemical Field Simulation

(1) Anodic reaction

Gutman [21] revealed that elastoplastic strain affects the anodic equilibrium potential, as shown in Equations (19) and (20).

Elastic deformation:

$$\Delta\varphi_{a,eq}^{e} = -\frac{\Delta P V_M}{zF} \tag{19}$$

Plastic deformation:

$$\Delta\varphi_{a,eq}^{p} = -\frac{TR}{zF} ln\left(\frac{v\alpha}{N_0}\varepsilon_p + 1\right) \tag{20}$$

The anodic equilibrium potential affected by continuous elastoplastic tension is shown in Equation (21):

$$\varphi_{a,eq}^{p} = \varphi_{a,eq}^{0} - \frac{\Delta P_m V_M}{zF} - \frac{TR}{zF} ln\left(\frac{v\alpha}{N_0}\varepsilon_p + 1\right) \tag{21}$$

where $\Delta\varphi_{a,eq}^{0}$ is the standard anodic equilibrium potential, ΔP and ΔP_m are the excess pressure (equal to 1/3 of the uniaxial tensile stress) and the excess pressure-to-elastic deformation limit (equal to 1/3 of the yield strength of the steel), respectively. $\Delta\varphi^{e}_{a,eq}$ and $\Delta\varphi^{p}_{a,eq}$ represent the shifts in the equilibrium potential of the anodic reaction under elastic and plastic deformation, respectively. P_m is the molar volume of the steel (7.13×10^6 m^3), z is the number of charges (2 for steel), t is an orientation-dependent factor (0.45), N_0 is the initial dislocation density of the steel (1×10^8 cm^{-2}) [22], ε_p is the equivalent plastic strain, and α is a constant (1.67×10^{11} cm^{-2}) [21]. R is the gas constant (8.314 J/mol K), T is the absolute temperature (298.15 K), and F is the Faraday's constant (96,485 C/mol).

(2) Cathodic reaction

Plastic deformation enhanced the cathodic reaction by facilitating the redistribution of electrochemical heterogeneities and expanding the cathodic reaction area [21,23]. Simultaneously, during the plastic deformation process, the activation energy of the cathodic reaction was reduced due to the formation of slip bands, microcracks, and other defects [15]. To simplify the FE simulation, a semi-empirical expression that only considered the increase in the exchange current density during plastic deformation was used to describe the M-E effect of the cathodic reaction [15]:

$$i_c = i_{0,c} \times 10^{\frac{\sigma_{Mises} V_M}{6F(-b_c)}} \tag{22}$$

where $i_{0,c}$ is the exchange current density of hydrogen evolution in the absence of external stress/strain, σ_{Mises} is the Von Mises stress, and b_c is the cathodic Tafel slope.

Distribution of the Electrical Fields in a Solution

The electrical field distribution in a solution for electrochemical reactions can be determined using electrical field theory [24]:

$$\nabla i_k = Q_k \tag{23}$$

$$i_k = -\sigma_k \nabla \varphi_k \tag{24}$$

where Q_k is a general source term, and σ_k and φ_k are the conductivity and potential, respectively. The conductivities of the NS4 solution and steel electrode were set as 0.096 S/m and 106 S/m [15].

Multifield Coupling Simulation

The multifield coupled FE simulation of the X80 pipeline assessed three factors: the influence of the M-E effect on hydrogen distribution in the pipeline immersed in the solution, the electrochemical corrosion analysis of the steel/solution interface, and the elastoplastic solid mechanics analysis.

This paper combined five modules in the COMSOL software package: solid mechanics, secondary current distribution, deformation geometry, diluted species transport, and a general-form partial differential equation module. The solid mechanics module was solved first to obtain the relevant mechanical parameters, which were used as inputs in the secondary current distribution module to determine the size changes in the corrosion defects over time. Then, the relevant mechanical parameters and corrosion defect size changes were entered into the diluted species transport module to obtain the hydrogen diffusion behavior and concentration, as well as the deformation geometry and a general-form partial differential equation. Table 2 shows the other parameters of the three-field coupling simulation.

Table 2. Parameters of the three-field coupling simulation.

Parameter	Value
The density of steel, ρ (g/cm^3)	7.85 [25]
Molar mass of steel, M (g/mol)	56
Partial molar volume of hydrogen, Vh (m^3/mol)	2×10^{-6} [26,27]

3. Results

3.1. Tensile, Hydrogen Permeation, and Potentiodynamic Polarization Characteristics

Figure 4a illustrates the engineering stress–strain curve of the X80 steel with a yield strength (σ_y) of 638 MPa. The true stress–strain curve, which was converted from the engineering stress–strain curve, was used to establish the material hardening function in the COMSOL software. Figure 4b shows the transient hydrogen permeation current density curve of the X80 steel, with a steady-state current density (i_∞) of 29 $\mu A/cm^2$, an initial current density (i_0) of 2.54 $\mu A/cm^2$, a D_{eff} of 7.73×10^{-10} m^2/s, and a C_{app} of 1.55 mol/m^3.

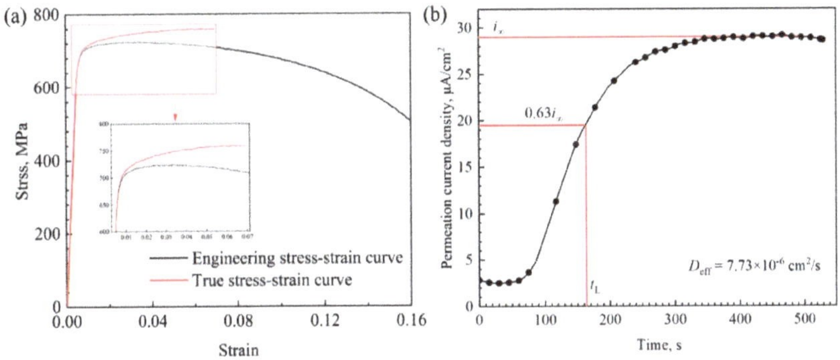

Figure 4. (**a**) Engineering and true stress–strain curves of the X80 steel. (**b**) Transient hydrogen permeation current density curve of the X80 steel.

Figure 5 shows the potentiodynamic polarization curve. Various electrochemical corrosion parameters, including the corrosion potential (E_{corr}), corrosion current density (i_{corr}), anode Tafel slope (b_a), cathodic Tafel slope (b_c), anode exchange current density (i_{0a}), cathodic exchange current density (i_{0c}), anodic equilibrium potential (φ_{0a}), and cathodic equilibrium potential (φ_{0c}), were derived from the polarization curves and used as initial conditions for the FE calculations. Table 3 shows the values of these electrochemical corrosion parameters.

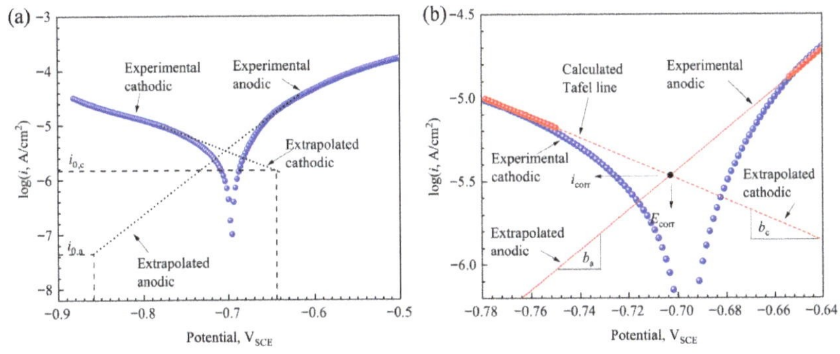

Figure 5. Potentiodynamic polarization curve of the X80 steel. (**b**) is the enlarged image of (**a**).

Table 3. Values of the electrochemical corrosion parameters.

E_{corr} V	i_{corr} A/cm^2	φ^0_{eqa} V	φ^0_{eqc} V	b_a V/dec	b_c V/dec	i_{0a} A/cm^2	i_{0c} A/cm^2
−0.704	3.373×10^{-6}	−0.859	−0.644	0.162	−0.083	4.508×10^{-8}	1.496×10^{-6}

3.2. Stress and Strain Distributions

The numerical simulations using the hydrogen diffusion coefficient derived from the hydrogen permeation test indicated that approximately 9 d were required for hydrogen to reach a steady-state distribution in the pipeline model. Figures 6–8 show the Von Mises stress, static water stress, and equivalent plastic strain distribution at the corrosion defect of the X80 pipeline under different tensile strains (0.1–0.4%) after simulation for 9 d, where the color legend indicates the stress (MPa). Consistent maximum Von Mises stress localization was evident at the center of the corrosion defect (Figure 6). The local Von Mises stress increased at a higher applied tensile strain. At applied tensile strains of 0.1% and 0.2%, maximum Von Mises stress values of 350 MPa and 600 MPa were present at the center of the corrosion defect, respectively, which were slightly higher than the stress level of the pipeline body. At this juncture, the corrosion defect was subjected to elastic deformation. A tensile strain of 0.3% significantly increased the localized stress at the defect site to 800 MPa, resulting in localized plastic deformation. The Von Mises stress levels in the root and subjacent regions of the defect were equal. At a tensile strain of 0.4%, the maximum Von Mises stress at the corrosion defect approached 900 MPa. However, the Von Mises stress level at the root of the defect was lower than that in the lateral and subjacent regions.

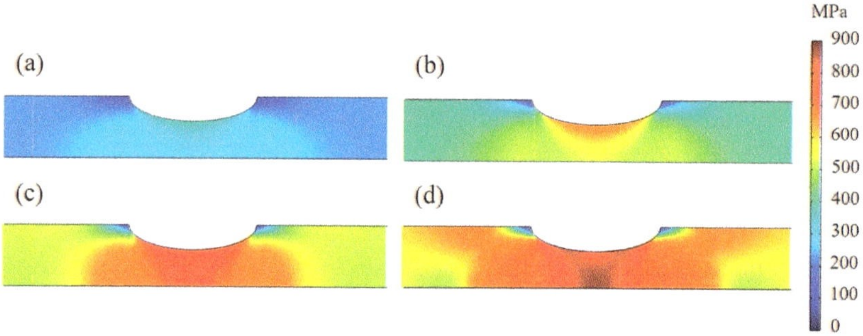

Figure 6. Von Mises stress distribution at the corrosion defect at different tensile strains of 0.1%, 0.2%, 0.3%, and 0.4% (**a**–**d**).

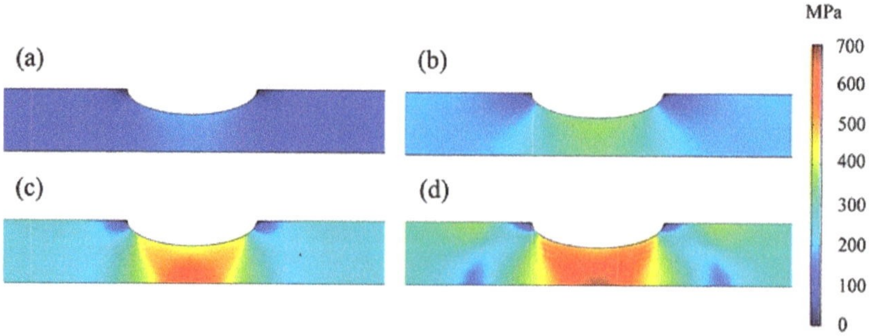

Figure 7. Hydrostatic stress distribution at the corrosion defect at different tensile strains of 0.1%, 0.2%, 0.3%, and 0.4% (**a**–**d**).

As shown in Figure 7, the magnitude of the local hydrostatic stress increased at a higher tensile strain, with the color legend referring to the hydrostatic stress (MPa). The hydrostatic stress at the corrosion defect (i.e., 200 MPa) was slightly higher than that of the pipeline body at an applied tensile strain of 0.1% and reached approximately 400 MPa at a tensile strain of 0.2%, which was below the yield strength of the steel. At tensile strains below 0.2%, the hydrostatic stress levels in the root and subjacent regions of the defect were essentially equal. When the tensile strain was increased to 0.3%, the local hydrostatic stress rose to 560 MPa, reaching a maximum subjacent to the corrosion defect. At a tensile strain of 0.4%, the maximum hydrostatic stress localized below the defect and along the inner wall of the pipeline reached 670 MPa.

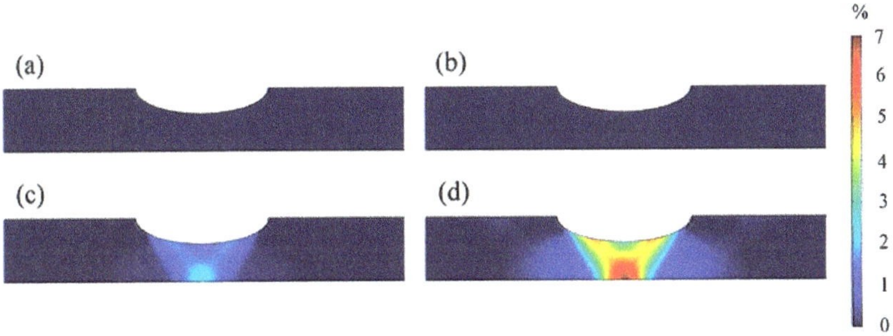

Figure 8. Equivalent plastic strain distribution at the corrosion defect at different tensile strains of 0.1%, 0.2%, 0.3%, and 0.4% (**a–d**).

The stress concentration at the corrosion defect significantly altered the localized strain. As shown in Figure 8, the equivalent plastic strain at the corrosion defect remained negligible (0%) at externally applied tensile strains of 0.1% and 0.2%, which were consistent with the strain level of the pipeline body. However, a tensile strain of 0.3% caused a localized concentration of equivalent plastic strain at the defect site, attaining a maximum strain of approximately 2.7%. When the applied tensile strain was further elevated to 0.4%, the equivalent plastic strain at the defect site exhibited a pronounced increase, reaching 7%. Furthermore, the maximum equivalent plastic strain was localized subjacent to the defect and along the inner wall of the pipeline.

3.3. Anodic Equilibrium Potential and Cathodic Exchange Current Density Distributions

Figures 9 and 10 show the distribution of the anodic equilibrium potential and cathodic exchange current density at the corrosion defect of the X80 pipeline at different tensile strains (0.1–0.4%) after simulation for 9 d. As shown in Figure 9, at external tensile strains of 0.1% and 0.2%, the anodic equilibrium potential at the corrosion defect site was equivalent to that on the pipeline body (−0.867 V). When the tensile strain reached 0.3%, the anodic equilibrium potential at the defect shifted to the negative side, with the most negative value approximating −0.871 V. When the applied tensile strain was increased to 0.4%, the negative shift of the anode equilibrium potential at the defect site became more pronounced, reaching −0.876 V.

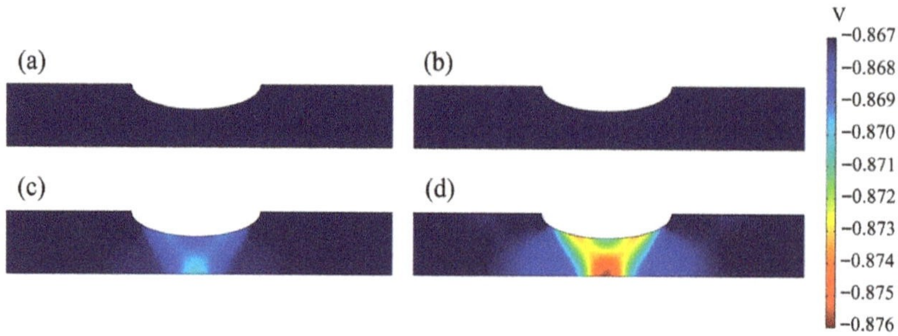

Figure 9. Anodic equilibrium potential distribution at the corrosion defect at different tensile strains of 0.1%, 0.2%, 0.3%, and 0.4% (**a**–**d**).

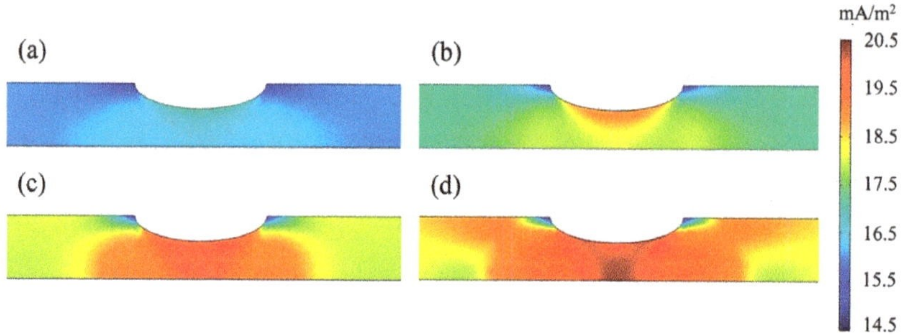

Figure 10. Cathodic exchange current density distribution at the corrosion defect at different tensile strains of 0.1%, 0.2%, 0.3%, and 0.4% (**a**–**d**).

As illustrated in Figure 10, the cathodic exchange current density exhibited a progressive increase as the tensile strain rose. At tensile strains of 0.1% and 0.2%, the exchange current density at the periphery of the corrosion defect was higher than in other regions of the pipeline. The maximum exchange current density was localized at the defect root at 0.1% and 0.2% tensile strains, reaching 17 mA/m^2 and 19 mA/m^2, respectively. The cathodic exchange current density increased significantly at a tensile strain of 0.3%, with values in the root and subjacent regions of the defect exceeding 19.5 mA/m^2. When the tensile strain was further increased to 0.4%, the maximum cathodic exchange current density at the corrosion defect surpassed 20 mA/m^2, localized subjacent to the defect and along the inner wall of the pipeline.

3.4. Hydrogen Distribution

Figure 11 shows the stable-state distribution and maximum position of the hydrogen concentration at the corrosion defect site of the X80 pipeline at different tensile strains (0.1–0.4%) after a 9 d diffusion simulation. The application of tensile strain altered the hydrogen distribution in the pipeline, resulting in hydrogen accumulation at the corrosion defect. At a tensile strain of 0.1%, the hydrogen concentration at the root of the corrosion defect was slightly higher than that in the pipeline body. When the tensile strain reached 0.2%, the maximum hydrogen concentration exhibited lateral displacement along the corrosion defect periphery. The hydrogen enrichment at the corrosion defect became more pronounced as the tensile strain increased, while the locus of maximum hydrogen concentration migrated both externally and subjacent to the defect root.

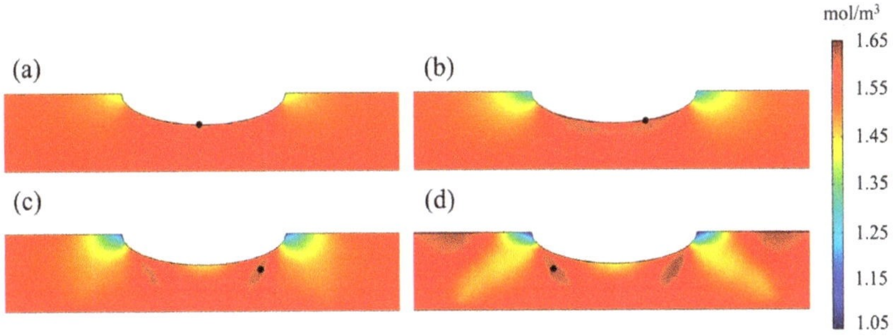

Figure 11. Hydrogen concentration distribution at the corrosion defect in the Mechano-electrochemical–hydrogen diffusion field at different tensile strains of 0.1%, 0.2%, 0.3%, and 0.4% (**a**–**d**).

Figure 12 shows the coupled hydrogen concentration results of the stress hydrogen diffusion in the same conditions, which are consistent with those in Figure 11. To facilitate comparative analysis, the maximum and minimum hydrogen concentration values are listed in Table 4.

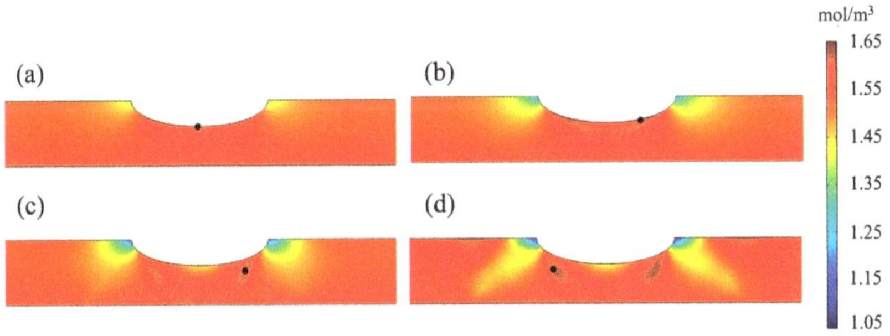

Figure 12. Hydrogen concentration distribution at the corrosion defect in the stress–hydrogen diffusion field at different tensile strains of 0.1%, 0.2%, 0.3%, and 0.4% (**a**–**d**).

Table 4. Maximum and minimum hydrogen concentrations (mol/m^3).

Physical Field	Hydrogen Concentration				
	Tensile Strain	**0.1%**	**0.2%**	**0.3%**	**0.4%**
mechano-chemical–H atom diffusion	minimum value	1.392	1.258	1.165	1.116
	maximum value	1.592	1.622	1.627	1.635
stress–H atom diffusion	minimum value	1.382	1.248	1.157	1.107
	maximum value	1.587	1.617	1.621	1.632

3.5. Linear Distribution of the Current Density and Hydrogen Concentration

Figure 13 illustrates the linear distributions of the anodic current density, cathodic current density, and hydrogen concentration along the periphery of the corrosion defect in the X80 pipeline at tensile strains ranging between 0% and 0.4% after a 9 d simulation. The anodic and cathodic current density values at the defect edge increased at a higher tensile strain and were comparable in the absence of tensile strain (Figure 13a,b). However, the values of current densities displayed divergent trends with applied strain. At 0.1% and 0.2% tensile strain, the anodic current density increased, while that of the cathodic current decreased. Despite these changes, the current density distribution remained uniform across the defects. As the tensile strain increased to 0.3% and 0.4%, the anodic and cathodic current densities were symmetrically distributed on both sides of the defect root. Although the anodic and cathodic current density increased significantly at the defect center, only a slight decrease was evident at the defect sides. The cathodic current density exhibited a substantial increase at the defect center and a marginal reduction on both sides. The peak anode current densities reached 19.1 mA/m^2 and 21.4 mA/m^2 at tensile strains of 0.3% and 0.4%, respectively. Furthermore, the most negative cathodic current densities were -17.2 mA/m^2 and -19.4 mA/m^2 at 0.3% and 0.4% tensile strains, respectively.

The linear hydrogen concentration distribution in Figure 13c aligns with the pattern observed in Figure 11. The hydrogen concentration along the periphery of the defect remained nearly uniform across the defect. The hydrogen concentration at the defect center increased at tensile strains of 0.1% and 0.2%, while the side values decreased to below that of the center. At tensile strains of 0.3% and 0.4%, the hydrogen concentration presented a pronounced non-uniform distribution. The disparity between the hydrogen concentration in the central and peripheral regions exhibited a substantial augmentation.

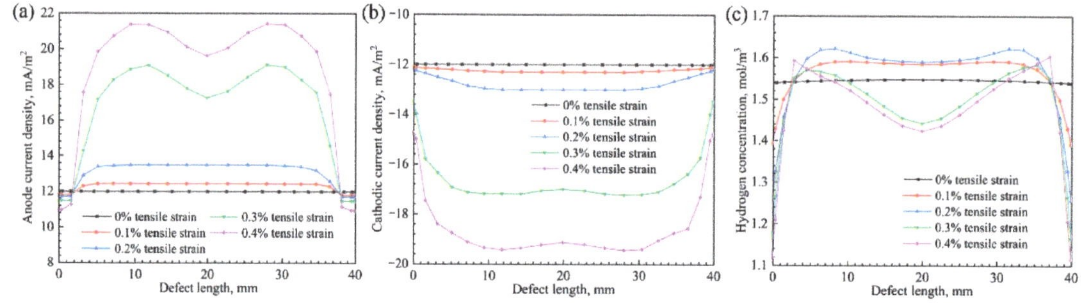

Figure 13. (**a**) Linear distribution of the anodic current density along the corrosion defect edge. (**b**) Linear distribution of the cathodic current density along the corrosion defect edge. (**c**) Linear distribution of the hydrogen concentration along the corrosion defect edge.

4. Discussion

4.1. The Effect of Tensile Strain on the Anodic Equilibrium Potential and Cathodic Exchange Current Density

Higher tensile strains increased the local Von Mises stress, hydrostatic stress, and equivalent plastic strain at the corrosion defect of the X80 pipeline (Figures 6–8). Similar to previous research findings, the tensile strain had varying effects on the stress–strain distribution [10]. Although both the von Mises and hydrostatic stress exhibited an upward trend at higher global tensile strains on the pipeline, the stress concentration at the corrosion defect was more pronounced than in the intact areas. Unlike global stress evolution, equivalent plastic strain only increased locally at the defect while remaining zero elsewhere in the pipeline. At lower tensile strains, the stress concentration was more pronounced at the defect root. As the tensile strain increased, the maximum stress and strain values shifted to the area below the defect and the inner wall of the pipeline.

The anodic and cathodic reactions of steel in near-neutral solutions were influenced by stress and strain, causing a negative shift in the anodic equilibrium potential and promoting hydrogen evolution reactions [21]. According to Equations (21) and (22), the anodic equilibrium potential and cathodic exchange current density varied in accordance with the equivalent plastic strain and Von Mises stress, respectively. This is because plastic deformation and stress can lead to the formation of defects in the material that increase the density of electrochemically active sites [28], and also increase the surface roughness of the metal, thereby altering its electronic properties [29]. Ultimately, these changes affect both the cathodic and anodic reactions.

Therefore, the distribution of the anodic equilibrium potential in Figure 9 and the cathodic exchange current density in Figure 10 correspond to the strain and stress distributions in Figures 6 and 8, respectively. Therefore, it can be concluded that low tensile strains (<0.3%), which induced elastic deformation in the pipeline, resulted in an equivalent plastic strain of zero and no alteration to the anodic equilibrium potential of the X80 steel in the NS4 solution. Contrarily, high tensile strains (≥0.3%) promoted localized plastic deformation at the corrosion defect, which is consist with previous research [30]. These phenomena could induce equivalent plastic strain amplification and a subsequent negative shift in the anodic equilibrium potential. Similarly, cathodic kinetics demonstrated strain-dependent localization. The cathodic exchange current density was concentrated at the defect midpoint in the elastic regime. However, in the plastic regime, correlated with the Von Mises stress redistributions, the maximum value of the cathodic exchange current density also migrated away from the defect root.

4.2. The Influence of the M-E Effect on the Anodic and Cathodic Current Densities

Due to the influence of stress on the potential distribution across the X80 pipeline, the high-stress region at the root of the defect became the anode, while the low-stress regions at both ends of the defect acted as the cathode. Electrons from the anodic reaction migrated through the pipeline to the cathodic regions, while Fe^{2+} ions diffused via the electrolyte. This further accelerated the anodic dissolution at the root of the defect while reducing the anodic dissolution at both ends. In the absence of tensile strain, the X80 steel underwent uniform dissolution, with an anodic current density of 12.0 mA/m^2 (Figure 13) However, at a tensile strain of 0.4%, the maximum anodic current density at the edge of the defect reached 21.4 mA/m^2, corresponding to the high-stress region, while the current density at both low-stress ends measured 11 mA/m^2. The current density at the root of the defect was 19.1 mA/m^2, indicating that this region did not represent the location of the fastest steel dissolution.

As shown in Figure 13, the M-E effect promoted hydrogen evolution, while the local corrosion reaction at the defect edge disrupted the linear distribution of the cathodic current density. In the absence of tensile strain, the defect underwent self-corrosion, with equal anodic and cathodic current densities at −12.0 mA/m^2. The application of a 0.4% tensile strain yielded maximum anodic and cathodic current densities of 21.4 mA/m^2 and −19.4 mA/m^2, respectively, creating an imbalance between the two. The algebraic difference between these values is referred to as the net current density. When considering the influence of the M-E effect on the electrochemical corrosion reaction, the polarization in the high-stress region was more pronounced, resulting in a positive net current density, indicating that this region was in an anodic polarization state, which enhanced corrosion. Contrarily, the ends of the defect were in a cathodic polarization state. For example, a 0.3% tensile strain resulted in a cathodic current density of −13.4 mA/m^2 and an anodic current density of 11.5 mA/m^2, yielding a net current density of −1.9 mA/m^2 at the ends, suggesting corrosion mitigation.

4.3. The Influence of the M-E Effect on the Hydrogen Distribution

In the coupled stress–hydrogen diffusion simulation, hydrogen diffusion was driven by both the hydrogen concentration and hydrostatic stress gradients in the pipeline. The combined effect of the hydrogen concentration and hydrostatic stress determined the distribution state. In this case, the magnitude of hydrostatic stress depended directly on the applied tensile strain. When considering the M-E effect, the size of the defect changed continuously as the corrosion reaction progressed, affecting the hydrostatic stress concentration at the defect. Therefore, even if the applied tensile strain remained constant, the hydrogen concentration distribution still underwent change.

As shown in Figures 11 and 12, the hydrogen distribution states and the locations of the maximum hydrogen concentration were the same for both the M-E effect–hydrogen diffusion and stress–hydrogen diffusion coupled simulations. The hydrogen enrichment at the corrosion defect was insignificant at a tensile strain of 0.1%, with only a slightly higher concentration at the defect root compared to the surrounding areas. Hydrogen enrichment occurred at the corrosion defect at a tensile strain of 0.2%, while the location of the maximum hydrogen concentration shifted to the right along the defect edge. Although higher tensile strains intensified hydrogen enrichment, the enriched region and peak concentration site shifted away from the defect root, moving toward the sides and downward. As shown in Figure 13, the maximum hydrogen concentration on the defect edge also shifts from the center to the sides under the influence of tensile strain.

As indicated in Table 4, the maximum and minimum hydrogen concentrations differed when considering the M-E effect compared to the stress–hydrogen diffusion coupling, reflecting the influence of the dynamic evolution of the corrosion defect on hydrogen diffusion distribution. At different tensile strains, the difference in the minimum hydrogen concentration was around 0.01 mol/m^3, while that of the maximum hydrogen concentration did not exceed 0.006 mol/m^3, which was approximately 0.39% of the initial hydrogen concentration. This suggests that the M-E effect has a relatively minor influence on the hydrogen concentration in the simulated conditions compared to the impact of the applied tensile strain. This was because the change range in the anodic equilibrium potential (−0.876 to −0.867 V) and the cathodic exchange current density (14.5 to 20.5 mA/m^2) in this simulation environment was small, indicating a relatively low corrosion rate. Therefore, the difference between the hydrogen concentration of the M-E effect and stress–hydrogen diffusion coupling was minimal. In this case, the hydrogen diffusion behavior was still dominated by the hydrostatic stress and hydrogen concentration gradients. However, the stress–hydrogen diffusion coupling did not account for the progression of the corrosion reaction over time. The simulation time in this chapter was 9 d, while in practical scenarios, predictions often cover multiple years. The continued corrosion reaction over time changed the shape and size of the corrosion defect, altering the related stress and strain, which in turn affected the anodic equilibrium potential and the hydrogen concentration and its distribution. When considering the long-term development of corrosion defects and hydrogen distribution, the model combining the M-E effect with hydrogen diffusion remains applicable. Existing research shows that extended time results in a higher hydrogen concentration enriched at the corrosion defect, increasing the risk of hydrogen-induced cracking.

5. Conclusions

This study employed an FE model to simulate hydrogen diffusion and distribution in an X80 pipeline with corrosion defects in a near-neutral solution. It analyzed the mechanism of the M-E effect behind hydrogen diffusion and enrichment behavior by coupling the stress field in the steel and the electrochemical field at the steel/solution interface to ensure that the simulation process and results were more consistent with the actual conditions of field pipelines. The main conclusions are as follows:

(1) The tensile strain influenced the cathodic and anodic reactions by altering the stress–strain distribution on the X80 pipeline. Under elastic deformation, the anodic equilibrium potential was uniform across the pipeline, while the maximum cathodic exchange current density was evident at the defect root. Under plastic deformation, the anodic equilibrium potential displayed a negative shift, while the maximum cathodic exchange current density deviated from the center.

(2) The M-E effect induced anodic polarization in high-stress regions, which enhanced corrosion, while cathodic polarization occurs in low-stress regions, resulting in corrosion mitigation.
(3) In a near-neutral solution environment, the hydrogen distribution state on the X80 pipeline was consistent with that of the stress–hydrogen diffusion coupling when considering the M-E effect. However, the presence of the M-E effect slightly increased the hydrogen concentration at the corrosion defect.

Author Contributions: Conceptualization, Z.D. and J.T.; methodology, Z.D.; software, J.T.; validation, S.Z., F.W. and Q.B.; formal analysis, S.Z.; data curation, F.W.; writing—original draft preparation, S.G. and J.T.; writing—review and editing, L.C. and Z.D.; visualization, X.W.; supervision, X.W.; project administration, P.K.; funding acquisition, P.K. All authors have read and agreed to the published version of the manuscript.

Funding: Scientific Research Projects of PetroChina Southwest Oil & Gasfield Company, 20230312-08.

Data Availability Statement: The data presented in this study are available on request from the corresponding authors.

Conflicts of Interest: Zhixiang Dai, Sijia Zheng, Feng Wang, Qin Bie, Pengcheng Kang and Xinyi Wang were employed by the Gathering and Transportation Engineering Technology Research Institute, PetroChina Southwest Oil & Gasfield Company, The remaining authors declare that the research was conducted in the absence of any commercial or financial relationships that could be construed as a potential conflict of interest.

References

1. Faye, O.; Szpunar, J.; Eduok, U. A critical review on the current technologies for the generation, storage, and transportation of hydrogen. *Int. J. Hydrogen Energy* **2022**, *47*, 13771–13802. [CrossRef]
2. Abdalla, A.M.; Hossain, S.; Nisfindy, O.B.; Azad, A.T.; Dawood, M.; Azad, A.K. Hydrogen production, storage, transportation and key challenges with applications: A review. *Energy Convers. Manag.* **2018**, *165*, 602–627. [CrossRef]
3. Zhu, Y.-Q.; Song, W.; Wang, H.-B.; Qi, J.-T.; Zeng, R.-C.; Ren, H.; Jiang, W.-C.; Meng, H.-B.; Li, Y.-X. Advances in reducing hydrogen effect of pipeline steels on hydrogen-blended natural gas transportation: A systematic review of mitigation strategies. *Renew. Sustain. Energy Rev.* **2023**, *189*, 113950. [CrossRef]
4. Yang, H.; Zhang, H.; Liu, C.; Wang, C.; Fan, X.; Cheng, Y.F.; Li, Y. Effects of defect on the hydrogen embrittlement behavior of X80 pipeline steel in hydrogen-blended natural gas environments. *Int. J. Hydrogen Energy* **2024**, *58*, 158–173. [CrossRef]
5. Chen, X.; Li, X.; Du, C.; Cheng, Y. Effect of cathodic protection on corrosion of pipeline steel under disbonded coating. *Corros. Sci.* **2009**, *51*, 2242–2245. [CrossRef]
6. Perdomo, J.; Song, I. Chemical and electrochemical conditions on steel under disbonded coatings: The effect of applied potential, solution resistivity, crevice thickness and holiday size. *Corros. Sci.* **2000**, *42*, 1389–1415. [CrossRef]
7. Wu, W.; Liu, Z.; Li, X.; Du, C. Electrochemical characteristic and stress corrosion behavior of API X70 high-strength pipeline steel under a simulated disbonded coating in an artificial seawater environment. *J. Electroanal. Chem.* **2019**, *845*, 92–105. [CrossRef]
8. Vanboven, G.; Chen, W.; Rogge, R. The role of residual stress in neutral pH stress corrosion cracking of pipeline steels. Part I: Pitting and cracking occurrence. *Acta Mater.* **2007**, *55*, 29–42. [CrossRef]
9. Zhang, Z.; Guo, L.; Cheng, Y.F. Interaction between internal and external defects on pipelines and its effect on failure pressure. *Thin-Walled Struct.* **2021**, *159*, 107230. [CrossRef]
10. Guo, S.; Xu, L.; Dong, S.; Cheng, Y.F. Finite element modeling of hydrogen atom diffusion and distribution at corrosion defect on aged pipelines transporting hydrogen. *Int. J. Hydrogen Energy* **2023**, *48*, 13566–13577. [CrossRef]
11. Li, W.; Guo, T.; Xu, L.; Chen, L.; Jiang, B.; Wang, X.; Wu, H.; Qiao, L. Promotion of pitting corrosion at hydrogen-enriched α/γ phase boundaries in austenitic stainless steel weld joints. *Acta Mater.* **2022**, *227*, 117728. [CrossRef]
12. Devanathan, M.A.V.; Stachurski, Z. The mechanism of hydrogen evolution on iron in acid solutions by determination of permeation rates. *J. Electrochem. Soc.* **1964**, *111*, 619–623. [CrossRef]
13. Doyle, D.; Palumbo, G.; Aust, K.; El-Sherik, A.; Erb, U. The influence of intercrystalline defects on hydrogen activity and transport in nickel. *Acta Met. Mater.* **1995**, *43*, 3027–3033. [CrossRef]
14. Cheng, Y.F. Fundamentals of hydrogen evolution reaction and its implications on near-neutral pH stress corrosion cracking of pipelines. *Electrochim. Acta* **2007**, *52*, 2661–2667. [CrossRef]

15. Xu, L.; Cheng, Y. Development of a finite element model for simulation and prediction of mechanoelectrochemical effect of pipeline corrosion. *Corros. Sci.* **2013**, *73*, 150–160. [CrossRef]
16. Hafsi, Z.; Mishra, M.; Elaoud, S. Hydrogen embrittlement of steel pipelines during transients. *Procedia Struct. Integr.* **2018**, *13*, 210–217. [CrossRef]
17. Krom, A.; Koers, R.; Bakker, A. Hydrogen transport near a blunting crack tip. *J. Mech. Phys. Solids* **1999**, *47*, 971–992. [CrossRef]
18. Wijmans, J.; Baker, R. The solution-diffusion model: A review. *J. Membr. Sci.* **1995**, *107*, 1–21. [CrossRef]
19. Sofronis, P.; McMeeking, R. Numerical analysis of hydrogen transport near a blunting crack tip. *J. Mech. Phys. Solids* **1989**, *37*, 317–350. [CrossRef]
20. McLellan, A.G.; Fyfe, W.S. Non-hydrostatic thermodynamics of chemical systems. *Proc. R. Soc. Lond. A Math. Phys. Sci.* **1970**, *314*, 443–455.
21. Gutman, E.; Solovioff, G.; Eliezer, D. The mechanochemical behavior of type 316L stainless steel. *Corros. Sci.* **1996**, *38*, 1141–1145. [CrossRef]
22. Wu, X.; Teng, M.; Jia, W.; Cai, J. Study on the mechanical properties of X80 pipeline steel under pre-charged high-pressure gaseous hydrogen. *Int. J. Hydrogen Energy* **2024**, *84*, 39–52. [CrossRef]
23. Xu, L.; Cheng, Y. Corrosion of X100 pipeline steel under plastic strain in a neutral pH bicarbonate solution. *Corros. Sci.* **2012**, *64*, 145–152. [CrossRef]
24. Gutman, E.M. *Mechanochemistry of Solid Surfaces*; World Scientific Publishing Company: Singapore, 1994.
25. Celikin, M. The role of dynamic transmission electron microscopy on the development of next generation magnesium alloys. *Mater. Today* **2015**, *18*, 472. [CrossRef]
26. Saxena, A.; Findley, K. Modelling the Effects of Hydrogen Pressure on Fatigue Crack Growth Behavior in SA372 Pressure Vessel Steels. In Proceedings of the ASME 2022 Pressure Vessels & Piping Conference, Las Vegas, Nevada, 17–22 July 2022. [CrossRef]
27. Toribio, J.; González, B.; Matos, J.-C. Numerical modeling of hydrogen embrittlement of pearlitic steel in the presence of blunt notches. *Procedia Struct. Integr.* **2021**, *33*, 1215–1218. [CrossRef]
28. Sahal, M.; Creus, J.; Sabot, R.; Feaugas, X. The effects of dislocation patterns on the dissolution process of polycrystalline nickel. *Acta Mater.* **2006**, *54*, 2157–2167. [CrossRef]
29. Dai, Y.Z.; Chiang, F.P. On the Mechanism of Plastic Deformation Induced Surface Roughness. *J. Eng. Mater. Technol.* **1992**, *114*, 432–438. [CrossRef]
30. Huang, H.; Jia, C.; Guo, L. Effect of local corrosion on tensile behavior of steel plates. *Structures* **2022**, *43*, 977–989. [CrossRef]

hydrogen

MDPI

Article

An Efficient Finite Element Model to Predict the Mechanical Response of Metallic-Reinforced Pressure Vessels

Ana Lucía León Razo, Miguel Ernesto Gutierrez Rivera *, Carlos Enrique Valencia Murillo, Elias Rigoberto Ledesma Orozco and Israel Martinez Ramirez

Department of Mechanical Engineering, University of Guanajuato, Salamanca 36885, Guanajuato, Mexico; al.leonrazo@ugto.mx (A.L.L.R.); ce.valenciamurillo@ugto.mx (C.E.V.M.); elias@ugto.mx (E.R.L.O.); israel.martinez@ugto.mx (I.M.R.)
* Correspondence: miguel.gutierrez@ugto.mx

Abstract

In the design of pressure vessels for hydrogen storage, the durability and robustness of the designs are tested by using experimental methods, numerical simulations, or both. However, in the initial design phase, it is widely known that using numerical simulation tools reduces the cost of performing experiments; therefore, models that provide accurate and reliable results must be developed. This work presents an axisymmetric finite element model to predict the mechanical response of reinforced wire pressure vessels of type II. The main contribution of the present model is the use of equivalent properties and a minor number of contact elements to simulate the behavior of the wire reinforcement, which reduces the computational effort compared to a model with a solid-based mesh. The accuracy of the proposed model is tested against solid elements with very good agreement and experimental results with reasonable agreement. A parametric study was conducted to test the influence of the number of layers of reinforcement, and it was concluded that there is a limit to increasing the number of layers, which does not increase the vessel's strength considerably, but it does with its mass.

Keywords: pressure vessel; finite element method; experimental results; hydrogen storage

1. Introduction

Emissions from fossil fuel combustion have produced harmful substances that threaten human health and the environment, demanding the search for cleaner and more sustainable energy sources for the industrial and transportation sectors [1]. One of the most promising alternatives is hydrogen due to its zero-carbon emission and high energy density [2,3]. However, there are still challenges to overcome, such as its transportation and storage [4]. In this regard, pressure vessels (PVs) play a crucial role, allowing for safe and efficient hydrogen storage at high pressure. For mobile applications, the most mature and accessible way to store hydrogen has been through its compression in PVs.

To date, compressed hydrogen can be stored in four types of PVs: type I (fully metallic), type II (metallic liner reinforced partially with fibers), type III (metallic liner with a fully composite reinforcement), and type IV (polymer-based vessel) [5,6]. Emerging applications such as hydrogen-powered vehicles require PVs with minimal weight, capable of containing a sufficient volume of hydrogen to maximize travel range. Then, the development of numerical models may prove advantageous for designers in reducing the manufacturing cost of prototypes in the initial design phase.

Considering the types previously discussed, several investigations have sought to determine optimal design parameters that improve storage capacity and optimize vessels' weight. The pressure vessels designs have been tested using experimental techniques and/or numerical tools. Batul et al. [7] applied the similitude theory to establish a method to reduce the cost of testing thin-walled PVs used in aerospace applications. Zhang et al. [8] proposed specific methods for designing the dome thickness and the load pressure of a liner in type IV PVs. They evaluated the design's safety and performance through failure criteria and finite element analysis. Kumar et al. [9] designed and analyzed an airborne cryogenic liquid hydrogen tank; the numerical analyses were performed by means of ANSYS. The work of Kwak et al. [10] presented a design method for improving failure resistance and inner capacity of the seamless compressed natural gas pressure vessel (PV) type II through finite element analysis. They modeled a small axisymmetric section to reduce computation time, with the model ensuring structural reliability for an autofrettage pressure. Park et al. [11] sought to improve structural design to maximize fuel efficiency and reduce the vessel's weight. Nguyen et al. [12] developed a modeling tool to analyze damage and design composite PVs for hydrogen storage. Celaya et al. [13] presented a method to evaluate equivalent properties of different metallic reinforcements of PVs by means of finite element analysis, which optimizes the use of computational resources. Błachut et al. [14] investigated the influences of the fiber tension of the filament winding on the mechanical properties of composite PVs; they manufactured and tested three series of samples and performed a numerical investigation to support the experimental results. For the numerical model, they used a representative volume element (RVE) cell to homogenize the carbon fiber filament and epoxy resin. They reported that the induced compressive stresses on the steel liner due to an increment in fiber tension leads to an increase in burst pressure, while the thickness of the composite layer is reduced. Wu et al. [15] designed a structure model of a fiber-wound composite gas cylinder following the classical grid theory and related regulations; they also generated a finite element model to perform strength and failure analysis of the designed structure. For the simulation, they used symmetrical geometry and modeled one-quarter of the geometry with solid elements. Azzem et al. [16] studied the influences of the winding angles on hoop stress in composite type IV PVs by means of finite element analysis. Katsumata et al. [17] designed and experimentally tested a type III tank with a dome–cylinder-split molded carbon-fiber-reinforced plastic composite (CFRP) structure. They illustrated the damage and burst mechanisms by using the finite element method (FEM). In the work presented by Reda et al. [18], different hydrogen PV types were analyzed by means of FEM. They evaluated the effects of high internal pressures, extreme temperatures, and absolute vacuum on the PV performance. Koutsawa and Bouhala [19] presented an uncertainty analysis in the design of composite type IV PVs for hydrogen storage. For the analysis, they developed a finite element model of the tank in ABAQUS; only 1/16 of the geometry was modeled, and periodic symmetry was used. Syed et al. [20] presented a design of a hydrogen pressure vessel type III; their study focused on identifying a safer dome shape that effectively reduces stress in the composite layers while retaining the same winding pattern of a prior model. They found that the stress distribution is reduced with the elliptical dome geometry.

In addition, it is worth mentioning that during recent years, there has been an increase in the number of reviews exploring different aspects of energy storage systems in the literature: for instance, the hydrogen storage types, the forthcoming challenges in this technology, manufacturing processes of storage vessels, design optimization and challenges in the theoretical modeling of storage mechanisms [2,3,6,21–25].

Despite the improvements in computational performance in recent decades and due to the increasing demand for numerical models capable of capturing the complex behavior of multi-physics models of structures, further development of efficient and accurate models is still required to overcome these issues. Moreover, regarding the use of numerical techniques for optimization in hydrogen storage research, there is still a need to address the balance between scability, accuracy, and computational cost [3]. Considering this and motivated by the works on hydrogen storage reported in the literature, this work presents the development and validation of a reinforced wire PV type II finite element model to evaluate the reinforcement influence on the mechanical response under static loading conditions. The reinforcement of the cylindrical section is made of a steel wire, which improves the tank's load-bearing capacity and reduces its weight compared to type I PVs, thus representing an improvement in structural efficiency and a reduction in manufacturing costs [26,27]. Finally, the reinforcement wires are modeled using equivalent properties in the present approach, and only contact conditions between layers are defined, thereby reducing the number of contacts commonly required in a model where wires are individually modeled.

2. Materials and Methods

This section describes the finite element model developed to predict the mechanical response of metallic-reinforced pressure vessels. In addition, a brief description of the prototype fabrication process and the experimental setup used to validate the model is presented. Finally, the numerical and experimental results are compared, and the conclusions are drawn.

2.1. Pressure Vessel Geometry

For the geometry of the finite element model and to simplify the experimental validation, in general, it is considered that the pressure vessel (the prototype) is composed of a cylindrical section and two caps. The cylinder section is a 2-by-6-inch galvanized steel pipe nipple, one cap is a 2-1/2-inch threaded bell reducer and the other is a threaded cap. Figure 1 presents a schematic representation of the prototype (without reinforcement), where the main components of the system are clearly identified.

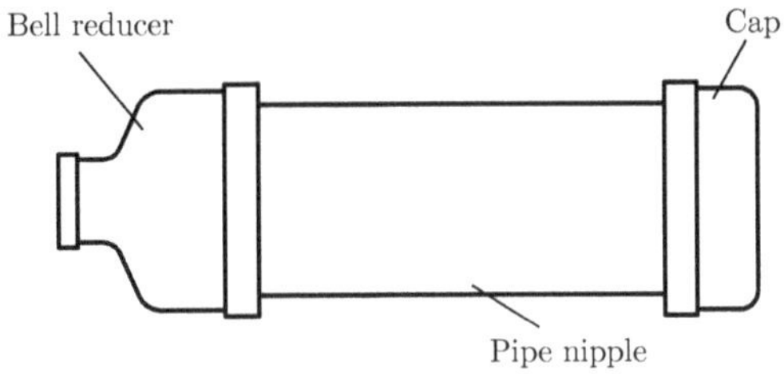

Figure 1. Schematic representation of prototypes.

2.2. Finite Element Models

In the finite element analysis of pressure vessels, according to the simplifications made, it is common to model the vessel geometry using shell, axisymmetric plane, or solid elements, which yield similar outcomes. However, computation time may vary considerably. For this reason, two finite element models are proposed in this work: one using axisymmetric plane elements and the other using solid elements. Both models were developed using the commercial software ANSYS 2024. The nipple pipe, bell reducer, and cap were modeled as isotropic materials using the mechanical properties of structural steel (A36), which are listed in Table 1 [28].

Table 1. Mechanical properties of ASTM A36 steel [28].

Mechanical Property	Value	Units
Modulus of elasticity	200	GPa
Poisson's ratio	0.26	-

An axisymmetric finite element model was developed using the element PLANE183, which is an 8-node bidimensional element. The material of the steel wire reinforcement was modeled using the orthotropic equivalent properties reported by Celaya et al. [13], shown in Table 2. Each reinforcement layer was modeled as a separated region in contact with the adjacent surfaces. No detachment and friction between surfaces are considered to simulate their interaction. For illustration purposes, Figure 2 shows the two-dimensional mesh used to model the tank reinforced with three layers of wire. The mesh shown is the result of the mesh sensitivity analysis made, with a 0.1% maximum variation convergence criterion for the von Mises stress, and the results for this study are presented in Table 3. Regarding boundary conditions, a uniform internal pressure was applied to the inner surface of the vessel, and axial displacement was constrained by applying line supports at the vessel's top (represented as green triangles in Figure 2).

Table 2. Equivalent mechanical properties for the steel wire reinforcement [13].

Mechanical Property	Value	Units
Elastic modulus E_x	160.0	GPa
Elastic modulus E_y	21.0	GPa
Elastic modulus E_z	21.0	GPa
Shear modulus G_{xy}	26.2	GPa
Shear modulus G_{xz}	26.2	GPa
Shear modulus G_{yz}	25.4	GPa
Poisson's ratio ν_{xy}	0.16	-
Poisson's ratio ν_{xz}	0.16	-
Poisson's ratio ν_{yz}	0.07	-

Table 3. Mesh independence study for the axisymmetric model.

Iteration	von Mises Stress (MPa)	Change	Nodes	Elements
1	41.973		725	146
2	42.338	0.94%	2241	530
3	42.568	−0.11%	7199	1978
4	42.566	0.16%	11,394	3271
5	42.654	−0.33%	25,388	7633
6	42.706	−0.14%	29,338	8861
7	42.709	0.04%	37,592	11,497

The solid model was developed using a 20-node solid element (SOLID186). In this case, symmetry was exploited, and only one degree of the whole geometry was modeled. The reinforcement was represented by cylinders that were in contact with each other and the tank. Contact elements were defined under the assumption of permanent bonding between surfaces, which prevents separation or relative sliding. Figure 3 shows the mesh used to model the tank reinforced with three layers of wire. Again, the mesh shown is the result of the mesh sensitivity analysis made, with a 0.1% maximum variation convergence criterion for the von Mises stress, and the results for this study are presented in Table 4. Even though the convergence is achieved earlier, compared with the plane model, the number of elements is more than double. Since solid elements are used, pressure boundary conditions were applied in the internal areas of the tank, while displacement at the top face of the tank was restricted in the Z-axis direction (see the green triangles in Figure 3). Finally, the mechanical properties of the reinforcement were taken from [13] and are presented in Table 5.

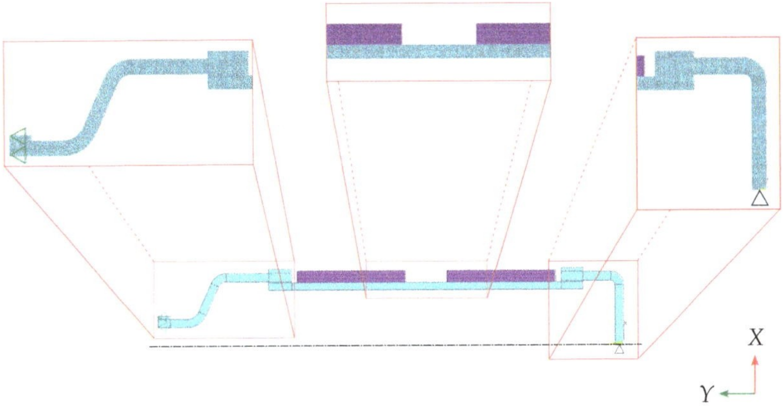

Figure 2. Axisymmetric mesh for a prototype with three layers of reinforcement.

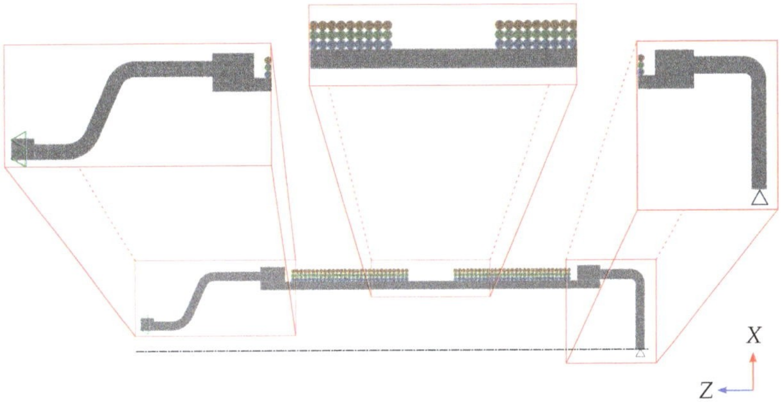

Figure 3. Solid mesh for a prototype with three layers of reinforcement.

Table 4. Mesh independence study for the solid model.

Iteration	von Mises Stress (MPa)	Change	Nodes	Elements
1	42.875		43,090	6220
2	42.611	−0.92%	48,938	7212
3	42.670	−0.35%	68,349	10,888
4	42.629	−0.15%	82,669	13,696
5	42.644	−0.05%	140,868	25,340

Table 5. Mechanical properties of the steel wire [13].

Mechanical Property	Value	Units
Modulus of elasticity	200	GPa
Poisson's ratio	0.3	-

2.3. Prototype Fabrication

In order to validate the finite element model, small prototypes of metallic-reinforced pressure vessels were manufactured. The reinforcement is achieved by winding successive layers of steel wire around the cylindrical section; a layer of steel wire is partially shown in Figure 4a.

The fabrication began by welding the wire to the pipe nipple mounted on a conventional lathe. Then, the assembly was rotated at a low speed while the wire was stretched with constant tension, ensuring a uniform and controlled winding of the reinforcement layers around the pipe nipple (see Figure 4b). A central 2 cm wide region was deliberately left without reinforcement to allow for the placing of strain gauges, which will be used to measure the strains in the central region, to see the effect of the reinforcement in the cylindrical region and to validate the model. The fabrication process is completed by closing the ends of the reinforced pipe nipple, one with the bell reducer and the other with the threaded cap.

(a) (b)

Figure 4. Prototype fabrication: (**a**) wire layer and (**b**) assembly.

After finishing the manufacture of prototypes, they were instrumented by positioning strain gauges to measure both circumferential and axial strains. Figure 5 presents an instrumented prototype of a metallic-reinforced pressure vessel.

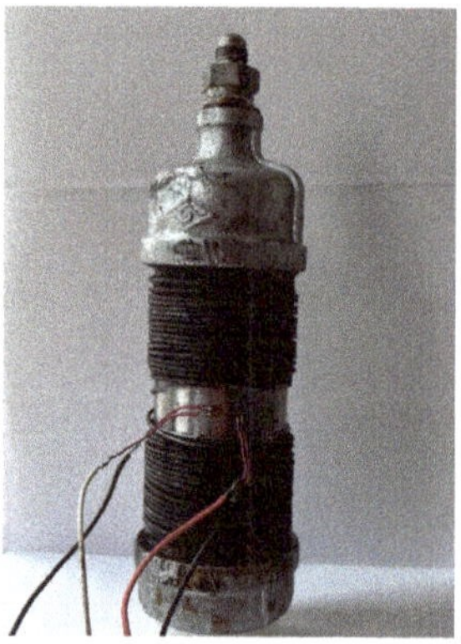

Figure 5. Prototype of a metallic-reinforced pressure vessel.

2.4. Experimental Setup

To validate the structural behavior of the present finite element model, the manufactured prototypes were subjected to pressure tests. For this purpose, a pressure washer with a maximum capacity of 5000 psi was used to generate the load. The pressure was regulated by means of a hydraulic circuit, which includes a control valve to maintain a specific pressure during testing, a pressure gauge, and a flow valve to regulate the water supplied to the system.

For safety reasons, the prototype was placed inside an empty cylindrical container during testing. The generated data during testing were collected by using a StudentDAQ MM01 data acquisition device from Micro-Measurement, optimized for 350 Ω strain gauges, and stored on a computer for subsequent analysis. Figure 6 shows an overview of the experimental setup.

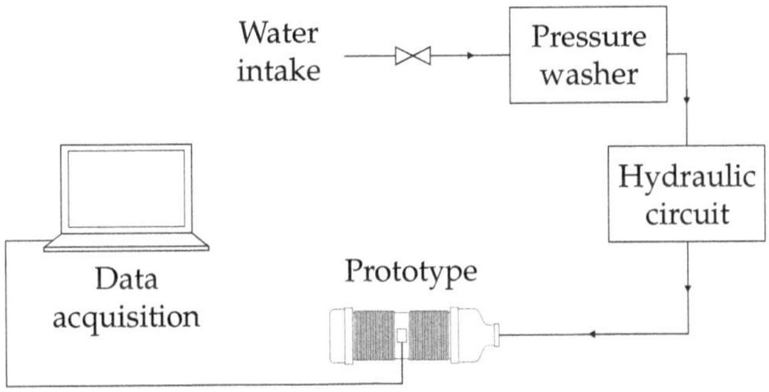

Figure 6. Overview of the experimental setup.

2.5. Validation

In order to validate the axisymmetric finite element model presented, a comparison was made between the numerical results obtained using solid elements and the experimental results. Also, to appreciate the influence of the reinforcement, a prototype without it was first tested and modeled. The results for the hoop and axial strains for this case are presented in Figure 7. An excellent agreement is noted for the hoop strain at the beginning of the test. However, after 500 psi, the numerical deformations start to differ from the experimental results, with a maximum relative error of 7%. On the other hand, for the axial strain, a slightly bigger difference is presented between the numerical results (which are almost identical) and the experimental results during all the comparisons, having a difference of 13% for the final load. Figure 8 presents the contours for the (a) hoop and (b) axial strains for the prototype without reinforcement at 1000 psi. It is worth noting that the strains are uniform in the cylindrical part of the prototype.

Figure 9 compares the experimental results and the finite element models for a prototype with one reinforcement layer for the hoop and axial strains. It is observed that the experimental hoop and axial strains are higher than those predicted for the finite element models, as the last two are almost identical. The difference between the experimental data and the finite element models, at the maximum applied pressure, is in the order of 12% and 13% for the hoop and axial strains, respectively.

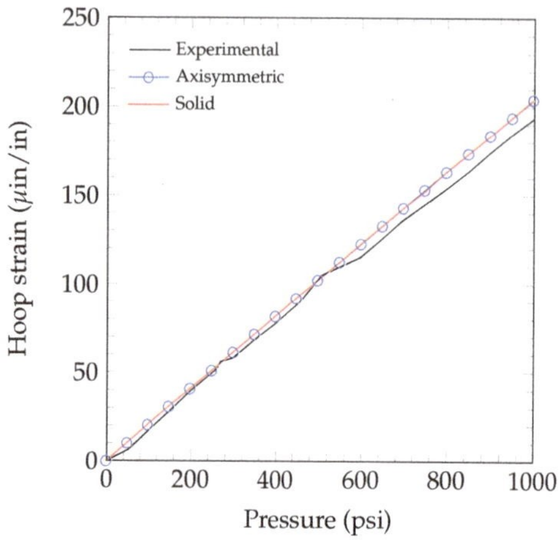

(**a**)

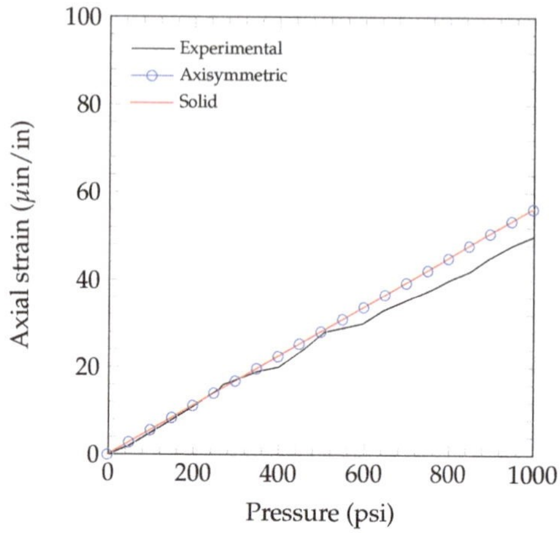

(**b**)

Figure 7. Internal pressure vs. (**a**) hoop and (**b**) axial strains for the prototype without reinforcement.

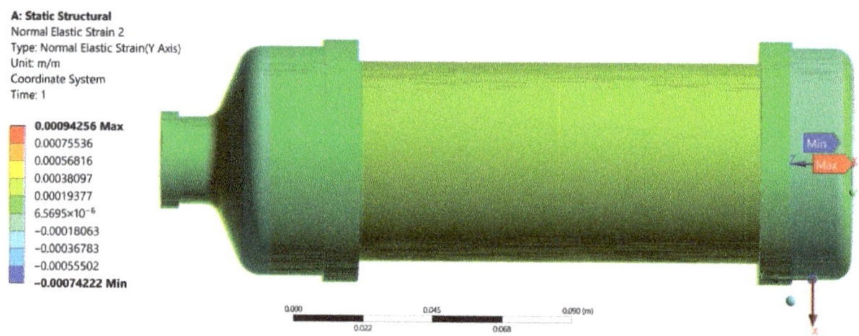

(**a**)

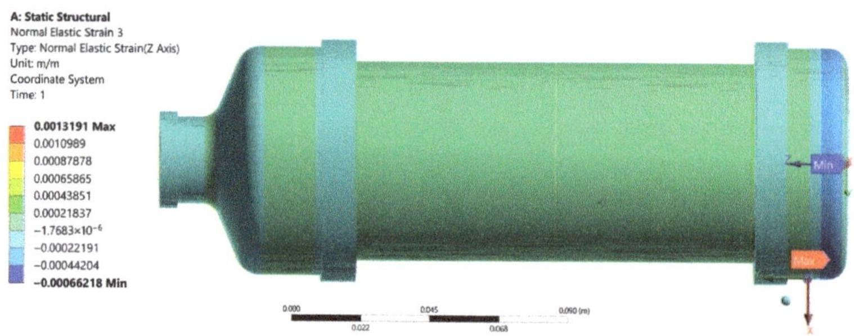

(**b**)

Figure 8. Components of the strain for the prototype without reinforcement at 1000 psi: (**a**) Hoop. (**b**) Axial.

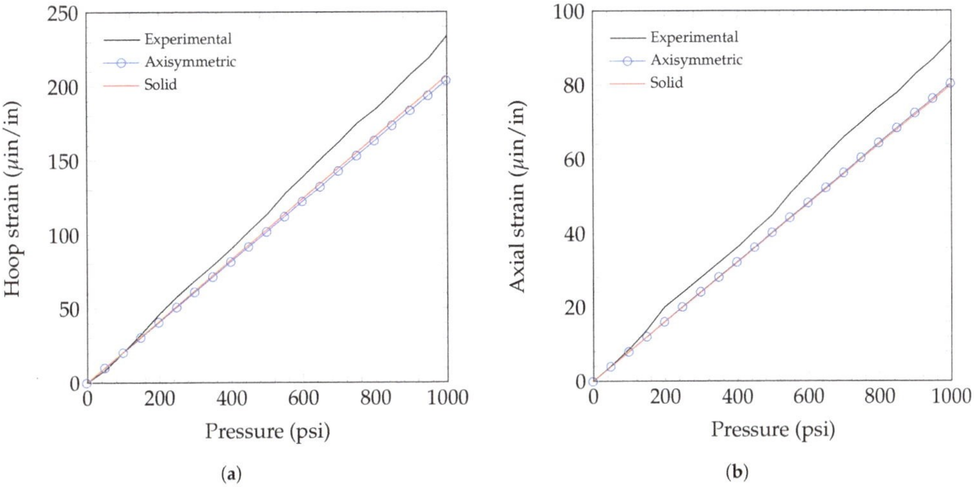

Figure 9. Internal pressure vs. (**a**) hoop and (**b**) axial strains for the prototype with one layer of reinforcement.

Finally, a comparison between the experimental results and the finite element models for the hoop and axial strains for a prototype with three layers of reinforcement is presented in Figure 10. For this case, the experimental hoop strains are higher than those predicted for the finite element models, while the axial strains are lower. The difference between the experimental data and the axisymmetric finite element model is 18% and 6% for the hoop and axial strains, respectively. Figure 11 presents the contours for the (a) hoop and (b) axial strains for the cylindrical part of the prototype with three layers of reinforcement at 1000 psi. Here, the effect of the reinforcement can be seen in the middle of the prototype, which has higher deformations since this part was left without reinforcement to place the strain gauges.

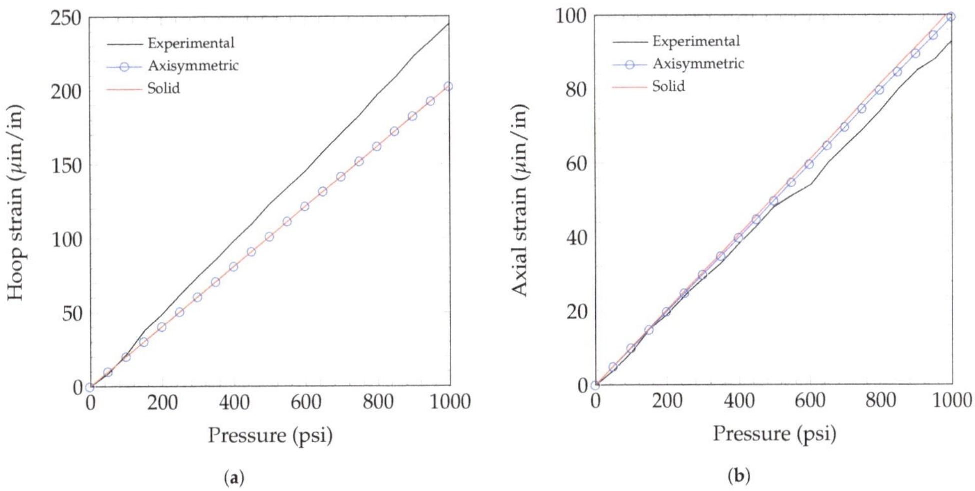

Figure 10. Internal pressure vs. (**a**) hoop and (**b**) axial strains for the prototype with three layers of reinforcement.

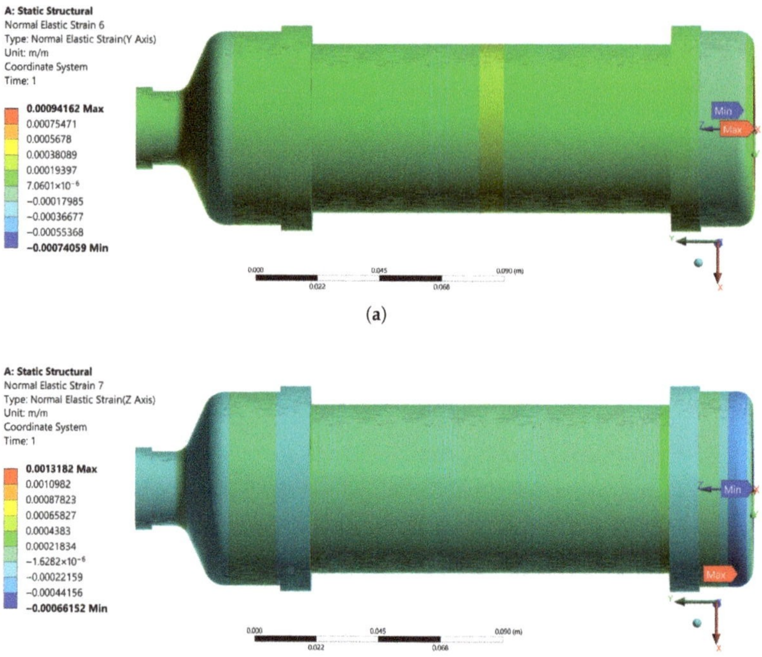

(a)

(b)

Figure 11. Components of the strain in the cylindrical part of the prototype with three layers of reinforcement at 1000 psi: (**a**) Hoop. (**b**) Axial.

3. Numerical Analysis

Since results from the solid and axisymmetric models are very close, but with considerably less computational effort used by the latter model, a study of the influence of the number of reinforcement layers on strain is performed using the axisymmetric model. Figure 12 shows the hoop and axial strains for different numbers of layers of reinforcement at 1000 psi. On the one hand, a significant reduction is observed for the hoop strain until five turns, decreasing from 204 μin/in to 93 μin/in, with differences higher than 5% between each layer increment. After that, the reduction is lower, with a difference between 13 and 14 layers of 0.33%. On the other hand, axial strains start increasing until two layers, reaching a maximum of 60 μin/in, and then decreasing slowly, having differences lower than 5%, and stabilizing at 0.70% between 13 and 14 layers.

For completeness, hoop and axial stresses are presented in Figure 13. It is observed that both stresses decrease when the number of layers is increased. For the first layer, the reductions in the hoop and axial stresses are in the order of 12.81 MPa and 3.04 MPa, respectively, compared to the model without reinforcement, while the mass increases by 290 g. For the case with 14 layers, the reductions of the stresses are 0.08 MPa in both directions, with respect to the model with 13 layers. However, the mass increases by 506 g for the same case.

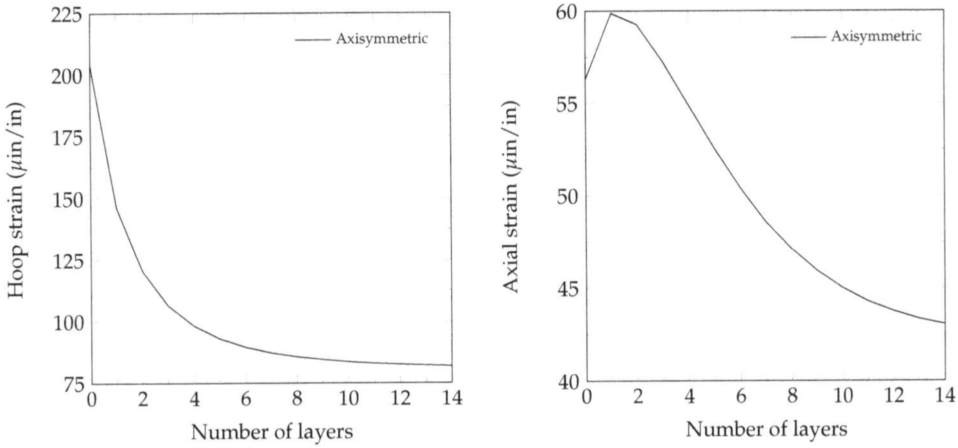

(**a**) (**b**)

Figure 12. Number of layers of reinforcement vs. (**a**) hoop and (**b**) axial strains for the prototype.

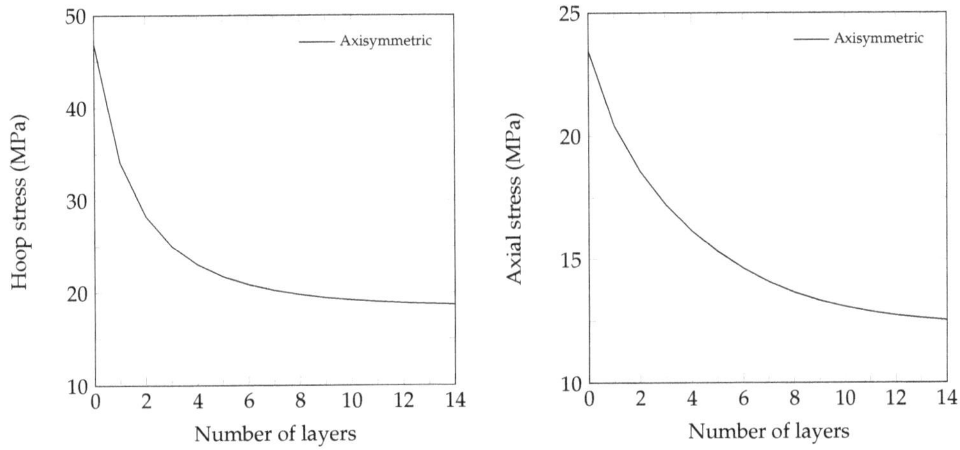

(**a**) (**b**)

Figure 13. Number of layers of reinforcement vs. (**a**) hoop and (**b**) axial stresses for the prototype.

Finally, Table 6 presents how the mass, the hoop, and axial stresses vary, with respect to the base model, when different numbers of layers of reinforcement are added. The mass added for each layer is higher than the previous one, while the stress reduction is lower. For example, from 1 to 2 layers, the mass increases by 19.44% and the hoop and axial stresses decrease by 12.51% and 7.86%. In comparison, from 13 to 14 layers, the mass increases by 33.98%, and the hoop and axial stresses only reduce by 0.17% and 0.36%, respectively. Based on these observations and the ASME Boiler and Pressure Vessel Code [29], it is recommended that the reinforcement has at least 10 layers. After that, the relative stress reduction should be monitored, and if it is lower than 1%, it is suggested that the number of layers should not be increased and the design should be tested. If it does not meet the requirements, another wire should be used as reinforcement.

Table 6. Percentage of mass increase and stress reduction for layers of reinforcement with respect to the base model.

Layers	% of Mass Added	% of Hoop Stress Reduction	% of Axial Stress Reduction
1	18.2	27.3	13.0
2	37.7	39.8	20.8
3	58.3	46.7	26.6
4	80.2	50.9	31.1
5	103.3	53.7	34.7
6	127.6	55.6	37.6
7	153.1	56.9	40.0
8	179.8	57.9	41.8
9	207.7	58.6	43.2
10	236.8	59.1	44.3
11	267.2	59.5	45.1
12	298.7	59.8	45.8
13	331.5	60.0	46.2
14	365.5	60.2	46.6

4. Conclusions

This work presents an axisymmetric finite element model that can predict the mechanical response of a wire-reinforced pressure vessel (type II). This model uses equivalent properties and contact elements between layers to model the wire reinforcement. The proposed model showed very good agreement with the results obtained through solid elements and reasonable agreement with the experimental results. The latest difference in results is mainly attributed to errors inherent in the experimental work, such as variations in mechanical properties, inaccuracies in measurement devices, fluctuations in input pressure, dimensional deviations of the final parts due to the manufacturing process, and misalignment of the strain gauge.

Also, the numerical study showed that there is a limit to increasing the number of layers, which does not increase the vessel's strength considerably but does increase its mass. Furthermore, the proposed model could be used in the optimization process in the early stages of the wire-reinforced pressure vessel type II design for hydrogen storage.

It is important to note that the present model only works when the geometry, loads, and boundary conditions are axisymmetric. This model should not be used if a load or boundary condition makes the problem asymmetric. Furthermore, it is worth mentioning that the manufacturing process for the reinforcement should be monitored and comply with the recommendations from the ASME Boiler and Pressure Vessel Code [29]. Specifically, the helix angle of the winding should be less than 1 degree, the maximum gap between wires in the longitudinal direction of the vessel should be less than 5% of the wire width, and neither the inner cylinder nor the wire should yield.

Author Contributions: Conceptualization, M.E.G.R. and E.R.L.O.; methodology, M.E.G.R. and A.L.L.R.; software, M.E.G.R. and A.L.L.R.; validation, M.E.G.R. and A.L.L.R.; formal analysis, M.E.G.R., A.L.L.R. and E.R.L.O.; investigation, M.E.G.R. and A.L.L.R.; writing—original draft preparation, M.E.G.R., A.L.L.R. and C.E.V.M.; writing—review and editing, M.E.G.R., E.R.L.O. and I.M.R.; visualization, A.L.L.R. and C.E.V.M.; supervision, M.E.G.R., E.R.L.O. and I.M.R.; project administration, M.E.G.R.; funding acquisition, M.E.G.R. All authors have read and agreed to the published version of the manuscript.

Funding: This research was funded by the University of Guanajuato, grant number CIIC 241/2024.

Data Availability Statement: The data presented in this study can be made available upon request from the corresponding author. The data are not publicly available.

Acknowledgments: The authors acknowledge the support of the University of Guanajuato for providing experimental, computational and software facilities. SECIHTI's assistance with graduate student and senior researcher fellowships is also appreciated.

Conflicts of Interest: The authors declare no conflicts of interest.

References

1. Cheng, Q.; Zhang, R.; Shi, Z.; Lin, J. Review of common hydrogen storage tanks and current manufacturing methods for aluminium alloy tank liners. *Int. J. Lightweight Mater. Manuf.* **2024**, *7*, 269–284. [CrossRef]
2. Ahad, M.T.; Bhuiyan, M.M.H.; Sakib, A.N.; Becerril Corral, A.; Siddique, Z. An overview of challenges for the future of hydrogen. *Materials* **2023**, *16*, 6680. [CrossRef]
3. Kadirgama, K.; Samylingam, L.; Aslfattahi, N.; Kiai, M.S.; Kok, C.K.; Yusaf, T. Advancements and challenges in numerical analysis of hydrogen energy storage methods: Techniques, applications, and future direction. *Int. J. Hydrogen Energy* **2025**, *125*, 67–85. [CrossRef]
4. Ma, N.; Zhao, W.; Wang, W.; Li, X.; Zhou, H. Large scale of green hydrogen storage: Opportunities and challenges. *Int. J. Hydrogen Energy* **2024**, *50*, 379–396. [CrossRef]
5. Abdalla, A.M.; Hossain, S.; Nisfindy, O.B.; Azad, A.T.; Dawood, M.; Azad, A.K. Hydrogen production, storage, transportation and key challenges with applications: A review. *Energy Convers. Manag.* **2018**, *165*, 602–627. [CrossRef]
6. Nachtane, M.; Tarfaoui, M.; Abichou, M.A.; Vetcher, A.; Rouway, M.; Aâmir, A.; Mouadili, H.; Laaouidi, H.; Naanani, H. An overview of the recent advances in composite materials and artificial intelligence for hydrogen storage vessels design. *J. Compos. Sci.* **2023**, *7*, 119. [CrossRef]
7. Batul, B.; Sohail, A.; Aizaz, A.; Jamil, Z. Application of structural similitude for scaling of a pressure vessel. In Proceedings of the IOP Conference Series: Materials Science and Engineering, Kuala Lumpur, Malaysia, 29–30 August 2019; IOP Publishing: Bristol, UK, 2019; Volume 642, p. 012004.
8. Zhang, Q.; Xu, H.; Jia, X.; Zu, L.; Cheng, S.; Wang, H. Design of a 70 MPa type IV hydrogen storage vessel using accurate modeling techniques for dome thickness prediction. *Compos. Struct.* **2020**, *236*, 111915. [CrossRef]
9. Kumar, S.S.; Bibin, C.; Ramachandran, M. Design and analysis of hydrogen storage tank with different materials by ansys. In Proceedings of the IOP Conference Series: Materials Science and Engineering, Mumbai, India, 23–24 December 2019; IOP Publishing: Bristol, UK, 2020; Volume 810, p. 012016.
10. Kwak, H.S.; Park, G.Y.; Kim, C. Design of compressed natural gas pressure vessel (Type II) to improve storage efficiency and structural reliability. *J. Press. Vessel Technol.* **2020**, *142*, 011303. [CrossRef]
11. Park, G.; Jang, H.; Kim, C. Design of composite layer and liner for structure safety of hydrogen pressure vessel (type 4). *J. Mech. Sci. Technol.* **2021**, *35*, 3507–3517. [CrossRef]
12. Nguyen, B.N.; Roh, H.S.; Merkel, D.R.; Simmons, K.L. A predictive modeling tool for damage analysis and design of hydrogen storage composite pressure vessels. *Int. J. Hydrogen Energy* **2021**, *46*, 20573–20585. [CrossRef]
13. Celaya Garcia, L.; Gutierrez Rivera, M.; Ledesma Orozco, E.; Aceves, S.M.; Martinez Ramirez, I. Equivalent properties of pure metallic reinforcement for pressure vessels using the finite element method. *J. Braz. Soc. Mech. Sci. Eng.* **2021**, *43*. [CrossRef]
14. Błachut, A.; Wollmann, T.; Panek, M.; Vater, M.; Kaleta, J.; Detyna, J.; Hoschützky, S.; Gude, M. Influence of fiber tension during filament winding on the mechanical properties of composite pressure vessels. *Compos. Struct.* **2023**, *304*, 116337. [CrossRef]
15. Wu, X.; Yang, B.; Zhou, S. Strength and Failure Analysis of Fiber-Wound Composite Gas Cylinder via Numerical Simulation. *Materials* **2024**, *17*, 717. [CrossRef]
16. Azeem, M.; Hamdan, H.Y.; Alam, M.A.; Kumar, M.; Sajid, Z.; Gohery, S.; Maziz, A.; Gemi, L.; Abdullah, S.S.; Khan, S.H. Influence of winding angles on hoop stress in composite pressure vessels: Finite element analysis. *Results Eng.* **2024**, *21*, 101667. [CrossRef]
17. Katsumata, S.; Ogasawara, T.; Uchino, T.; Hirayama, N.; Sakata, K.; Uzawa, K. Experimental and analytical study of a high-pressure hydrogen storage tank made of CFRP with dome–cylinder split molding structure for fuel cell vehicles. *Int. J. Hydrogen Energy* **2025**, *101*, 269–279. [CrossRef]
18. Reda, R.; Ataya, S.; Ashraf, A. Finite Element Modeling of Different Types of Hydrogen Pressure Vessels Under Extreme Conditions for Space Applications. *Processes* **2025**, *13*, 1429. [CrossRef]
19. Koutsawa, Y.; Bouhala, L. Uncertainty analysis in the design of Type-IV composite pressure vessels for hydrogen storage. *Compos. Part C Open Access* **2025**, *16*, 100544. [CrossRef]
20. Syed, Z.A.H.; Che, J.l.; Chang, S.H. Design of a Type-III hydrogen pressure vessel with an elliptical dome for stress reduction in the composite layer. *Int. J. Hydrogen Energy* **2025**, *103*, 937–950. [CrossRef]

21. Barthélémy, H.; Weber, M.; Barbier, F. Hydrogen storage: Recent improvements and industrial perspectives. *Int. J. Hydrogen Energy* **2017**, *42*, 7254–7262. [CrossRef]
22. Hassan, I.; Ramadan, H.S.; Saleh, M.A.; Hissel, D. Hydrogen storage technologies for stationary and mobile applications: Review, analysis and perspectives. *Renew. Sustain. Energy Rev.* **2021**, *149*, 111311. [CrossRef]
23. Wang, Z.; Wang, Y.; Afshan, S.; Hjalmarsson, J. A review of metallic tanks for H2 storage with a view to application in future green shipping. *Int. J. Hydrogen Energy* **2021**, *46*, 6151–6179. [CrossRef]
24. Zhou, W.; Wang, J.; Pan, Z.B.; Liu, J.; Ma, L.H.; Zhou, J.Y.; Su, Y.F. Review on optimization design, failure analysis and non-destructive testing of composite hydrogen storage vessel. *Int. J. Hydrogen Energy* **2022**, *47*, 38862–38883. [CrossRef]
25. Li, J.; Chai, X.; Gu, Y.; Zhang, P.; Yang, X.; Wen, Y.; Xu, Z.; Jiang, B.; Wang, J.; Jin, G.; et al. Small-scale high-pressure hydrogen storage vessels: A review. *Materials* **2024**, *17*, 721. [CrossRef] [PubMed]
26. Bohorquez Rico, H.A. Diseño y Simulación de un Tanque Tipo II Revestido con Alambre de Acero SA-905. Master's Thesis, Universidad de Guanajuato, Guanajuato, Mexico, 2018. Available online: http://repositorio.ugto.mx/handle/20.500.12059/392 (accessed on 19 June 2025).
27. Saxena, A.; Prakash, A.; Sharp, G.; Thomson, W.; David, T.; Harless, F.; Vain, J.; Miller, I.; Nagpal, V.; Peters, D.T.; et al. *Low Cost Hydrogen Storage at 875 Bar Using Steel Liners and Steel Wire Wraps*; Technical report; WireTough Cylinders, LLC: Bristol, VA, USA, 2019.
28. MatWeb, LLC. ASTM A36 Steel, Plate. 2025. Available online: https://www.matweb.com/search/DataSheet.aspx?MatGUID=afc003f4fb40465fa3df05129f0e88e6 (accessed on 26 June 2025).
29. American Society of Mechanical Engineers. *ASME Boiler and Pressure Vessel Code*, 2021st ed.; ASME: New York, NY, USA, 2021.

hydrogen

MDPI

Article

Classification Framework for Hydrological Resources for Sustainable Hydrogen Production with a Predictive Algorithm for Optimization

Mónica Álvarez-Manso [1], Gabriel Búrdalo-Salcedo [2,*] and María Fernández-Raga [1,*]

1 Department of Applied Physics and Chemistry, University of León, 24071 León, Spain; tmimam00@estudiantes.unileon.es

2 Department of Mechanical Engineering, Computer Science and Aerospace, University of León, 24071 León, Spain

* Correspondence: gabriel.burdalo@unileon.es (G.B.-S.); maria.raga@unileon.es (M.F.-R.); Tel.: +34-987-291-000-5343 (G.B.-S.); +34-987-291-000-5342 (M.F.-R.)

Abstract

Given the urgent need to decarbonize the global energy system, green hydrogen has emerged as a key alternative in the transition to renewables. However, its production via electrolysis demands high water quality and raises environmental concerns, particularly regarding reject water discharge. This study employs an experimental and analytical approach to define optimal water characteristics for electrolysis, focusing on conductivity as a key parameter. A pilot water treatment plant with reverse osmosis and electrodeionization (EDI) was designed to simulate industrial-scale pretreatment. Twenty water samples from diverse natural sources (surface and groundwater) were tested, selected for geographical and geological variability. A predictive algorithm was developed and validated to estimate useful versus reject water based on input quality. Three conductivity-based categories were defined: optimal (0–410 μS/cm), moderate (411–900 μS/cm), and restricted (>900 μS/cm). Results show that water quality significantly affects process efficiency, energy use, waste generation, and operating costs. This work offers a technical and regulatory framework for assessing potential sites for green hydrogen plants, recommending avoidance of high-conductivity sources. It also underscores the current regulatory gap regarding reject water treatment, stressing the need for clear environmental guidelines to ensure project sustainability.

Keywords: green-hydrogen; waste water; conductivity; hydrogen generation; algorithm

1. Introduction

The collapse of the current energy model has resulted in a drive for renewable energies [1]. This results in a need to identify products that minimize the negative impacts of the carbon emissions [2] generated by energy systems based on fossil fuels such as coal, then oil, and now natural gas. Hydrogen has emerged as an alternative fuel to enable decarbonization by maintaining energy generation capacity while reducing CO_2 emissions. In this regard, on 7 August 2021, the Intergovernmental Panel on Climate Change (IPCC) indicated that the point of zero emissions must be reached by the year 2050 [3]. The International Energy Agency (IEA) then established a roadmap to reach the point where all hydrogen consumed will be generated from renewable sources, as well as increasing its use in new applications [4]. This results in a prediction that hydrogen consumption will increase from the current value of 70 Mt/year to about 435 Mt/year by 2050, but with no associated emissions due to either its production or consumption. The European Union accepted this challenge by establishing the European Green Pact [5] and the EU Hydrogen Strategy [6], the REPowerEU Plan [7], and the Paris Agreement [8].

Investors and producers (especially large fossil fuel and energy companies) have thus begun a race to develop projects for the renewable production of hydrogen to contribute to this energy transition process. However, despite the tools currently in place, growth is not being achieved at the desired rate owing to technical problems with production, since the installation of production plants does not meet the environmental requirements specified in national regulations. Specifically, one of the most critical problems is to ensure that the waste produced by such plants based on water is acceptable for release into the natural environment. Hydrogen production is based on the extraction of water from a natural system followed by an electrolysis process to separate the hydrogen and oxygen by applying external energy. The electrolyzers used to achieve this rely on water of very high quality, which thus requires previous preparation. This step involves the generation of large amounts of reject water that must be discharged back into natural water cycles. Herein lies the problem, as reject water does not have characteristics suitable for its incorporation into the natural environment, which is not permitted when it exceeds certain parameter values.

A further problem is the additional energy cost required for the electrolysis process, which is increased if the water requires further preparation [9]. Moreover, to achieve green hydrogen production as driven and subsidized by the European Union, this energy must only come from renewable sources. All of these conditions limit the establishment of green hydrogen production plants, but they have not yet been studied sufficiently owing to the relatively recent introduction of these processes [10].

The aim of this work is to identify the most appropriate characteristics and establish a range of water parameter values that will enable the installation of a 2.8-MW microplant for green hydrogen production as recommended by Lima in 2025 [11].

Several phases are required to achieve this final objective:

- A pilot water treatment plant will be designed and built to achieve the conditions required for the electrolyzer of a green hydrogen generation plant.
- A series of experiments will be carried out using this plant to purify water from different natural sources with varying conductivity and hardness values and different origins (groundwater or surface water) to classify water from different sources according to its potential for renewable hydrogen production.
- The characteristics of the reject water produced when using each type of water will be evaluated to assess whether it could be discharged directly into the natural environment or will require previous treatment, which would drastically reduce the profitability of such a plant.
- On the basis of all these results, we aim to establish the water quality limits corresponding to the feasible production of renewable hydrogen on an industrial scale.

2. Materials and Methods

2.1. Construction of a Test Plant to Evaluate Different Types of Water for the Production of Green Hydrogen

2.1.1. Plant Description and Water Requirements

First, an exhaustive literature review of existing techniques for the production of green hydrogen was carried out, revealing that electrolysis is the most widely used and environmentally friendly process [12–16]. The underlying mechanism of production based on water electrolysis is the controlled breakdown of water molecules into hydrogen and oxygen by applying energy obtained from a renewable source. Two partial reactions thereby occur: (1) hydrogen production and (2) oxygen release. Through this process, molecular hydrogen is obtained in a gaseous state in the cathodic zone or at the negative electrode, while oxygen is produced in a gaseous state in the anodic zone or at the positive electrode, under the same conditions.

The water required for such green hydrogen production must undergo a series of processes. The key point regarding the impact of large-scale production is to evaluate the water treatment stage applied to make it suitable for use in the electrolyzer. Only water that is adequately pretreated and that meets the characteristics required by the electrolyzer can be used. The energy required and the number of process steps, and thus the amount of reject water produced, depend greatly on the type of water available as input to the system. Determining the water requirements of the electrolyzer thus becomes very important, to ensure that these requirements can be met.

2.1.2. Electrolyzer Requirements

Four types of commercial electrolyzer exist for such electrolysis processes and are classified according to the type of electrolyte they use [17]: alkaline electrolyzers (AEL/AWE) [18], which use an alkaline solution; polymeric exchange membrane (PEM) electrolyzers [19,20], which rely on a polymeric electrolyte; anion electrolysis membrane (AEM) electrolyzers [21], which use an anionic membrane and combine the principles of alkaline and PEM technologies; and solid oxide electrolyzer cells (SOECs) [22,23], distinguished by the fact that they are composed entirely of ceramic components, with electrolytes made from yttrium-stabilized zirconia (ZrO_2) and scandium oxides [24] (Table 1). The operating parameters depend on the type of electrolyzer used, differentiating between those operating at low (AEL/AWE, PEM, and AEM devices) versus high temperature (SOECs). Electrolyzers that require operation at high temperature are less efficient, as resources must be consumed to reach such temperatures. Moreover, note that, among these four types of electrolyzer, the technology for two (SOEC and AEM) is less well developed, as they were introduced only recently, thus requiring further studies to confirm their feasibility [25]. The use of SOEC and AEM/AWE electrolyzers in the pilot plant studied herein was thus ruled out (Figure 1). Another important parameter to consider is the feasibility in terms of the average durability in operating hours, which is longer for PEM and AEL/AWE equipment. Analyzing all these aspects, it was decided to use a PEM-type electrolyzer, which is also the most commercially developed and frequently applied in plants today, as well as offering the lowest water consumption.

The operation of all of these technologies requires water, in either liquid (AEL/AWE, PEM, and AEM) or gaseous state (SOEC). Electrolyzers consume an average of 15–20 L of water per kg of H_2 produced, typically with conductivity values below 0.1 μS/cm. In this work, we mainly focus on the conductivity, since the other parameters analyzed, e.g., the amount of sulfates and chlorides, temperature, suspended solids, hardness, and pH, are indirectly related to the conductivity value.

The conductivity thus becomes a limiting factor, as it leads to the most difficult requirement to meet among all the parameters considered (Table 1). Using the consumption of the electrolyzer to calculate the total amount of water required to produce green hydrogen is difficult because it also depends on the purity of the initial input water [26]. However, these calculations suggest that, to achieve the objectives set by Europe for 2030 for hydrogen production corresponding to 40 GW, that is, ten million metric tons of renewable hydrogen generation [27], the amount of water required is equivalent to that used in France in eight years, even with optimal water input. This result underlines the importance of studying water in areas of plant implantation, to avoid possible environmental problems due to the availability of water, especially in countries in Southern Europe, which are considered to suffer from high water stress and in terms of the quality of this raw material [28].

Table 1. A description and comparison of the different types of electrolyzer.

Title	Type of Electrolyzer			
Characteristics of the required input water	AEL/AWE	PEM	AEM	SOEC
Water phase at entry	Liquid	Liquid	Liquid	Gaseous
Amount of water per kW produced, as indicated by the manufacturer	15–20 L	11–17 L	13–18 L	10–16 L
Recommended water conductivity	<0.2 μS/cm	<0.1 μS/cm	0.1 μS/cm	0.1 μS/cm
Hardness	No hardness allowed at input			
Acceptable pH range	4–10	7–8	7	7–9
Operating temperature	Low 100–150 °C	Low 70–90 °C	Low 30–70 °C	High 800–1000 °C
Chloride content	<5 ppm	Maximum 2 ppm	Maximum 2 ppm	Maximum 2 ppm
Turbidity	<1 ppm	<1 ppm	1 ppm	1 ppm
Total organic carbon (TOC)	<30 ppb			
Resistivity	>10 MΩ cm			
Technology development status	Since the nineteenth century	Since the twentieth century	Under study	Under study
Device lifetime	60,000 h	30,000 h	10,000 h	<20,000 h
Energy consumption of the electrolyzer	4.6 kWhel/m^{3}H$_2$	4.8 kWhel/m^{3}H$_2$ or 49.9 kWh/kg H$_2$	4.8 kWhel/m^{3}H$_2$ or 53.3 kWh/kgH$_2$	3.8 kWhel/m^{3}H$_2$

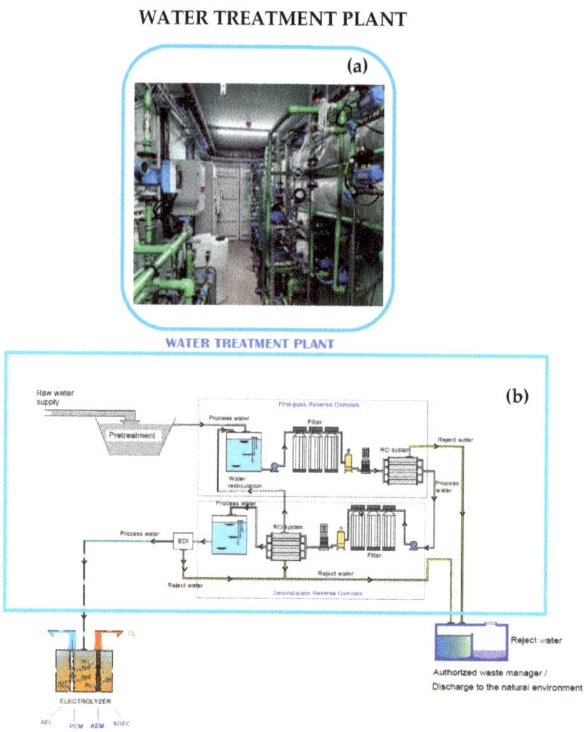

Figure 1. The technical design of the pilot treatment plant: (**a**) a photograph of the treatment plant built at the University of León and (**b**) its technical design.

A water footprint analysis of the hydrogen production process reveals that the water entering in a plant can be divided into three parts, one recoverable and two non-recoverable. The recoverable part of the water used corresponds to that effectively converted into hydrogen since the consumption of the hydrogen produced will naturally result in the production of water again. The two remaining, non-recoverable portions of water are the cooling water and the reject water. The cooling water part is not recoverable owing to the additives used to increase its performance, but this is used in a closed cycle, so the amount of water that cannot be recovered for this reason is not significant compared with the total water consumed by the plant [29]. But the total amount of reject water is a much more significant quantity, which will depend on the quality of the input water and will be directly proportional to the production volume throughout the entire lifetime of the plant. The fraction of reject water versus that actually converted into hydrogen will depend on the water quality. However, owing to a lack of studies, the specific water volume fractions that will end up as reject water or that are effectively converted into hydrogen are not currently known. Studies of plant efficiency depending on various water characteristics can thus provide critical information for environmental protection by contributing to the optimization of the amount of water consumed during production and minimizing the discharge that must be managed, which is essential to ensure plant viability. The objective is to select the optimal raw or input water so as to minimize the amount of reject water discharged into the river while also enabling cost savings by avoiding the management of large volumes of effluent, as this water would not meet the requirements for unrestricted discharge into the river. Herein lies the importance of this highly practical work, which is imperative owing to the legal vacuum and the lack of information and specific studies on this question of determining the optimal water characteristics for sustainable green hydrogen production. During the water treatment stage prior to entering the electrolyzer, the water is separated into reject water and water for use in production.

2.1.3. Preparation of Water for Entry into the Electrolyzer

This treatment stage, consisting of a sequence of reverse osmosis followed by electrodeionization (EDI), ensures that water with the required characteristics enters the EDI (Figure 1). The osmosis process must be repeated until the water meets the requirements described in Table 1 for the (most common) PEM electrolyzer. The amount of reject water will increase as the required number of osmosis processes increases. Therefore, not all water sources are suitable for hydrogen production, since quality below certain limiting values will result in the production of reject water in quantities that prevent its feasible use. To determine the amount of reject water produced when using different water samples, an experimental treatment plant was designed and constructed in-house (Figure 1). This plant was built at the facilities of the H-4 laboratory at the University of León (León, Spain) and consists of an osmosis + EDI skid unit.

2.1.4. Construction of the Pilot Water Treatment Plant In-House

The design of the prototype treatment microplant to prepare water suitable for hydrogen production followed basic industrial procedures. The most basic industrial osmosis systems always include a double osmosis process. However, for the prototype, a reverse osmosis plant was designed, enabling recirculation before entering EDI equipment. After performing empirical tests using this plant, it was observed that the water recirculated after the second osmosis process could be used, while that recirculated after the first osmosis process never showed improvements in water quality beyond 0.02% of conductivity, so it was not considered for the tests with the different types of water to analyze the optimal characteristics for sustainable H_2 green production. Unlike an industrial plant, water samples were monitored before and after each of the reverse osmosis processes in this experimental plant. The ultrapure water needed for electrolyzers requires an initial demineralization pretreatment of the input water (Figure 1 and Table 2) to avoid the formation of metal hydroxide precipitates in subsequent stages [10,30]. Becker et al. and Zhang et al. emphasize that these metallic precipitates are formed during electrolysis because the Ca^{2+} and Mg^{2+} ions react with the OH^- generated and are deposited on the electrode surface, reducing their efficiency and service life. Subsequently, the water was subject to a demineralization process, for which osmosis was chosen among all the existing technologies, owing to the input volume required. Currently, membrane technologies are used in the water purification sector [31], differing in that some (such as ultrafiltration, microfiltration, nanofiltration, and reverse osmosis) require the application of a pressure gradient, while others (such as electrodialysis) use electricity as the driving force. A combination of reverse osmosis with EDI was chosen here because this also eliminates turbidity (corresponding to organic compounds in Table 2) and retains a greater quantity of salts by eliminating monovalent ions as well as organic compounds of very low molecular weight, as filters have a pore size of approximately 0.0001 microns, enabling the production of high-quality drinking water (Table 2).

Table 2. Evaluation of the different membrane techniques according to the requirements of the electrolyzer.

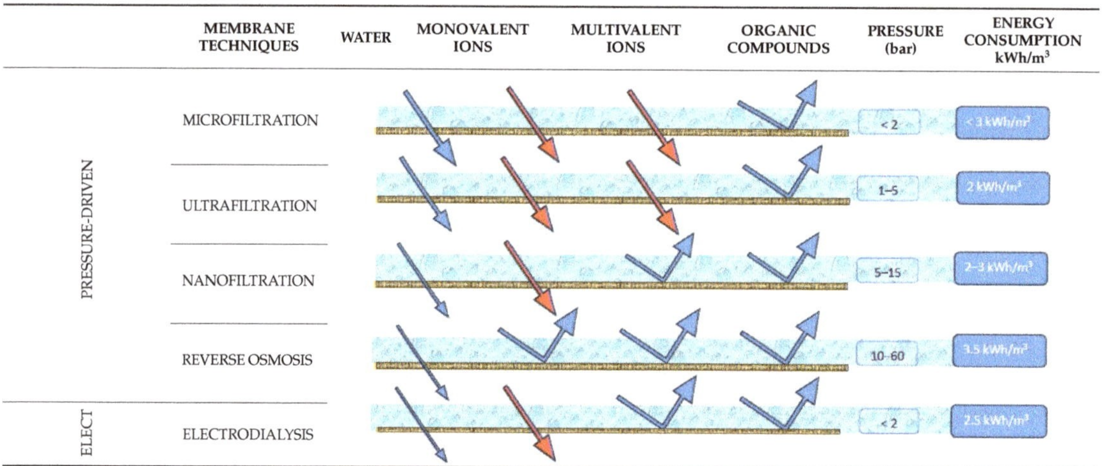

Notes: The permeability of desired substances is indicated by a blue arrow, while that to be avoided is indicated by a red arrow, resulting in the selection of reverse osmosis technology.

The prototype plant (Figure 2) starts with the collection of water in two 100-L tanks (A), where it is also analyzed before undergoing purification treatment. Initially, these 100 L of water are subject to physical pretreatment (B) by fine filtration using a cartridge filter with absolute selectivity of 5 microns. Chemical pretreatment (B) is then carried out using a dosing pump and a tank with 100-L capacity, using an average antifoulant dose of 1.5–3 mg per liter of input or raw water.

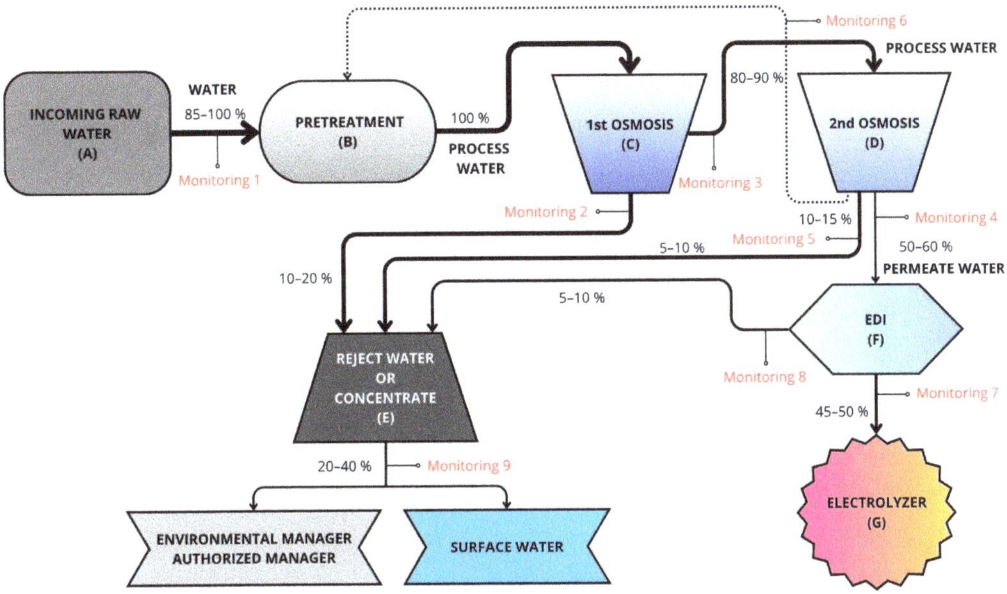

Figure 2. Water flow before entering the electrolyzer. Gray indicates incoming water, blue implies that it is more deionized, and darker colors indicate higher salt concentration. Therefore, at the end of the treatment, the water appears light blue (deionized water).

The water then undergoes the first reverse osmosis stage (C), a process that combines ion exchange resins and membranes to transfer ionic impurities into a waste or concentrated water stream. The water inlet pressure for these membranes is regulated by using a pump with a frequency controller. This process results in the loss of ions, causing a drop in pH. A pH value of 7.5 is required for the second osmosis stage (D), so bicarbonates are added to increase the pH, then the second reverse osmosis stage is applied. This second reverse osmosis stage results in a very low percentage of reject water. The reject water from both reverse osmosis stages is called the concentrate (E) owing to its high salt content (with almost all the salts that entered in the feed water being concentrated therein). The remaining water is then called demineralized water and is passed to the final, fine-processing stage based on EDI (F). This water treatment technology uses electricity (requiring continuous electrical supply) to remove ions from the water. This results in fully purified water with conductivity below 0.1 µS/cm, which can be fed to the inlet of the electrolyzers (G) to produce green H_2. The whole process is monitored so that, if limit values are exceeded, automatic purges and draining can be applied to remove water from the process.

To evaluate the different types of water, a monitoring system was installed (Figure 2). The nine installed monitors are used both to measure parameters and to control the inflow of water into the subsequent tanks, up to a maximum of 100% capacity. An example of this can be seen in Figure 2 (Monitoring 1); the Monitoring 1 line allows 100% water inflow only when the recirculation line from the second reverse osmosis stage does not supply any water. Otherwise, the Monitoring 1 line allows the entry of complementary water (85–90%) so as to reach 100% of the pretreatment tank's capacity.

On the basis of the hydraulic circuit of the process, this indicates the characteristics of the deionized water in terms of the flow rate, pressure, conductivity, pH, hardness, and dissolved oxygen content, with two sensors for each measurement variable (Figure 2). Two additional channels are provided in each transmitter to read flow and pressure variables.

2.2. *Water Performance Analysis Phase*

2.2.1. Selection of Different Natural Water Sources for the Production of Green Hydrogen

To consider a broad spectrum of possible water sources that might feed future green hydrogen generators, water from different origins were selected, including 13 from groundwater and 7 from surface sources (Figure 3 and Table 3). In addition to taking the type of source into consideration, these 20 sites were selected by considering the most favorable locations in terms of their proximity to renewable energy sources and access to water catchments, and their water was used in the plant. These sites represent the greatest variability in terms of conductivity, pH, and hardness values. The map shows that, in addition to these 20 sites, 57 other sites have been sampled, but the water from these sites has not been used in the plant.

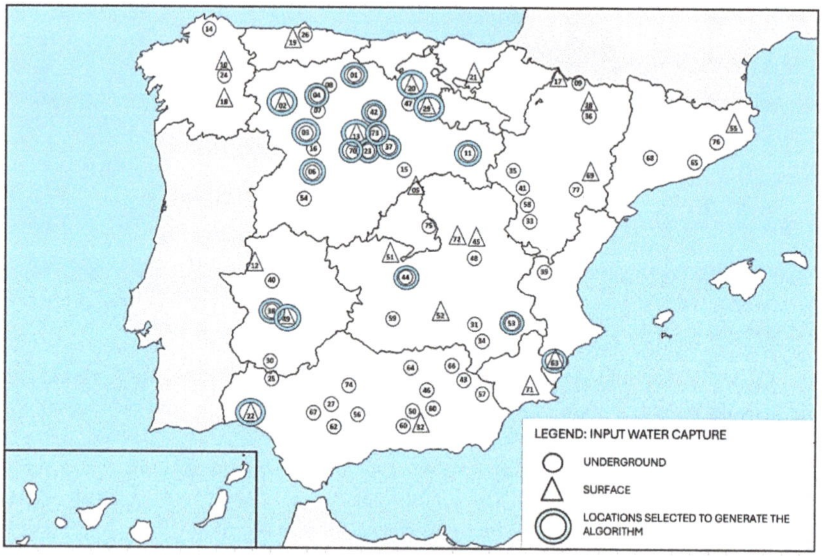

Figure 3. Location map of water sources used for the raw water treatment pilot plant. Triangles indicate surface springs, while circles indicate underground sources. The numbers indicate each location as indicated in Table 3.

Water analysis was performed by an authorized laboratory using samples collected on site at each of the 20 locations in blue (Figure 3 and Table 3). These 20 locations were selected on the basis of a proportional distribution according to the Köppen–Geiger climate classification, ensuring the inclusion of both groundwater and surface water samples. From each location, 175 L was collected, of which 25 L was used for analysis, 100 L for purification by the treatment process in the pilot water treatment plant, and 50 L for recirculation in the circuit. The water was collected in three 50-L polyethylene containers for input into the treatment plant described above (Point A in Figure 2). In the first pretreatment phase, relying on decantation and filtration (Point B in Figure 2), visible impurities such as stones, plant debris, and microorganisms were removed. Subsequently, chlorine was added and left to act for two hours. Then, the free chlorine particles and colloids were mainly removed. The water flowed through the double-pass reverse osmosis stages (Point C and Point D in Figure 2), while the permeate and intermediate water was monitored at monitoring points 2 (reject water) and 3 (water continuing in the process to the second reverse osmosis stage). After the second reverse osmosis stage, the water was divided into three streams: 10–15% was sent to the pretreatment tank (monitoring point 6), 5–10% was directed to reject water or concentrate (monitoring point 5), and 50–60% continued in the process toward EDI (monitoring point 6). During EDI, the water was treated by electrodeionization to remove the remaining ions, again generating reject water (monitoring point 8). Finally, the water leaving the EDI passed by an ultraviolet lamp for bacteria reduction treatment, where a water sample was also taken (monitoring point 7), before passing through to the electrolyzer to confirm that it had the desired composition according to the requirements of the electrolyzer.

Table 3. Characteristics of water samples used to evaluate the treatment efficiency, type of soil, type of clays in the area, and conductivity of the water.

	Water Characteristics	Köppen–Geiger Climate Classification * [32]	Soil Types	Presence of Clay	Input Water Conductivity (μs/Cm)
01	Cervera de Pisuerga (Palencia)	Csb	Chalky/rocky	Low	32.00
02	Bembibre (León)	Csb	Slate/clays	None	39.00
03	Santibáñez de Vidriales (Zamora)	Csb	Slate	Low	55.00
04	Villadangos del Páramo (León)	Cfb	Clay–loam	None	83.00
05	Altazar (Madrid)	Csb	Entisols and Inceptisols over shales and sandstones	Low	94.00
06	Villamayor (Salamanca)	Csb	Sandy and sandy loam soils	None	116.00
07	Onzonilla (León)	Csb	Alluvial soils (Fluvisols) and sandy loam	Moderate	153.50
08	Almanza (León)	Csb	Brown chalky and clayey soils	Low	156.00
09	Panticosa (Huesca)	Cfb	Leptosols and poorly evolved mountain soils	None	183.00
10	Lugo (Lugo)	Csb	Umbrisols and Cambisols	None	214.00
11	Almazán (Soria)	Cfb	Fluvisols and Cambisols over alluvial terraces	Moderate	228.00
12	Alcántara (Cáceres)	Csa	Leptosols and Regosols over slate and quartzite	Low	260.00
13	Arroyo de la Encomienda (Valladolid)	Csb	Fluvisols and chalky Cambisols	Moderate	266.00
14	As Pontes (A Coruña)	Csb	Metamorphic soils with the presence of alluvium	None	275.00
15	Muñoveros (Segovia)	Csb	Sandy soils and sandy loam	Low	288.00
16	Manganeses de la Lampreana (Zamora)	Csb	Sandy and silty	Low	302.00
17	Sabiñanigo (Huesca)	Cfb	Alluvial and Quaternary terraces	None	303.00

Table 3. *Cont.*

	Water Characteristics	Köppen–Geiger Climate Classification * [32]	Soil Types	Presence of Clay	Input Water Conductivity (µs/Cm)
18	Orense (Orense)	Csb	Acid soils (Umbrisols and Cambisols) and sandy loam.	None	304.00
19	Llanera (Asturias)	Cfb	Deep soils, loamy to clayey loam in texture	Low or moderate	314.00
20	Pineda Trasmonte (Burgos)	Cfb	Brown limestone and clayey soils	None or low	357.00
21	Salvatirra (Álava)	Csb	Soils over alluvial deposits	Low	375.00
22	Palos de la Frontera (Huelva)	sa	Sandy and sandy loam, with alluvial deposits	High	375.00
23	Bercero (Valladolid)	Csb	Luvisols and alluvial terrace soils	Moderate	387.00
24	Lugo (Lugo)	Csb	Shallow with silty or sandy-loam texture	None	397.00
25	Cala (Huelva)	Csa	Regosols and Leptosols over shales and quartzite	High	418.00
26	Silvota (Asturias)	Cfb	Deep soils, loamy to clayey loam in texture	Low or moderate	430.00
27	Marismas (Sevilla)	Csa	Loamy/clayey	High	441.00
28	Jaca (Huesca)	Cfb	Rocky/pebbly	None	446.00
29	Miranda de Ebro (Burgos)	Csb	Claystones, limestones, and sandstones	Low	463.00
30	Cabeza de Vaca (Badajoz)	Csa	Brown limestone and clayey soils	Low	473.00
31	El Bonillo (Albacete)	Csa	Limestone and dolomitic soils	Low	474.00
32	Iznajar (Córdoba)	Csa	Clayey and limestone soils	Moderate	500.00
33	Villarquemado (Teruel)	Cfb	Brown limestone and clayey soils	Moderate	510.00
34	Yeste (Albacete)	Csa	Mediterranean rendzina, brown limestone, and alluvial	Low	515.00
35	Berrueco (Zaragoza)	Cfb	Xerophytic (desert), loose, dusty soils	Low	516.00
36	Jaca (Huesca)	Cfa	Fluvisoles, Leptosoles, Cambisoles	None	533.00
37	Rueda (Valladolid)	Cfa	Sandy loam and limestone soils	Moderate	572.00
38	La Venta Usagre (Badajoz)	Csa	Brown limestone and clayey soils	Low	576.00
39	Requena (Valencia)	Csa	Fluvisols and Cambisols with clay accumulation	Low	617.00
40	Torico (Cáceres)	Csa	Leptosols and Regosols developed over shales	Low	636.00
41	Used (Zaragoza)	Cfb	Ordovician quartzites and dolomitic limestones	Low	664.00

Table 3. *Cont.*

	Water Characteristics	Köppen–Geiger Climate Classification * [32]	Soil Types	Presence of Clay	Input Water Conductivity (µs/Cm)
42	Villaviudas (Palencia)	Cfb	Rendzinas with brown limestone soils in higher areas	Moderate	667.00
43	Puebla de Don Fabrique (Granada)	BSk	Chalky and clayey soils over Mesozoic materials	Low	670.00
44	Fuensalida (Toledo)	BSk	Fluvisols and Cambisols over alluvial materials	Low	706.00
45	Nohales (Cuenca)	Csa	Chalky and clayey soils	Moderate	712.00
46	La Guardia de Jaén (Jaén)	BSk	Clayey and limestone soils	Moderate	720.00
47	Miranda de Ebro (Burgos)	Cfb	Clayey, limestone, and sandstone soils	Low	788.00
48	Villarez de Saz (Cuenca)	Csa	Limestone and clayey soils over Mesozoic materials	Moderate	823.00
49	Torremejia (Badajoz)	Csa	Southern brown earth and clayey red soils	Low	927.00
50	Colomera (Granada)	Csa	Limestone and clayey soils over Mesozoic materials	Low	970.00
51	Villamiel de Toledo (Toledo)	BSk	Fluvisols and Cambisols over alluvial materials	Moderate	1060.00
52	Argamasilla de Alba (Ciudad Real)	BSk	Clayey loam and chalky clay soils	Moderate	1080.00
53	Chinchilla de Montearagón (Albacete)	BSk	Limestone, clayey, and gypsiferous soils	Moderate	1087.00
54	Calzada de Don Diego (Salamanca)	Csb	Sandy and sandy loam over granitic peneplains	Low	1100.00
55	Santa Julia de Ramis (Girona)	Cfa	Granitic and alluvial materials	None	1160.00
56	Puente Genil (Córdoba)	BSk	Clayey and limestone soils	Moderate	1240.00
57	Orce (Granada)	BSk	Calcium Cambisols and chalky Regosols	Low	1250.00
58	Moreal del Campo (Teruel)	Cfb	Brown limestone and clayey soils over Tertiary soils	Moderate	1263.00
59	Almagro (Ciudad Real)	BSk	Clayey and limestone soils with Miocene sediments	Moderate	1431.00
60	Montejicar (Granada)	BSk	Limestone, clayey, and marly limestone Subbetic mountain ranges	Moderate	1480.00
61	Vega de Granada (Granada)	BSk	Fluvisols with alluvial soils, with clay–loam texture	Moderate	1677.00
62	Sierra de los Caballos (Sevilla)	Csa	Limestone, clayey and stony with Mesozoic material	High	1786.00
63	Santa Cruz De La Muela (Alicante)	BSh	Clayey and limestone soils	Moderate	1800.00
64	Bailen (Linares)	BSk	Strongly to slightly chalky soils	Moderate	1826.00
65	L'Hospitalet de Llobregat (Barcelona)	Csa	Alluvial soils with Fluvisols	Low	1880.00

Table 3. *Cont.*

	Water Characteristics	Köppen–Geiger Climate Classification * [32]	Soil Types	Presence of Clay	Input Water Conductivity (µs/Cm)
66	Navas de San Juan (Jaén)	Csa	Marls, clays, and gypsum	High	1887.00
67	Carmona (Sevilla)	Csa	Alluvial and chalky soils	High	1975.00
68	Verdu (Lleida)	BSk	Chalky and sandy loam soils over Tertiary soils	Moderate	2010.00
69	Caspe (Zaragoza)	BSk	Chalky, stony, and loose soils over Tertiary soils	Moderate	2300.00
70	Peñafiel (Valladolid)	Csb	Sandy loam and limestone soils	High	2580.00
71	Mazarrón (Murcia)	BSk	Soils with chalky crust and ferruginization	High	2800.00
72	Torrejoncillo Del Rey (Cuenca)	Cfb	Chalky and clayey soils over Mesozoic materials	Moderate	2834.00
73	Villafuerte (Valladolid)	Csb	Chalky moor and sandy or sandy loam soils	High	3126.00
74	La Luisiana (Sevilla)	Csa	Alluvial Fluviosols and clayey soils	High	3285.00
75	Mejorada del Campo (Madrid)	Csa	Soils with chalky crust and ferruginization	High	3380.00
76	Girona (Girona)	Cfa	Sandy loam to clay–loam soils	Low	3563.00
77	Caspe (Zaragoza)	BSk	Chalky, stony, and loose soils over marls	High	3857.00

Notes: Based off of a geological map and map of risk prevention due to clay expansivity in Spain (Geological and Mining Institute of Spain (IGME)). Numbers in blue indicate that water samples taken at these points were used for algorithm development. * BSh: hot steppe; BSk: cold steppe; Csa: temperate with a dry or hot summer; Csb: temperate with a dry or temperate summer; Cfa: temperate with a dry season and hot summer; Cfb: temperate with a dry season and temperate summer.

2.2.2. Analysis of Existing Water Legislation

In addition to meeting the minimum requirements for the feed water at the electrolyzer input, it is also essential to define the minimum parameters for discharge of the water generated by the hydrogen generation plant so as to determine whether the reject water can be discharged directly into the environment.

Water legislation defines discharge as the act of introducing water with parameter values that exceed certain limits or contain pollutants by mixing it with inland water or with the rest of the public water system [33]. Hydrogen production, being an industrial activity, must comply with these parameter limits. Currently, there is no specific Spanish or European legislation defining the limits and specific parameters in relation to discharges generated during renewable hydrogen production. Given this lack of an established standard to define maximum allowable values, we studied the limit values found in both legislation and literature to determine the limiting conductivity value that might affect the receiving system. Conductivity values above this would require further treatment by an authorized waste manager or the application of other systems such as forced vacuum evaporation to reduce the conductivity.

The results presented in Table 4 indicate that conductivity values below 1000 µS/cm do not affect crops, human health, or freshwater species. Meanwhile, water with conductivities above 6000 µS/cm would not be appropriate for any use. Also, water with higher conductivity can only be used for certain crops.

Table 4. Conductivity limit values, the sector affected, and information source.

Limit Values	Sector Affected	Source
>6000 μS/cm	Crops	[34] Ismayilov et al. (2021)
>5000 μS/cm	Surface water	[35] Degree 141/2012, of June 21 (Galicia, Spain)
	Human health	[36] Martín and Ángel (2023)
>4000 μS/cm	Surface water	[37] R.D 38/2004, BOA 10 March 2004, (Spain)
	Irrigation	[34] Ismayilov, A., et al. (2021)
>3000 μS/cm	Irrigation	[34] Ismayilov, A., et al. (2021) [38] Álvarez, M., et al. (2004)
	Crops	[39] Pérez-Vázquez et al. (2020)
>2500 μS/cm	Human consumption	[40] European Directive 2020/2184 [41] European Directive 98/83/EC [42] European Council Directive 98/83/EC, 1998 [43] R.D. 1138/90 of 14 September 1990 [44] R.D. 140/2003
>2000 μS/cm	Freshwater species	[45] Flores and Jara (2017) [46] Cañedo-Argüelles et al. (2016)
>1800 μS/cm	Discharges into the natural environment	[47] R.D. 606/2003 (CNAE 3519)
>1000 μS/cm	Human consumption	[31] Solís-Carvajal (2017) [39] European Council Directive 75/440/EEC of 16 June 1975
	Crops	[48] Porta and Herrero (1996)

2.2.3. Trial Monitoring and Data Collection

The water samples of the 20 sites were introduced into the pilot pretreatment plant while monitoring the most important parameters (conductivity, alkalinity, hardness, and contents of different minerals) in each of the stages. All of the parameters were found to be related to conductivity. As it was impossible to perform a continuous process in the pilot plant, the analysis was performed by introducing 150 L and analyzing a volume of 100 L per sample.

2.2.4. Analysis of Results

After each stage of the pilot plant process, the following parameters were measured: temperature, flow rate, pressure, conductivity, pH, hardness, and dissolved oxygen content.

The following monitoring equipment was used, with two sensors per variable: for conductivity, Hamilton Conducell in-line sensors with a range of 0.1 μS/cm to 200 mS/cm; for pH, Mettler Toledo glass electrode sensors; for flow rates, Endress + Hauser Promag electromagnetic flowmeters; for pressure, WIKA S-20 transducers for input values between 10 and 15 bar; for temperature, a PT100 resistance temperature detector; for dissolved oxygen, Mettler Toledo optical sensors; and for hardness, HACH HQ440D automatic titrators. All these parameters were monitored by using a centralized system for data acquisition and alarms (SCADA, Siemens WinCC, Loveland, Colorado–EEUU).

2.3. Development of an Algorithm to Estimate the Amounts of Reject Water and Usable Water

With the detailed results from the pilot pretreatment plant, with 13 groundwater samples and 7 surface samples, an algorithm was developed (Figure 3 and Table 3). After introducing 100 L of each of these water samples into the constructed pretreatment plant, we studied the values of temperature, flow rate, conductivity, pH, and hardness obtained after each stage of the process. These data were input into the ChatGPT Version 1.2025.203 artificial intelligence system, asking it to integrate them into Python version 3.13.2.

The algorithm is based on the following main formulas (Equation (1)):

$$\begin{array}{l}\text{Conductivity of filtered water, process water, or output water: C\_filtered} = \text{C\_input} \times (1 - \text{salt\_rejection}) \\ \text{Reject water conductivity (by mass balance): (inflow\_flow} \times \text{inflow\_conductivity)} = \\ (\text{filtered\_flow} \times \text{filtered\_conductivity}) + (\text{reject\_flow} \times \text{reject\_conductivity}) \\ \text{Filtered or process water flow rate: filtered\_flow} = \text{input\_flow} \times \text{recovery} \\ \text{Reject water flow rate: reject\_flow} = \text{input\_flow} - \text{filtered\_flow}\end{array} \quad (1)$$

On the basis of this analysis and simulations, a program was generated (registered as intellectual property with code 00765-02622024) to predict, using data for the conductivity of water streams, the potential production of reject and output water when used in a hydrogen plant.

This algorithm was subsequently validated via analysis of 26 other new water samples (13 surface and 13 groundwater) (Table 8) that were also introduced and monitored after putting in the treatment plant to compare the results obtained by the algorithm with the final results obtained when using the plant.

Finally, the algorithm was applied to 46 water samples that had not been tested in the plant to calculate the amount of water that would theoretically have to be managed to implement such a hydrogen generation plant with water from each source (Figure 4).

2.4. Determination of the Limit Values Corresponding to Optimum, Moderate, and Restricted Water Quality for Green Hydrogen Production

On the basis of the analysis of our 20 tests monitored in the experimental treatment plant (Table 5), it was possible to determine the conductivity and volume of the reject water that would be generated by such a plant when installed. There are legal and health requirements, as recommended and described in Section 2.2.2 (Table 4), which demand the management and treatment of water with values above certain limits. Combining these two concepts, one can determine the maximum conductivity of the input water that will result in reject water exceeding established limit values. Using this information, three grades of water quality for hydrogen production were determined with the following values:

1. **Optimum**: when the incoming raw water has a conductivity between 0 and 410 µS/cm.
2. **Moderate**: when the incoming raw water has a conductivity between 411 and 900 µS/cm.
3. **Restricted**: when the incoming raw water has a conductivity above 900 µS/cm.

These parameters, together with the algorithm, allowed us to classify the 77 water samples whose conductivity was measured into qualities and to analyze the geographical distribution and its relationship with groundwater and surface water locations (Figure 3).

Table 5. Conductivity values of the reject water measured for the 20 water sampling points selected for the algorithm design.

	Origin	Input Conductivity (μS/cm)	Reject Water Conductivity (μS/cm)	Percentage of Reject Water Flow (%)
01	Underground	32.00	74.34	9.02
02	Surface	39.00	90.60	9.42
03	Underground	55.00	128.58	10.12
04	Underground	83.00	197.75	10.42
06	Underground	116.00	282.51	10.83
11	Underground	228.00	594.10	12.92
13	Surface	266.00	706.12	15.05
20	Surface	357.00	984.11	17.66
22	Surface	375.00	1042.16	17.85
23	Underground	387.00	1079.96	19.38
29	Surface	463.00	1308.25	20.96
37	Underground	572.00	1622.25	28.87
38	Underground	576.00	1633.71	29.12
42	Underground	667.00	1896.75	30.98
44	Underground	706.00	2008.01	34.62
49	Surface	927.00	2683.76	47.15
53	Underground	1087.00	3156.65	51.32
63	Surface	1800.00	5364.90	55.98
70	Underground	2580.00	7898.67	61.71
73	Underground	3126.00	9767.50	68.63

Notes:

Optimum	Raw water conductivity: 0–410 μS/cm
Moderate	Raw water conductivity: 411–900 μS/cm
Restricted	Raw water conductivity: >900 μS/cm

3. Results and Discussion

3.1. Analysis of the Limit Values in Legislation

Our search of current legislation showed that high water conductivity is a critical indicator of contamination and salinization, having adverse effects on health and the environment, so its control is essential to ensure the sustainability and viability of renewable hydrogen production plants.

There is no harmonized standard, and limits vary according to the Water Framework Directive (WFD), which prioritizes the non-alteration of ecosystems. The absence of a single conductivity standard requires case-by-case analysis, prioritizing the receiving medium and subsequent uses of the water.

3.2. Flow Analysis and Water Analysis at the Pretreatment Plant

The analysis of the flow rate, hardness, conductivity, and pH for water from the 20 water sampling points showed that all these parameters are related to conductivity, so the conductivity is the parameter with the highest influence on the possible behavior of the water for generation of hydrogen [49,50]. According to the flow analyses throughout the pretreatment plant, shown in Figure 2, the reject water has a higher flow rate after the first osmosis, but this decreases after the second osmosis and after EDI.

It is observed from these results that surface water is slightly more optimal than groundwater (46% versus 57%), considering that surface water benefits from dilution of salts via precipitation. However, this difference is minimal, and industrial uses of surface water are always subject to the requirement to maintain natural flows and supply urban centers. Therefore, when possible, it is preferable to obtain water from groundwater sources rather than directly from surface water, owing to the greater flow stability, reduced susceptibility to short-term seasonal and climatic variations, and less exposure to immediate regulatory restrictions, especially for an industrial process where operational continuity, predictability of supply, and minimization of risks from external factors are paramount.

The viability of green hydrogen production plants is directly related to the quality of the water used, mainly its conductivity, but no reference values are available to help delimit the suitability of water when installing such a plant. This value is of great interest since the amount of reject water produced by a plant, as well as the cost of the maintenance and membranes in the plant, will increase strongly when water with high conductivity is used. After empirically analyzing these reference values, it was found that water with conductivity values below 410 µS/cm would be optimal. Such water will allow the generation of reject water with salinity suitable for discharge directly into the environment, and always at quantities of less than 30% of the input water entering the plant, which also reduces the consumption of raw materials as well as the cost of membranes (Table 5). Conductivity values between 410 and 900 µS/cm mean managing volumes of reject water that approach 50% of the input water volume (Table 5), associated with much higher maintenance costs. In addition, the electricity consumption will be almost three times higher than in green areas (consumption of ~1.32 kWh/m^3 in green areas versus ~3.88 kWh/m^3 in red areas).

In addition, hydrogen production in such areas may suffer from interruptions at certain times of the year when environmental restrictions prohibit discharge above certain regulatory limits. The results of this study also show that input water with conductivity values above 900 µS/cm will make the process nonviable, owing to the strong increase in the cost of membranes used in the plant as well as the management of reject water volumes that could reach up to 80% (Table 5).

Raw water with conductivity values below 410 µS/cm has quality suitable for hydrogen production, generating reject water with salinity values suitable for direct discharge into the receiving medium. Meanwhile, the consumption of raw materials necessary for the operation of the renewable hydrogen plant is also reduced.

Several conclusions can be drawn from Table 5. The first and most obvious is that the conductivity of the input water is closely related to the conductivity of the reject water, increasing by approximately threefold while passing through the plant.

However, the table also reveals that, when using input water from locations with the highest water quality, that is, optimum quality, the reject water volume is always less than 20% of the input water. Meanwhile, when the water quality is moderate, this reject water can reach 20–35% of the input water, and for restricted water levels, at least 47% of the input volume will end up as reject water. Indeed, for a plant where the conductivity of the input water exceeds 1800 µS/cm, the reject water would account for almost 56% of the total. Moreover, this reject water would have such an increased conductivity that it would require prior treatment before discharge into the receiving medium, via an authorized waste manager.

This means that, to produce 1 kg of green hydrogen with a PEM-type electrolyzer, the most widely used today, about 21 L of input water classified as optimal quality would be required, while up to about 38 L of water in the restricted category would be needed. For hydrogen production plants using other types of electrolyzers, a theoretical estimate of the water consumption required for the production of 1 kg of green hydrogen is presented in Table 6, revealing that AEL/AEW-type electrolyzers would require the most water, reaching almost 44.44 L in the worst conditions.

Since the water consumption depends on the electrolyzer type and water quality, we do not propose using only the amount of water available as a viable indicator for selecting a particular location for a hydrogen generation plant. This study shows that, as pointed out in other works [51], water consumption for green hydrogen production is not very high, since the water is partly returned to the receiving medium. The analysis reveals that, if current energy demands were to be met only through the production of green hydrogen, the water consumption required would be only a very small percentage of the volume of water consumed worldwide for energy production currently, namely 0.6%.

The results of this study argue instead for a dual indicator for the location of hydrogen generation plants that considers both the quantity and quality of water available, with the quality of the available water being the most important indicator. The main contribution of this work is the proposed classification of input water quality, which will make it possible to minimize risks related to increased production costs and predict feasibility as early as the implementation and performance study phases of projects under development.

Table 6. Average amount of input ultrapure water required (in liters) to produce 1 kg of green hydrogen depending on the water quality and the type of electrolyzer (AEL/AWE, PEM, AEM, or SOEC).

Water Quality Category	Percentage Average Water Use	Ultrapure Water Required to Generate 1 kgH$_2$ [52]							
		AEL/AWE [53]		PEM [53,54]		AEM [55]		SOEC [56]	
		15 l	22.3 l	11 l	17 l	13 l	18 l	10 l	16 l
Optimum	81%	18.52	24.69	13.58	21.01	16.05	22.22	12.35	19.75
Moderate	75%	20.00	26.67	14.66	22.68	17.33	23.98	13.33	21.33
Restricted	45%	33.3	44.44	24.44	37.74	28.88	40.00	22.22	35.56

Therefore, when designing a green hydrogen production plant, raw water being of poor quality, especially in areas with water conductivity values above 900 µS/cm, means an increase in the raw water flow consumption, an increase in the energy required for the reverse osmosis process, and an increase in discharges with high conductivity requiring treatment by an authorized waste manager. This entails a high production cost. Normally, costs corresponding to wastewater treatment after the osmosis and EDI stages are not calculated for projected renewable hydrogen production plants, since it is assumed that the wastewater will have properties appropriate for direct discharge into the receiving environment or the existing sewage system, but this is not the case. The costs of replacing the membranes when they are clogged, which will occur much more frequently in the case of raw water with low quality, are also not taken into account. Hydrogen production with water in the restricted category would face frequent interruptions, in addition to being exposed to the risk of uncertainties surrounding environmental regulations regarding discharge management and obtaining combined environmental authorization, in extreme cases making the plant unfeasible. This classification could also be useful for the design of plants of different sizes and powers by scaling the flow requirements.

Intensive research is currently being conducted on both fuel cell and electrolyzer components to reduce costs and make hydrogen technology more competitive. However, a technically and economically viable technology or system that can meet the limits for the dissolved salt concentration in the reject water discharged into the receiving environment has yet to be developed.

In addition, to study the temporal stability and the effect of the different seasons, in case of changes in the water table through the year, we selected 11 random points from the 20 blue points in Table 3 (Table 7) representing the most frequent Köppen–Geiger climate types (BSk, Csa, Csb, Cfb). Water from these 11 points was resampled after a few months and reintroduced into the plant to analyze the conductivity of the reject water. These results revealed that the water quality classification remained constant for all values except site 44. In the first sampling, site 44 was placed in the moderate category but was very close to the cut-off for the restricted group, to which it switched when the water table dropped.

Table 7. Comparison of conductivity values measured in water taken at 11 points on two occasions over time, showing the temporal variability of conductivity over time.

Location	Input Conductivity (µS/cm)	Reject Water Conductivity (µS/cm)	Date of First Measurement	Aquifer Water Level	Input Conductivity (µS/cm)	Reject Water Conductivity (µS/cm)	Date of Second Measurement	Aquifer Water Level	Water TABLE over Time
42	667	1896.75	15 April 2023	High	653	1855.76	10 April 2024	High	Constant
23	387	1079.96	15 April 2024	High	394	1099.50	22 May 2024	High	
22	375	1042.16	19 June 2023	Low	342	923.54	19 June 2024	Low	
44	706	2008.01	15 April 2023	High	1200	3499.08	13 September 2024	Low	Strong variation
3	55	128.58	22 May 2024	High	57	133.49	25 September 2024	Low	
4	83	197.75	24 June 2024	High	81	196.51	25 September 2024	Low	
73	3126	9767.50	23 February 2023	High	3201	10,039.62	15 September 2024	Low	
2	39	90.60	16 June 2023	High	41	95.63	15 September 2024	Low	
53	1087	3156.65	12 December 2023	High	1118	3249.02	10 April 2024	Moderate	Moderate variation
29	463	1308.25	8 May 2023	High	503	1423.34	28 November 2024	Moderate	
6	116	282.51	16 June 2023	Moderate	116	282.51	15 September 2024	Low	

Having established that the quality and quantity of the incoming raw water are the indicators necessary to determine the optimal location of the plant, it became essential to develop a tool that, on the basis of an analysis of the input water and flow rate, could determine the hydrogen production achievable with a given electrolyzer. Under this premise, a validation algorithm was designed and developed.

3.3. Algorithm Development and Validation

On the basis of a very simple and efficient analysis of conductivity, the designed algorithm estimates the costs of an installation in terms of the water to be treated as well as the electricity and membrane consumption.

The validation of the algorithm revealed that the relative error between the predicted values and those obtained by the algorithm was a maximum of 4.54% with respect to the real values (Table 8), confirming that the algorithm is valid to justify the location of a specific power plant. This validation is suitable for both groundwater and surface water sources.

Following its validation, the algorithm was extended to locations of the Iberian landmass representing different geologies and edaphological characters (Table 3). The results revealed no relationship between the data calculated using the algorithm and the geology at each site, or with the different edaphological characteristics. This lack of relationship makes sense considering that the water will have passed through different agricultural or industrial lands that could significantly alter its conductivity by mixing with certain byproducts; in addition, changes could occur owing to hydromorphological alterations due to channeling or intensive extraction, or the presence of pathogenic microorganisms such as *E. coli* and *Streptococci* sp. associated with poor health practices and variations in pH and hardness due to the dissolution of minerals or the introduction of chemical residues. In addition, the breadth of the geologic map, which includes many different geologic characters in each square, greatly hinders any comparison with specific water sampling points, yielding inconclusive results. However, an effect on the quality of water is found when correlating the classification of the sampling points with the map of clays (Figure 4). It appears that, in those locations where there is no clay or low-expansivity clay, the water seems to have better quality for use in green hydrogen production. The relationship between water conductivity and clays is determined by the physicochemical and hydrodynamic mechanisms that influence the expansivity and electrical properties of these materials. In the case of sulfate-rich clays, the concentration of sulfates ($SO_4{}^{2}$) would be directly linked to their electrical conductivity since these chemicals dissociate in water, releasing mobile ions that contribute significantly to the water's ability to carry electric current.

Table 8. Validation of the designed algorithm.

Location	Source	Input Conductivity (μS/cm)	Reject Water Conductivity (μS/cm)	Reject Water Conductivity According to the Algorithm (μS/cm)	Relative Error (%)
San Román de Bembibre (León)	Surface	41.00	95.63	97.14	1.58
Quintana Dueñas (Burgos)	Underground	86.00	205.25	213.27	3.91
Robleda (Salamanca)	Underground	137.00	335.68	340.44	1.42
Cerdeña (A Coruña)	Surface	180.00	459.00	456.37	0.57
Aboño (A Coruña)	Surface	220.00	572.46	598.46	4.54
Rozas de Valdearroyo (Cantabria)	Surface	254.00	667.44	670.30	0.43
Guareña (Badajoz)	Surface	270.00	716.74	716.40	0.05
Fuentidueña (Segovia)	Underground	365.00	1007.40	1007.00	0.04
Talavera la Real (Badajoz)	Surface	406.00	1141.14	1147.70	0.57
Alquezar (Huesca)	Surface	413.00	1161.44	1162.00	0.05
Villasalla (Soria)	Underground	422.00	1194.39	1198.00	0.30
Mérida (Badajoz)	Underground	442.00	1248.56	1257.02	0.68
Cabezas Rubias (Huelva)	Underground	501.00	1417.68	1422.00	0.30
Leza (Álava)	Surface	560.00	1587.26	1589.60	0.15
Lupiana (Guadalajara)	Underground	562.00	1593.27	1598.77	0.35
Caravaca de la Cruz (Murcia)	Underground	565.00	1601.94	1604.00	0.13
Brihuega (Guadalajara)	Underground	600.00	1701.96	1697.14	0.28
Martín del Rio (Teruel)	Surface	602.00	1708.11	1709.00	0.05
Alcázar de San Juan (Ciudad Real)	Underground	625.00	1775.50	1776.71	0.07
Alfaro (La Rioja)	Surface	693.00	1972.35	1978.50	0.31
Toro (Zamora)	Surface	746.00	2124.68	2119.00	0.27
Valdepeñas (Ciudad Real)	Surface	869.00	2483.69	2471.79	0.48
Daroca (Zaragoza)	Surface	940.00	2725.81	2733.00	0.26
Villamediana (Palencia)	Underground	1170.00	3408.68	3391.13	0.51
Calzada de Calatrava (Ciudad Real)	Underground	2200.00	6644.22	6640.10	0.06
Azanjuez (Madrid)	Underground	2520.00	7671.89	7669.33	0.03

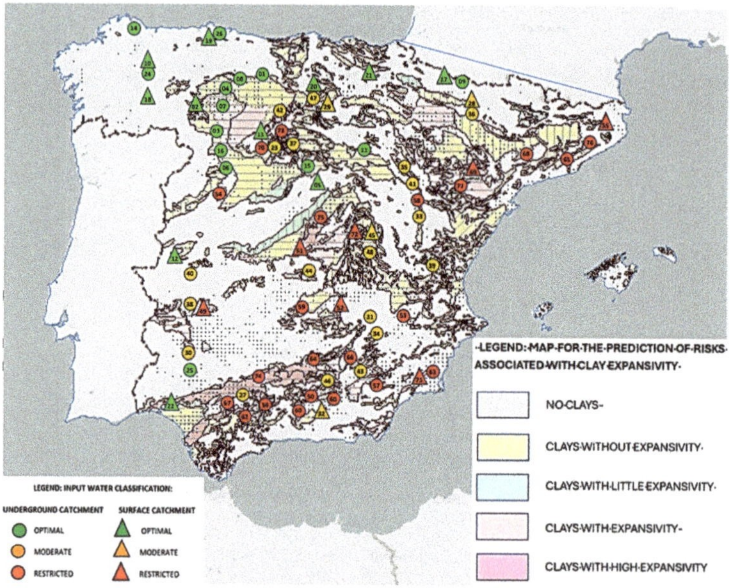

Figure 4. Predictive Map of Risks due to Clay Expansivity (Geological and Mining Institute of Spain (IGME)), with the location of the selected [57] water sampling points overlaid. Optimal or higher-quality water locations are indicated in green, moderate or medium-quality water sources in orange, and restricted-quality water sources in red. Circles indicate underground sources, while triangles indicate surface sources. Numbers indicate each location as indicated in Table 3.

Researchers such as Coitiño López [58] have indicated that water sources located on the margins of surface water bodies (rivers, reservoirs, etc.), especially those with shrub vegetation, as well as areas with detrital aquifers and granitic (not chalky) soils, usually have more suitable conditions in terms of conductivity, so these would be more suitable sites for the production of green hydrogen. Installations on sedimentary rock are less beneficial since this implies higher conductivity [57]. Our analysis of the sampling data herein confirms these results.

Finally, it is interesting to note that this analysis in the Iberian Peninsula could be extrapolated to the European Union, as it includes locations with very different geomorphological and catchment characteristics. This could contribute to the development of a more intelligent management policy and the promotion of green hydrogen production in Europe. However, such management would require the establishment of standards to regulate the quality of input water and associated conductivity limits to ensure the sustainability of the renewable hydrogen industry.

The current results align with the sixth Sustainable Development Goals related to efficient water management and water stress reduction, indicating that sustainable water management is essential for energy production and the preservation of aquatic ecosystems [59]. Indeed, the sustainability of green hydrogen production plants is more related to adequate water quality in terms of conductivity than to being located in areas with abundant renewable resources, which is the parameter used currently.

4. Future Research

This study highlights the issues related to the installation of green hydrogen production plants and the establishment of water quality values to draft the necessary compliance standard specifying the limiting parameters for the quality of the raw input water, as well as the annual volumes of extraction permits that regulate this type of activity in a sustainable manner. We look forward to contributing to the development of this legislation in the future.

5. Conclusions

The following conclusions can be drawn from the work carried out in this research:

1. There is no legislation or literature describing established limit values for hydrogen production. At present, a legal vacuum exists regarding hydrogen generation plants with respect to protocols for which parameters should be analyzed when it comes to sampling reject water. Instead, one must refer to general regulations applicable to industrial production, which do not take into account the specific characteristics of electrolysis processes, which result in an increase in the concentration of salts and very high conductivity. Therefore, we recommend the development of specific regulatory frameworks to define such parameters and help drive hydrogen production, reducing the uncertainty faced by companies in this sector; this would be highly beneficial for society, as it would guarantee a sustainable and responsible process.

2. From the analysis of the parameters of the water produced during generation of hydrogen, it can be concluded that conductivity is the most important parameter when it comes to classifying the viability of using a given water source for green hydrogen production, as this affects both the technical and economic viability of the electrolysis process. A criterion has thereby been established to classify raw water from different locations according to its potential for green hydrogen production on the basis of conductivity.

3. Tests in a pilot plant confirmed that, for input water conductivities above 900 μS/cm, the production of reject water and waste from the filters would compromise the viability of the process by greatly increasing the operating costs, the consumption of raw materials, and the generation of waste not suitable for direct discharge into the receiving environment, making such green hydrogen production plants unfeasible. A comparative calculation of the average amount of water required to produce 1 kg of renewable hydrogen using water of different quality and different types of electrolyzer revealed values between 21 and 58 L. We have proposed a classification of natural water from various locations based on their conductivity to determine the potential usefulness of water from these locations for green hydrogen production.

4. An algorithm was developed and validated in Python to enable the automatic determination of the amount of reject water production, as well as the consumption of membranes required during the production process. The results of the algorithm indicated the flow behavior and the flow rates and conductivities in each part of the plant, with a maximum difference of 5% with respect to the real values, which will facilitate decision-making regarding location selection and the initial stage of project evaluation. This study involved the selection of 20 (algorithm development) + 26 (validation of the algorithm) water sampling points on the basis of their variability and proximity to renewable sources and water access. This underlines the importance of prior detailed hydrogeological analyses to ensure adequate water supply and minimize the impact of water stress, especially in vulnerable regions.

5. This paper has demonstrated that the quality of the input water used at the pretreatment plant and the efficiency of the purification process (reverse osmosis and EDI) are crucial not only to the performance of the PEM electrolyzer but also when it comes to minimizing the volume of reject water. The need to manage large volumes of reject water, especially when not meeting the parameter requirements for direct discharge, can drastically reduce the profitability and sustainability of such plants.

6. The pilot plant designed in this study includes a detailed monitoring system, and an algorithm to estimate the amount of reject and useful water was developed. These tools are vital for real-time decision-making, the fine-tuning of treatment processes, and prediction of performance on the basis of the input water quality.

6. Patents/Records

The algorithm resulting from this research has been registered under file number 00765-02622024 on 6 June 2025.

Author Contributions: Conceptualization, M.Á.-M. and M.F.-R.; Methodology: M.Á.-M., M.F.-R. and G.B.-S.; Investigation, M.Á.-M. and M.F.-R.; Resources, M.Á.-M.; Writing—Original Draft Preparation, M.F.-R. and G.B.-S.; Writing—Review & Editing, M.Á.-M., G.B.-S. and M.F.-R. Supervision, G.B.-S. and M.F.-R. Project administration, M.F.-R.; All authors have read and agreed to the published version of the manuscript.

Funding: This research received no external funding.

Data Availability Statement: The data presented in this study are available on request from the corresponding author.

Acknowledgments: We thank Global Virtualizza Ingenieros for their technical support and expertise in green hydrogen projects and water resource management, which were essential for the design and implementation of the pilot plant and the experimental validation in this study. Their specialized collaboration was key to the quality and relevance of the results obtained. This work also benefited from institutional support and laboratory facilities that enabled the experimental analyses required to characterize the water samples and validate the predictive algorithm developed.

Conflicts of Interest: The authors state that they have no conflicts of interest.

Abbreviations

The following abbreviations are used in this manuscript:

EDI	Electrodeionization
IEA	International Energy Agency
AEL/AWE	Alkaline electrolyzers
PEM	Polymeric exchange membrane
AEM	Anion electrolysis membrane
SOEC	Solid oxide electrolyzer cells

References

1. Jacobson, M.Z. *100% Clean, Renewable Energy and Storage for Everything*; Cambridge University Press: Cambridge, UK, 2020.
2. Kharissova, A.B.; Kharissova, O.; Kharisov, B.; Méndez, Y.P. Carbon negative footprint materials: A review. *Nano-Struct. Amp Nano-Objects* **2024**, *37*, 101100. [CrossRef]
3. Intergovernmental Panel on Climate Change. *Climate Change 2021—The Physical Science Basis: Working Group I Contribution to the Sixth Assessment Report of the Intergovernmental Panel on Climate Change*, 1st ed.; Cambridge University Press: Cambridge, UK, 2023. [CrossRef]
4. Bouckaert, S.; Pales, A.F.; McGlade, C.; Remme, U.; Wanner, B.; Varro, L.; D'Ambrosio, D.; Spencer, T. *Net Zero by 2050—A Roadmap for the Global Energy Sector*; International Energy Agency: Paris, France, 2021.
5. Consejo Europeo, 12 y 13 de Diciembre de 2019—Consilium. Available online: https://www.consilium.europa.eu/es/meetings/european-council/2019/12/12-13/ (accessed on 10 January 2024).
6. TA-9-2021-0241_ES.pdf. Available online: https://www.europarl.europa.eu/doceo/document/TA-9-2021-0241_ES.pdf (accessed on 10 January 2024).
7. REPowerEU. Available online: https://commission.europa.eu/topics/energy/repowereu_en (accessed on 8 July 2025).
8. United Nations. Adoption of the Paris Agreement. 2015. Available online: https://unfccc.int/resource/docs/2015/cop21/eng/l09r01.pdf (accessed on 26 January 2024).
9. Terlouw, T.; Bauer, C.; McKenna, R.; Mazzotti, M. Large-scale hydrogen production via water electrolysis: A techno-economic and environmental assessment. *Energy Environ. Sci.* **2022**, *15*, 3583–3602. [CrossRef]
10. Becker, H.; Murawski, J.; Shinde, D.V.; Stephens, I.E.L.; Hinds, G.; Smith, G. Impact of impurities on water electrolysis: A review. *Sustain. Energy Fuels* **2023**, *7*, 1565–1603. [CrossRef]
11. Lima, A.; Torrubia, J.; Valero, A.; Valero, A. Non-Renewable and Renewable Exergy Costs of Water Electrolysis in Hydrogen Production. *Energies* **2025**, *18*, 1398. [CrossRef]
12. Lim, T.; Kim, S.-K. Non-precious hydrogen evolution reaction catalysts: Stepping forward to practical polymer electrolyte membrane-based zero-gap water electrolyzers. *Chem. Eng. J.* **2022**, *433*, 133681. [CrossRef]
13. Marouani, I.; Guesmi, T.; Alshammari, B.M.; Alqunun, K.; Alzamil, A.; Alturki, M.; Abdallah, H.H. Integration of Renewable-Energy-Based Green Hydrogen into the Energy Future. *Processes* **2023**, *11*, 2685. [CrossRef]
14. Naqvi, S.R.; Kazmi, B.; Taqvi, S.A.A.; Chen, W.-H.; Juchelková, D. Techno economic analysis for advanced methods of green hydrogen production. *Curr. Opin. Green Sustain. Chem.* **2024**, *48*, 100939. [CrossRef]
15. Nasser, M.; Megahed, T.F.; Ookawara, S.; Hassan, H. A review of water electrolysis–based systems for hydrogen production using hybrid/solar/wind energy systems. *Environ. Sci. Pollut. Res.* **2022**, *29*, 86994–87018. [CrossRef]
16. Chi, J.; Yu, H. Water electrolysis based on renewable energy for hydrogen production. *Chin. J. Catal.* **2018**, *39*, 390–394. [CrossRef]
17. Oliveira, A.M.; Beswick, R.R.; Yan, Y. A green hydrogen economy for a renewable energy society. *Curr. Opin. Chem. Eng.* **2021**, *33*, 100701. [CrossRef]
18. Egert, F.; Ullmer, D.; Marx, S.; Taghizadeh, E.; Morawietz, T.; Gerle, M.; Le, T.A.; Schneider, L.P.C.; Biswas, I.S.; Wirz, R.E.; et al. Intensification of Alkaline Electrolyzer with Improved Two-Phase Flow. *Adv. Energy Mater.* **2025**, *15*, 2405285. [CrossRef]
19. Ceballos-Alvarez, C.; Maziar, J.; Moradizadeh, L.; Siaj, M.; Shahgaldi, S.; Izquierdo, R. Enhanced Graphene Oxide-Nafion® membranes for proton exchange membrane water electrolysis. *J. Membr. Sci.* **2025**, *734*, 124267. [CrossRef]
20. Ceballos-Alvarez, C.; Jafari, M.; Siaj, M.; Shahgaldi, S.; Izquierdo, R. Influence of Graphene Oxide on Mechanical and Morphological Properties of Nafion® Membranes. *Nanomaterials* **2025**, *15*, 68. [CrossRef] [PubMed]
21. Ma, W.; Tian, L.; Zhang, Q.; Shangguan, L.; Wang, F.; Zhu, H. Hydrophilic and hydrophobic dual-side-linked poly(carbazole-butanedione) anion exchange membranes for water electrolysis and fuel cells applications. *J. Membr. Sci.* **2025**, *722*, 123840. [CrossRef]
22. Li, L.; Watanabe, M.; Inada, M.; Ishihara, T. Planer type solid oxide cells using La0.9Sr0.1Ga0.8Mg0.2O3-δ thin-film electrolyte prepared by dip-coating method for high performance CO_2/H_2O co-electrolysis. *Appl. Catal. Gen.* **2025**, *694*, 120146. [CrossRef]

23. Magnino, A.; Marocco, P.; Santarelli, M.; Gandiglio, M. Economic viability and CO_2 emissions of hydrogen production for ammonia synthesis: A comparative analysis across Europe. *Adv. Appl. Energy* **2025**, *17*, 100204. [CrossRef]
24. Mosiałek, M.; Hanif, M.B.; Šalkus, T.; Kežionis, A.; Kazakevičius, E.; Orliukas, A.F.; Socha, R.P.; Łasocha, W.; Dziubaniuk, M.; Wyrwa, J.; et al. Synthesis of Yb and Sc stabilized zirconia electrolyte (Yb0.12Sc0.08Zr0.8O2–δ) for intermediate temperature SOFCs: Microstructural and electrical properties. *Ceram. Int.* **2023**, *49*, 15276–15283. [CrossRef]
25. Brey, J.J. La electrolisis como una herramienta clave de la descarbonización del siglo XXI. *Econ. Ind.* **2022**, *424*, 87–96.
26. Rey, J.; Segura, F.; Andújar, J. Green Hydrogen: Resources Consumption, Technological Maturity, and Regulatory Framework. *Energies* **2023**, *16*, 6222. [CrossRef]
27. Clerici, A.; Furfari, S. Future energy scenarios, impact on hydrogen development and EU green energy strategy. In Proceedings of the en 2021 AEIT International Annual Conference (AEIT), Milan, Italy, 4–8 October 2021; pp. 1–6. [CrossRef]
28. Garrote, L.; Iglesias, A.; Granados, A. Country-level assessment of future risk of water scarcity in Europe. In *Proceedings of IAHS*; Copernicus GmbH: Göttingen, Germany, 2018; pp. 455–462. [CrossRef]
29. Kim, Y.; Jung, S.; Kim, S.; Choi, S.T.; Kim, M.; Lee, H. Heat transfer performance of water-based electrospray cooling. *Int. Commun. Heat Mass Transf.* **2020**, *118*, 104861. [CrossRef]
30. Zhang, Y.; Zhang, Y.; Li, Z.; Yu, E.; Ye, H.; Li, Z.; Guo, X.; Zhou, D.; Wang, C.; Sha, Q.; et al. A Review of Hydrogen Production via Seawater Electrolysis: Current Status and Challenges. *Catalysts* **2024**, *14*, 691. [CrossRef]
31. Solís-Carvajal, C.A.; Pasos, C.A.V.; Ramírez-Navas, J.S. Tecnología de membranas: Ultrafiltración. *Entre Cienc. E Ing.* **2017**, *11*, 26–36. [CrossRef]
32. Cunha, S.; Silva, A.P.; Herráez, C.F.; Pires, V.C.; Chazarra, A.; Mestre, A.; Nunes, L.F.; Mendes, M.T.; Neto, J.; Marques, J.; et al. *Atlas Climático Ibérico—Iberian Climate Atlas*; Portugal Instituto de Meterologia: Lisbon, Portugal, 2011.
33. Real Decreto Legislativo 12001, de 20 de julio, p.pdf. Available online: https://www.boe.es/buscar/pdf/2001/BOE-A-2001-14276-consolidado.pdf (accessed on 14 April 2025).
34. Ismayilov, A.I.; Mamedov, A.I.; Fujimaki, H.; Tsunekawa, A.; Levy, G.J. Soil Salinity Type Effects on the Relationship between the Electrical Conductivity and Salt Content for 1:5 Soil-to-Water Extract. *Sustainability* **2021**, *13*, 3395. [CrossRef]
35. *DOG 129 del 6/07/2012—DECRETO 141/2012*; De 21 de Junio, Por el Que se Aprueba el Reglamento Marco del Servicio Público de Saneamiento y Depuración de Aguas Residuales de Galicia. Xunta de Galicia: Santiago, Spain, 2012. Available online: https://www.xunta.gal/dog/Publicados/2012/20120706/AnuncioCA02-270612-0001_es.html (accessed on 28 May 2024).
36. Martín, P.; Ángel, M. Modelo Distribuido de Simulación del Ciclo Hidrológico y Calidad del Agua, Integrado en Sistemas de Información Geográfica para Grandes Cuencas. Aportación al Análisis de Presiones e Impactos de la Directiva Marco del Agua. ene. Ph.D. Thesis, Universitat Politècnica de València, Valencia, Spain, 2023. Available online: https://riunet.upv.es/handle/10251/191462 (accessed on 2 July 2025).
37. DECRETO 38/2004; de 24 de Febrero, del Gobierno de Aragón, Por el Que se Aprueba el Reglamento de los ver-tidos de Aguas Residuales a las Redes Municipales de Alcantarillado. Available online: https://www.boa.aragon.es/cgi-bin/EBOA/BRSCGI?CMD=VEROBJ&MLKOB=376569325554 (accessed on 8 July 2025).
38. Olías, M.; Cerón, J.C.; Fernández, I. Sobre la utilización de la clasificación de las aguas de riego del U.S. Laboratory Salinity (USLS). *Geogaceta* **2005**, *37*, 111–113.
39. Pérez-Vazquez, E.L.; Gaucín-Delgado, J.M.; Ramírez-Rodríguez, S.C.; Sariñana-Aldaco, M.d.L.Á.; Sifuentes, G.Z.; Valenzuela, E.Z. Electrical conductivity of the nutrient solution its effect on the yield and nutraceutical quality of bell pepper. *Rev. Mex. Cienc. Agríc.* **2020**, *11*, 1669–1675. [CrossRef]
40. Directive—2020/2184—EN—EUR-Lex. Available online: https://eur-lex.europa.eu/eli/dir/2020/2184/oj/eng (accessed on 8 July 2025).
41. *Directiva 98/83/CE*; Del Consejo de 3 de Noviembre de 1998 Relativa a la Calidad de las Aguas Destinadas al Consumo Humano. European Union: Brussels, Belgium, 1998; Volume 330. Available online: http://data.europa.eu/eli/dir/1998/83/oj/spa (accessed on 8 July 2025).
42. Directive—98/83—EN—EUR-Lex. Available online: https://eur-lex.europa.eu/eli/dir/1998/83/oj/eng (accessed on 8 July 2025).
43. Ministerio de Relaciones con las Cortes y de la Secretaría del Gobierno. Real Decreto 1138/1990, de 14 de septiembre, por el que se aprueba la Reglamentación Técnico-Sanitaria para el abastecimiento y control de calidad de las aguas potables de consumo público. *BOE* **1990**, *BOE-A-1990-23231*, 27488–27497. Available online: https://www.boe.es/eli/es/rd/1990/09/14/1138 (accessed on 8 July 2025).
44. BOE-A-2003-3596 Real Decreto 140/2003; de 7 de Febrero, Por el que se Establecen los Criterios Sanitarios de la Calidad del Agua de Consumo Humano. Available online: https://www.boe.es/buscar/act.php?id=BOE-A-2003-3596 (accessed on 8 July 2025).

45. Guerra Jara, D.A. Alteración Física y Fisiológica en un Macroinvertebrado Acuático (Género Smicridea) Como Efecto de Cambios en la Salinidad. Bachelor's Thesis, Universidad de las Américas, TPuebla, Mexico, 2017. Available online: https://www.researchgate.net/publication/315701621_ALTERACION_FISICA_Y_FISIOLOGICA_EN_UN_MACROINVERTEBRADO_ACUATICO_GENERO_SMICRIDEA_COMO_EFECTO_DE_CAMBIOS_EN_LA_SALINIDAD (accessed on 9 July 2025).
46. Cañedo-Argüelles, M.; Prat, N.; Ladrera, R. Impact of potash mining in streams: The Llobregat basin (northeast Spain) as a case study. *J. Limnol.* **2016**, *76*, 343–354. [CrossRef]
47. Ministerio de Medio Ambiente. Real Decreto 606/2003, de 23 de mayo, por el que se modifica el Real Decreto 849/1986, de 11 de abril, por el que se aprueba el Reglamento del Dominio Público Hidráulico, que desarrolla los Títulos preliminar, I, IV, V, VI y VIII de la Ley 29/1985, de 2 de agosto, de Aguas. *BOE* **2003**, *BOE-A-2003-11384*, 22071–22096. Available online: https://www.boe.es/eli/es/rd/2003/05/23/606 (accessed on 8 July 2025).
48. Porta, J.; Herrero, J. *Vulnerability of Soils Under Irrigation*; Springer: Dordrecht, The Netherlands, 1996; pp. 85–96. [CrossRef]
49. Yousefi, M.; Esrafili, A.; Gholami, M.; Mohammadi, A.A.; Khan, N.A.; Baziar, M.; Oskoei, V. Modeling and analysis of the groundwater hardness variations process using machine learning procedure. *Desalination Water Treat.* **2021**, *238*, 135–141. [CrossRef]
50. Casanova, J.F.; Figueroa, A. Determinación Del Caudal Ambiental Y Su Relación Con Variables Indicadoras De Calidad Del Recurso Hídrico. *Luna Azul* **2015**, *40*, 5–24. [CrossRef]
51. Jiao, Y.-J.; Jiang, K.-J.; Chen, S.; Xiang, P.-P.; He, C.-M. Water use in the industrial sector based on the IPAC model under the carbon-neutral transformation path in China. *Adv. Clim. Change Res.* **2024**, *15*, 1130–1146. [CrossRef]
52. Saulnier, R.; Minnich, K.; Sturgess, K. *Water for the Hydrogen Economy*; WaterSMART Solutions Ltd.: Calgary, AB, Canada, 2025; p. 13.
53. Lin, N.; Arzumanyan, M.; Calzado, E.R.; Nicot, J.-P. Water Requirements for Hydrogen Production: Assessing Future Demand and Impacts on Texas Water Resources. *Sustainability* **2025**, *17*, 385. [CrossRef]
54. Newborough, M.; Cooley, G. Green hydrogen: Water use implications and opportunities. *Fuel Cells Bull.* **2021**, *2021*, 12–15. [CrossRef]
55. Azam, A.M.I.N.; Ragunathan, T.; Zulkefli, N.N.; Masdar, M.S.; Majlan, E.H.; Yunus, R.M.; Shamsul, N.S.; Husaini, T.; Shaffee, S.N.A. Investigation of Performance of Anion Exchange Membrane (AEM) Electrolysis with Different Operating Conditions. *Polymers* **2023**, *15*, 1301. [CrossRef] [PubMed]
56. Harrison, S. Electrolyser innovations PEM, alkaline, SOEC, and AEM. In Proceedings of the Presentado en Hydrogen Tech World, Essen, Germany, 26–27 June 2024; Available online: https://hydrogentechworld.com/wp-content/uploads/sites/22/2024/07/1.7.-Stephen-B-Harrison-sbh4-consulting.pdf (accessed on 15 June 2025).
57. Grozier, R.U.; Land, L.F. *Water Resources Activities of the U.S. Geological Survey in Texas Fiscal Years 1982-84*; U.S. Geological Survey: Austin, TX, USA, 1984.
58. Coitiño-López, J.; Barbazán, M.; Ernst, O. Conductividad eléctrica aparente para delimitar zonas de manejo en un suelo agrícola con reducida variabilidad en propiedades físico-químicas. *Agrociencia Urug.* **2015**, *19*, 102–111. [CrossRef]
59. Maru, B.T. Sustainable Production of Hydrogen and Chemical Commodities from Biodiesel Waste Crude Glycerol and Cellulose by Biological and Catalytic Processes. Universitat Rovira i Virgili, Tarragona, Spain. 2013. Available online: https://dialnet.unirioja.es/servlet/tesis?codigo=84088 (accessed on 28 May 2025).

hydrogen

MDPI

Article

Comparative Designs for Standalone Critical Loads Between PV/Battery and PV/Hydrogen Systems

Ahmed Lotfy [1,*], Wagdy Refaat Anis [2], Fatma Newagy [2] and Sameh Mostafa Mohamed [3]

1 Faculty of Engineering, Ain Shams University, Cairo 11517, Egypt
2 Electronics and Communication Engineering Department, Faculty of Engineering, Ain Shams University, Cairo 11517, Egypt; wagdyanis51@yahoo.com (W.R.A.); fatma_newagy@eng.asu.edu.eg (F.N.)
3 Egyptian Chinese College for Applied Technology, Suez Canal University, Ismailia 41522, Egypt; s_m_elsayed@yahoo.com
* Correspondence: alotfy718@gmail.com

Abstract

This study presents the design and techno-economic comparison of two standalone photovoltaic (PV) systems, each supplying a 1 kW critical load with 100% reliability under Cairo's climatic conditions. These systems are modeled for both the constant and the night load scenarios, accounting for the worst-case weather conditions involving 3.5 consecutive cloudy days. The primary comparison focuses on traditional lead-acid battery storage versus green hydrogen storage via electrolysis, compression, and fuel cell reconversion. Both the configurations are simulated using a Python-based tool that calculates hourly energy balance, component sizing, and economic performance over a 21-year project lifetime. The results show that the PV/H_2 system significantly outperforms the PV/lead-acid battery system in both the cost and the reliability. For the constant load, the Levelized Cost of Electricity (LCOE) drops from 0.52 USD/kWh to 0.23 USD/kWh (a 56% reduction), and the payback period is shortened from 16 to 7 years. For the night load, the LCOE improves from 0.67 to 0.36 USD/kWh (a 46% reduction). A supplementary cost analysis using lithium-ion batteries was also conducted. While Li-ion improves the economics compared to lead-acid (LCOE of 0.41 USD/kWh for the constant load and 0.49 USD/kWh for the night load), this represents a 21% and a 27% reduction, respectively. However, the green hydrogen system remains the most cost-effective and scalable storage solution for achieving 100% reliability in critical off-grid applications. These findings highlight the potential of green hydrogen as a sustainable and economically viable energy storage pathway, capable of reducing energy costs while ensuring long-term resilience.

Keywords: PV system; battery size; reliability; electrolyzer; fuel cell; levelized cost of energy

1. Introduction

This paper presents a comparative study of two energy storage systems used in standalone photovoltaic (PV) setups: traditional lead-acid batteries and green hydrogen storage. Both the systems rely on solar energy generated during the day to power loads. This study focuses on two types of loads—the constant load and the night load—while excluding the daily load, as its minimal storage requirements have a negligible impact on the system performance.

To evaluate these storage solutions under challenging conditions, the system is evaluated based on a worst-case weather scenario in Egypt: three consecutive cloudy days, which require storage to cover four full days of demand.

The constant load refers to a 24 h uninterrupted demand, typical of critical applications such as military operations or field hospitals, where reliability is essential. The night load requires the storing of solar energy generated during the day for use at night.

This study includes a comparison of system sizing, battery state-of-charge (SOC), and hydrogen tank capacity, along with performance metrics such as cost and reliability. It emphasizes use cases that demand uninterrupted energy availability, especially in off-grid or critical infrastructure contexts.

Hydrogen is gaining attention across multiple sectors due to its potential to decarbonize energy systems. Green hydrogen is increasingly viewed as a clean energy carrier with applications spanning fuel cells, industrial heating, transportation, and electricity generation. Fuel cells powered by hydrogen offer a zero-emissions alternative to traditional combustion engines, particularly in heavy-duty transport and long-haul applications [1].

Recent technological advancements in green hydrogen production, storage, and distribution are accelerating its integration into modern energy systems. Improved methods for hydrogen storage, such as liquefaction, compression, and chemical storage, are making it more efficient and cost-effective [2,3]. Green hydrogen also addresses a key challenge of solar and wind energy: intermittency, helping to ensure a stable energy supply.

The global energy demand continues to drive fossil fuel use, contributing to climate change. As a result, alternative energy sources like solar and wind are increasingly adopted. However, their variability necessitates effective energy storage solutions to ensure grid reliability [4].

This paper presents a comprehensive mathematical model for both lead-acid battery and hydrogen-based systems. It also compares two energy management strategies: one using batteries for short-term storage, and another employing hydrogen for long-term storage. Among renewable sources, solar energy is advantageous for its year-round availability and zero emissions [5].

PV system output is highly dependent on sunlight and can fluctuate under cloudy conditions, leading to instability [6]. To mitigate this, robust energy storage is essential. While batteries are common in PV setups, they often fall short during prolonged cloudy periods. A promising alternative is the hydrogen fuel cell (HFC), which generates electricity from stored hydrogen and enhances long-term system efficiency.

A Proton Exchange Membrane (PEM) electrolyzer is modeled in this study due to its high efficiency, rapid dynamic response, and compatibility with variable solar input. This makes it well-suited for integration with standalone PV systems. Hydrogen can be produced through water electrolysis and stored in tanks for future use, unlike batteries, which are typically short-term solutions. In PV/HFC systems, excess daytime solar energy powers an electrolyzer to produce hydrogen, which is stored and later converted into electricity via a fuel cell during the night or cloudy periods [7].

PEM (Proton Exchange Membrane) fuel cells offer advantages like fast startup, compact size, and long-term stability [8]. Hydrogen storage is suited for long-term energy needs, while batteries serve medium- and short-term applications [9].

This paper uses a Python-based simulation tool that dynamically calculates the sizing of system components, solar panels, electrolyzers, fuel cells, and batteries. The tool also simulates solar radiation based on the location and the time, and adjusts the energy supply according to environmental inputs and user-defined parameters. It tracks the battery SOC and optimizes the energy management to improve the system performance. Previous studies have explored the techno-economic performance of standalone PV systems using either battery or hydrogen-based storage.

Nasser and Hassan compared the PV/Battery and PV/Hydrogen systems for powering street lighting as a night load in New Borg El-Arab City, Egypt. Although the battery system had a higher energy efficiency (17.8%), the hydrogen system proved more economical, with a lower Levelized Cost of Energy (LCOE) of 1.06 USD/kWh compared to 2.80 USD/kWh for the battery system [10]. The hydrogen system also achieved a shorter payback period of 6.44 years versus 11.7 years, highlighting its cost-effectiveness despite the lower efficiency.

Nousir and Anis developed GUI-based software to calculate the PV and battery sizes for various load types, including constant, daily, and night loads. Their tool provides system sizing, cost analysis, and reliability metrics under real climatic conditions in Egypt. For a 1 kW off-grid system, they found that the LCOEs for daily, night, and constant loads were 0.14 USD/kWh, 1.05 USD/kWh, and 0.60 USD/kWh, respectively [11]. These results underline the higher cost of systems designed for continuous energy supply under adverse weather.

Miled et al. [12] examined a PV–electrolyzer–fuel cell (PVEFC) system for an off-grid green villa in Tunisia. Using HOMER software version 3.14.2, they optimized a system consisting of a 20 kW PV array and a 6 kW fuel cell. The system produced 306 kg of hydrogen annually and met 87% of the villa's energy needs through solar energy. With a total LCOE of 97,577.6 EUR—73% of which was attributed to the fuel cell—the study emphasized the importance of optimal PV tilt angle and solar potential in cost reduction.

Farhani et al. [13] assessed a hybrid solar hydrogen system for an agricultural farm in Kairouan, Tunisia. Their 140 kW PV array generated 234,765 kWh/year and produced 1991 kg of hydrogen annually. The system's capital cost was 809,121,000 EUR, and the optimization process using HOMER Pro confirmed its feasibility for remote applications, despite the high initial investment.

In another study, Farhani et al. [14] evaluated a hybrid solar–wind–hydrogen system in the Tunisian Sahel. With a PV capacity of 3000 kWp and a hydrogen output reaching 101.8 kg/day, the region's favorable solar and wind conditions proved ideal for such hybrid systems. The study highlighted the feasibility of combining renewables with hydrogen storage to achieve a sustainable, off-grid energy supply.

Okonkwo et al. [15] investigated PV system sizing for hydrogen refueling stations in Oman. A 3 MWp grid-connected PV system produced 58,615 kg of hydrogen annually with a Levelized Hydrogen Cost (LHC) of 5.5 EUR/kg. Standalone PV systems with batteries or fuel cells resulted in slightly higher LHCs of 5.74 EUR/kg and 7.38 EUR/kg, respectively. The system also achieved significant CO_2 reductions, illustrating the environmental benefits of integrating green hydrogen.

The main comparison in this study focuses on green hydrogen and lead-acid battery storage systems. A supplementary cost analysis using lithium-ion batteries is also included to highlight the potential trade-offs of more advanced battery technologies.

While several studies have evaluated PV systems with either battery or hydrogen storage, most focus on a single load type or use generic sizing tools such as HOMER. This work introduces a custom-built Python simulation tool that dynamically calculates the system performance under both constant and night load scenarios using actual solar radiation data from Cairo, Egypt. Unlike earlier works, this study simulates a worst-case scenario of 3.5 consecutive cloudy days and compares the two systems in terms of reliability, storage behavior, and economic indicators, including the LCOE and the payback period. The dual comparison across two load profiles under the worst conditions provides a new layer of insight for critical standalone system planning in off-grid environments.

2. System Description

This section compares two standalone energy storage configurations: A PV/Hydrogen (PV/H_2) system and a PV/Battery system.

As shown in Figure 1a., the PV/H_2 system uses solar panels to generate electricity during the day. This electricity powers an electrolyzer, which splits water into hydrogen and oxygen. The hydrogen is then compressed and stored in a tank, which is one of the most widely accepted methods for on-site hydrogen storage.

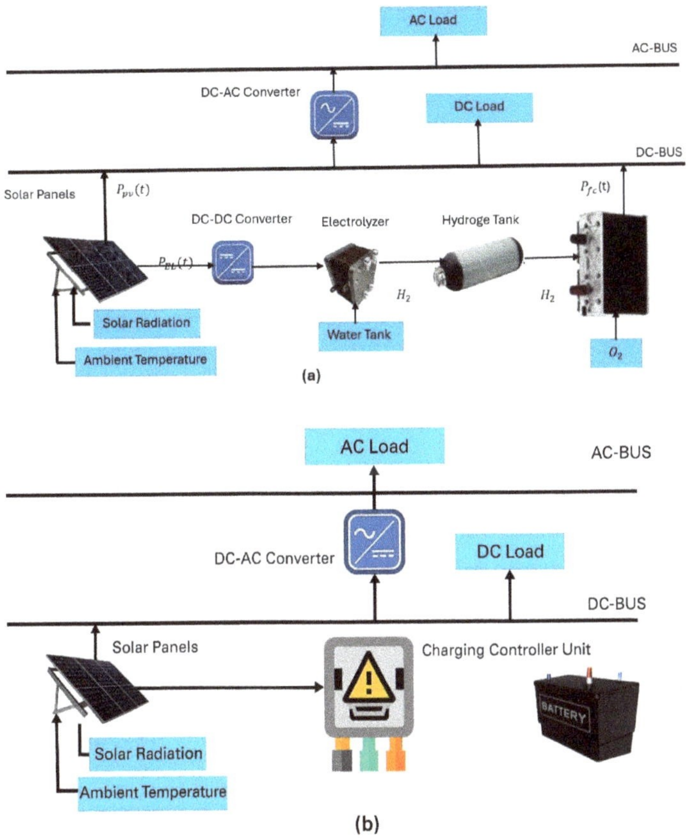

Figure 1. Schematic diagram of the comparison systems (**a**) off-grid PV/H_2 (**b**) off-grid PV/Batteries.

This configuration is designed to store hydrogen during surplus production periods (e.g., summer months) and utilize it during low-production periods (e.g., winter months). At night, or when solar radiation is unavailable, a fuel cell converts the stored hydrogen into electricity to meet both the constant and the night load demands.

As shown in Figure 1b., the PV/Battery system consists of PV panels, a battery storage bank, a charge controller, and a DC/AC inverter. During daylight hours, the PV array supplies power to the load and charges the battery bank. During the night, the stored electricity in the batteries powers the load through the DC/AC converter.

This system is commonly used for off-grid applications or backup power systems, especially where energy demand must be met consistently.

The solar time at the study location in Cairo, Egypt, ranges from 6 to 12 h per day, as shown in Figure 2a: solar radiation is the highest between April and September, leading to increased power generation. The excess energy produced during these periods is stored (in batteries or as hydrogen) for later use during the months with lower solar radiation.

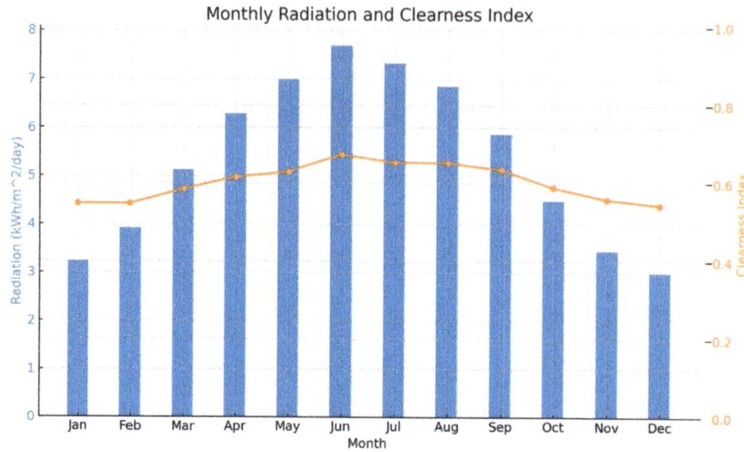

(**a**)

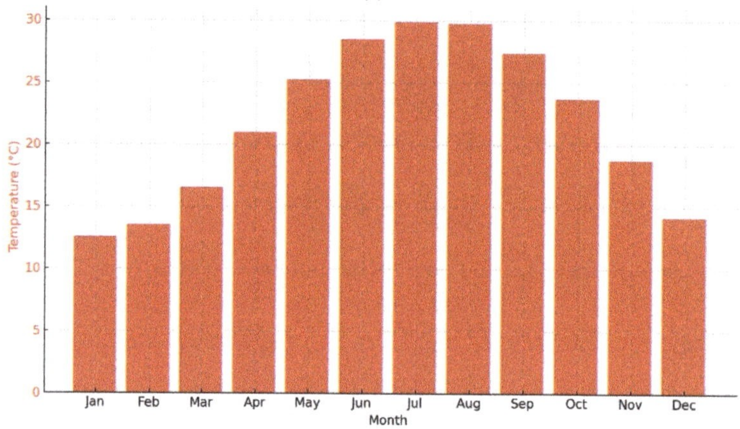

(**b**)

Figure 2. (**a**) Monthly solar global horizontal irradiance @Cairo, Egypt (30° N) (**b**) Monthly average ambient temperature.

Figure 2a,b presents the monthly average global solar radiation (H_T) and the ambient temperature for Cairo, Egypt, based on long-term meteorological data extracted from HOMER software using the Surface Meteorology and Solar Energy (SSE) database [16]. According to simulations using HOMER software, daily solar irradiation at the site reaches a peak of 7.69 kWh/m^2/day in June, while the lowest value is 3.0 kWh/m^2/day in December. A constant 1 kW load is assumed throughout the study.

The fuel cell is modeled based on the electrochemical reaction of hydrogen and oxygen to produce electricity and water. Its efficiency is influenced by several parameters, including: the operating pressure, the current density, the temperature, the fuel cell coefficient, and the hydrogen-to-electricity conversion ratio. These values are selected based on typical commercial fuel cell systems [2–17].

Both the electrolyzer and the fuel cell operate in a loop, where the electrolyzer produces hydrogen during surplus solar generation and the fuel cell consumes it to generate electricity during low or no solar periods. These dynamic supports both the constant and night load scenarios.

3. A Mathematical Model of the System

This study utilizes actual environmental data specific to Cairo, Egypt (Latitude: 30.0444° N, Longitude: 31.2357° E), to simulate and evaluate the performance of two energy storage systems. Three scenarios are considered, corresponding to different numbers of continuous cloudy days (NC = 1, 2, 3) occurring during January and December. These cases are used to investigate the impact of prolonged cloudy conditions on the required storage capacity and overall system cost.

This system is designed to supply a continuous 1 kW load throughout the year, under both constant and night load profiles. To optimize solar energy capture, photovoltaic (PV) panels are installed at a 30° tilt angle facing south, which is typical for this latitude [16].

The simulation is implemented using Python, performing an hourly energy balance analysis over a full year (8760 h). Component sizes, including PV arrays, batteries, electrolyzers, hydrogen storage tanks, and fuel cells, are determined based on the hourly power demand and solar availability.

The simulation process begins with input data: load profiles, solar irradiance, the temperature, and system duration. Control component efficiencies are included to ensure realistic power flow estimates. Both the systems are analyzed under identical load and weather conditions, and an economic model is integrated to evaluate the capital cost, Levelized Cost of Energy (LCOE), and payback period, enabling a comprehensive comparison between the PV/Battery and PV/Hydrogen configurations.

3.1. PV Panels Model

The photovoltaic (PV) array consists of multiple silicon-based modules connected in series and in parallel. Each module is rated at 200 W under Standard Test Conditions (STC), which correspond to solar radiation of 1000 W/m^2, a cell temperature of 25 °C, and an air mass (AM) of 1.5. Each panel generates 8.27 A at 24 V under these conditions [18]. To estimate the solar energy availability, the global radiation on a tilted surface (G_T) is calculated using standard solar geometry equations that account for the hour angle, declination, latitude, and panel tilt angle [19]:

$$G_T = \frac{\pi}{24} H_T\left(\frac{\cos w + \tan б\tan(\varnothing - ß)}{sinW_S' + W_S'\tan б\tan(\varnothing - ß)}\right) \quad (1)$$

where:

H_T Global radiation (Kwh/m^2/day) (direct and diffused radiation on tilted surface)

w hour angle

W_S' The sunset hour angle on tilted surface

б declination angle

Ø latitude of location

ß tilted angle

Figure 3 illustrates the monthly radiation and clearness index, calculated using HOMER-derived data for Cairo. The clearness index represents the ratio between global and extraterrestrial radiation and the seasonal variation of solar radiation on a tilted surface in Cairo. The radiation levels peak during the summer (June–August) and fall in the winter months (November–December), reflecting the impact of seasonal solar availability on energy production. The current, power, and energy generated by the PV array are determined as follows:

$$I = I_{sc0} \times G_T - I_0 \times e^{\frac{V_A}{V_T * \eta}} \quad (2)$$

where:

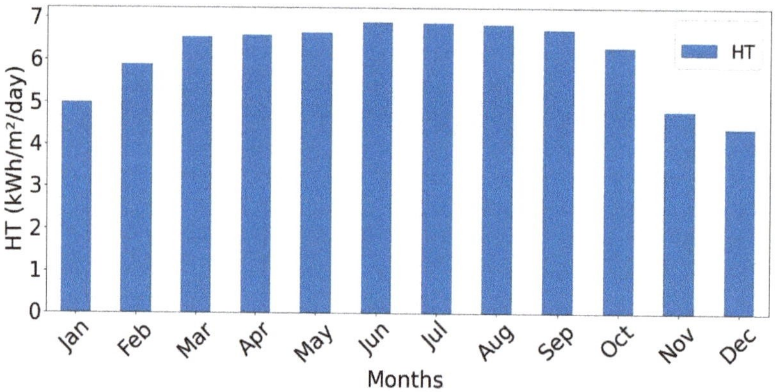

Figure 3. Global radiation @Cairo, Egypt (30° N) (direct and diffused radiation on tilted angle 30°).

I_{sc0} the short circuit current of PV array under standard test condition (STC)
V_A array voltage
ɳ perfection factor
V_T thermal voltage
I_0 reverse saturation current

$$P_A = I \times V_A \tag{3}$$

where: P_A is the array power

$$E = E_0 + P_A \times \Delta_T \tag{4}$$

where:

$\Delta_T = 0.1$ h
E_0 is the energy present in the array at previous (Δ_T)
E output energy from PV array

These equations allow hourly energy output to be tracked over time, accounting for changing solar conditions and system performance.

3.2. Battery Model

The batteries in the standalone PV system store energy for nighttime use. Battery capacity during charging and discharging can be expressed using the following equation:

$$I_B = I - I_L \tag{5}$$

where I_B is battery current and I_L is load current.

A critical parameter for battery sizing is the State of Charge (SOC), which represents the available energy in the battery relative to its capacity. This increases during charging and decreases during discharging. The SOC is updated using [20]:

If $I > I_L$, then the battery is charging, as:

$$SOC_{ch} = SOC_0 + \frac{I_B \times \eta ch \times \Delta_T}{AH} \tag{6}$$

where:

ɳch = Charging efficiency for battery is set at 85%.
AH is the amount of current a battery can provide over a while.
SOC_0 is the ratio of stored energy in the battery at previous moment.

If $I < I_L$, then battery is discharging.

$$SOC_{disch} = SOC_0 - \frac{I_B \times \Delta_T}{\eta_{disch} \times AH} \quad (7)$$

where η_{disch} is the discharging efficiency for the battery set at 85%, and $SOC = 0.8$ for January first.

To ensure reliable operation, the battery SOC is maintained between 20% and 95%, avoiding deep discharges and overcharging. A key design assumption is that the SOC at the start of each year equals the end-of-year SOC (0.8), ensuring energy continuity across annual cycles.

During the winter solstice (21 December), Cairo experiences its longest night, approximately 14 h. In a worst-case scenario of three consecutive cloudy days (NC = 3), the system must store 3.5 days of energy. During cloudy periods, solar radiation is assumed to drop to 15% of the normal values. The Battery Factor (BF)as shown in Equation (8), represents the number of storage hours required, following the formulation in [11].

$$\mathrm{BF} = \frac{Battery\ energy\ (wh)}{average\ load\ power\ (w)} \quad (8)$$

A Python program has been developed, and the result obtained is shown in Figure 4, showing the instantaneous SOC of the battery for one year, with the array size 10 KWp, assuming 6 cloudy days (3 consecutive days during December and 3 consecutive days during January). During cloudy days, the solar irradiance is supposed to be 15% of that during sunny days. It is shown that the SOC drops significantly during the cloudy days in January and December. However, the SOC is always higher than the minimum acceptable value (0.2).

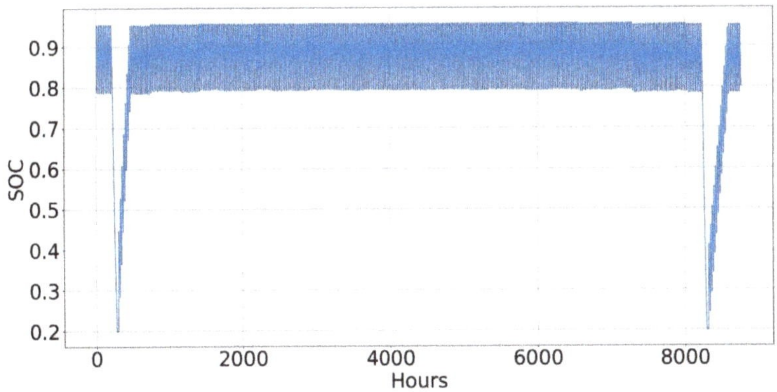

Figure 4. The SOC variation during the entire year @ NC = 3, BF = 140, Array Size 10 kWp, tilted angle 30°.

The battery design assumes the SOC at the beginning of each year remains consistent with that at the year's end (0.8), ensuring no energy loss at the start of the following year.

The design is made so that the SOC never decreases below 0.2, so that the power is always available for the load (i.e., 100% reliability). Also, the design discussed in this work considers the system reliability, which is defined as:

$$\mathrm{Reliability} = \frac{\left(L_d \times d - T_{off}\right)}{L_d * d} \quad (9)$$

L_d Load duration in hours per day

T_{off} Number of hours during which no power is supplied to the load during the year (unmet load)

d Number of days during the year

Figure 5 compares the reliability of the PV/Battery system under two load profiles, the constant and the night load, assuming three consecutive cloudy days (NC = 3). The results show that the reliability increases with the battery capacity in both cases. However, to achieve 97% reliability, the constant load system requires only 30 kWh of battery storage, whereas the night load system requires approximately 50 kWh. This is due to the misalignment between the energy production (daytime) and the energy consumption (nighttime), which makes the night load more demanding in terms of the storage capacity. These findings emphasize the importance of considering the load timing in the battery storage system design.

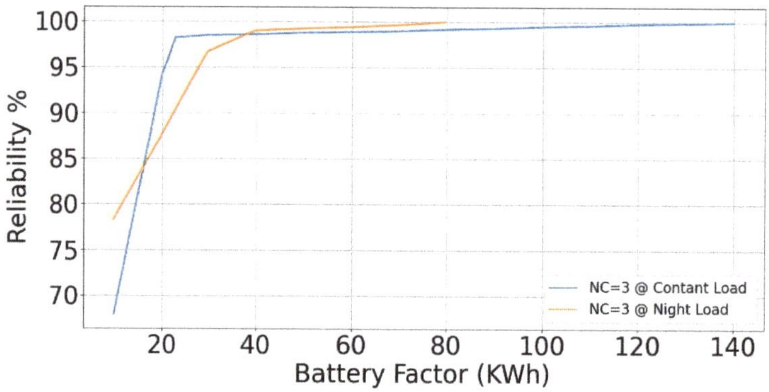

Figure 5. Battery Factor versus Reliability @ different number of cloudy days.

The simulation results show that, for a constant 1 kW load in Cairo (30° N latitude), increasing the battery capacity to 140 kWh with a 10 kWp PV array reduces the annual system downtime $\left(T_{off}\right)$ to zero hours per year. In comparison, under a night load scenario, only 80 kWh of the battery storage with an 8 kWp PV array is required to achieve the same zero-downtime condition. This demonstrates that the night load configuration requires nearly half the storage capacity to maintain full reliability, primarily due to the better alignment between the energy generation and the consumption.

As shown in Tables 1 and 2, the system reliability has a direct and significant impact on the overall cost. Even a small reduction in reliability can lead to substantial cost savings in the battery and PV sizing. The Python-based simulation tool developed in this study dynamically calculates the system reliability versus the cost by tracking the total number of hours during the year when the load is not served $\left(T_{off}\right)$.

Table 1. A 1 kW constant load standalone PV/Battery system (NC = 3).

Battery Factor (BF)	Array Size (kWp)	T_{off} (h)	Reliability	Initial Cost USD	Cost USD/kWh
20	10	516.84	94.1%	10,334.18	0.10
30	10	131.4	98.5%	13,938.78	0.14
40	10	122.64	98.6%	17,543.37	0.17
60	10	96.36	98.9%	24,752.55	0.24
80	10	70.08	99.2%	31,961.73	0.31
120	10	17.52	99.8%	46,380.12	0.45
140	10	0	100%	53,589.28	0.52

Table 2. A 1 kW night load standalone PV/Battery system (NC = 3).

Battery Factor (BF)	Array Size (kWp)	T_{off} (h)	Reliability	Initial Cost USD	Cost USD/kWh
20	8	257.5	97.06%	9709.18	0.25
30	8	151.6	98.27%	13,313.78	0.32
40	8	116.7	98.67%	16,918.37	0.39
50	8	84.1	99.04%	20,522.96	0.46
60	8	66.7	99.24%	24,127.55	0.53
70	8	15.3	99.65%	27,732.14	0.60
80	8	0	100%	31,336.73	0.67

To model realistic worst-case conditions, the simulation assumes three consecutive cloudy days in both January (days 10–12) and December (days 349–351). During these periods, the solar irradiance is reduced to 15% of the normal values, forcing the system to rely solely on stored energy. The tool helps the designers balance the critical load requirements and economic constraints by showing the cost and reliability trade off, depending on the intended application.

3.3. Electricity Controller Unit

The controller's primary function is to supply the electricity efficiently. It includes a DC/DC converter that maximizes the hydrogen production by matching the PV array and the electrolyzer characteristics (power point tracking), along with an inverter for the AC loads. The controller is assumed to operate with an efficiency of 95% [21].

3.4. Electrolyzer

The electrolyzer is directly connected to the DC bus and uses excess electricity generated by the PV panels to produce hydrogen through electrolysis. This hydrogen is then stored for later use in the fuel cell. The DC/DC converter ensures optimal power delivery to both the electrolyzer and the compressor, matching the PV array output via power point tracking [22]. The annual energy consumption of the compressor is approximately 1% of the total generated energy. The electrolysis process splits the water into hydrogen and oxygen, and the hydrogen is stored in a dedicated tank. The hydrogen production rate of the electrolyzer is modeled as:

$$mh_2(t)= \frac{P_{EL} \times \eta_{EL*3600}}{HHV_{H2}} \times \Delta_T \tag{10}$$

where P_{EL} is the rating of electrolyzer (kW), η_{EL} is the efficiency of electrolyzer and HHV_{H2} is the heating value of hydrogen in MJ/kg, respectively. In the mathematical modeling of an electrolyzer, the hydrogen is produced in kg/h.

For a 1 kW electrolyzer and 90% efficient electrolyzer with the heating value of the hydrogen being 143 MJ/kg, which equals 39.44 kWh/kg, the mass flow rate of the hydrogen will be 0.02268 kg/h/kW. Therefore, a 1 kW electrolyzer will produce 0.02268 kg/h hydrogen [2–17], taking into account Δ_T = 0.1 h. The electrolyzer produces hydrogen, which is stored in a proposed tank for night usage to produce electric power through the fuel cell.

3.5. Fuel Cells

The fuel cell subsystem is responsible for converting stored hydrogen into electricity through an electrochemical reaction with oxygen. In this study, a proton exchange membrane (PEM) fuel cell is used, which is known for its compact size, high energy density, and quick startup time [23].

The electrolytic process utilizes a thin membrane that allows positive ions to pass while preventing electrons and neutral gases. Hydrogen from the electrolyzer enters through the anode, while oxygen enters through the cathode, initiating a chemical reaction that dissociates into protons and electrons [24].

This creates a concentration gradient across the membrane, with protons diffusing toward the cathode and electrons flowing from the anode to the cathode, generating an electrical current. The fuel cell model used here is based on a linear approximation of the hydrogen consumption as a function of the power output, rather than a fixed efficiency percentage. The hydrogen consumption rate is calculated as:

$$R_{H2}= \Upsilon_1 \, P_{fc}^{r} + \Upsilon_2 \, P_{fc}(t) \quad (11)$$

where R_{H2} is the mass flow rate of the hydrogen, Υ_1 and Υ_2 are the fuel cell intercept coefficient in kg/h/kW rated and the fuel cell slope curve in kg/h/kW, respectively, and P_{fc}^{r} is the rated capacity of the fuel cell in kW. For a 1 kW FC rating, considering Υ_1 and Υ_2 to be 0.0003 and 0.058 kg/h/kW, respectively, these factors determining the efficiency of the fuel cell, R_{H2} is calculated as 0.059 kg/h. Therefore, 0.059 kg/h of hydrogen is required to generate 1 kW of power by FC [11–19]. The hydrogen low heating value (LHV) is 33.3 kWh/kg [25]. The coefficients Υ_1 and Υ_2 play a crucial role in modeling hydrogen consumption in fuel cell systems. Υ_1 represents the baseline hydrogen consumption rate per unit of rated fuel cell capacity, while Υ_2 accounts for the hydrogen required relative to the fuel cell's actual power output at any given time. These coefficients help to refine the system's hydrogen flow calculations, ensuring a more accurate estimation of hydrogen consumption during both steady-state and variable operating conditions.

This approach better reflects real fuel cell behavior, including idle consumption and dynamic load response. In contrast, fixed-efficiency models (e.g., 50–60%) do not account for changes in operating conditions. By using intercept and slope coefficients, this model enables more accurate sizing and hydrogen flow tracking, which improves the reliability and system optimization.

3.6. Hydrogen Tank

A constant mass flow of hydrogen is required for power generation by FC. Therefore, the size of the hydrogen tank becomes a critical parameter to ensure the reliability of the power generation system. The hydrogen tank stores the hydrogen produced by the electrolyzer and supplies it to the fuel cells for power generation. The hydrogen generated daily is stored to ensure a continuous supply during the periods when solar radiation is unavailable. The size of the hydrogen tank, measured in kilograms, is treated as a decision variable in the system model. The hydrogen storage system designed for this work operates on a small scale, meeting a 1 kW load and ensuring a continuous power supply over 24 h a day, 365 days a year, in constant load and from sunset to sunrise periods in night load. Unlike large-scale industrial hydrogen storage systems, this system operates at low pressure and stores a limited volume of hydrogen, which significantly reduces associated risks. Therefore, while hydrogen storage systems of any scale require careful safety considerations, the risks associated with this small-scale design are substantially lower compared to large-scale applications [26,27]. The modeling of hydrogen gas utilizes steady-state properties and the ideal gas law, assuming ideal behavior under standard conditions. This approach simplifies the analysis and allows for efficient estimation of hydrogen's thermodynamic behavior during storage and system operations. While the optimal conditions for hydrogen storage have been applied in this design, it is essential to consider that, in practical applications, safety factors such as pressure relief systems, leak detection, and material durability must be rigorously addressed to ensure the system's reliability and safe operation.

The Python-based simulation tool developed in this study is fully flexible and user-configurable. It allows users to input different values for key system parameters such as the tilt angle, the latitude, PV module characteristics, the battery depth of discharge (DOD), and various electrolyzer and fuel cell efficiencies. This makes the model highly adaptable to different locations, system scales, and operational strategies. It can be used not only for standalone systems under Cairo's climate but also for broader applications across diverse environmental and technical conditions.

3.7. Model Validation and Assumption Justification

The parameters used in this simulation are based on real data and technical specifications, most of which are described in detail in Section 3: hourly solar radiation and temperature data for Cairo. The PV module selected for this study has an open-circuit voltage of 33 V, a short-circuit current of 8.27 A, voltage at a maximum power point of 25 V, and a current at a maximum power point of 8 A [18]. The panel is installed with a fixed 30° tilt angle facing south. Battery efficiency (70% efficiency), electrolyzer sizing assumptions (90% efficiency), and PEM fuel cell behavior (using voltage coefficients) are based on published values and typical commercial systems. The hydrogen consumption was set at 0.058 kg/kWh. Economic inputs, including capital and replacement costs, follow recent literature values. The simulation assumes a worst-case scenario of 3.5 consecutive cloudy days (NC = 3), requiring four days of full autonomy. All the technical values and sources are listed in the relevant system description and model sections.

4. Energy Management Strategy

Efficient and reliable operation of a standalone PV/H_2 system requires a well-defined energy management strategy. This strategy ensures that the load demands are continuously met, either through direct PV generation or through stored energy in the form of hydrogen.

The energy management system monitors the following:

- Power supplied to the load
- Excess PV energy is used for hydrogen production via the electrolyzer
- Power generation by the fuel cell using stored hydrogen

The loads demands must be satisfied by solar PV generation $P_{pv}(t)$ during the day and fuel cells during the night to supply the load. The difference between power generation is calculated as:

$$\Delta P(t) = P_{pv}(t) - P_L(\mathrm{t}) \quad (12)$$

Therefore, depending on $\Delta P(t)$ three Scenarios are formed as follows:

[Scenario:1] If $\Delta P(t) > 0$, i.e., the generation is more than the load, the following steps have been implemented [17].

Step:1 The excess power ($P_{EL} = \Delta P(t)$) is fed to the electrolyzer. The amount of hydrogen generated by the electrolyzer per (Δ_T) as explained in Equation (10).

Step:2 The amount of hydrogen accumulated per Δ_T, ($G_{H2}(t)$) in the storage tank is calculated as:

$$G_{H2}(t) = G_{H2}(t-1) + mh_2(t) \quad (13)$$

where $G_{H2}(t-1)$ is the hydrogen present in the tank at the previous moment.

Step:3 In the case of storing hydrogen as shown in Equation (10), the volume of hydrogen becomes larger than the volume of the tank (G^r_{H2}), i.e., $G_{H2}(t) > G^r_{H2}$, Then the electrolyzer is not turned on because there is no volume available in the hydrogen tank.

[Scenario:2] If $\Delta P(t) = 0$, then no power exchange is there; the total demand is met by the solar power generation only.

[Scenario:3] If $\Delta P(t) < 0$, the load demand is more than the power generated by the PV cells, and the required power to meet the load is provided by fuel cells and the PV during the day; during the night all power is supplied by the fuel cell. The power is managed as follows:

Step:1 As a first step, check the hydrogen availability in the hydrogen tank. The power required to be produced by the fuel cell is calculated as:

$$P_{fc}(\mathrm{t}) = P_L(\mathrm{t}) - P_{pv}(t) \quad (14)$$

Step:2 The amount of hydrogen consumed by the FC from the hydrogen tank at a particular time to produce P_{fc}(t) is given by Equation (14). The hydrogen storage tank is updated after taking hydrogen; the remaining hydrogen in the tank is calculated as:

$$G_{H2}(t) = G_{H2}(t-1) - \mathrm{R}_{H2}(t) \quad (15)$$

where $G_{H2}(t-1)$ is the hydrogen in the tank at time $(t-1)$.

Step:3 In case the hydrogen tank does not have sufficient hydrogen i.e., $\mathrm{R}_{H2} \geq G_{H2}(t)$, this was taken into account in the design to avoid any energy loss because the load is critical.

To ensure efficient energy management and reduce the energy loss, careful sizing of all major system components is essential. These include photovoltaic panels (P_{pv}), fuel cells (P^r_{fc}), the electrolyzer (P_{EL}), and hydrogen storage tank size (G^r_{H2}).

These parameters were tuned to handle worst-case weather conditions, specifically three consecutive cloudy days on 10–12 January and December 349–351. These scenarios are embedded in the Python version 3.13.5-based simulation, which tracks the hydrogen levels, the battery SOC (for comparison), and the power flows hourly.

Figure 6 illustrates the full energy flow logic of the system, showing the generation, storage, and load management under dynamic environmental and load conditions.

Figure 7 illustrates the full-year hydrogen production and storage cycle, assuming three consecutive cloudy days in both January and December. At the start of the year, the hydrogen production and storage levels are low due to reduced solar radiation in the winter. As the spring and the summer arrive, increased solar energy enables greater hydrogen generation, allowing the system to accumulate surplus hydrogen. This reserve is then used in the autumn and the winter to maintain the load supply. The hydrogen storage level at the end of the year closely matches the starting value, ensuring energy continuity into the following year. This seasonal balance confirms the system's ability to meet load demand year-round with no energy loss.

The results clearly show that the hydrogen storage is at its minimum during the winter and reaches its maximum in the summer, due to higher solar energy availability in the summer months. Notably, the amount of stored hydrogen at the end of the year is approximately equal to the amount at the beginning of the next year, indicating that the hydrogen generated annually is fully utilized to meet the load demand.

Figures 7 and 8 present an analysis based on a system configuration under the assumption of three consecutive cloudy days. The system operates with a constant load of 1 kW. During the day, the photovoltaic (PV) array has a capacity of 10 kWp, and the electrolyzer is sized at 8 kW. For the night load, the array size is reduced to 8 kWp, and the electrolyzer is downsized to 5 kW. These sizing choices are made to ensure that the maximum current generated by the solar panels can be effectively utilized. Additionally, a 1 kW fuel cell is included to match the load demand during the periods without sunlight.

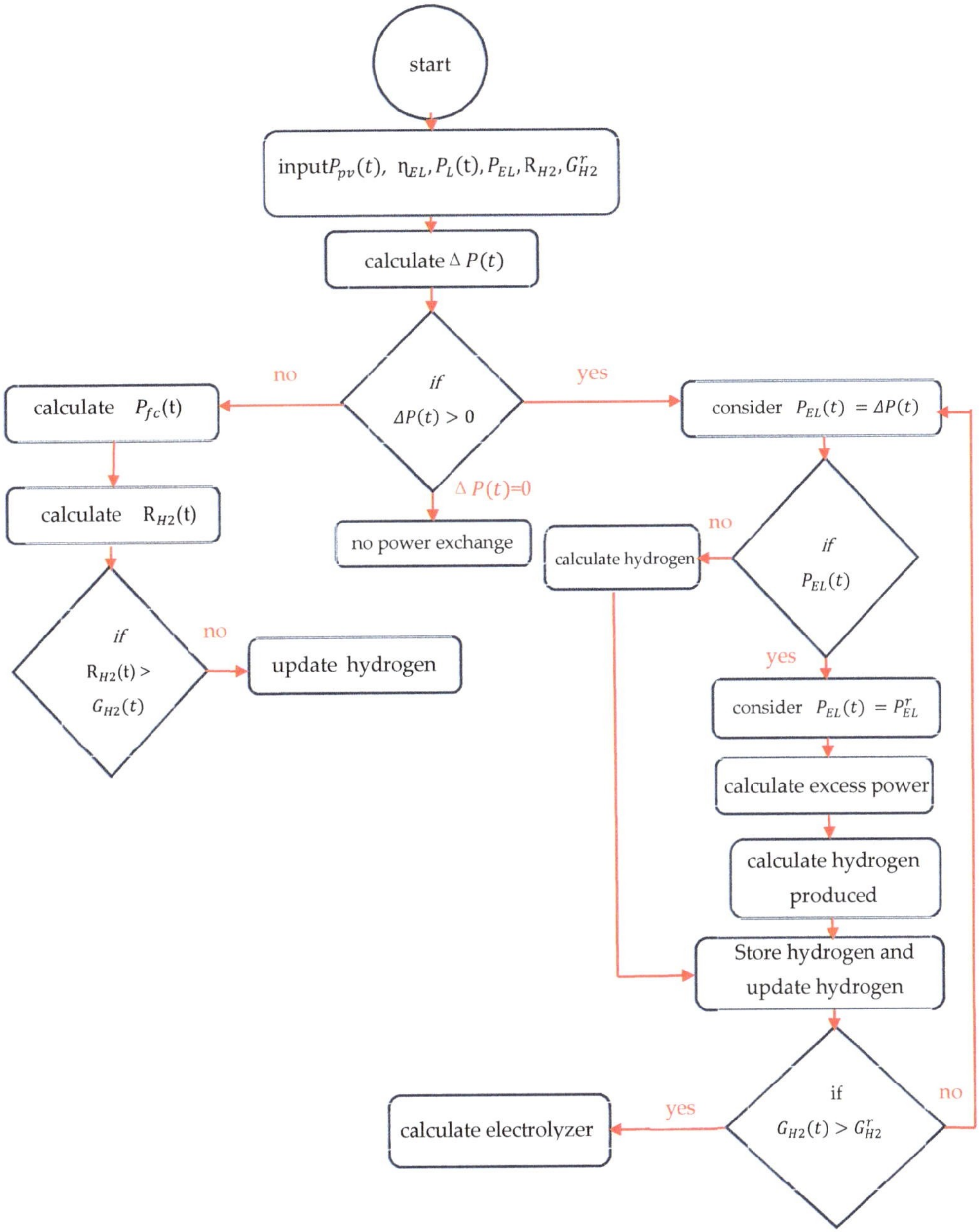

Figure 6. Flow chart for energy management of the proposed system.

The hydrogen tank begins the year with an initial capacity of 15 kg and reaches 19 kg by the end of the year for a constant load. For the night load, the tank starts at 11.5 kg and increases to 17.6 kg at the end of the year. The results show that the hydrogen production fluctuates across the four seasons, primarily influenced by the varying energy output of the PV system. This PV-generated energy plays a critical role in driving the electrolyzer and determining the overall hydrogen yield, sufficient hydrogen to meet the load demand during the solar radiation period.

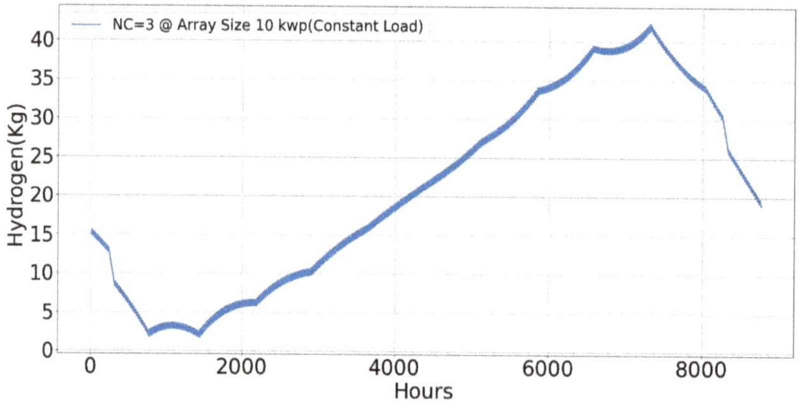

Figure 7. State of hydrogen entire year @ NC = 3, array size 10 kWp, tilted angle 30°.

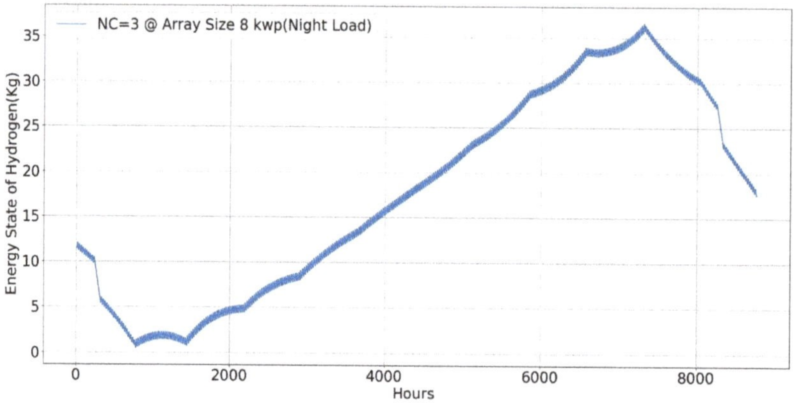

Figure 8. State of hydrogen entire year @ NC = 3, array Size 8 kWp night load, tilted angle 30°.

Figure 9a,b illustrates the accounts for three cloudy days in January (10th, 11th, and 12th) and three days in December (349th, 350th, and 351st), respectively; these figures highlight the impact of reduced solar irradiance on hydrogen production and storage. During these periods, solar energy generation drops significantly, leading to a complete halt in hydrogen production. Consequently, the system draws upon the existing hydrogen stored in the tank to supply the load through the fuel cell. This rapid decline in the hydrogen storage demonstrates the system's reliance on prior storage to maintain an uninterrupted energy supply under extended low-solar conditions. These periods represent the worst-case climatic scenarios considered in the system design to ensure reliability during prolonged cloudy days.

Figures 10 and 11 show the monthly hydrogen production under varying seasonal conditions, emphasizing the effect of cloudy days in winter. The hydrogen production drops significantly during the cloudy periods in January and December due to reduced solar input. However, the system's sizing ensures that enough hydrogen is stored in advance to handle these events. The design maintains a sufficient end-of-year hydrogen reserve to guarantee the energy availability at the beginning of the following year, demonstrating resilience against Cairo's most adverse solar conditions.

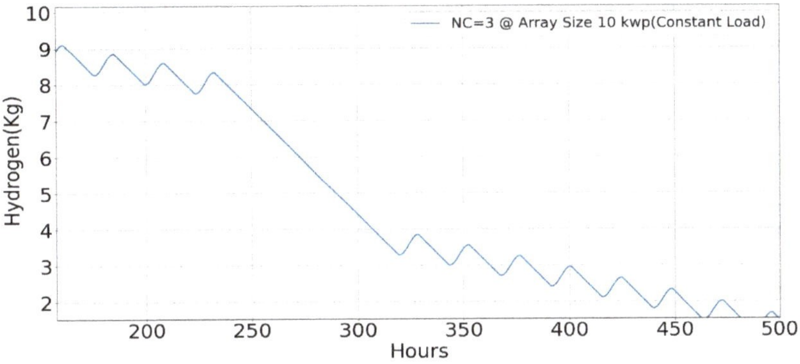

(**a**)

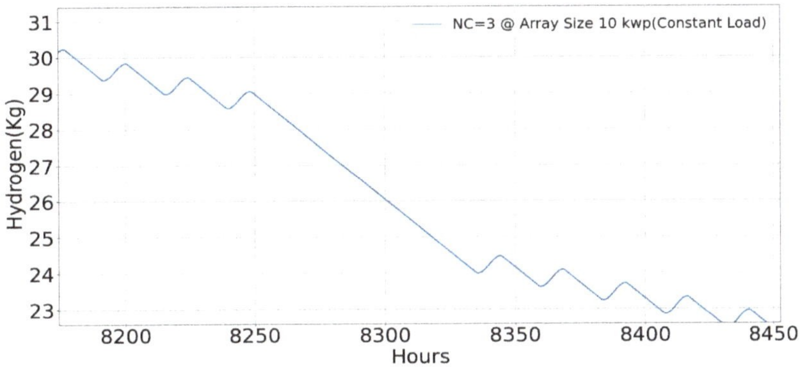

(**b**)

Figure 9. (**a**) Three successive cloudy days in January. (**b**) Three successive cloudy days in December.

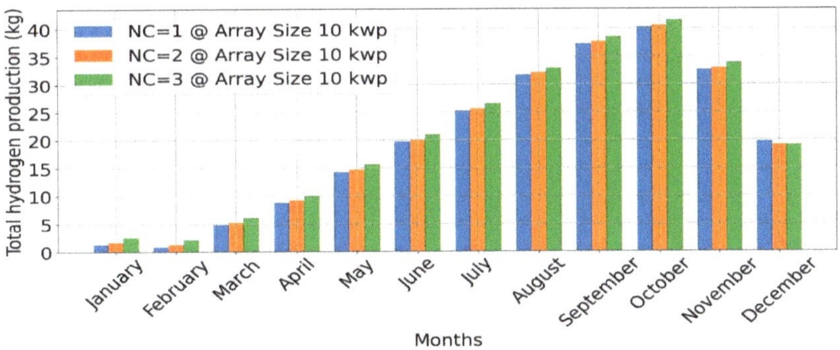

Figure 10. The hydrogen production during the solar radiation period by month through the year @ NC = 1&2&3, constant load 1 KW, array size 10 kWp, tilted angle 30°.

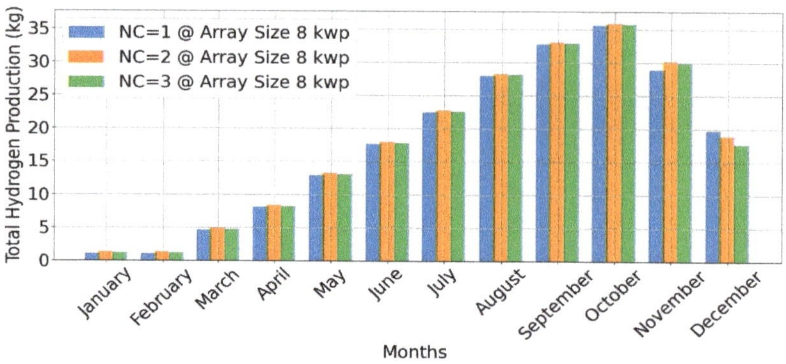

Figure 11. The hydrogen production during the solar radiation period by month through the year @ NC = 1&2&3, constant load 1 KW, array size 10 kWp night load, tilted angle 30°.

5. Economic Analysis

This section presents a cost-based comparison between the PV/H_2 and PV/Battery systems under constant and night load conditions. The reference system is designed to supply a 1 kW continuous load throughout the year, including worst-case weather conditions. All the initial costs include mounting and installation.

The total cost of each system includes the capital cost of the PV array and other major components such as batteries, Proton Exchange Membrane Fuel Cell (PEMFC), Proton Exchange Membrane Electrolyzer (PEMEZ), converters, and hydrogen tanks. Table 3 summarizes the initial costs and lifespans for all components. A 21-year project lifetime and an annual interest rate of 6% are assumed, with an operation and maintenance (O&M) cost of 5% of the component's initial cost as reported in [10].

Table 3. Initial cost and lifetime of both system components [18].

Component	Initial Cost	Lifetime (Years)
PV	312 USD/Kw	25
PEMEZ	1000 USD/Kw	20
PEMFC	1000 USD/Kw	10
Converter	146 USD/Kw	10
H_2 Tank	570 USD/Kg	20
Compressor	1800 USD/Kw	20
Battery	100 USD/Kw	3

The component requiring a replacement in the PV/Hydrogen system is the fuel cell and DC/DC Converter. The FC operates based on hours rather than years. The configuration results in 50,000 h of fuel cell operation throughout the year, necessitating replacement at the end of the 10th year. The effective interest over each 10-year period is 0.8% for the FC and DC/DC Converter.

For the PV/Battery system, the batteries require replacement every three years, with an effective interest rate of 0.19% per replacement cycle. Replacement costs are determined according to established equations; all other components share the system's full 21-year lifetime and are not replaced.

To incorporate the effect of photovoltaic (PV) performance degradation over time, the Levelized Cost of Energy (LCOE) was calculated using a present worth factor that includes an annual degradation rate. This is represented by Equation (16)

$$P_0 = E_0 \times C_0 \times (\frac{1-\left(\frac{1+g}{1+i}\right)^N}{i-g}) \tag{16}$$

where:

P_0 Total cost over 21 years
E_0 Average yearly generated energy (kWh)
C_0 Cost of Energy per kWh
i Yearly interest rate
g the annual degradation rate (assumed 2% for PV systems in Egypt)

This formula accounts for the typical 2% annual drop in the PV output, which is often caused by environmental factors such as dust accumulation and surface glass abrasion from sandstorms. If degradation is neglected (i.e., g = 0), the equation reduces to the conventional annuity factor used in Equation (17):

$$P_0 = E_0 \times C_0 \times (\frac{1-(1+i)^{-N}}{i}) \tag{17}$$

Tables 4 and 5 present the impact of adding three cloudy days to the capital investments for each system configuration. The impact of system reliability was factored in when comparing the two loads and their effect on the cost of energy. The Levelized Cost of Electricity (LCOE) is used to estimate the cost of providing 1 kW of electricity. The results show that the PV/H_2 system is more cost-effective than the PV/Battery configuration for critical applications. For a constant load with several cloudy days, NC = 3, the LCOE values are 0.23 USD/kWh for the PV/H_2 system and 0.52 USD/kWh for the PV/Battery system. In the case of the night load, the LCOE increases to 0.36 USD/kWh for the PV/H_2 system and 0.67 USD/kWh for the PV/Battery system.

Table 4. A 1 kW constant load standalone PV/hydrogen system (NC = 3).

Tank Size	Array Size (kWp)	T_{off} (h)	Reliability	Initial Cost USD/kWh	Cost USD/kWh
2	10	219	97.5%	13,255.37	0.13
4	10	148	98.3%	14,395.37	0.14
6	10	132	98.5%	15,535.37	0.15
8	10	74	99.16%	16,675.37	0.16
10	10	68	99.22%	17,815.37	0.17
13	10	7	99.99%	19,525.37	0.19
15	10	0	100%	20,665.37	0.20
19 (End of The First Year)	10	0	100%	22,945.37	0.23

Table 5. A 1 kW night load standalone PV/hydrogen system (NC = 3).

Tank Size	Array Size (kWp)	T_{off} (h)	Reliability	Initial Cost USD/kWh	Cost USD/kWh
2	8	164	96.6%	9630.37	0.19
4	8	115	97.37%	10,770.37	0.21
6	8	89	97.97%	11,910.37	0.23
8	8	54	98.77%	13,050.37	0.25
10	8	19	99.57%	14,190.37	0.28
11.5	8	0	100%	15,045.37	0.29
17.6 (End of The First Year)	8	0	100%	18,522.37	0.36

As a supplementary economic estimate, the performance of lithium-ion batteries was also considered. Using typical values from the recent literature, 400 USD/kWh capital cost, and a 13-year lifespan [20] calculated an approximate Levelized Cost of Electricity (LCOE), which was calculated using the same energy demand profiles as in the lead-acid simulation. The resulting LCOE was 0.41 USD/kWh for the constant load and 0.49 USD/kWh for the night load, representing a 21% and a 27% cost reduction, respectively, compared to lead-acid batteries.

Lithium-ion batteries offer clear advantages in terms of the energy density, longer life cycle, and reduced replacement frequency. However, they also require advanced thermal management, especially in hot climates, due to their sensitivity to high operating temperatures, which can reduce the performance and the safety over time [28,29]. While this estimate is not derived from a full simulation, it provides a realistic economic reference, reinforcing the potential of lithium-ion as an improved battery storage option, though hydrogen storage remains more cost-effective in this study's scenarios.

The total cost of the hydrogen-based system is primarily driven by the cost of the hydrogen storage tank, in addition to the costs of the electrolyzer, the fuel cell, and the photovoltaic (PV) components. The overall system cost can be expressed as the sum of these individual components. Among them, the tank cost increases linearly with the required storage size and tends to dominate the total cost, especially in scenarios with extended cloudy day requirements. Since the number of hydrogen tanks must be an integer, this introduces stepwise changes in the total cost, while the remaining cost variation is attributed to the continuous variation in the other system components.

The higher LCOE observed in the battery-based system is primarily due to the substantial number of batteries required to meet the energy demand, their relatively short operational lifespan, and the significant costs associated with battery replacement over the system's lifetime.

5.1. Payback Period

The Payback Period (PBP) reflects how quickly the system investment is recovered. It is calculated as:

$$P_0 = E_0 \times C_P \times \left(\frac{1-(1+i)^{-PBP}}{i}\right) \tag{18}$$

where:

C_P Cost of Energy multiplied by the profit of the system.
PBP Payback Period in years

5.2. Economic Sensitivity Analysis

A sensitivity analysis was conducted to assess how changes in solar irradiance, hydrogen storage capacity, and electricity costs affect the economic feasibility of the systems. This analysis revealed that the areas with lower solar irradiance may experience higher LCE values, suggesting the need for regional optimization and adaptation to local conditions. The criterion used to assess the economic feasibility of any system is the payback period. The shorter the payback period, assuming an annual interest rate of 6%, the better the investment.

Figures 12 and 13 show the constant load. The PBP is 7 years for the PV/H_2 system and 16 years for the PV/Battery system. In the case of the night load, the PBP increases to 7.2 years for the PV/H_2 system and 15 years for the PV/Battery system.

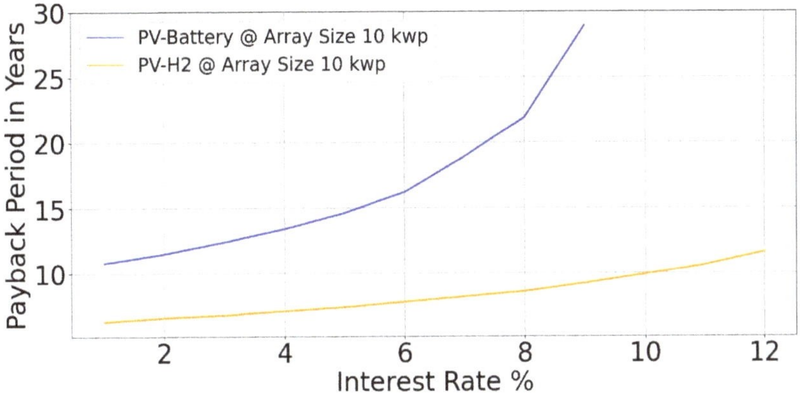

Figure 12. The payback period of the two considered systems for the constant load versus the interest rate.

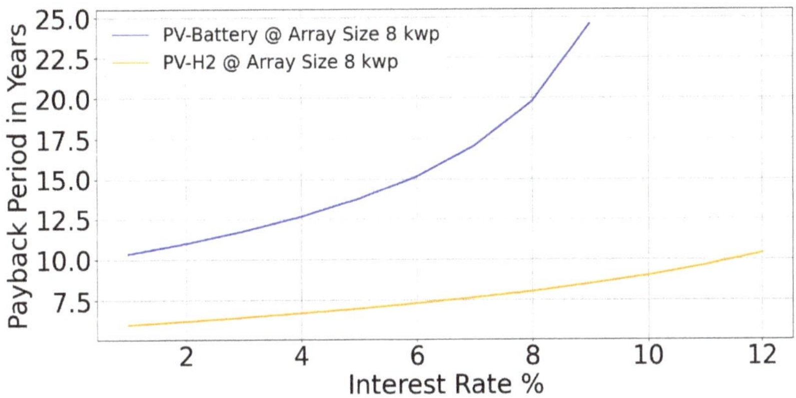

Figure 13. The payback period of the two considered systems for the night load versus the interest rate.

The payback period of the PV/Battery system is always longer than that of the PV/H_2 system. The main reason for this is the high cost and short lifetime of the batteries, as the battery cost is the dominant factor in the PV/Battery system. The short lifespan of the batteries requires their replacement every three years. On the other hand, in the PV/H_2 system, the hydrogen tanks are the primary cost, but the tank lifetime is much longer, with no need for replacement.

Figure 14 shows the effect of the selling price of one kWh on the payback period for both systems for a constant load. It is clear that as the selling price increases (i.e., an increase in profit), Figure 14 illustrates that the payback period (PBP) is a vital indicator of the project feasibility. The calculations were performed using established equations, with a profit of 20% selected to facilitate a comparison between the two systems under various load conditions. This percentage was chosen based on the highest observed LCOE, which corresponds to the battery-based configuration.

The results for the constant load are as follows: the PBP is 5.3 years for the PV/H_2 system and 9.7 years for the PV/Battery system. In the case of the night load, the PBP increases to 6.3 years for the PV/H_2 system and 12.2 years for the PV/Battery system.

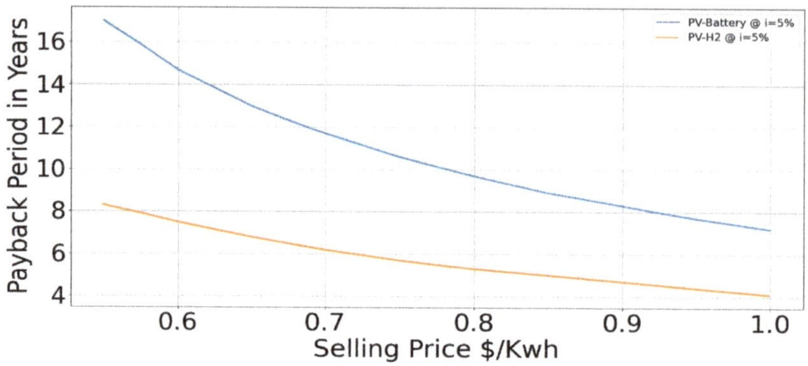

Figure 14. Payback period in years versus the selling price for the constant load @i = 5%.

As shown in Figure 15, the payback period the PV/H_2 is shorter than that of the PV/Batteries at different interest rates and different load profiles.

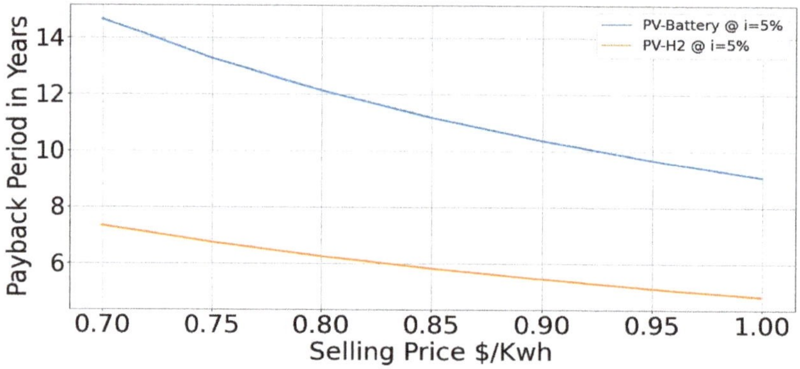

Figure 15. Payback period in years versus selling price for night load @i = 5%.

5.3. Inflation Causes

Inflation and interest rate changes affect the component replacement costs differently, depending on the system.

PV/Battery system: since the batteries are replaced multiple times, the cost increases with inflation. The energy cost rises to 0.30 USD/kWh (the constant load) and 0.40 USD/kWh (the night load) at an average battery replacement interest rate.

PV/H_2 system: the hydrogen tank, with its long lifetime, is less sensitive to inflation. Cost estimates are 0.37 USD/kWh (the constant load) and 0.60 USD/kWh (the night load), assuming a 6% interest rate and a 50 kg tank.

Figures 16 and 17 illustrate these economic sensitivities. The PV/Battery system cost is closely tied to battery inflation, while the PV/H_2 system is more capital-heavy upfront but less sensitive over time. While this study focuses on small-scale, low-pressure hydrogen storage suitable for standalone systems, the results offer an insight into how scaling affects economic performance. At the grid scale, the economies of scale would likely reduce the capital cost per unit of hydrogen storage, particularly in tank fabrication and system integration. As a result, the LCOE and the payback period could improve, especially for systems with large energy demand and long-duration storage needs. However, such systems introduce additional factors such as high-pressure storage and advanced safety systems, which were not modeled in this study. Future work may expand the analysis to grid-scale applications, incorporating those considerations.

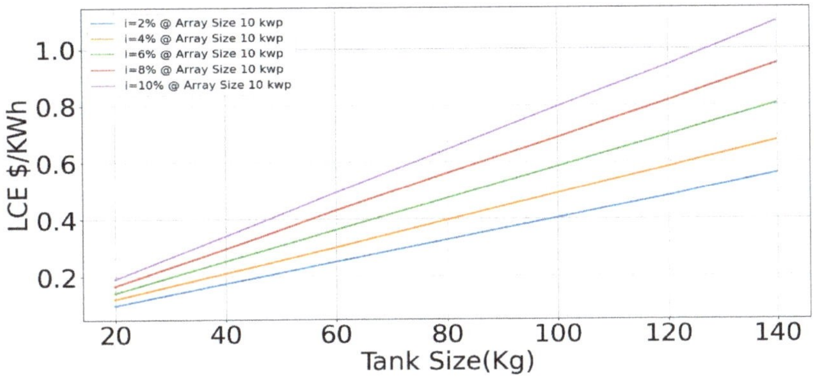

Figure 16. The Levelized Cost of Energy versus the tank size for the constant load.

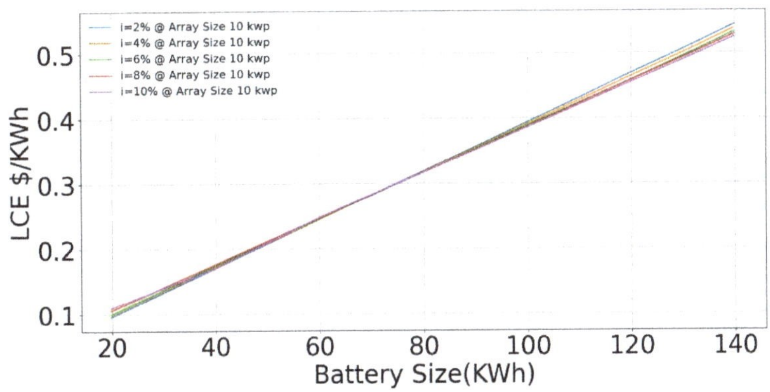

Figure 17. The Levelized Cost of Energy versus the battery size for the constant load.

6. Conclusions

This study presents a comparative techno-economic analysis of two standalone photovoltaic (PV) systems designed to supply critical loads, one using lead-acid batteries and the other using green hydrogen as storage. Both the systems were modeled and simulated under Cairo's climatic conditions, assuming worst-case scenarios of 3.5 consecutive cloudy days. The systems were evaluated for two load profiles: the constant load and the night load, to achieve 100% reliability and reduce energy loss.

The simulation results show that the PV/H_2 system consistently outperforms the PV/Battery system in both cost and performance. For the constant load, the Levelized Cost of Electricity (LCOE) is reduced from USD 0.52/kWh (Battery) to USD 0.23/kWh (Hydrogen). For the night load, the LCOE improves from 0.67 USD/kWh to 0.36 USD/kWh. This confirms that the night load is inherently more expensive to support due to its reliance on stored energy during solar-off hours.

Payback period analysis further supports the advantage of hydrogen. At a 6% interest rate, the PV/H_2 system achieves a payback period of 7 years for the constant load, compared to 16 years for the PV/Battery system. With a 20% profit margin included, this drops to 5.3 years for hydrogen, versus 9.7 years for batteries. Night load scenarios follow a similar trend, with hydrogen offering significantly shorter cost recovery times.

In terms of reliability–cost trade-offs, the analysis shows that small compromises in reliability can significantly reduce the system cost. For instance, increasing the tank size from 2 kg to 15 kg (the constant load) improves the reliability from 97.5% to 100%, but raises the initial cost by 55% and the LCOE by 0.13 USD/kWh. A similar pattern is observed in night load scenarios, emphasizing the need to balance the system cost with the criticality of the application.

The economic sensitivity analysis reveals that hydrogen systems are more resilient to inflation and interest rate changes. While battery systems suffer from frequent replacements and compounding costs, hydrogen systems front-load most expenses, benefiting from longer component lifespans, especially the hydrogen tank. Even under extended cloudy day conditions and variable inflation rates, the PV/H_2 system maintains a lower LCE than its battery-based counterpart.

In the current simulation, energy loss is defined as the duration or amount of unmet load (T_{off}), which is set to zero in the final optimized scenarios. However, the real-world losses occur through system inefficiencies (e.g., inverter loss, DC-DC converter loss, electrolyzer/fuel cell conversion, and curtailment when hydrogen storage is full). These losses are embedded in the efficiency parameters of each component (e.g., 90% electrolyzer efficiency, 95% inverter, and 95% DC-DC converter), but excess PV energy that cannot be stored is not explicitly tracked as curtailed energy. Future work could quantify annual energy loss due to curtailment and other inefficiencies more explicitly.

This study compares hydrogen storage with traditional lead-acid batteries due to their lower initial cost and continued use in small-scale off-grid systems. In addition to the core comparison between hydrogen storage and lead-acid batteries, a supplementary economic estimate was conducted for lithium-ion batteries to provide a broader perspective on storage alternatives. The results showed that lithium-ion storage reduces the Levelized Cost of Electricity (LCOE) compared to lead-acid by approximately 21% for the constant load and 27% for night-only load scenarios. While lithium-ion offers improved the energy density, higher efficiency, and longer lifespan, it also introduces operational challenges, particularly in hot climates like Egypt, where thermal management becomes critical. These findings suggest that lithium-ion technology may offer a viable alternative for certain applications. Still, green hydrogen remains the most cost-effective and scalable solution across both load profiles in this study.

While this study focuses on the techno-economic performance of hydrogen and battery storage systems, we recognize the importance of assessing their environmental impacts. Battery manufacturing, particularly for Li-ion systems, involves the mining and processing of rare materials, which can result in significant environmental burdens. Hydrogen produced via electrolysis (green hydrogen) is cleaner during operation, but still requires energy and water input. A full life-cycle environmental analysis comparing hydrogen and battery systems is an important area for future research. Such analysis could complement the economic findings and provide a more comprehensive sustainability perspective.

The PV/H_2 system offers a more reliable, economically viable, and future-resilient solution for standalone critical-load applications. Its superior long-term performance under adverse weather and economic conditions makes it a strong candidate for off-grid, high-reliability energy systems.

Author Contributions: Conceptualization, A.L. and W.R.A.; methodology, A.L.; software, A.L.; validation, A.L., W.R.A. and S.M.M.; formal analysis, A.L.; investigation, A.L.; resources, S.M.M.; data curation, A.L.; writing—original draft preparation, A.L.; writing—review and editing, W.R.A.; visualization, F.N.; supervision, W.R.A.; project administration, A.L. and S.M.M. All authors have read and agreed to the published version of the manuscript.

Funding: This research received no external funding.

Data Availability Statement: The original contributions presented in the study are included in the article; further inquiries can be directed at the corresponding author.

Conflicts of Interest: The authors declare that they have no known competing financial interests or personal relationships that could have appeared to influence the work reported in this paper.

References

1. Angelico, R.; Giametta, F.; Bianchi, B.; Catalano, P. Green Hydrogen for Energy Transition: A Critical Perspective. *Energies* **2025**, *18*, 404. [CrossRef]
2. Kumar, N.; Lee, S.Y.; Park, S.J. Advancements in hydrogen storage technologies: A comprehensive review of materials, methods, and economic policy. *Nano Today* **2024**, *56*, 102302. [CrossRef]
3. Alam, S.N.; Khalid, Z.; Singh, B.; Guldhe, A. Integration of government policies on the global level for green hydrogen production. In *Green Hydrogen Economy for Environmental Sustainability*; American Chemical Society: Washington, DC, USA, 2024; Volume 1: Fundamentals and Feedstocks, pp. 1–28.
4. Masters, G.M. *Renewable and Efficient Electric Power Systems*, 2nd ed.; John Wiley & Sons: Hoboken, NJ, USA, 2013.
5. Ozden, E.; Tari, I. Energy–exergy and economic analyses of a hybrid solar–hydrogen renewable energy system in Ankara, Turkey. *Appl. Therm. Eng.* **2016**, *99*, 169–178. [CrossRef]
6. Rakhshani, E.; Rouzbehi, K.; Sánchez, A.J.; Tobar, A.C.; Pouresmaeil, E. Integration of large-scale PV-based generation into power systems: A survey. *Energies* **2019**, *12*, 1425. [CrossRef]
7. Rahman, S.; Tam, K.S. A feasibility study of photovoltaic-fuel cell hybrid energy system. *IEEE Trans. Energy Convers.* **1988**, *3*, 50–55. [CrossRef] [PubMed]
8. Sun, L.; Jin, Y.; Pan, L.; Shen, J.; Lee, K.Y. Efficiency analysis and control of a grid-connected PEM fuel cell in distributed generation. *Energy Convers. Manag.* **2019**, *195*, 587–596. [CrossRef]
9. Hassan, Q.; Jaszczur, M.; Hafedh, S.A.; Abbas, M.K.; Abdulateef, A.M.; Hasan, A.; Abdulateef, J.; Mohamad, A. Optimizing a microgrid photovoltaic-fuel cell energy system at the highest renewable fraction. *Int. J. Hydrogen Energy* **2022**, *47*, 13710–13731. [CrossRef]
10. Nasser, M.; Hassan, H. Assessment of standalone streetlighting energy storage systems based on hydrogen of hybrid PV/electrolyzer/fuel cell/desalination and PV/batteries. *J. Energy Storage* **2023**, *63*, 10698. [CrossRef]
11. Nousir, S.E.; Anis, W.R.; Abouelatta, M.; Adly, G.S.; Abdelaal, M.A. Developing Techno-economic GUI-. based Software for On-grid, and Off-grid Photovoltaic Systems Sizing. *Ain Shams Eng. J.* **2024**, *15*, 102508. [CrossRef]
12. Miled, A.; Farhani, S.; Askri, F. Theoretical modeling and economic analysis of PV and fuel cell energy system for an off-grid green villa in Sidi Bouzid, Tunisia-A case study. *Int. J. Hydrogen Energy* **2025**, *105*, 932–945. [CrossRef]
13. Farhani, S.; Barhoumi, E.M.; Islam, Q.U.; Becha, F. Optimal design and economic analysis of a stand-alone integrated solar hydrogen water desalination system case study agriculture farm in Kairouan Tunisia. *Int. J. Hydrogen Energy* **2024**, *63*, 759–766. [CrossRef]
14. Farhani, S.; Grissa, H.; Ouda, M.; Bacha, F. A techno-economic feasibility study of electricity and hydrogen production in hybrid solar-wind energy park. The case study of Tunisian Sahel. *Eng. Technol. Appl. Sci. Res.* **2024**, *14*, 15154–15160. [CrossRef]
15. Barhoumi, E.M.; Okonkwo, P.C.; Ben Belgacem, I.; Zghaibeh, M.; Tlili, I. Optimal sizing of photovoltaic systems based green hydrogen refueling stations case study Oman. *Int. J. Hydrogen Energy* **2022**, *47*, 31964–31973. [CrossRef]
16. Stackhouse, P.W.; Whitlock, C.H.; Huang, D.; Mikovitz, J.C. *Surface Meteorology and Solar Energy, Version 6.0 Methodology*; NASA Technical Report NASA/TM—2008-215961; NASA Langley Research Center: Hampton, VA, USA, 2008. Available online: https://ntrs.nasa.gov/citations/20080012200 (accessed on 25 June 2025).
17. Singh, S.; Chauhan, P.; Singh, N. Capacity optimization of grid connected solar/fuel cell energy system using hybrid ABC-PSO algorithm. *Int. J. Hydrogen Energy* **2020**, *45*, 10070–10088. [CrossRef]
18. Kazem, H.A.; Chaichan, M.T.; Al-Waeli, A.H.A.; Sopian, K.; Alnaser, W.E.; Kazmerski, L.; Alnaser, N.W. A review of photovoltaic/thermal (PV/T) incorporation in the hydrogen production process. *Glob. Energy Interconnect.* **2025**, *8*, 363–393. [CrossRef]
19. Duffie, J.A.; Beckman, W.A.; Blair, N. *Solar Engineering of Thermal Processes, Photovoltaics and Wind*; John Wiley & Sons: Hoboken, NJ, USA, 2020.
20. Han, X.; Garrison, J.; Hug, G. Techno-economic analysis of PV-battery systems in Switzerland. *Renew. Sustain. Energy Rev.* **2022**, *158*, 112028. [CrossRef]
21. Tebibel, H. Methodology for multi-objective optimization of wind turbine/battery/electrolyzer system for decentralized clean hydrogen production using an adapted power management strategy for low wind speed conditions. Energy Conversion and strategy for low wind speed conditions. *Energy Convers. Manag.* **2021**, *238*, 114125.

22. Nasser, M.; Megahed, T.F.; Ookawara, S.; Hassan, H. A review of water electrolysis–based systems for hydrogen production using hybrid/solar/wind energy systems. *Environ. Sci. Pollut. Res.* **2022**, *29*, 86994–87018. [CrossRef]
23. Ahmadi, M.H.; Mohammadi, A.; Pourfayaz, F.; Mehrpooya, M.; Bidi, M.; Valero, A.; Uson, S. Thermodynamic analysis and optimization of a waste heat recovery system for proton exchange membrane fuel cell using transcritical carbon dioxide cycle and cold energy of liquefied natural gas. *J. Nat. Gas Sci. Eng.* **2016**, *34*, 428–438. [CrossRef]
24. EG&G Technical Services, Inc. *Fuel Cell Handbook*, 7th ed.; U.S. Department of Energy Office of Fossil Energy National Energy Technology Laboratory: Morgantown, WV, USA, 2004.
25. Al-Orabi, A.M.; Osman, M.G.; Sedhom, B.E. Analysis of the economic and technological viability of producing green hydrogen with renewable energy sources in a variety of climates to reduce CO_2 emissions: A case study in Egypt. *Appl. Energy* **2023**, *338*, 120958. [CrossRef]
26. Magliano, A.; Perez Carrera, C.; Pappalardo, C.M.; Guida, D.; Berardi, V.P. A comprehensive literature review on hydrogen tanks: Storage, safety, and structural integrity. *Appl. Sci.* **2024**, *14*, 9348. [CrossRef]
27. Emma, D.; Ehrmann, A.; Schwenzfeier-Hellkamp, E. Safety of Hydrogen Storage Technologies. *Processes* **2024**, *12*, 2182. [CrossRef]
28. Liu, J.; Zhou, J.; Li, M.; Shao, W.; Cao, X.; Ding, J. Lithium-ion battery combustion with different state of charge and combustion prevention based on flame arrester. *Therm. Sci. Eng. Prog.* **2025**, *61*, 103550. [CrossRef]
29. Ortiz, Y.; Arévalo, P.; Peña, D.; Jurado, F. Recent advances in thermal management strategies for lithium-ion batteries: A comprehensive review. *Batteries* **2024**, *10*, 83. [CrossRef]

hydrogen

MDPI

Review

Thermo-Fluid Dynamics Modelling of Liquid Hydrogen Storage and Transfer Processes

Lucas M. Claussner [1,*], Giordano Emrys Scarponi [2] and Federico Ustolin [1,*]

1 Department of Mechanical and Industrial Engineering, Norwegian University of Science and Technology NTNU, Richard Birkelands vei 2b, 7034 Trondheim, Norway
2 Department of Civil, Chemical, Environmental, and Materials Engineering—DICAM, Alma Mater Studiorum—University of Bologna, Via Terracini 28, 40131 Bologna, Italy
* Correspondence: lucas.claussner@ntnu.no (L.M.C.); federico.ustolin@ntnu.no (F.U.)

Abstract

The use of liquid hydrogen (LH_2) as an energy carrier is gaining traction across sectors such as aerospace, maritime, and large-scale energy storage due to its high gravimetric energy density and low environmental impact. However, the cryogenic nature of LH_2, with storage temperatures near 20 K, poses significant thermodynamic and safety challenges. This review consolidates the current state of modelling approaches used to simulate LH_2 behaviour during storage and transfer operations, with a focus on improving operational efficiency and safety. The review categorizes the literature into two primary domains: (1) thermodynamic behaviour within storage tanks and (2) multi-phase flow dynamics in storage and transfer systems. Within these domains, it covers a variety of phenomena. Particular attention is given to the role of heat ingress in driving self-pressurization and boil-off gas (BoG) formation, which significantly influence storage performance and safety mechanisms. Eighty-one studies published over six decades were analyzed, encompassing a diverse range of modelling approaches. The reviewed literature revealed significant methodological variety, including general analytical models, lumped-parameter models (0D/1D), empirical and semi-empirical models, computational fluid dynamics (CFD) models (2D/3D), machine learning (ML) and artificial neural network (ANN) models, and numerical multidisciplinary simulation models. The review evaluates the validation status of each model and identifies persistent research gaps. By mapping current modelling efforts and their limitations, this review highlights opportunities for enhancing the accuracy and applicability of LH_2 simulations. Improved modelling tools are essential to support the design of inherently safe, reliable, and efficient hydrogen infrastructure in a decarbonized energy landscape.

Keywords: liquid hydrogen; thermodynamics; modelling; multi-phase flow; cryogenic storage

1. Introduction

In response to growing global energy demands and the urgent need for low-carbon solutions, hydrogen has emerged as a pivotal energy carrier in the transition toward sustainable systems. Hydrogen (H_2) offers distinct advantages due to its high gravimetric energy density (120 MJ/kg) and relatively low environmental impact [1]. Compared to compressed gaseous hydrogen (CGH_2 at 700 bar), LH_2 provides superior volumetric energy density (8 MJ/L vs. 5.6 MJ/L), making it particularly attractive for applications in aerospace, maritime transport, and large-scale energy storage [2]. Despite these benefits, the cryogenic nature of LH_2—requiring storage at approximately 20 K (−253 °C)—introduces significant thermodynamic and safety challenges [3]. These include complex heat and mass transfer processes, phase transitions, and pressure dynamics, all of which are further complicated by LH_2's chemical properties such as high diffusivity, low ignition energy, and wide flammability limits. As a result, the safe and efficient handling, storage, and transfer of LH_2 demand a deep understanding of interdependent physical phenomena and robust modelling approaches. Table 1 provides a comparison of key properties of different fuels, including CGH_2 at 700 bar, LH_2, natural gas (NG) at 250 bar, liquefied natural gas (LNG), and gasoline.

Table 1. Comparison of safety- and efficiency-relevant properties of some energy carriers [1,2,4].

Properties	CGH_2 (700 bar)	LH_2	NG (250 bar)	LNG	Gasoline
Storage temperature	Ambient	Ca. −253 °C	Ambient	Ca. −162 °C	Ambient
Gravimetric energy density (MJ/kg)	120		53.6		44
Volumetric energy density (MJ/L)	5.6	8	9	22.2	32.0
			At atm. conditions		
Flammability limits in air (vol%)	4–74		5–15		1–7
Minimum ignition energy in air (mJ)	0.02		0.30		0.30
Stoichiometric flame speed (m/s)	2.1		0.4		0.3
Diffusion coefficient in air (cm^2/s)	0.61		0.16		0.05

To date, no comprehensive review has systematically examined the modelling techniques used to simulate LH_2 behaviour across its storage and transfer lifecycle. Existing studies often focus on isolated phenomena—such as heat ingress, boil-off gas (BoG) formation, or sloshing—without fully addressing the interconnected nature of these processes. This fragmentation limits the development of integrated models capable of supporting the design of inherently safe and efficient LH_2 infrastructure.

This literature review aims to fill that gap by consolidating and critically analyzing modelling approaches across two principal domains:

1. Thermodynamic behaviour within LH_2 storage tanks;
2. Multi-phase flow dynamics in storage and transfer systems.

Within these domains, the review explores seven focus areas: heat transfer, mass transfer (evaporation/condensation), thermal stratification and pressurization, venting and depressurization, sloshing, LH_2 transfer operations, and flash boiling with pressure recovery. These phenomena are not independent; rather, they interact dynamically, influencing system performance and safety outcomes. Recognizing and modelling these interdependencies is essential for accurate simulation and practical application.

The objectives of this review are as follows:

- Map the current landscape of LH_2 modelling techniques;
- Evaluate the extent of model validation against experimental data;
- Identify persistent knowledge gaps;
- Provide guidance for future research and model development.

By synthesizing insights from over six decades of research, this review contributes to the advancement of LH_2 technologies and supports the development of safer, more reliable, and economically viable hydrogen systems in a decarbonized energy future.

2. State of the Art

This section provides an overview of insulation techniques of LH_2 storage and transfer systems, followed by an introduction to the focus areas that are subject to the reviewed literature and the corresponding simulation models. Each focus area and the connected phenomena are thereafter presented in more detail. One of the main concerns when handling LH_2 is the evaporation of the liquid phase due to unavoidable heat ingress. The evaporated LH_2 is called boil-off gas (BoG) and is a driving factor of pressurization. Cryogenic tanks, commonly used to store substances such as LH_2 or LNG, are typically constructed as double-walled containers. In the case of LH_2 transfer pipes, this concept extends to double-walled piping. The space between the inner and outer wall (annular space) is filled with insulation material and, depending on the insulation material, maintained under vacuum. An overview of the different insulation materials is provided in a work by Claussner et al. [5]. The performance of the various insultation materials can be supplemented with additional efforts. A popular method to further improve the insulation characteristics is the deployment of vapour-cooled shields (VCS). A VCS is a thermal shield placed inside the multi-layer insulation (MLI) of a cryogenic tank [6]. It uses cold hydrogen vapour, which is already evaporating from the stored LH_2, to intercept and absorb some of the incoming heat before it reaches the inner wall of the cryotank. Studies show VCS can cut heat flux by up to 70%. While multiple VCS layers can improve performance, the benefits diminish beyond two shields [6]. A practical drawback of VCS is increased complexity of tank design. This will affect developmental and manufacturing costs. Further investigation is needed to determine whether this additional cost could be offset—or even surpassed—by savings resulting from improved dormancy behaviour, including a review of experimental studies. Active cooling of an LH_2 tank can be achieved through a two-stage cryocooler setup. A prototype by NASA uses a first stage, a 90 K cryocooler, that is paired with a tube-on-shield (TOS) broad-area cooling (BAC) heat exchanger that intercepts and removes incoming heat before it reaches the propellant tank [7]. The second stage employs a 20 K cryocooler and a tube-on-tank (TOT) heat exchanger to extract heat directly from the LH_2 liquid and ullage space, maintaining stable pressure and preventing boil-off [7].

Seven focus areas were identified during the review process. These individual areas were not strictly equivalent to one specific thermodynamic phenomenon but rather defined by the objective of the authors. In some cases, the area of focus incorporates many thermodynamic phenomena to explain a more complex condition or operational scenario.

The focus areas were divided into two main groups, thermodynamic behaviour of the storage tank and multi-phase flow in storage tank and pipelines:

- Thermodynamic behaviour of the storage tank:
  - Heat transfer;
  - Mass transfer including evaporation/condensation;
  - Thermal stratification and pressurization;
  - Venting and depressurization.
- Multi-phase flow:
  - Sloshing;
  - LH_2 transfer;
  - Flash boiling and pressure recovery.

Figure 1 shows the thermodynamic phenomena inside of LH_2 storage tanks corresponding to the focus areas.

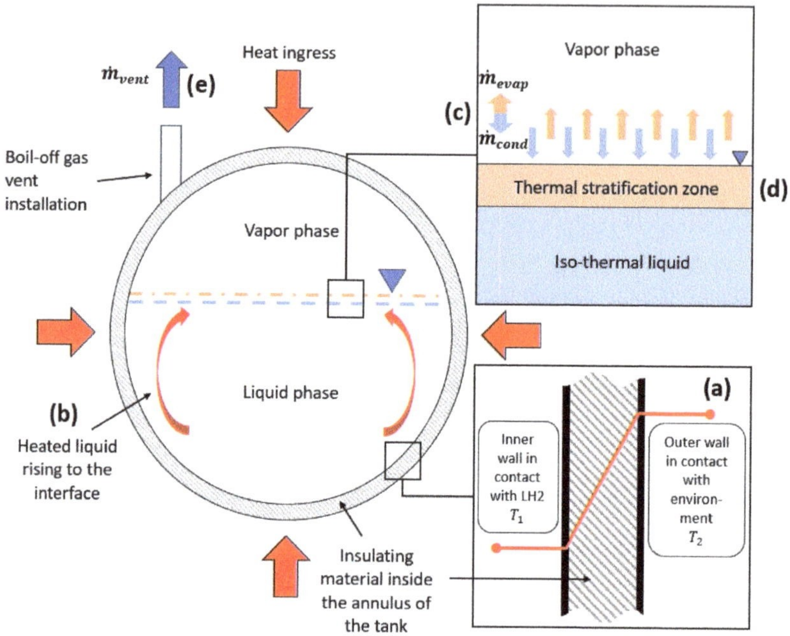

Figure 1. Thermodynamic behaviour of LH_2 storage tank and corresponding phenomena: (**a**) heat transfer into the tank through conduction; (**b**) heat transfer through natural convection; (**c**) evaporation and condensation at the interface; (**d**) thermal stratification, (**e**) venting and tank depressurization.

2.1. Heat Transfer

Heat transfer into the LH_2 storage and transfer equipment plays a central role in leading to adverse effects limiting operability. Even though heat transfer is omnipresent in LH_2 applications, this section focuses on the heat transfer into a storage tank. Heat transfer occurs through conduction, radiation, and convection. Heat conduction is the transfer of heat through direct molecular interactions. In fluids, it occurs via inelastic collisions in nonconducting solids through molecular (lattice) vibrations, and in metals via electron movement (conduction electrons) and molecular collisions [8]. It is the primary heat transfer mechanism in opaque solids, while transparent solids may also transfer heat via radiation. Conduction is quantified using Fourier's law, with thermal conductivity determining efficiency. The law is based on two key quantities: temperature and heat flow, expressed as

$$\Delta Q = \kappa \Delta T \tag{1}$$

$$q = -k \nabla T \tag{2}$$

where Q represents the rate of heat transfer across a temperature gradient T, determined by thermal conductivity κ [9]. To calculate heat transfer by conduction through a plane wall, Fourier's law of heat conduction applies where L is the thickness of the wall:

$$Q = \frac{\kappa A (T_1 - T_2)}{L} \tag{3}$$

The surface area of heat transfer is given by A, and the wall temperatures at the inner and outer side are given by T_1 and T_2, respectively. L refers to the wall thickness. Figure 1a shows a schematic of the heat ingress into an LH_2 tank, mainly through conduction. For LH_2 applications, conduction heat transfer occurs through the wall of a tank or a pipe, even though the annulus contains insulation material. Furthermore, the support structure of the inner tank and cables for level and temperature sensors act as thermal bridges. Insulating materials like foams and fibres can reduce heat transfer by increasing path length and decreasing cross-sectional area [8]. In VMLI (vacuum multi-layer insulation) of LH_2 tanks, spacers between the layers of insulation material are made from materials with low conductivity.

Convection is described as heat transfer involving both molecular motion and fluid transport. It is governed by Newton's law of cooling and depends on the heat transfer coefficient, which is influenced by fluid properties, velocity, and flow shape [9]. Convection occurs through diffusion (due to gradients) and advection (bulk fluid motion). It can be natural (driven by density differences) or forced (using fans or pumps). Convection plays a crucial role in heating, cooling, and phase-change processes [9]. Newton's law of cooling is described as

$$Q = h_c A_w \Delta T \tag{4}$$

where h_c represents the convection heat transfer coefficient [10], A_w denotes the surface area exposed to the fluid, influencing the total heat exchange, and ΔT is the temperature difference between the surface and the surrounding fluid, which drives the heat transfer process [10]. Convection is part of the heat transfer inside the tank between the fluid in different sections of the tank, especially between the liquid and the vapour phases as interface-to-liquid and interface-to-vapour heat transfer [11]. The liquid in contact with the inner tank wall increases in temperature, decreases in density, and flows up toward the liquid–vapour interface as consequence of the gained buoyancy. The internal convection can be minimized if the insulating properties of the tank shell (inner wall, outer wall, and insulation inside the annulus) are enhanced. Figure 1b shows the schematic of an LH_2 storage tank in which heat ingress resulted in natural convection from the bulk liquid to the liquid–vapour interface. Heat transfer through natural convection [12] may provoke thermal stratification (see Section 2.3).

Radiation heat transfer occurs as all materials emit energy in all directions, requiring no medium, unlike conduction and convection [9]. This allows energy transfer through a vacuum. Governed by the Stefan–Boltzmann law, radiation depends on surface properties like emissivity, absorptivity, transmissivity, and reflectivity. The heat transfer rate (power) by emitted radiation is determined by

$$Q = \sigma \varepsilon A T^4 \tag{5}$$

where σ represents the Stefan–Boltzmann constant, the variable A denotes the surface area of the object, T is its absolute temperature in K, and ε represents the emissivity of the object, indicating how efficiently it emits radiation. Radiation spans different wavelengths, shifting to shorter, higher-energy wavelengths as temperature increases [9]. In LH_2 applications, radiation manifests through the tank shell. External heat sources, such as the sun or accidental fires, can heat up the outer wall of an LH_2 tank which then starts emitting heat radiation towards the inner wall through the insulation. The insulation plays a crucial role in minimizing heat transfer through radiation. Common practices to reduce heat ingress through radiation include the use of reflecting materials in the insulation. In the case of cryogenic LH_2 tanks, radiation and conduction are the dominant forms of heat ingress, since heat transfer through convection is limited by the vacuum inside the annulus of the tank. The heat transfer mechanisms described here are not isolated; they directly influence mass transfer, thermal stratification, and pressurization, underscoring the need for integrated modelling approaches.

2.2. Mass Transfer Including Evaporation and Condensation

Depending on the filling level of the LH_2 tank, as well as on the internal temperature and pressure, the LH_2 is usually present in two phases: a gaseous and a liquid phase. The gaseous phase is often described as vapour phase or ullage. Due to thermodynamic processes, mainly resulting from heat ingress into the tank, condensation and evaporation at the vapour–liquid interface occur. This mass transfer happens simultaneously in both directions. Figure 1c shows a schematic of mass transfer at the vapour–liquid interface of an LH_2 tank. D. Lee [13] developed a model for evaporation and condensation, assuming that phase change occurs based on a relaxation process. In this model, the mass transfer rate for evaporation is described as [13]:

$$\dot{m}_{lv} = C \cdot \rho_l \cdot \frac{T - T_{sat}}{T_{sat}} \; for\; T > T_{sat} \tag{6}$$

$$\dot{m}_{lv} = 0 \; for\; T < T_{sat} \tag{7}$$

The mass transfer rate for condensation is described as [13]:

$$\dot{m}_{vl} = C \cdot \rho_v \cdot \frac{T_{sat} - T}{T_{sat}} \; for\; T < T_{sat} \tag{8}$$

$$\dot{m}_{vl} = 0 \; for\; T > T_{sat} \tag{9}$$

where $\dot{m}_{lv}$ and $\dot{m}_{vl}$ represent the mass transfer rates due to evaporation and condensation, respectively. The parameter C is the mass transfer coefficient, expressed with unit $\frac{1}{s}$, and is often referred to as the relaxation parameter [13]. Furthermore, ρ_l and ρ_v denote the densities of the liquid and vapour phases. T is the local temperature, while T_{sat} is the saturation temperature at which phase change occurs [13].

2.3. Thermal Stratification and Tank Pressurization

Thermal stratification in a cryogenic tank refers to the temperature distribution within the liquid column and the ullage space [14]. Thermal stratification and pressurization are emergent phenomena resulting from the interplay of heat and mass transfer, making their accurate prediction dependent on understanding these underlying processes. Thermal stratification of LH_2 in cryogenic tanks occurs due to heat ingress into the tank over time. As heat enters, it forms a natural convection boundary layer in the liquid near the tank wall. This causes warmer liquid to rise from the bulk region near the tank walls to the liquid–vapour interface due to buoyant forces, creating a distinct warm liquid zone at the top of the iso-thermal bulk liquid [14]. The gradual increase in the thickness of this warm zone leads to an vertical temperature gradient in the liquid, known as thermal stratification [15]. Figure 1d shows a schematic of a thermal stratification layer build-up in an LH_2 tank. The warmer upper layer of the liquid phase starts to evaporate to reach thermodynamical equilibrium, transferring mass and heat into the vapour phase. However, heat and mass are transferred in parallel from the vapour phase to the liquid in the form of condensation. LH_2 tanks for large-scale storage can be considered closed systems when all valves and ports are closed, which results in pressurization when the contents of the tank are warming up and starting to expand. Furthermore, pressurization leads to more condensation at the interface where the vapour is colder than the rest of the vapour space. Due to the vertical temperature gradient, it is very difficult to accurately predict temperature and pressure evolution over time. In stationary tanks, temperature stratification cannot be avoided. However, the formation can be slowed down by limiting heat ingress by improving the tanks insulation. Tanks with installations for active cooling could utilize spray-cooling to constantly introduce subcooled droplets of LH_2 at the top of the tank into the vapour phase. LH_2 is subcooled when the temperature is below the saturation point at a certain pressure [16]. Mobile tank systems could take advantage of the effects of sloshing, constantly mixing the fluid inside of the tank to avoid the formation of stratified layers. However, the efficiency of measures to avoid stratification should be subject to investigations with the purpose of evaluating economic viability.

2.4. Venting and Tank Depressurization

Venting (releasing hydrogen from its containment into the atmosphere) is initiated when the internal tank pressure has increased above the set pressure of the release device, which can be achieved by an actively controlled or a passively activated pressure-release valve (PRV). Venting can be the result of self-pressurization or active pressurization through the introduction of gas or through heating [17]. Filling procedures can make active venting necessary when the filled hydrogen is not subcooled or the receiving tank's initial internal temperature is too high, thus making excessive evaporation unavoidable. In both cases, excess gas needs to be released to lower the internal pressure and temperature and to continue with the filling procedure. In summary, discharge of LH_2 from the tank leads to depressurization and to cooling. Figure 1e shows a schematic of the upper section of an LH_2 tank with a vent mast and the corresponding mass flow during venting operation. Understanding venting is relevant for the operation of LH_2 tanks, not just for safety reasons, but also for the efficient transfer process when receiving or supplying LH_2. This concept extends to the depressurization behaviour, which is relevant during outflow when LH_2 consumers, such as fuel cells or internal combustion engines, are actively draining the supplying tanks content.

2.5. Sloshing

Sloshing is the movement of fluids, such as cryogenic propellants, within a storage tank, which becomes more pronounced in low-gravity environments. Sloshing can be caused by slight vibrations due to the low viscosity of LH_2 and it disturbs the thermodynamic equilibrium within the tank by mixing high-temperature vapour with subcooled fluid. This can cause a rapid pressure drop and ullage collapse [18]. In space travel, it affects the control of pressurants. A pressurant is intentionally evaporated cryogenic fuel that is used to move the liquid propellant and to keep the internal pressure of a tank high enough to operate the consumers, such as rocket engines [19]. Sloshing, however, can lead to changes in temperature and pressure due to gas condensation on the vapour–liquid interface and increases in BoG formation [20]. Sloshing increases BoG formation in two main ways:

1. Heat transfer—Sloshing dissipates kinetic energy into the tank system, transferring heat and accelerating fuel evaporation.
2. Increased surface area—The motion increases the liquid–vapour interface, enhancing evaporation.

Sloshing effects in LH_2 are not well studied compared to LNG, where previous research has linked sloshing to pressure and temperature variations. To limit the negative effects of sloshing, the movement of the tank should be kept minimal. However, for mobile applications, this is not feasible. Instead, so-called baffles are installed inside the tank to slow down fluid movement. Baffles can, under certain circumstances, act as thermal bridges, accelerating heat ingress into the tank and therefore promoting stratification, particularly in operational scenarios in which the tank is not moving. However, this depends on the insulation design of the tank. Furthermore, the initial chill-down of a warm tank can be more time- and energy-intensive due to the additional thermal capacity of the baffles. The thermodynamic effects of sloshing on LH_2 are underexplored, making it a critical area for further research, especially in designing fuel-efficient and cost-effective LH_2 carriers [21]. Figure 2 shows a schematic cross section of an LH_2 tank with baffles to mitigate sloshing of the liquid phase inside.

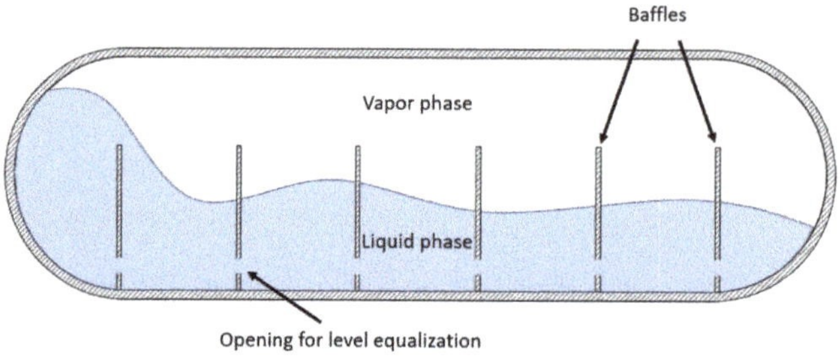

Figure 2. Schematic cross section of an LH_2 tank with sloshing liquid and installed baffles.

2.6. LH_2 Transfer Operations

LH_2 transfer operations refer to the processes of flowing LH_2 from one container to another, typically involving pipelines, storage tanks, or transport vessels. Transfer operations involve complex multi-phase flow, where heat ingress, evaporation, and pressure changes interact dynamically, requiring holistic modelling to capture these interdependencies. There are several methods for transferring LH_2 efficiently. Pump transfer relies on cryogenic pumps to move LH_2 under controlled pressure, while pressure-driven transfer utilizes differential pressure to push LH_2 from one vessel to another. If the transferred fluid consists of a gas phase and a liquid phase, the flow is considered as multi-phase. Especially at the beginning of a transfer process, when the pipe is not entirely cooled down yet, multi-phase flow occurs due to evaporation. This formation of gas has a significant negative impact on the filling procedure of a receiving tank, since the presence of a gas phase leads to necessary venting or energy consumption through active cooling. Furthermore, the geometry of LH_2-containing transfer components influences vapour formation. Generally, turbulence and thereby friction can be promoted by certain geometries, causing pressure drop and heat ingress and therefore enhancing evaporation. To what extend existing models underpredict the influence of pipe geometry should be subject to separate analysis. Figure 3 shows a schematic cross section of an LH_2 transfer pipe with a developed multi-phase-flow.

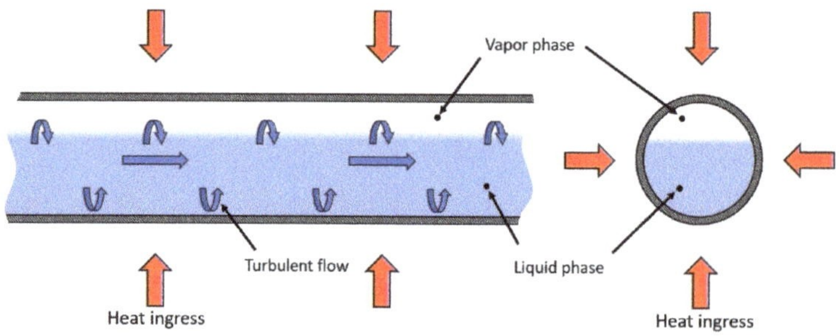

Figure 3. Schematic cross section of an insulated pipe with a turbulent two-phase flow of LH_2 inside.

2.7. Flash Boiling and Pressure Recovery

Flashing, or flash evaporation, occurs when a liquid boils due to a sudden reduction in pressure. Several key factors influence this phenomenon, including the initial temperature, overheating temperature, pressure reduction rate, and liquid level [22]. Thermal stratification plays a significant role in flashing, with the hot gas–liquid interface boiling first, followed by vapour generation at the bottom of the vessel. This process pushes up the liquid surface, ultimately leading to pressure recovery [22]. Figure 4 shows the phase diagram of H_2, with the red arrow indicating a sudden pressure drop at equilibrium leading to evaporation.

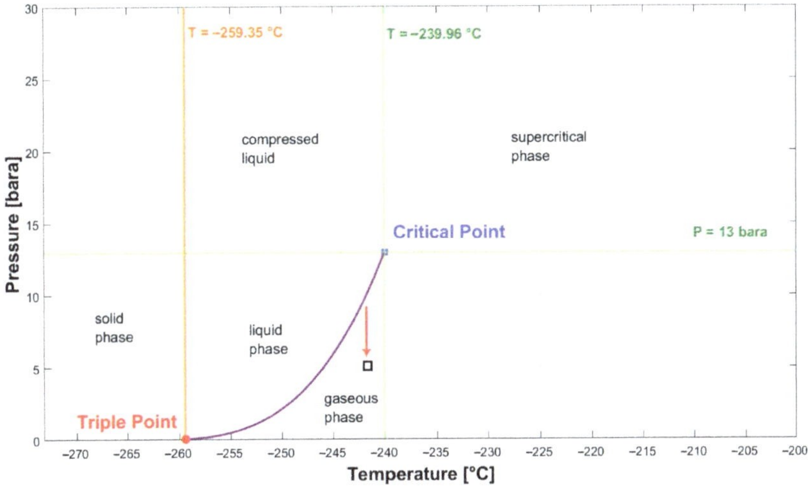

Figure 4. Phase diagram of H_2.

3. Methodology

A narrative literature review approach was selected in this study to provide an overview of existing modelling approaches to simulate LH_2 thermo- and fluid dynamics in storage and transfer components. A narrative literature review is a versatile methodology widely used in research to synthesize existing literature, provide overviews, and critique findings in a coherent manner. It is particularly effective when addressing complex or multifaceted topics that require flexibility and interpretative depth [23]. This literature review was carried out following the key steps introduced in Figure 5.

The key steps were derived from Pickering and Byrne [24] and adapted for the purpose of conducting a structured narrative review. As a first step, a topic was selected and related research questions were considered. The topic was defined as "Review of Modelling Approaches for LH_2 Thermodynamics during Storage and Transfer". The objective was to provide an overview of relevant thermodynamic phenomena, problems, and different modelling approaches for LH_2 dynamics in the corresponding systems and components. This review addresses the following research questions:

1. What thermodynamic phenomena are covered by the existing literature for LH_2?
2. What are the proposed modelling approaches to simulate these phenomena?
3. Are the proposed approaches validated?
4. Why are certain models not validated/is there an identifiable knowledge gap or need for data?

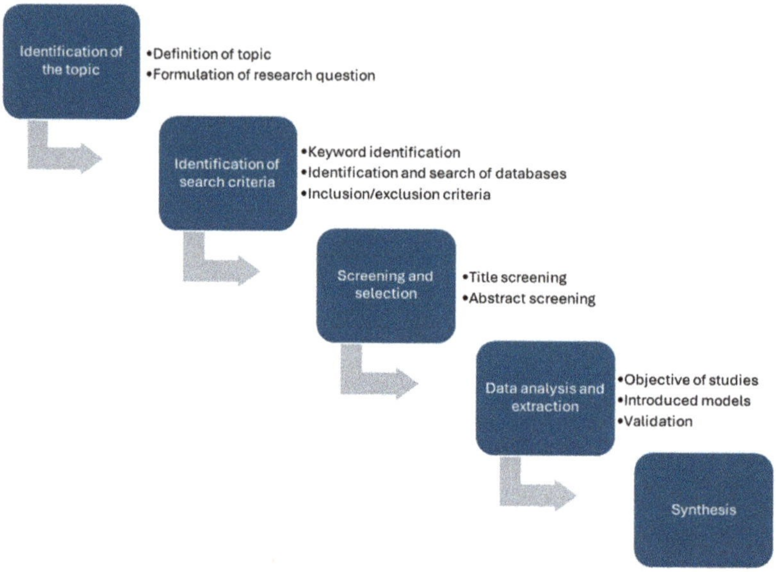

Figure 5. Key steps adapted to conduct a narrative review.

The literature review utilized a comprehensive search strategy across multiple scientific databases to ensure a broad overview. Key sources included Scopus, Google Scholar, and the NASA archive, the latter offering access to an extensive collection of studies dating back to the 1960s, many of which are not available elsewhere. The search terms and keywords focused on topics central to the research:

- Liquid hydrogen or LH_2;
- Model or Modelling;
- Simulation;
- Pressurization;
- Storage;
- Tank;
- Thermodynamic;
- Fluid dynamics.

Studies were included or excluded based on specific criteria:

- Publications in languages other than English and German were excluded;
- Models referring to substances other than LH_2 were excluded;
- Excluded subject areas included the following terms:
    a. Medicine;
    b. Chemistry;
    c. Computer science;
    d. Social science;
    e. Pharmacology;
    f. Environmental science;
    g. Biology.
- Excluded keywords included the following terms:
    a. Combustion;
    b. Diffusion;
    c. Oxygen;

d. Helium;
e. Rocket engine;
f. Ballistics;
g. Viscosity;
h. Cavitation;
i. Catalyst.

The criteria presented in this section applied to the use of Scopus as search tool exclusively. Literature found in the other databases or sources was screened for exclusion criteria via title and abstract screening. The screening and selection process involved several systematic steps. Initially, title screening was conducted to filter out unrelated studies. This was followed by an abstract screening to determine the purpose and scope of the studies, assessing whether they introduced simulation models and to make sure that none of the exclusion criteria were met. The full-text analysis, extraction, and synthesis process began with an initial document overview, where each paper was scanned to identify its key focus by reviewing the abstract, introduction, and conclusion. Following this, a focused reading was conducted, analyzing specific sections relevant to the research question. After focused reading, the synthesis stage categorized studies based on the thermodynamic phenomena they addressed. For each phenomenon, relevant papers were assigned, and their modelling approaches were examined in detail, including simulation techniques, software used, and LH_2 storage components such as tanks or pipes. Lastly, the models were screened for whether they were validated or not, and if a model lacked validation, an effort was made to determine the reason. Identified knowledge gaps, theoretical advances, and consequently recommendations for future research were included.

4. Modelling of LH_2 Thermal Dynamic Behaviour

In total, 81 papers that focus on one or more of the seven topics described in Section 2 were identified and included in this study. The analyzed sources have been published over a period of 64 years, with the oldest work dating back to 1961. In Figure 6, an overview of the distribution of the analyzed papers over the years is provided. The literature review was concluded in January 2025. In recent years, there has been a trend of an increasing number of publications connected to the topic of this review. This growing interest in LH_2 modelling underscores the relevance of LH_2 thermodynamics and corresponding modelling approaches.

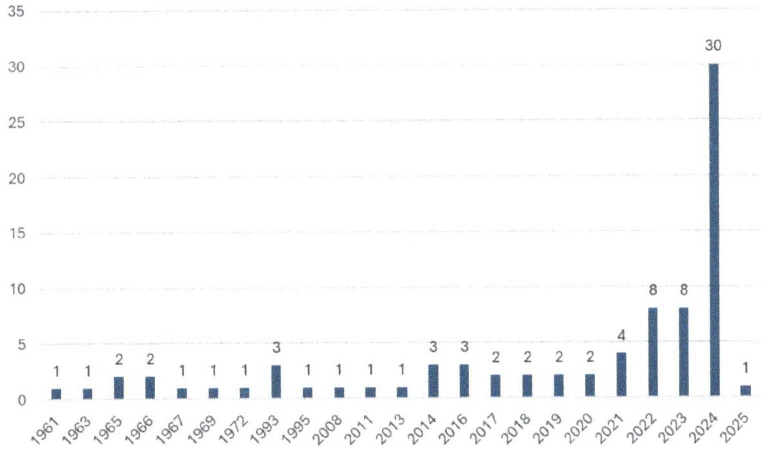

Figure 6. Number of papers over the years falling within the seven focus areas described in Section 2.

Most of the papers (56) focused on the modelling of thermodynamic behaviour of LH_2 in storage tanks; 24 papers focused on the modelling of the behaviour during multiphase flow and in pipelines. Figure 7 presents the numbers of papers for each focus area, indicating how many of the corresponding models were validated.

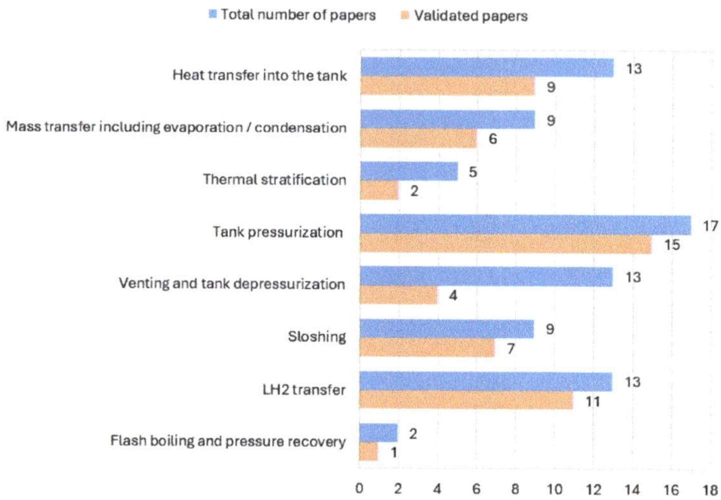

Figure 7. Number of models and validated models per focus area.

The analyzed literature revealed a wide range of modelling approaches:

- General analytical models;
- Lumped-parameter models (0D/1D models);
- Empirical and semi-empirical models;
- Computational fluid dynamics (CFD) models (2D/3D);
- Machine learning (ML) and artificial neural network (ANN) models;
- Numerical multidisciplinary simulation models.

Figure 8 provides an overview of the amount of different modelling approaches that were found in the analyzed papers.

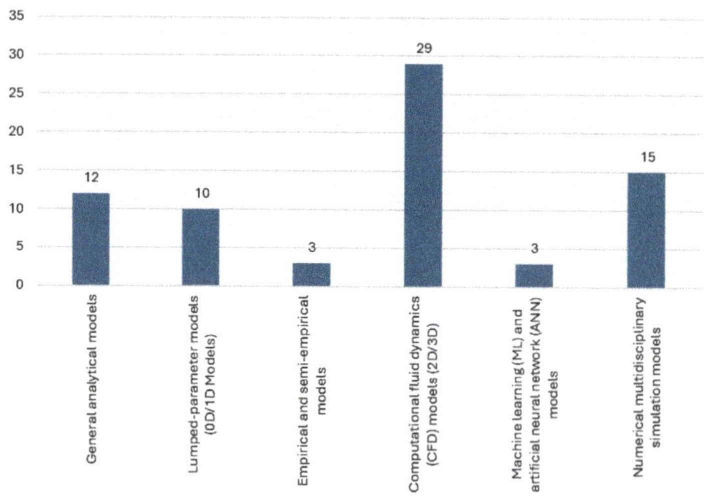

Figure 8. Amount of different modelling approaches.

Some of the presented models are combinations or the paper introduces multiple approaches. Therefore, the summed-up number of modelling approaches does not fit the total number of papers.

4.1. Models Focusing on Thermodynamic Behaviour in Storage Tanks

4.1.1. Models Focusing on Heat Transfer

Huntley et al. [25] conducted research on the thermal behaviour of LH_2 in a tank when subjected to heating from the wall and bottom. They introduced an experimental and analytical model combined with dimensionless parameterization. The models were compared with each other. This experimental study on LH_2 behaviour in a heated tank at constant pressure revealed several key findings. A predictive analysis successfully estimated temperature stratification trends with increasing wall heat flux [25]. L. Wang et al. [26] conducted a axisymmetric CFD investigation on how varying gravity conditions influence the thermodynamic characteristics of LH_2 tanks. The developed CFD model incorporates multicomponent effects and phase-change dynamics using the Schrage equation and Dalton's law. Due to lack of data, the model was not validated [26]. J. W. Jiang et al. [6] simulated the transient behaviour of a MLI LH_2 tank coupled with a VCS. The model accounted for heat transfer into the tank, pressurization, and the effects of active cooling. The model was validated using experimental data from NASA's Multipurpose Hydrogen Test Bed (MHTB) [27]. A CFD model based on the VoF method (volume of fluid method) to determine phase-change coefficients for evaporation and condensation was developed by R. Lv et al. [28]. The study explores scale effects on self-pressurization in LH_2 tanks under normal- and microgravity. Grid-independence tests confirm grid selection for 16 cases, ensuring numerical reliability and efficiency [28]. F. Ustolin et al. [29] developed an analytical model based on thermodynamic equations to estimate the time to failure for a cryogenic LH_2 tank exposed to fire [29]. The model accounts for the heat transfer into the tank, the pressurization behaviour, the mass transfer at the liquid–vapour interface, and the release rate of the BOG. It was validated against experimental results. Due to missing experimental details, assumptions were made regarding the pressure-relief valve section area and MLVI thermal conductivity [29]. Furthermore, F. Ustolin et al. [30] developed a 2D CFD model to analyze the thermal behaviour of an LH_2 tank exposed to an external fire. The heat transfer into the tank, the mass transfer at the liquid–vapour interface, and the pressurization behaviour were investigated [30]. The model was validated against experimental data, showing reasonable agreement despite measurement uncertainties [31]. Sun et al. [32] developed a model to analyze the heat transfer into an LH_2 storage tank including mass transfer at the liquid–vapour interface, tank self-pressurization, and the evaporation rate. The model is a numerical quasi-2D model, which is essentially a hybrid approach where heat transfer is modelled as one-dimensional (1D) in certain directions and two-dimensional (2D) in others, simplifying full 3D modelling. The validation was conducted by comparing the model predictions with experimental data from four distinct cryogenic storage tanks [32]. Wang et al. [11] modelled and compared the thermodynamic behaviour of LH_2 tanks for maritime transport with different insulation types at varying filling levels. Therefore, an analytical model was presented. The model was validated through comparisons with experimental data from LH_2 and liquefied natural gas (LNG) tanks. Key parameters affecting pressure rise and vapour temperature predictions were adjusted to match experimental results. The model showed good agreement with vapour temperature measurements across different filling levels, though initial pressure rise predictions were slightly overestimated at higher fill levels [11]. For heat transfer in insulation, the model was compared against a chill-down experiment conducted on a 208 m^3 LH_2 spherical tank at Los Alamos National Laboratory [11]. Parello et al. [33]

conducted research on design and integration of LH_2 tanks for aircraft by developing a multi-phase thermodynamic model using phyton, which accounts for heat transfer into the tank and the resulting boil-off. The model integrates thermodynamic and structural analyses to assess tank performance during mid-range missions. The model was not validated because the components that were simulated were not built yet [33]. Alipour Bonab and Yazdani-Asrami [16] utilized a cascading feed-forward neural network (CFNN) approach to estimate the heat transfer coefficient of subcooled liquid hydrogen (sLH_2) for transportation applications. The study employs 4-fold cross-validation for the CFNN model, which features five hidden layers with 20 neurons each, the Levenberg–Marquardt optimization algorithm, and Logsig-Purelin activation functions [16]. Z. Xu et al. [34] presented a numerical multipurpose simulation model that accounts for phase change at the liquid–vapour interface, pressurization, and heat transfer of cryogenic tanks for LH_2-powered hybrid trains. The study examines a thermal management strategy to ensure stable hydrogen flow while maintaining tank safety. The findings provide insights into optimizing hydrogen storage and supply [34]. The model was not validated. Numerical formulations were presented by H. Lv et al. [35] to analyze heat transfer into a composite multilayer insulated tank with a vapour-cooled shield and the dormancy behaviour of the stored LH_2. The model was not validated [35]. K. Li et al. [36] developed a numerical heat and mass transfer model of the fluid domain and a heat conduction model of VMLI coupled with a VCS in an LH_2 tank. The heat and mass transfer model accounts for active cooling, mass transfer at the liquid–vapour interface, and the overall heat ingress into the tank. It identifies optimal positions and operating conditions for single and double VCS to maximize dormancy extension [36]. The heat conduction model was validated through experimental comparisons and conservation verification for mass and energy. The numerical model of VMLI coupled with VCS and the heat and mass transfer model were verified using NASA's experimental data [27]. Table 2 provides an overview of models focusing on heat transfer.

Table 2. Overview of models focusing on heat transfer.

Authors	Year	Type of Model	Validated
S. C. Huntley et al. [25]	1966	Experimental and analytical	Yes
L. Wang et al. [26]	2019	FLUENT-CFD	No
W. Jiang et al. [6]	2021	Numerical model	Yes
R. Lv et al. [28]	2021	CFD, VoF method	Yes
F. Ustolin et al. [29]	2021	Analytical equations	Yes
F. Ustolin et al. [30]	2022	2D CFD	Yes
Z. Sun et al. [32]	2023	Numerical, 2D	Yes
J. Wang et al. [11]	2024	Analytical equations	Yes
R. Parello et al. [33]	2024	Multi-phase thermodynamic	No
S. Alipour Bonab and M. Yazdani-Asrami [16]	2024	CFNN	Yes
Z. Xu et al. [34]	2024	Numerical model	No
H. Lv et al. [35]	2024	Numerical model	No
K. Li et al. [36]	2024	Numerical model	Yes

4.1.2. Models Focusing on Mass Transfer Including Evaporation and Condensation

W. A. Olsen [37] conducted an analytical investigation into heat and mass transfer at the liquid–vapour interface inside of an LH_2 tank. It was found that the rapid pressurization of a two-phase hydrogen tank without mixing leads to significant heat transfer to the liquid due to condensation, which must be considered in stratification studies [37]. The analytical model was not validated. C. Lin et al. [38] conducted research on mixing and transient interface condensation inside LH_2 tanks. An empirical model was developed to account for heat transfer into the tank and mass transfer at the liquid–vapour interface. No validation was carried out. R. B. Schweickart [39] analyzed a demonstration concept for CPST (Cryogenic Propellant Storage and Transfer). Simulations of CPST operations, including microgravity cryogenic fuel transfers and long-term storage, were conducted using a validated thermal and thermodynamic analysis tool. A model running in SINDA/FLUINT software and accounting for heat transfer into the tank and the mass transfer at the liquid–vapour interface was introduced. At the time of publication, no validation was carried out [39]. A 1D model equation that emphasizes heat and mass transfer at the interface was developed by M. Stewart [40]. The equation was incorporated into a CFD simulation, and the results were compared with experimental data. The numerical challenge addressed in the paper is the physics of the liquid–vapour interface, particularly focusing on the interfacial temperature gradient and heat transfer mechanisms. The paper resolves this challenge by utilizing a sub-grid interface model [40]. The validation against experimental data revealed that the model accurately predicted interface temperature jumps, heat flux distributions, and condensation effects. H. Wang et al. [41] investigated heat leakage distribution between the liquid and the vapour phases in LH_2 tanks. Therefore, a heat leakage distribution model (HDM) was developed under consideration of experimental data. The model accounts for mass and heat transfer at the liquid–vapour interface, as well as for heat leakage into the tank and for self-pressurization. For validation purposes, experimental data from NASA were used. The HDM was validated using three experimental datasets, showing improved pressure prediction accuracy. Compared to TMZM (thermodynamic multi-zone model) and CFD models, the HDM-coupled TMZM had better agreement with experiments. A CFD simulation model utilizing the VoF method was developed by Y. Jiang [42]. The objective was to account for heat transfer and phase-change characteristics of LH_2 in cryogenic tanks under microgravity conditions. Multi-phase flow, phase-change dynamics, turbulence, and evaporation and condensation processes were simulated. The phase-change model for an LH_2 storage tank was validated by comparing simulated and experimental results. To compare the thermal performance of cylindrical and spherical LH_2 tanks, Z. Wang and W. Mérida [43] developed a numerical non-equilibrium thermodynamic model. The model accounts for heat transfer into the tank, mass transfer at the liquid–vapour interface, and boil-off-rates [43]. The numerical validation of the non-equilibrium thermodynamic model was conducted using experimental data from NASA on both cylindrical and spherical LH_2 tanks. D. Choi [44] developed a numerical, physical-based model and conducted an energy–mass balance analysis to model heat transfer into an LH_2 fuel tank, the phase change at the liquid–vapour interface, and pressure changes due to fuel consumption and self-pressurization. The proposed thermodynamic model was validated by comparing its predictions with experimental data for LH_2 vaporization and fuel emission [44]. H. Wang et al. [45] conducted a literature review on the fluid-dynamic behaviour of LH_2 during non-vented storage. The review highlights experimental and computational studies on LH_2 storage [45]. A simulation approach was presented, which utilizes the VoF method for CFD modelling of phase change at the liquid–vapour interface and heat transfer into the tank. NASA test data aided CFD model validation [45]. Table 3 provides an overview of models focusing on mass transfer.

Table 3. Overview of models focusing on mass transfer.

Authors	Year	Type of Model	Validated
W. A. Olsen [37]	1966	Analytical equations	No
C. Lin et al. [38]	1993	Empirical model	No
R. B. Schweickart [39]	2014	SINDA/FLUINT	No
M. Stewart [40]	2017	1D CFD	Yes
H. Wang et al. [41]	2023	Analytical, HDM	Yes
Y. Jiang [42]	2023	CFD, VoF method	Yes
Z. Wang and W. Mérida [43]	2024	Numerical, non-equilibrium	Yes
D. Choi [44]	2024	Numerical model	Yes
H. Wang et al. [45]	2024	2D CFD, VoF method	Yes

4.1.3. Models Focusing on Thermal Stratification

To simulate thermal stratification and self-pressurization in a closed LH_2 tank, R. W. Arnett and R. O. Voth [46] proposed mathematical formulation of boundary layer equations, particularly for turbulence. The model was not validated. A 3D fluid-dynamic model using Flow-3D (1992) software was presented by Grayson [47]. The objective was to present a coupled thermodynamic–fluid-dynamic solution for liquid-hydrogen tank behaviour with a focus on temperature stratification. The model enabled combined propellant sloshing and thermal stratification analysis [47]. There was no validation carried out due to a lack of data. M. J. Daigle et al. [48] developed a dynamical model of a cryogenic fuel tank that accounts for temperature stratification and non-equilibrium condensation–evaporation at the liquid–vapour interface [48]. The model was not validated. To model the effects of insulation thickness on self-pressurization rate and thermal stratification in LH_2 tanks, J. Joseph et al. [15] used a transient two-phase thermodynamic lumped model [15]. The numerical model was validated against experimental data from a cryogenic tank pressurization test with liquid nitrogen. Simulations accurately captured pressure evolution, liquid stratification, and temperature changes, closely matching experimental data [15]. Z. Liu et al. [49] conducted a study on thermal stratification in LH_2 tanks under different gravity levels. A CFD model, utilizing the VoF method, was developed to study the thermal processes in a cryogenic LH_2 tank, analyzing fluid temperature distribution, thermal stratification, and surface tension effects [49]. The model was validated against experimental data through parameter calibration. Table 4 provides an overview of models focusing on thermal stratification.

Table 4. Overview of models focusing on thermal stratification.

Authors	Year	Type of Model	Validated
R. W. Arnett and R. O. Voth [46]	1972	Analytical equations	No
G. D. Grayson [47]	1995	CFD, Flow-3D	No
M. J. Daigle et al. [48]	2013	Dynamical, low-dimensional approach	No
J. Joseph et al. [15]	2017	Transient two-phase thermodynamic lumped model	Yes
Z. Liu et al. [49]	2018	CFD, VoF method	Yes

4.1.4. Models Focusing on Tank Pressurization

A model for predicting LH_2 self-pressurization proposed by Aydelott [50] based on analytical equations was validated with experimental data. A series of 21 self-pressurization tests were conducted on a 9-inch spherical dewar partially filled with LH_2, ending at 100 psi (6.89 bar) absolute pressure. The tests varied filling degree, heat transfer rate, and heater configuration [50]. Gursu et al. [51] introduced three different pressure rise models to analyze LH_2 thermal stratification, self-pressurization, and boil-off-rates in cryogenic vessels. The models are a homogeneous model, a surface-evaporation model, and a thermal stratification model. The homogeneous and surface-evaporation models assumed uniform temperature within the cryogenic vessel, while the thermal stratification model accounted for temperature gradients. The modelling approach is validated in the second part of the paper [52]. To simulate the self-pressurization of a partially filled LH_2 tank in normal gravity, Barsi and Kassemi [53] developed a two-phase lumped-vapour CFD model. Validation against experimental data showed reasonable accuracy in predicting pressure rise across various fill levels. O. V. Kartuzova et al. [54] conducted a simulation of self-pressurization and spray-cooling, a process where droplets of subcooled liquid are introduced into the vapour phase, in LH_2 storage tanks. They accounted for heat transfer into the tank, as well as for heat and mass transfer at the liquid–vapour interface. A two-phase CFD compressible VOF model was developed to study self-pressurization and pressure control in cryogenic storage [54]. The model was validated using pressurization and pressure-control test results from the MHTB at NASA Marshall Space Flight Centre. Kassemi and Kartuzova [55] presented two-phase CFD models for cryogenic tank self-pressurization and pressure control. Pressure control, mass transfer at the liquid–vapour interface, heat ingress into the tanks, active cooling, and multi-phase flow were also subject of the research. Therefore, a sharp-interface CFD model was developed. The model utilized the VoF method and was validated against experimental data obtained from the NASA MHTB [55]. With the help of a CFD model using ANSYS Fluent version 16.0, Stewart and Moder [56] described the self-pressurization of a flight-weight LH_2 tank. The model accounts for self-pressurization, mass transfer at the liquid–vapour interface, and the heat ingress into the tank. The paper highlights the challenges posed by experimental measurement accuracy, saturation pressure uncertainty, and velocity variations at the interface [56]. The simulation results were compared with experimental data for validation purposes. A notable pressure difference was observed between iso-thermal and steady boil-off cases [56]. A multimode model for simulating the self-pressurization of a cryogenic tank was developed by Majumdar et al. [57] using the Generalized Fluid System Simulation Program (GFSSP). Additionally, the model accounted for heat transfer through MLI, SOFI (spray-on foam insulation), and the metal wall into the tank, heat and mass transfer at the liquid–vapour interface, and pressure regulation through a vent system. Validation was not conducted [57]. Y. Liu et al. [58] modelled the heat transfer and the resulting boil-off gas formation analytically to optimize a type-C tank for LH_2 marine transport. The study evaluated insulation strategies and materials. Three insulation strategies, involving varying positions of the VCS, were analyzed, showing significant reductions in heat transfer and boil-off rate [58]. The thermodynamic model developed in this research was validated using results from other publications. Ahluwalia et al. [59] investigated para-ortho hydrogen conversion, the resulting dormancy behaviour, and fractions of LH_2 stored in a cryogenic tank for heavy-duty trucks. The objective was to develop a model accounting for the BOG formation in the tank [59]. A numerical model utilizing the Adams–Bashford–Molton method and the Benedict–Webb–Rubin (BWR) equation of state was developed using REFPROP. The approach was not validated [59]. Metveev and Leachman [60] simulated the effect of the tank size on self-pressurization and venting rate of an LH_2 tank. A simplified lumped-element model was developed to simulate thermal processes in stationary LH_2 tanks. The model efficiently simulates multi-day processes and was validated for self-pressurization [60]. The model was validated using experimental data from the MHTB tank's self-pressurization process [60]. S. Jeong et al. [61] described a 3D CFD model incorporating the VoF method to simulate self-pressurization, heat transfer and phase change at the liquid–vapour interface inside of a partially filled LH_2 tank. The model, validated against experimental and computational data, accurately captured pressurization rates, interfacial mass transfer, and heat flow [61]. H. R. Wang et al. [62] modelled the pressurization rate, the heat ingress, the mass transfer at the interface, and thermal stratification in LH_2 tanks. A modified TMZM model was developed to predict pressure build-up in LH_2 tanks and validated using MHTB experimental data [62]. The TMZM model was validated using experimental data from a MHTB tank under four conditions showing a maximum relative error of 4.67%. Furthermore, the TMZM outperformed a CFD model in predicting pressure rise, confirming its reliability [62]. Anas. A. Rahman et al. [63] developed a prediction model for pressure evolution in non-venting LH_2 tanks, accounting for heat flux into the tank and thermal stratification. Therefore, an artificial neural network (ANN) model was configured as a four-layered feed-forward network with a back-propagation algorithm [63]. The ANN model was validated using regression and error histogram plots, showing a strong fit between predicted and actual results [63]. H. Wang et al. [64] describe the use of thermal models for predicting the self-pressurization process in LH_2 tanks: TEMs (thermodynamic equilibrium models), SEM (surface-evaporation model), TMZM, TSM (thermal stratified model), TMNM (thermal multi-node model), and a CFD model. These models are assessed based on their comprehensiveness, calculation speed, and precision [64]. The validation was conducted by comparing the predictions of various thermal models, including TEMs, SEM, TMZM, TSM, TMNM, and CFD, against experimental data [65]. J. Wang et al. [66] conducted modelling of LH_2 thermodynamics for maritime export. The study assesses the use of double-wall vacuum insulation in pressurized spherical tanks, with a new analytical model predicting heat ingress, pressure rise, and BOG generation [66]. Validation was undertaken against experimental data. It predicted a pressure rise of 3.95 kPa/h, closely matching experimental results. Insulation heat transfer validation aligned with NASA data, predicting a boil-off rate (BOR) of 1.48 m^3/day. Boil-off validation showed general agreement with cumulative BOR but underpredicted late-stage boiling due to liquid stratification simplifications [66]. T. Yu and Y. Lim [67] developed a lumped-element model to predict changes in holding time and to assess the influence of LH_2 filling ratio, heat transfer, and the shape of the tank. Four case studies showed that low liquid levels lead to faster pressure build-up during laden voyages, while ballast voyages generate more boil-off gas [67]. Validation was conducted against experimental data by NASA with the MHTB [67]. Table 5 provides an overview of models focusing on tank pressurization.

Table 5. Overview of models focusing on tank pressurization.

Authors	Year	Type of Model	Validated
J. C. Aydelott [50]	1967	Analytical equations	Yes
S. Gursu et al. [51]	1993	Homogeneous, SEM, TSM	Yes
S. Barsi and M. Kassemi [53]	2008	Lumped-vapour CFD	Yes
O. V. Kartuzova et al. [54]	2014	CFD, VoF method	Yes
M. Kassemi and O. Kartuzova [55]	2016	Sharp-interface CFD, VoF method	Yes
M. Stewart and J. P. Moder [56]	2016	CFD, ANSYS Fluent	Yes
A. Majumdar et al. [57]	2016	Multinode model in GFSSP	No
H. R. Wang et al. [62]	2022	TMZM	Yes
Y. Liu et al. [58]	2023	Analytical equations	Yes
R. K. Ahluwalia et al. [59]	2023	Numerical, Adams–Bashford–Molton method, Benedict–Webb–Rubin (BWR) equation	No
K. I. Metveev and J. W. Leachman [60]	2023	Lumped-element model	Yes
S. Jeong et al. [61]	2023	3D CFD, VoF method	Yes
Anas. A. Rahman et al. [63]	2024	ANN	Yes
H. Wang et al. [64]	2024	TEM, SEM, TMZM, TSM, TMNM, CFD	Yes
J. Wang et al. [66]	2024	Analytical equations	Yes
T. Yu and Y. Lim [67]	2024	Lumped-element model	Yes

4.1.5. Models Focusing on Venting and Tank Depressurization

To investigate the LH_2 tank pressurization during outflow in space travel, Mandell and Roudebush [68] developed a dimensionless, simplified model, relying on modified Stanton numbers [68]. Roudebush and Mandell [69] furthermore conducted research focused on important parameters relevant for the outflow problem in LH_2 tanks. Experimental data from Lewis and Lockheed-Georgia experiments were used to validate the model [69]. Aydelott and Spuckler [17] presented an analytical equation to account for BOG formation and the losses occurring during H_2 venting. In a series of venting tests on a 22-inch spherical tank partially filled with LH_2, the mass vented was analyzed as a function of energy input. The model was not validated [17]. Winters and Houf [70] simulated small-scale releases from LH_2 storage systems. The study uses turbulent entrainment models to predict characteristics of hydrogen leaks. The COLDPLUME code models the leak stream's behaviour, accounting for both thermal and solutal buoyancy effects [70]. The model was not validated. To accurately predict and minimize the losses that occur due to the cryogenic nature of LH_2, Petitpas [71] presented a thermodynamic model. The model is a modified 0D thermodynamic code for simulating the LH_2 pathway from liquefaction to refuelling stations. BOG losses were estimated, with key findings including negligible venting during liquefaction, significant losses during transfer to receiving vessels, and varying losses depending on vessel pressure and filling methods [71]. BOG losses at refuelling stations were found to decrease with higher delivery capacities. The approach was not validated due to a lack of data [71]. Zuo et al. [72] conducted a numerical investigation into the venting process of LH_2 in an on-orbit storage tank. The authors utilized the VoF method to develop a full procedural thermodynamic venting system model which additionally accounts for evaporation and condensation at the liquid–vapour interface, pressurization, and depressurization. The model was validated through mesh independence testing and by comparing simulated self-pressurization rates against experimental tank data [72]. A 6×10^5 elements mesh was chosen for its balance between accuracy and computational efficiency. Self-pressurization simulations were compared to experimental data, showing good agreement with an evaporation intensity factor of 8.5 s^{-1}. Lower values led to unrealistic evaporation resistance [72]. Al Ghafri et al. [73] estimated the BOG losses from LH_2 storage as well as the heat transfer into the tank. Therefore, a superheated vapour (SHV) model implemented in BoilFAST using a non-equilibrium approach was developed. The BoilFAST simulation tool shows strong agreement with experimental data across various conditions. The SHV model was validated against experimental data for liquid nitrogen, LNG mixtures, and self-pressurization tests, showing good agreement with the literature sources [73]. Hamacher et al. [74] conducted modelling of the thermodynamic behaviour of CcH_2 (cryo-compressed hydrogen) tanks for trucks. The tank behaviour regarding dormancy, phase change, pressurization during discharge, and refuelling was investigated. The model utilized a DAE approach and was not validated [74]. H. Chen et al. [75] researched venting systems of LH_2 tanks under microgravity conditions. Of central interest was the control strategy to manage pressurization and depressurization during venting of LH_2. For simulation purposes, a lumped-vapour model was coupled with a CFD simulation in Ansys FLUENT [75]. The model was not validated. Stops et al. [76] conducted modelling of LH_2 storage tanks for truck applications. The developed model accounted for heating requirements, dormancy behaviour, and depressurization at discharge. The presented model is a Differential-Algebraic System of Equations (DAE) which is implemented into a decision logic for different operational modes. Validation was not carried out. Dutta and Mukhopadhyay [77] developed a thermodynamic model of storage and discharge of LH_2 and LOX (liquid oxygen). They simulated the heat transfer into the tank, the phase change, and primarily the behaviour during outflow. The numerical formulations utilized the fourth-order Runge–Kutta method [77]. The model was not validated [77]. Okpeke et al. [78] presented an analytical modelling approach to assess BOG losses and their cost during LH_2 tank filling with and without precooling. The study analyzed LH_2 tank filling with and without precooling, considering two tank materials. The model was not validated [78]. Kumar and Sleiti [79] investigated modelling and analysis of the effect of LH_2 tank geometries and relief pressure on BOG losses. Therefore, the heat transfer into the tank, phase change at the vapour–liquid interface, and BOG dynamics were modelled. The objective was to optimize the design and operational efficiency of LH_2 tanks [79]. The numerical model was implemented using BoilFAST [79]. The simulation results were not validated. Table 6 provides an overview of models focusing on venting and tank depressurization.

Table 6. Overview of models focusing on venting and tank depressurization.

Authors	Year	Type of Model	Validated
D. A. Mandell and W. H. Roudebush [68]	1965	Dimensionless, Stanton numbers	Yes
W. H. Roudebush and D. A. Mandell [69]	1965	1D equations	Yes
J. C. Aydelott and C. M. Spuckler [17]	1969	Analytical equations	No
W. S. Winters and W. G. Houf [70]	2011	Turbulent entrainment model, COLDPLUME code	No
G. Petitpas [71]	2018	0D MATLAB code	No
Z. Zuo et al. [72]	2020	Numerical, VoF method	Yes
S. Z. S. Al Ghafri et al. [73]	2022	SHV, BoilFAST	Yes
J. Hamacher et al. [74]	2023	DAE	No
H. Chen et al. [75]	2024	Lumped-vapour model, Ansys FLUENT	No
L. Stops et al. [76]	2024	DAE	No
J. Dutta and A. Mukhopadhyay [77]	2024	Numerical, fourth-order Runge–Kutta method	No
B. E. Okpeke et al. [78]	2024	Analytical equations	No
L. Kumar and A. Sleiti [79]	2024	Numerical, BoilFast	No

4.2. Models Focusing on Multi-Phase Flow in Storage Tanks and Pipelines

4.2.1. Models Focusing on Sloshing

Z. Liu et al. [18] presented a numerical model with a coupled motion–mesh and VoF method to simulate fluid sloshing inside an LH_2 tank [18]. The model was validated against experimental data. The validation study compared three k-ε turbulence models for predicting fluid sloshing, using experimental data from other sources. To simulate the heat transfer into an LH_2 tank and the mass transfer at the liquid–vapour interface, as well as the BOG formation rate, Smith et al. [21] proposed a numerical thermodynamic model utilizing the first law of thermodynamics. Modelling of sloshing is relevant for the design of future LH_2 carriers [21]. Therefore, thermodynamic modelling with empirical data on ship motion and sloshing effects was integrated. Initially calibrated for LNG carriers, the model was adapted for LH_2. Due to a lack of data, the model was not validated [21]. S. Li et al. [80] conducted numerical simulation of the thermodynamic behaviour of LH_2 tanks for trailers focusing on multi-phase flow, phase change, and sloshing. The model in use is a multi-phase turbulent flow CFD model utilizing the VoF method [80]. The numerical model was validated using NASA's 1960s self-pressurization experiment data of a spherical

LH_2 tank. Initial conditions and self-pressurization simulations were performed in two steps. The results showed a 0.74% temperature difference for bulk LH_2, with temperature and pressure trends closely matching experimental data, confirming the model's reliability within a 10% error margin for pressure [80]. A CFD model for tank pressurization with gaseous hydrogen and subsequent sloshing was developed by Kartuzova et al. [20]. The model first simulated autogenous pressurization. Sloshing simulations followed, comparing two cases: one described as moderate sloshing with a slight pressure drop and one described as violent sloshing causing a sharp pressure decrease [20]. The model was validated against the beforementioned two cases [20]. W. Kang et al. [81] conducted research on the effect of sloshing on the properties of LH_2 inside the tank of an LH_2 carrier. Therefore, a multi-phase 3D CFD model utilizing the VoF method was developed. The model accounts for the sloshing dynamics, heat ingress into the tank, and the phase change [81]. The model was validated against experimental data. Validation against other oscillatory flow experiments showed strong agreement, with a pressure drop delay but a relative error below 5% after stabilization [81]. To analyze the thermodynamic behaviour of LH_2 under sinusoidal sloshing, H. Lv et al. [82] developed a numerical simulation model, examining the effects of baffles, amplitude, frequency, and storage pressure. The model was developed in Ansys FLUENT 2022. It was found that baffles reduce fluid fluctuations but accelerate evaporation by acting as thermal bridges [82]. The CFD model was validated by comparing pressurization rates and sloshing behaviour in LH_2 tanks with reference studies. Pressurization rates under stationary conditions matched well with an average absolute error of 339 Pa and a relative error of 0.32% [82]. Sloshing behaviour was also verified, showing strong agreement. A mesh independence study determined that a 2.96×10^6 elements mesh ensures accuracy while optimizing computational resources [82]. Furthermore, H. Lv et al. [83] developed a CFD code with implemented fluid–structure-interaction (FSI) to simulate the thermo-hydrodynamic response of LH_2 in tankers equipped with baffles during the braking process. The model accounts for phase change, heat transfer into the tank, and pressurization. The model's pressurization predictions were validated against experimental data. Sloshing behaviour was further verified through comparisons with experimental results [83]. J. Zhang et al. [84] developed a numerical model utilizing the VoF method together with a SST k-omega turbulence model to account for sloshing behaviour of LH_2 on floating storage platforms. C-type and Moss-type LH_2 storage tanks under extreme sea conditions were compared [84]. Self-pressurization, heat transfer into the storage tank, and phase change were subject to the investigation [84]. Validation was not carried out. Z. Xinjia [85] conducted research on condensation heat transfer and sloshing in LH_2 tanks. A modified thermal stratified model was developed to account for thermal stratification, mass transfer at the liquid–vapour interface, and heat transfer into the tank, with the focus being on sloshing behaviour [85]. The model was validated against experimental data [85]. Table 7 provides an overview of models focusing on sloshing.

Table 7. Overview of models focusing on sloshing.

Table Authors	Year	Type of Model	Validated
Z. Liu et al. [18]	2019	Numerical, VoF method	Yes
J. R. Smith et al. [21]	2022	Numerical	No
S. Li et al. [80]	2022	CFD-VoF	Yes
O. V. Kartuzova et al. [20]	2024	CFD	Yes
W. Kang et al. [81]	2024	3D CFD	Yes
H. Lv et al. [82]	2024	CFD, Ansys FLUENT	Yes
J. Zhang et al. [84]	2024	Numerical, VoF method	No
H. Lv et al. [83]	2024	CFD	Yes
Z. Xinjia [85]	2024	Thermal stratified model	Yes

4.2.2. Models Focusing on Multi-Phase Flow During LH_2 Transfer Operations

To investigate the requirements for the pressurized transfer of LH_2, Gluck and Kline [86] presented an analytical model based on system parameters and experimentally determined interfacial mass transfer and gas-phase heat transfer [86]. A developed correlation accurately predicted gas-phase heat transfer within the experimental range. The analytical model successfully estimated gas requirements and provided qualitative insights into the effects of system parameters [86]. Heat transfer correlations were developed and compared with experimental data, yielding reasonable agreements [86]. Campi et al. [87] developed a numerical lumped-parameter model to account for two-phase flow characteristics, chill-down, and heat transfer of LH_2. The model was validated against experimental results [87]. Rame et al. [88] investigated the flow behaviour of LH_2 during line chill-down tests. The focus was on modelling the heat transfer through the pipe wall and the fluid dynamics inside of the pipe. By analyzing video recordings alongside stream temperature data, they identified correlations between temperature variations and flow regime changes [88]. A simple plug flow model was proposed to explain temperature humps in a fluid-filled pipe. It assumes uniform velocity, negligible radial conduction, and axial conduction in the wall. The model suggests that as liquid fraction increases, the heat transfer coefficient rises, briefly raising fluid temperature before it decreases to a steady state [88]. The model was not validated. Darr and Hartwig [89] developed a numerical lumped-node heat transfer model to simulate the chill-down and mass and heat transfer during two-phase convection inside LH_2 pipes. The model outputs were compared to experimental data [89]. The results validated the correlation for actual quality, heat transfer across all boiling regimes, and dividing points like Leidenfrost and DNB temperatures (departure from nucleate boiling temperature) [89]. Kang et al. [90] developed a transient thermal fluid simulation model that utilized the VoF method. The model was developed to investigate the loading process of LH_2 tanks for mobile applications. The model accounts for multi-phase flow, but also for phase-change dynamics and losses through BoG venting [90]. The simulation analyzed variations in volume fraction, pressure, mass flow rate, and temperature. Validation was conducted through a grid-dependency test, where simulations with different mesh densities were compared to ensure consistent LH_2 volume fraction trends, leading to the selection of a sufficiently fine and computationally efficient grid [90]. Mangold et al. [91] investigated the refuelling process of LH_2-powered aircraft, emphasizing safety and economic efficiency. The model accounted for heat transfer into an LH_2 tank and between the liquid and vapour phase, pressurization, and the chill-down of the tank. The model consists of analytical equations and was not validated [91]. To investigate multi-phase thermal flow dynamics in vacuum-insulated LH_2 pipelines for hydrogen-fuelled vessels, Seo et al. [92] developed a numerical multi-phase thermal flow simulation model focusing on phase-change phenomena. It evaluates insulation performance using a composite vacuum system and compares it to polyurethane foam [92]. Results closely matched experimental data, with errors between 0.3% and 1.4%, confirming the model's reliability for LH_2 multi-phase thermal flow simulations [92]. Chung et al. [93] presented a numerical FVM (finite volume method) approach for analyzing fluid flow in an LH_2 tank for liquefied hydrogen carriers, using a CFD model implemented in STAR-CCM+. The model incorporates fluid dynamics, turbulence, multi-phase flow, heat transfer, and BOR [93]. The multi-phase thermal analysis was validated by comparing BOR estimations for a type-C LN2 (liquefied nitrogen) storage tank with experimental data. Results showed that the simulations provided accurate BOR predictions, with good agreement to experimental results, highlighting the effectiveness of the FVM approach for cryogenic storage analysis [93]. Y. Liu et al. [94] simulated the flow dynamics of LH_2 in pipelines under the influence of external forces, focusing on both the fluid behaviour and the structural response of the pipe using a 3D CFD model [94]. The developed model is validated using experimental data for LH_2 flow under superheat conditions. Simulation results closely match experiments, with a maximum relative error below 15%. Deviations stem from complex two-phase flow behaviours and minor sensor errors [94]. Numerical models of flow boiling in heated tubes were developed by LeClair et al. [95] using a GFSSP approach utilizing a finite-volume flow network solver to evaluate fluid flow, heat transfer into the pipe, and pressure drop correlations [95]. The model was validated. Pesich et al. [96] conducted a CFD validation study simulating the no-vent filling and chilling process of LH_2 tanks using three different commercial CFD codes: STAR-CCM+, Fluent, and Flow3D. The models were validated using LH_2 no-vent fill test data from Moran et al. [97], specifically, Test 9093G with upward pipe discharge. Due to missing injector and tank lid details, assumptions were made about pipe dimensions. Experimental uncertainties included inlet conditions, trapped helium, and time-dependent mass flow rate variations [96]. Molkov et al. [98] investigated the behaviour of sLH_2 during the refuelling process at a hydrogen refuelling station (HRS). The study develops a two-phase CFD model, incorporating a new correlation for the Nusselt number that accounts for condensation heat transfer, filling ratio, and HLR (heat loss rate). The results have been validated against data from a conceptual refuelling procedure [98]. Alipour Bonab and Yazdani-Asrami [99] focused on predicting the heat transfer coefficient (HTC) in flow boiling of LH_2, which is intended to be used as both fuel and cryogenic coolant for future hydrogen-powered airplanes [99]. They developed a cascade-forward neural network (CFNN) model, achieving 99.88% accuracy. Cross-validation was applied to assess the CFNN model's performance by dividing the dataset into three folds. The model was trained and tested on each fold, averaging results to reduce overfitting [99]. Table 8 provides an overview of models focusing on multi-phase flow during transfer operations.

Table 8. Overview of models focusing on multi-phase flow during transfer operations.

Table Authors	Year	Type of Model	Validated
D. F. Gluck and F. Kline [86]	1961	Analytical model, system- and experimentally determined parameters	Yes
F. A. Campi et al. [87]	1963	Numerical lumped-parameter model	Yes
E. Rame et al. [88]	2014	Analytical plug flow model	No
S. R. Darr and J. W. Hartwig [89]	2020	Numerical lumped-node model	Yes
Y. Liu et al. [94]	2022	3D CFD	Yes
D. Kang et al. [90]	2022	CFD-VoF	Yes
J. Mangold et al. [91]	2022	Analytical equations	No
Y. S. Seo et al. [92]	2024	Numerical multi-phase thermal flow model	Yes
S.-M. Chung et al. [93]	2024	FVM model	Yes
A. LeClair et al. [95]	2024	Numerical model using a GFSSP and finite-volume flow network solver	Yes
J. M. Pesich et al. [96]	2024	CFD (STAR-CCM+, Fluent, and Flow3D)	Yes
V. Molkov et al. [98]	2024	CFD-based modified thermal stratified model	Yes
S. Alipour Bonab and M. Yazdani-Asrami [99]	2025	CFNN model	Yes

4.2.3. Models Focusing on Flash Boiling and Pressure Recovery

A CFD analysis was undertaken by K. Tani et al. [22] to investigate into the pressure recovery and flashing phenomenon during a pressure reduction experiment in large-scale LH_2 tanks. A flashing experiment at a high liquid level in a 30 m^3 tank was conducted [22]. The model was not validated. Kangwanpongpan et al. [100] presented a two-phase LES turbulent model of flash boiling and pressure recovery for release scenarios from large-scale LH_2 tanks. The model was validated against a large-scale LH_2 flash boiling experiment at JAXA, and successfully predicted pressure recovery and gaseous hydrogen temperature trends [100]. Table 9 provides an overview of models focusing on flash boiling and pressure recovery.

Table 9. Overview of models focusing on flash boiling and pressure recovery.

Authors	Year	Type of Model	Validated
K. Tani et al. [22]	2021	CFD analysis	No
T. Kangwanpongpan et al. [100]	2024	CFD, two-phase LES	Yes

5. Discussion

Thermodynamic phenomena related to LH_2 storage and transfer are often interdependent. Therefore, a comprehensive understanding of multiple phenomena is essential for accurate modelling. Experimental data are either scarce or entirely missing in several key areas of LH_2 thermodynamics research, limiting model validation and accuracy. For instance, studies on gravity's influence on LH_2 tanks have relied on FLUENT-CFD simulations, but due to a lack of experimental validation, the accuracy of these models remains uncertain. Similarly, aircraft LH_2 tank design models could not be validated because the simulated components have not yet been constructed. Thermal behaviour under fire exposure is another area requiring further experimental investigation. A CFD model analyzing LH_2 tanks subjected to external fire indicated the need for additional experiments to refine understanding and improve validation. Likewise, research on cryogenic tank venting and depressurization relies on assumptions due to missing experimental data for insulation thermal conductivity. In terms of boiling heat transfer, some LH_2 thermodynamic models do not incorporate this phenomenon due to the absence of experimental data. This limitation extends to large-scale LH_2 storage, where models assessing pressurization and boil-off gas formation for maritime export lack sufficient large-scale validation. Similarly, thermal stratification in low-gravity conditions presents challenges, as CFD models investigating LH_2 tank behaviour under varying gravitational forces have shown significant deviations from experimental pressure data, requiring parameter calibration due to missing data. Lastly, models for cryogenic hydrogen transfer in space remains largely unvalidated, as experimental testing under relevant space conditions has not yet been conducted. To support validation efforts, the generation of more experimental data should be pursued. This concerns mainly data for pressure and temperature inside LH_2 storage tanks during filling or phases of self-pressurization, and inside LH_2 pipes during transfer operations.

Improved insulation extends dormancy, reducing heat ingress, pressure build-up, and subsequently energy loss through H_2 venting. Advancements in thermal management, including active cooling and optimized venting, enhance operational reliability. Insights gained through multi-phase flow modelling contribute to improving pipeline and transfer line design to optimize flow conditions and to reduce phase separation. Models that are backed through validation against experimental or real-life data particularly support safety and effectiveness. By providing predictive insights into thermodynamic phenomena relevant for LH_2 storage and transfer, validated modelling approaches directly inform the design of safety systems and operational protocols. In this way, modelling is not only a tool for efficiency optimization but also a critical enabler of inherently safer LH_2 storage and transfer infrastructure. The most common type of model found in the literature was the CFD model. CFD models provide detailed insights but are computationally expensive, limiting real-time application. The trade-off between accuracy and feasibility requires further analysis. This constraint highlights the importance of developing advanced reduced-order models (ROMs) to achieve a balance between accuracy and computational efficiency, as emphasized in the future research agenda.

While the narrative review approach provides flexibility to synthesize diverse modelling studies, it also carries the risk of selection bias. This limitation underscores the importance of complementing narrative reviews with systematic meta-analyses in the future to ensure balanced coverage and to strengthen the reliability of safety- and efficiency-related insights. Additionally, limited discussion of safety risks, including high-pressure transfer and leaks, highlights a need for further exploration in practical LH_2 deployment. Future research should prioritize large-scale experimental validation to enhance model reliability. Collaboration with industry stakeholders can establish standardized testing protocols, while dedicated sensor technologies could provide real-time heat transfer and pressurization data. Computational efficiency improvements, including reduced-order models (ROMs) and artificial intelligence (AI) integration, could balance prediction speed and accuracy. Further research is needed on LH_2 storage in extreme environments, such as microgravity and maritime transport, to understand phase behaviour and sloshing effects. CFD studies of scenarios under microgravity should furthermore be investigated regarding sensitivity to mesh refinement. Venting and BOG management should be optimized through re-liquefaction technologies and predictive strategies to minimize hydrogen losses. Systematic meta-analyses of LH_2 modelling approaches can assess accuracy and efficiency, reducing potential biases. Addressing these gaps through interdisciplinary research will improve the safety, efficiency, and economic feasibility of LH_2 storage and transport.

Models, whether mathematical or computational, typically serve three fundamental purposes: prediction, design, and control. Predictive models aim to anticipate system behaviour under specific conditions or project future states. Design-oriented models help explore design spaces, evaluate concepts, and guide innovation, while control models support decision-making and operational strategies for managing complex systems. In this context, the interpretation of "lack of validation" requires careful nuance, particularly for models developed for foresight or exploratory analysis. Although validation is essential when experimental data exists, some models are intentionally created for purposes where validation is not yet feasible. These include exploratory prediction of prospective system behaviour, concept evaluation for technologies still in early development, and design-space mapping to identify promising directions before prototypes or empirical datasets become available. Therefore, the absence of validation does not necessarily indicate a methodological weakness. Rather, it reflects the epistemic nature of the model's intended purpose: to inform thinking, guide research priorities, and frame possibilities, rather than to provide definitive, empirically confirmed outcomes.

6. Conclusions

This structured narrative review has examined 81 studies spanning over six decades, focusing on modelling approaches for LH_2 thermodynamics during storage and transfer. By categorizing the literature into two primary domains—thermodynamic behaviour within storage tanks and multi-phase flow dynamics in transfer systems—the review identified seven key focus areas and evaluated the extent of model validation across them. The analysis reveals that while a wide range of modelling techniques have been developed—including analytical, CFD, and machine learning-based models—many remain unvalidated due to the scarcity of experimental data, especially under extreme or large-scale conditions. This limits their predictive reliability and practical applicability. In mapping the current research landscape, this review contributes a consolidated reference for future model developers and researchers. It highlights the need for integrated modelling approaches that account for the interdependence of heat transfer, mass transfer, stratification, pressurization, and flow dynamics. It also underscores the importance of validation against experimental or operational data to ensure safety and efficiency in LH_2 infrastructure.

Despite the progress made, several limitations persist. These include insufficient modelling of coupled phenomena, limited data on accident scenarios, and a lack of large-scale experimental validation.

Future research should prioritize the following:

- Development of reduced-order and hybrid models for real-time applications;
- Experimental campaigns under microgravity and maritime conditions;
- Integration of AI and sensor technologies for dynamic system monitoring;
- Systematic meta-analyses to assess model accuracy and generalizability.

In the frame of this review and due to the high number of different approaches found in the literature, it is not feasible to present each modelling approach in greater detail. For future reviews, the scope should therefore be limited to one focus area, possibly an operational scenario, highlighting key differences, innovations, or contradictions between models. The analyzed models should be discussed regarding reliability and applicability by comparing accuracy, modelling assumptions, and computational efficiency.

By addressing these gaps, the scientific community can advance the design of inherently safe, efficient, and scalable LH_2 systems and support the broader transition to a clean hydrogen economy.

Author Contributions: Conceptualization, L.M.C. and F.U.; Methodology, L.M.C.; Writing—Original Draft Preparation, L.M.C.; Writing—Review & Editing, F.U. and G.E.S.; Visualization, L.M.C.; Supervision, F.U. and G.E.S.; Project Administration, F.U. All authors have read and agreed to the published version of the manuscript.

Funding: This research was funded by the ELVHYS project grant number 101101381.

Data Availability Statement: No new data were created or analyzed in this study.

Acknowledgments: This work was undertaken as part of the ELVHYS project No. 101101381 supported by the Clean Hydrogen Partnership and its members and the European Union. UK participants in Horizon Europe Project ELVHYS are supported by UKRI grant numbers 10063519 (University of Ulster) and 10070592 (Health and Safety Executive). Funded by the European Union. Views and opinions expressed are, however, those of the authors only and do not necessarily reflect those of the European Union or Clean Hydrogen JU. Neither the European Union nor the granting authority can be held responsible for them. During the preparation of this manuscript, the authors used Microsoft Copilot (M365 Copilot, Version number: bizchat.20251204.45.4) for the purposes of language editing. The authors have reviewed and edited the output and take full responsibility for the content of this publication.

Conflicts of Interest: The authors declare no conflict of interest.

References

1. Hydrogen Storage. Available online: https://www.energy.gov/eere/fuelcells/hydrogen-storage (accessed on 2 May 2025).
2. Møller, K.T.; Jensen, T.R.; Akiba, E.; Li, H. Hydrogen—A Sustainable Energy Carrier. *Prog. Nat. Sci. Mater. Int.* **2017**, *27*, 34–40. [CrossRef]
3. NIST Chemistry WebBook Hydrogen. Available online: https://webbook.nist.gov/cgi/cbook.cgi?Name=hydrogen&Units=SI (accessed on 10 December 2023).
4. Vudumu, S.K. Experimental and Computational Investigations of Hydrogen Safety, Dispersion and Combustion for Transportation Applications. Ph.D. Thesis, Missouri University of Science and Technology, Rolla, MI, USA, 2010.
5. Claussner Lucas, M.; Ustolin, F.; Scarponi, G.E. Design and Operation of Liquid Hydrogen Storage Tanks. *Chem. Eng. Trans.* **2024**, *111*, 31–36. [CrossRef]
6. Jiang, W.; Sun, P.; Li, P.; Zuo, Z.; Huang, Y. Transient Thermal Behavior of Multi-Layer Insulation Coupled with Vapor Cooled Shield Used for Liquid Hydrogen Storage Tank. *Energy* **2021**, *231*, 120859. [CrossRef]

7. Stephens, J.R.; Belcher, T.; Furby, J.; Giddens, P.; Hamill, B.; Nguyen, S.; Pedersen, K.; Smith, J.; Tashakkor, S.; Valenzuela, J.G.; et al. Long Duration Storage of Liquid Hydrogen via Two-Stage Active Cooling Hardware Characterization and Test Planning. In Proceedings of the AIAA SCITECH 2024 Forum, Orlando, FL, USA, 8–12 January 2024; American Institute of Aeronautics and Astronautics: Orlando, FL, USA, 2024.
8. Wooldridge, M.; Luebbers, R. *Heat Transfer*; McGraw Hill: Columbus, OH, USA, 2020. Available online: https://www.accessscience.com/content/article/a311100 (accessed on 30 November 2025).
9. Matsuno, K.; Nemoto, A. Quantum as a Heat Engine—The Physics of Intensities Unique to the Origins of Life. *Phys. Life Rev.* **2005**, *2*, 227–250. [CrossRef]
10. Barron, R.F.; Nellis, G.F. *Cryogenic Heat Transfer*; CRC Press: Boca Raton, FL, USA, 2016.
11. Wang, J.; Webley, P.A.; Hughes, T.J. Thermodynamic Modelling of Low Fill Levels in Cryogenic Storage Tanks for Application to Liquid Hydrogen Maritime Transport. *Appl. Therm. Eng.* **2024**, *256*, 124054. [CrossRef]
12. Zheng, L.; Zhang, X. Numerical Methods. In *Modeling and Analysis of Modern Fluid Problems*; Elsevier: Amsterdam, The Netherlands, 2017; pp. 361–455, ISBN 978-0-12-811753-8.
13. Lee, W.H. A Pressure Iteration Scheme for Two-Phase Flow Modeling. *Tech. Rep. LA-UR 79-975* **1980**, *1*, 407–431.
14. Xavier, M.; Edwin Raj, R.; Narayanan, V. Thermal Stratification in LH_2 Tank of Cryogenic Propulsion Stage Tested in ISRO Facility. *IOP Conf. Ser. Mater. Sci. Eng.* **2017**, *171*, 012063. [CrossRef]
15. Joseph, J.; Agrawal, G.; Agarwal, D.K.; Pisharady, J.C.; Sunil Kumar, S. Effect of Insulation Thickness on Pressure Evolution and Thermal Stratification in a Cryogenic Tank. *Appl. Therm. Eng.* **2017**, *111*, 1629–1639. [CrossRef]
16. Alipour Bonab, S.; Yazdani-Asrami, M. Investigation on the Heat Transfer Estimation of Subcooled Liquid Hydrogen for Transportation Applications Using Intelligent Technique. *Int. J. Hydrogen Energy* **2024**, *84*, 468–479. [CrossRef]
17. Aydelott, J.C.; Spuckler, C.M. *Venting of Liquid Hydrogen Tankage*; National Aeronautics and Space Administration: Washington, DC, USA, 1969.
18. Liu, Z.; Feng, Y.; Lei, G.; Li, Y. Fluid Sloshing Dynamic Performance in a Liquid Hydrogen Tank. *Int. J. Hydrogen Energy* **2019**, *44*, 13885–13894. [CrossRef]
19. Inman, A.E.; Muehlbauer, J.G. *Shuttle Performance Augmentation with the Titan Liquid Boost Module*; National Aeronautics and Space Administration: Washington, DC, USA, 1980.
20. Kartuzova, O.V.; Kassemi, M.; Hauser, D. Validation of a Two-Phase CFD Model for Predicting Propellant Tank Pressurization and Pressure Collapse in The Ground-Based K-Site Hydrogen Slosh Experiment. *NASA Glenn Res. Cent.* **2024**, *15*, 0547.
21. Smith, J.R.; Gkantonas, S.; Mastorakos, E. Modelling of Boil-Off and Sloshing Relevant to Future Liquid Hydrogen Carriers. *Energies* **2022**, *15*, 2046. [CrossRef]
22. Tani, K.; Himeno, T.; Sakuma, Y.; Watanabe, T.; Kobayashi, H.; Toge, T.; Unno, S.; Kamiya, S.; Muragishi, O.; Kanbe, K. Pressure Recovery During Pressure Reduction Experiment with Large-Scale Liquid Hydrogen Tank. *Int. J. Hydrogen Energy* **2021**, *46*, 29583–29596. [CrossRef]
23. Xiao, Y.; Watson, M. Guidance on Conducting a Systematic Literature Review. *J. Plan. Educ. Res.* **2019**, *39*, 93–112. [CrossRef]
24. Pickering, C.; Byrne, J. The Benefits of Publishing Systematic Quantitative Literature Reviews for PhD Candidates and Other Early-Career Researchers. *High. Educ. Res. Dev.* **2014**, *33*, 534–548. [CrossRef]
25. Huntley, S.C.; Gauntner, J.W.; Anderson, B.H. *Wall and Bottom Heating of Liquid Hydrogen in a Propellant Tank*; National Aeronautics and Space Administration: Washington, DC, USA, 1966.
26. Wang, L.; Yan, T.; Wang, J.; Ye, S.; Li, Y.; Zhuan, R.; Wang, B. CFD Investigation on Thermodynamic Characteristics in Liquid Hydrogen Tank During Successive Varied-Gravity Conditions. *Cryogenics* **2019**, *103*, 102973. [CrossRef]
27. Martin, J.J.; Hastings, L. *Large-Scale Liquid Hydrogen Testing of a Variable Density Multilayer Insulation with a Foam Substrate*; National Aeronautics and Space Administration: Washington, DC, USA, 2001.
28. Lv, R.; Huang, Y.; Wu, J. Thermodynamic Analysis of Partially Filled Hydrogen Tanks in a Wide Scale Range. *Appl. Therm. Eng.* **2021**, *193*, 117007. [CrossRef]
29. Ustolin, F.; Iannaccone, T.; Cozzani, V.; Jafarzadeh, S.; Paltrinieri, N. Time to Failure Estimation of Cryogenic Liquefied Tanks Exposed to a Fire. In Proceedings of the 31st European Safety and Reliability Conference (ESREL 2021), Angers, France, 19–23 September 2021; Research Publishing Services: Singapore, 2021; pp. 935–942.
30. Ustolin, F.; Scarponi, G.E.; Iannaccone, T.; Cozzani, V.; Paltrinieri, N. Cryogenic Hydrogen Storage Tanks Exposed to Fires: A CFD Study. *Chem. Eng. Trans.* **2022**, *90*, 535–540. [CrossRef]
31. Pehr, K. Experimental Examinations on the Worst-Case Behaviour of LH_2/LNG Tanks for Passenger Cars. In Proceedings of the 11th World Hydrogen Energy Conference, Stuttgart, Germany, 23–28 June 1996; 1996; pp. 2168–2187.
32. Sun, Z.; Li, M.; Qu, Z.; Tian, D.; Zhang, J. A Quasi-2D Thermodynamic Model for Performance Analysis and Optimization of Liquid Hydrogen Storage System with Multilayer Insulation and Vapor-Cooled Shield. *J. Energy Storage* **2023**, *73*, 109128. [CrossRef]

33. Parello, R.; Defoort, S.; Benard, E.; Gourinat, Y. Design and Integration of a Liquid Hydrogen Tank on an Aircraft. In Proceedings of the AIAA SCITECH 2024 Forum, Orlando, FL, USA, 8–12 January 2024; American Institute of Aeronautics and Astronautics: Orlando, FL, USA, 2024.
34. Xu, Z.; Zhang, Y.; Zhao, N.; Li, S.; Yan, Y.; Hillmansen, S. Optimal Thermal Management on the Storage Vessle for Cryogenic Hydrogen-Powered Hybrid Train. In Proceedings of the 2024 IEEE Vehicle Power and Propulsion Conference (VPPC), Washington DC, USA, 7–10 October 2024; IEEE: Washington, DC, USA, 2024; pp. 1–6.
35. Lv, H.; Zhang, Z.; Chen, L.; Zhang, Z.; Chen, S.; Hou, Y. Thermodynamic Analysis of Vapor-Cooled Shield with Para-to-Ortho Hydrogen Conversion in Composite Multilayer Insulation Structure for Liquid Hydrogen Tank. *Int. J. Hydrogen Energy* **2024**, *50*, 1448–1462. [CrossRef]
36. Li, K.; Wen, J.; Xin, B.; Zhou, A.; Wang, S. Transient-State Modeling and Thermodynamic Analysis of Self-Pressurization Liquid Hydrogen Tank Considering Effect of Vacuum Multi-Layer Insulation Coupled with Vapor-Cooled Shield. *Energy* **2024**, *286*, 129450. [CrossRef]
37. Olsen, W.A. *Experimental and Analytical Investigation of Interfacial Heat and Mass Transfer in a Pressurized Tank Containing Liquid Hydrogen*; National Aeronautics and Space Administration: Washington, DC, USA, 1966.
38. Lin, C.; Hasan, M.; Nyland, T. Mixing and Transient Interface Condensation of a Liquid Hydrogen Tank. In Proceedings of the 29th Joint Propulsion Conference and Exhibit, Monterey, CA, USA, 28–30 June 1993; American Institute of Aeronautics and Astronautics: Monterey, CA, USA, 1993.
39. Schweickart, R.B. Thermodynamic Analysis of a Demonstration Concept for the Long-Duration Storage and Transfer of Cryogenic Propellants. *Cryogenics* **2014**, *64*, 283–288. [CrossRef]
40. Stewart, M. Pressurization of a Flightweight, Liquid Hydrogen Tank: Evaporation & Condensation at the Liquid/Vapor Interface. In Proceedings of the 53rd AIAA/SAE/ASEE Joint Propulsion Conference, Atlanta, GA, USA, 10–12 July 2017; American Institute of Aeronautics and Astronautics: Atlanta, GA, USA, 2017.
41. Wang, H.; Wang, B.; Li, R.; Shen, X.; Wu, Y.; Quanwen, P.; Yuanxin, H.; Weiming, Z.; Zhihua, G. Theoretical Investigation on Heat Leakage Distribution between Vapor and Liquid in Liquid Hydrogen Tanks. *Int. J. Hydrogen Energy* **2023**, *48*, 17187–17201. [CrossRef]
42. Jiang, Y.; Yu, Y.; Wang, Z.; Zhang, S.; Cao, J. CFD Simulation of Heat Transfer and Phase Change Characteristics of the Cryogenic Liquid Hydrogen Tank Under Microgravity Conditions. *Int. J. Hydrogen Energy* **2023**, *48*, 7026–7037. [CrossRef]
43. Wang, Z.; Mérida, W. Thermal Performance of Cylindrical and Spherical Liquid Hydrogen Tanks. *Int. J. Hydrogen Energy* **2024**, *53*, 667–683. [CrossRef]
44. Choi, D.; Lee, S.; Kim, S. A Thermodynamic Model for Cryogenic Liquid Hydrogen Fuel Tanks. *Appl. Sci.* **2024**, *14*, 3786. [CrossRef]
45. Wang, H.; Wang, B.; Sun, J.; Pan, Q.; Luo, G.; Tao, X.; He, Y.; Pfotenhauer, J.; Jin, T.; Gan, Z. Experimental and Computational Fluid Dynamic Investigation on Thermal Behaviors of Liquid Hydrogen During the No-Vented Storage Process: A Literature Review. *Int. J. Hydrogen Energy* **2024**, *57*, 822–843. [CrossRef]
46. Arnett, R.W.; Voth, R.O. *A Computer Program for the Calculation of Thermal Stratification and Self-Pressurization in a Liquid Hydrogen Tank*; National Aeronautics and Space Administration: Washington, DC, USA, 1972.
47. Grayson, G.D. Coupled Thermodynamic-Fluid-Dynamic Solution for a Liquid-Hydrogen Tank. *J. Spacecr. Rockets* **1995**, *32*, 918–921. [CrossRef]
48. Daigle, M.J.; Smelyanskiy, V.N.; Boschee, J.; Foygel, M. Temperature Stratification in a Cryogenic Fuel Tank. *J. Thermophys. Heat Transf.* **2013**, *27*, 116–126. [CrossRef]
49. Liu, Z.; Li, Y.; Zhou, G. Study on Thermal Stratification in Liquid Hydrogen Tank Under Different Gravity Levels. *Int. J. Hydrogen Energy* **2018**, *43*, 9369–9378. [CrossRef]
50. Aydelott, J.C. Self Pressurization of Liquid Hydrogen Tankage. Master's Thesis, Cornell University, Ithaca, NY, USA, 1967.
51. Gursu, S.; Sherif, S.A.; Veziroglu, T.N.; Sheffield, J.W. Analysis and Optimization of Thermal Stratification and Self-Pressurization Effects in Liquid Hydrogen Storage Systems—Part 1: Model Development. *J. Energy Resour. Technol.* **1993**, *115*, 221–227. [CrossRef]
52. Gursu, S.; Sherif, S.A.; Veziroglu, T.N.; Sheffield, J.W. Analysis and Optimization of Thermal Stratification and Self-Pressurization Effects in Liquid Hydrogen Storage Systems—Part 2: Model Results and Conclusions. *J. Energy Resour. Technol.* **1993**, *115*, 228–231. [CrossRef]
53. Barsi, S.; Kassemi, M. Numerical and Experimental Comparisons of the Self-Pressurization Behavior of an LH_2 Tank in Normal Gravity. *Cryogenics* **2008**, *48*, 122–129. [CrossRef]
54. Kartuzova, O.V.; Kassemi, M.; Moder, J.P.; Agui, J.H. Self-Pressurization and Spray Cooling Simulations of the Multipurpose Hydrogen Test Bed (MHTB) Ground-Based Experiment. In Proceedings of the 50th AIAA/ASME/SAE/ASEE Joint Propulsion Conference, Cleveland, OH, USA, 28–30 July 2014; American Institute of Aeronautics and Astronautics: Cleveland, OH, USA, 2014.

55. Kassemi, M.; Kartuzova, O. Effect of Interfacial Turbulence and Accommodation Coefficient on CFD Predictions of Pressurization and Pressure Control in Cryogenic Storage Tank. *Cryogenics* **2016**, *74*, 138–153. [CrossRef]
56. Stewart, M.; Moder, J.P. Self-Pressurization of a Flightweight, Liquid Hydrogen Tank: Simulation and Comparison with Experiments. In Proceedings of the 52nd AIAA/SAE/ASEE Joint Propulsion Conference, Salt Lake City, UT, USA, 25–27 July 2016; American Institute of Aeronautics and Astronautics: Salt Lake City, UT, USA, 2016.
57. Majumdar, A.; Valenzuela, J.; LeClair, A.; Moder, J. Numerical Modeling of Self-Pressurization and Pressure Control by a Thermodynamic Vent System in a Cryogenic Tank. *Cryogenics* **2016**, *74*, 113–122. [CrossRef]
58. Liu, Y.; Zhou, P.; Jeong, B.; Wang, H. Design and Optimization of a Type-C Tank for Liquid Hydrogen Marine Transport. *Int. J. Hydrogen Energy* **2023**, *48*, 34885–34896. [CrossRef]
59. Ahluwalia, R.K.; Peng, J.-K.; Roh, H.-S.; Papadias, D.; Wang, X.; Aceves, S.M. Liquid Hydrogen Storage System for Heavy Duty Trucks: Capacity, Dormancy, Refueling, and Discharge. *Int. J. Hydrogen Energy* **2023**, *48*, 34120–34131. [CrossRef]
60. Matveev, K.I.; Leachman, J.W. The Effect of Liquid Hydrogen Tank Size on Self-Pressurization and Constant-Pressure Venting. *Hydrogen* **2023**, *4*, 444–455. [CrossRef]
61. Jeong, S.; Lee, S.; Moon, S. CFD Thermo-Hydraulic Evaluation of a Liquid Hydrogen Storage Tank with Different Insulation Thickness in a Small-Scale Hydrogen Liquefier. *Fluids* **2023**, *8*, 239. [CrossRef]
62. Wang, H.R.; Wang, B.; Pan, Q.W.; Wu, Y.Z.; Jiang, L.; Wang, Z.H.; Gan, Z.H. Modeling and Thermodynamic Analysis of Thermal Performance in Self-Pressurized Liquid Hydrogen Tanks. *Int. J. Hydrogen Energy* **2022**, *47*, 30530–30545. [CrossRef]
63. Rahman, A.A.; Wang, H.; Wang, B.; Xu, T.; Li, J.; He, Y.; Jin, T.; Gan, Z. Prediction of Pressure Evolution in Non-Venting Self-Pressurized Liquid Hydrogen Tanks Using Artificial Neural Network Approach. *Int. J. Hydrogen Energy* **2024**, *68*, 428–439. [CrossRef]
64. Wang, H.; Wang, B.; Xu, T.; Shen, X.; He, Y.; Zhou, W.; Pfotenhauer, J.; Jin, T.; Gan, Z. Thermal Models for Self-Pressurization Prediction of Liquid Hydrogen Tanks: Formulation, Validation, Assessment, and Prospects. *Fuel* **2024**, *365*, 131247. [CrossRef]
65. Blatt, M.H. Impirical Correlations for Pressure Rise in Closed Cryogenic Containers. *JSR Spacecr. Rockets* **1968**, *5*, 733–735. [CrossRef]
66. Wang, J.; Webley, P.A.; Hughes, T.J. Thermodynamic Modelling of Pressurised Storage and Transportation of Liquid Hydrogen for Maritime Export. *Int. J. Hydrogen Energy* **2024**, *62*, 1273–1285. [CrossRef]
67. Yu, T.; Lim, Y. Changes in Holding Time Depending Liquid Filling Ratio, Heat Transfer Parameters, and Shape of the Tank. *Int. J. Hydrogen Energy* **2024**, *82*, 836–852. [CrossRef]
68. Mandell, D.A.; Roudebush, W.H. *Parametric Investigation of Liquid Hydrogen Tank Pressurization During Outflow*; NASA: Washington, DC, USA, 1965.
69. Roudebush, W.H.; Mandell, D.A. *Analytical Investigation of Some Important Parameters in the Pressurized Liquid Hydrogen Tank Outflow Problem*; NASA: Washington, DC, USA, 1965.
70. Winters, W.S.; Houf, W.G. Simulation of Small-Scale Releases from Liquid Hydrogen Storage Systems. *Int. J. Hydrogen Energy* **2011**, *36*, 3913–3921. [CrossRef]
71. Petitpas, G. *Boil-Off Losses Along LH_2 Pathway*; Lawrence Livermore National Laboratory: Livermore, CA, USA, 2018; p. LLNL-TR-750685.
72. Zuo, Z.; Jiang, W.; Qin, X.; Huang, Y. Numerical Investigation on Full Thermodynamic Venting Process of Liquid Hydrogen in an On-Orbit Storage Tank. *Int. J. Hydrogen Energy* **2020**, *45*, 27792–27805. [CrossRef]
73. Al Ghafri, S.Z.S.; Swanger, A.; Jusko, V.; Siahvashi, A.; Perez, F.; Johns, M.L.; May, E.F. Modelling of Liquid Hydrogen Boil-Off. *Energies* **2022**, *15*, 1149. [CrossRef]
74. Hamacher, J.; Stary, A.; Stops, L.; Siebe, D.; Kapp, M.; Rehfeldt, S.; Klein, H. Modeling the Thermodynamic Behavior of Cryo-Compressed Hydrogen Tanks for Trucks. *Cryogenics* **2023**, *135*, 103743. [CrossRef]
75. Chen, H.; Li, X.; Zhang, H.; Yang, P.; Liu, Y.; Ye, W.; Yan, C.; Wang, X. Control Strategy Optimization of Thermodynamic Venting System in Liquid Hydrogen Storage Tank Under Microgravity. *Microgravity Sci. Technol.* **2024**, *37*, 1. [CrossRef]
76. Stops, L.; Siebe, D.; Stary, A.; Hamacher, J.; Sidarava, V.; Rehfeldt, S.; Klein, H. Generalized Thermodynamic Modeling of Hydrogen Storage Tanks for Truck Application. *Cryogenics* **2024**, *139*, 103826. [CrossRef]
77. Dutta, J.; Mukhopadhyay, A. Thermodynamic Model of Storage and Discharge of Liquid Hydrogen and Oxygen in Pressure Vessels Under Cryogenic Conditions. In *Advances in Energy and Sustainability*; De, S., Roy, P.C., Eds.; Lecture Notes in Mechanical Engineering; Springer Nature: Singapore, 2024; pp. 171–184, ISBN 978-981-97-7307-7.
78. Okpeke, B.E.; Ait Aider, C.; Baetcke, L.; Ehlers, S. Assessment of Boil-Off Losses and Their Cost Implication during Liquid Hydrogen Tank Filling with and without Precooling. *Energies* **2024**, *17*, 4092. [CrossRef]
79. Kumar, L.; Sleiti, A. Modeling and Analysis of Effect of Various Tank Geometries and Relief Pressure on Liquid Hydrogen (LH_2) Boil-Off Losses. *Energy Proc.* **2024**, *44*, 1–7. [CrossRef]
80. Li, S.; Yan, Y.; Wei, W.; Wang, Z.; Ni, Z. Numerical Simulation on the Thermal Dynamic Behavior of Liquid Hydrogen in a Storage Tank for Trailers. *Case Stud. Therm. Eng.* **2022**, *40*, 102520. [CrossRef]

81. Kang, W.; Gao, B.; Fan, X. Numerical Study of the Effect of Sloshing on the Properties of Fluids in Tanks of Liquid Hydrogen Carriers. In Proceedings of the Ninth International Symposium on Energy Science and Chemical Engineering (ISESCE 2024), Nanjing, China, 22–24 March 2024; Ensafi, A., Abdullah, A.Z., Aruna, K.K., Eds.; SPIE: Nanjing, China, 2024; p. 30.
82. Lv, H.; Chen, L.; Zhang, Z.; Chen, S.; Hou, Y. Numerical Study on Thermodynamic Characteristics of Large-Scale Liquid Hydrogen Tank with Baffles Under Sloshing Conditions. *Int. J. Hydrogen Energy* **2024**, *57*, 562–574. [CrossRef]
83. Lv, H.; Zhang, Z.; Chen, L.; Chen, S.; Yang, T.; Zhang, K.; Hou, Y. Study on Thermohydrodynamic Responses of Liquid Hydrogen in Baffled Tankers During Braking Process. *Int. J. Heat Fluid Flow* **2024**, *110*, 109612. [CrossRef]
84. Zhang, J.; Shi, S.; Yuan, H.; Yang, K. Study on Coupled Motion of Floating Hydrogen Storage Platform with Liquid Hydrogen Sloshing. *China Ocean Eng.* **2024**, *38*, 999–1011. [CrossRef]
85. Xinjia, Z.; Zhang, X.; Xu, C. Thermodynamic Study on Condensation Heat Transfer and Sloshing of the Liquid Hydrogen Storage. In Proceedings of the Fifth International Conference on Green Energy, Environment, and Sustainable Development (GEESD 2024), Mianyang, China, 28–30 June 2024; Aghaei, M., Ren, H., Zhang, X., Eds.; SPIE: Mianyang, China, 2024; p. 93.
86. Gluck, D.F.; Kline, F. *Gas Requirements in Pressurized Transfer of Liquid Hydrogen*; NASA: Washington, DC, USA, 1961.
87. Campi, F.A.; Chi, J.W.H.; DeZubay, E.A.; Holmgren, J.D.; Vetere, A.M. *Transient Two-Phase Heat Transfer and Flow Characteristics of Liquid Hydrogen*; Westinghouse Electric Corporation Astronuclear Laboratory: Large, PA, USA, 1963.
88. Rame, E.; Hartwig, J.W.; McQuillen, J.B. Flow Visualization of Liquid Hydrogen Line Chill Down Tests. In Proceedings of the 52nd Aerospace Sciences Meeting, National Harbor, MD, USA, 13–17 January 2014; American Institute of Aeronautics and Astronautics: National Harbor, MD, USA, 2014.
89. Darr, S.R.; Hartwig, J.W. Two-Phase Convection Heat Transfer Correlations for Liquid Hydrogen Pipe Chilldown. *Cryogenics* **2020**, *105*, 102999. [CrossRef]
90. Kang, D.; Yun, S.; Kim, B.; Kim, J.; Kim, G.; Lee, H.; Choi, S. Numerical Investigation of the Initial Charging Process of the Liquid Hydrogen Tank for Vehicles. *Energies* **2022**, *16*, 38. [CrossRef]
91. Mangold, J.; Silberhorn, D.; Moebs, N.; Dzikus, N.; Hoelzen, J.; Zill, T.; Strohmayer, A. Refueling of LH2 Aircraft—Assessment of Turnaround Procedures and Aircraft Design Implication. *Energies* **2022**, *15*, 2475. [CrossRef]
92. Seo, Y.S.; Chung, S.-M.; Park, J.-C. Multiphase-Thermal Flow Simulation in a Straight Vacuum-Insulated LH_2 Pipe: Fuel Gas Supply System in a LH_2-Fueled Ship. *J. Mar. Sci. Eng.* **2024**, *12*, 914. [CrossRef]
93. Chung, S.-M.; Ahn, H.-J.; Park, J.-C. Numerical Approach to Analyze Fluid Flow in a Type C Tank for Liquefied Hydrogen Carrier (Part 2: Thermal Flow). *J. Energy Storage* **2024**, *76*, 109599. [CrossRef]
94. Liu, Y.; Qiu, Y.; Liu, Z.; Lei, G. Modeling and Analysis of the Flow Characteristics of Liquid Hydrogen in a Pipe Suffering from External Transient Impact. *Energies* **2022**, *15*, 4154. [CrossRef]
95. LeClair, A.; Baldwin, M.; Majumdar, A.; Hartwig, J.; Ganesan, V.; Mudawar, I. Modeling of Cryogenic Heated-Tube Flow Boiling Experiments of Hydrogen and Helium with the Generalized Fluid System Simulation Program. *Cryogenics* **2024**, *143*, 103926. [CrossRef]
96. Pesich, J.M.; Hauser, D.M.; Yamashita, K.G.; Baker, M.C.; Sakowski, B.A. Liquid Hydrogen Tank Chill and No-Vent Fill Prediction Using Computational Fluid Dynamics. *IOP Conf. Ser. Mater. Sci. Eng.* **2024**, *1301*, 012065. [CrossRef]
97. Moran, M.E.; Nyland, T.W. *Liquid Transfer Cryogenic Test Facility-Initial Hydrogen and Nitrogen No-Vent Fill Data*; NASA: Washington, DC, USA, 1990.
98. Molkov, V.; Ebne-Abbasi, H.; Makarov, D. Liquid Hydrogen Refuelling at HRS: Description of sLH2 Concept, Modelling Approach and Results of Numerical Simulations. *Int. J. Hydrogen Energy* **2024**, *93*, 285–296. [CrossRef]
99. Alipour Bonab, S.; Yazdani-Asrami, M. Artificial Intelligence-Based Model to Predict the Heat Transfer Coefficient in Flow Boiling of Liquid Hydrogen as Fuel and Cryogenic Coolant in Future Hydrogen-Powered Cryo-Electric Aviation. *Fuel* **2025**, *381*, 133323. [CrossRef]
100. Kangwanpongpan, T.; Makarov, D.; Cirrone, D.; Molkov, V. LES Model of Flash-Boiling and Pressure Recovery Phenomena During Release from Large-Scale Pressurised Liquid Hydrogen Storage Tank. *Int. J. Hydrogen Energy* **2024**, *50*, 390–405. [CrossRef]

hydrogen

MDPI

Review

Multilevel Mechanisms of Magnetic Nanoparticles in Enhancing Dark Fermentative Hydrogen Production: From Pure to Mixed Cultures

Junwei Yan [1] and Zhangzhang Xie [2,*]

1 School of Engineering, Guangzhou College of Technology and Business, Guangzhou 510850, China; yanjunwei@gzgs.edu.cn
2 National-Regional Joint Engineering Research Center for Soil Pollution Control and Remediation in South China, Guangdong Key Laboratory of Integrated Agro-Environmental Pollution Control and Management, Institute of Eco-Environmental and Soil Sciences, Guangdong Academy of Sciences, Guangzhou 510650, China
* Correspondence: zzxie@soil.gd.cn

Abstract

Dark fermentative hydrogen production is constrained by challenges including low hydrogen yield and operational instability. Magnetic nanoparticles (MNPs) have emerged as promising additives for enhancing biohydrogen production due to their unique physicochemical characteristics, such as high specific surface area, excellent electrical conductivity, and inherent magnetic recyclability. This review systematically compares the enhancement mechanisms of MNPs in two distinct microbial systems: pure cultures and mixed cultures. In pure cultures, MNPs function primarily at the cellular and molecular levels through the following: (1) serving as sustained-release sources of essential metallic cofactors like Fe and Ni to promote hydrogenase synthesis and activation; (2) acting as efficient electron carriers that facilitate intracellular and extracellular electron transfer; and (3) redirecting central carbon metabolism toward high-hydrogen-yield acetate-type fermentation. In mixed cultures, which are more representative of practical applications, MNPs operate at the ecological level through the following: (1) modifying microenvironmental niches to exert selective pressure that enriches hydrogen-producing bacteria, such as *Clostridium*; (2) forming conductive networks that promote direct interspecies electron transfer and strengthen syntrophic metabolism; and (3) enhancing system robustness via toxin adsorption and pH buffering. Despite promising phenomenological improvements, critical knowledge gaps remain, including unclear structure–activity relationships of MNPs, insufficient quantification of electron transfer pathways, unknown genetic regulatory mechanisms, and overlooked magnetobiological effects. Future research should integrate electrochemical monitoring, multi-omics analyses, and advanced characterization techniques to deepen the mechanistic understanding of nanomaterial–microbe interactions. This review aims to provide theoretical insights and practical strategies for developing efficient and sustainable MNP–microorganism hybrid systems for scalable biohydrogen production.

Keywords: nanoadditives; hydrogenase activity; electron shuttling; metabolic flux analysis; community dynamics; syntrophy

1. Introduction

With the increasingly severe global energy crisis and environmental pollution, the development of green and sustainable alternative energy sources has become urgent. Hy-

 https://doi.org/10.3390/hydrogen6040120

drogen is hailed as an ideal energy carrier for the 21st century, due to its high energy density (141.9 MJ/kg) [1]. Among various hydrogen production technologies, dark fermentative hydrogen production stands out due to its ability to utilize diverse organic wastes and biomass as substrates under mild operational conditions, offering both environmental and economic benefits [2,3]. Dark fermentation is an anaerobic process where carbohydrates are converted into hydrogen, carbon dioxide, and volatile fatty acids (VFAs), such as acetate and butyrate, by microorganisms [4]. The process can be carried out by defined pure cultures (e.g., *Clostridium* spp. [5], *Enterobacter* spp. [6]) or by mixed microbial consortia [7]; each system presents distinct metabolic advantages and operational challenges. The key to the process is the activity of hydrogenase enzymes [8]. Its efficiency is governed by critical parameters, including substrate concentration and complexity [9], microbial community structure [10], pH and temperature [11], and hydraulic retention time [12]. Dark fermentation could be applied to a variety of feedstocks, including lignocellulosic materials, industrial effluents, and municipal organic wastes, thereby contributing to waste valorization and circular economy principles [13–15]. Moreover, dark fermentation operates under anaerobic conditions and does not require light energy, making it more adaptable and easier to scale compared to photobiological methods [16].

However, the practical application of dark fermentation is hindered by several bottlenecks, including low substrate conversion efficiency, slow hydrogen production rates, and process instability. The theoretical maximum hydrogen yield from glucose is 12 mol H_2/mol glucose, yet actual yields typically range only from 1 to 2.5 mol H_2/mol glucose due to branched metabolic pathways, hydrogen partial pressure inhibition, and the activity of hydrogen-consuming bacteria [17]. These limitations are further exacerbated by the accumulation of soluble metabolites, such as volatile fatty acids and alcohols, which could inhibit microbial activity and reduce the system pH, leading to process failure [18,19]. Traditional optimization strategies, such as operational condition tuning and strain improvement, often yield limited enhancements and involve complex processes. For instance, maintaining strict anaerobic conditions and optimizing pH and temperature only marginally improve hydrogen yields, while genetic modification of hydrogen-producing strains is often time-consuming and not always transferable to mixed-culture systems [20,21].

The emergence of nanotechnology offers a promising pathway to address these limitations. Numerous studies have shown that various nanoparticles (NPs) could stimulate microbial growth and enhance enzymatic activity during dark fermentation, thereby improving hydrogen productivity and yield [22,23]. Nanoparticles, due to their high surface area-to-volume ratio, unique catalytic properties, and ability to facilitate electron transfer, have been demonstrated to significantly enhance the activity of key enzymes involved in hydrogen production, such as hydrogenases and nitrogenases [24,25].

For instance, silver nanoparticles (AgNPs) at an optimal concentration of 20 nM were found to increase hydrogen production from glucose by 67.6% in a mixed culture dominated by *Clostridium butyricum*, while also shifting the metabolic flux toward more favorable acetate and butyrate pathways [26]. Similarly, gold nanoparticles (AuNPs) at 5 nM were reported to improve hydrogen yield from sucrose by 62.3% in anaerobic sludge systems [27]. The incorporation of silica nanoparticles (SBA-15) at 120 mg/L in a high-loading industrial wastewater treatment system resulted in a remarkable 66.6% increase in hydrogen production (7 mol H_2/kg COD) through enhanced microbial entrapment and substrate degradation efficiency [28]. The addition of zinc oxide nanoparticles (ZnO NPs) at 50 mg/L to a mixed microbial consortium fermenting glucose was shown to enhance hydrogen yield by 33%, accompanied by a significant shift in metabolic pathways toward acetate-type fermentation and improved electron transfer efficiency [29]. Moreover, palladium nanoparticles (PdNPs) at 5 mg/L were shown to reduce the lag phase and enhance the hydrogen yield by *Enterobacter cloacae*, demonstrating their potential as efficient electron mediators [30].

The operational feasibility of nanoparticles frequently surpasses that of more intricate and energy-intensive strategies. However, the high cost and non-renewable nature of many nanomaterials present significant economic barriers to their widespread adoption [31,32]. Moreover, issues such as potential nanotoxicity, aggregation, and long-term stability in biological systems require careful evaluation before large-scale implementation [33,34].

In this context, magnetic nanoparticles (MNPs)—including Fe, Ni, Co, and their oxides, as well as emerging classes of magnetic composite materials—offer a promising alternative. MNPs exhibit distinctive properties, such as high specific surface area, elevated surface reactivity, excellent electrical conductivity, and favorable biocompatibility [35,36]. Their incorporation into dark fermentation systems has been shown to significantly enhance hydrogen yield and process efficiency. For example, Fe_3O_4 NPs have been reported to increase hydrogen production by up to 41% in sugarcane bagasse fermentation systems, while NiO NPs boosted the hydrogen yield by 55.6% in glucose-fed anaerobic sludge systems [37,38]. Moreover, due to their magnetic properties, MNPs could be efficiently recovered and reused post-fermentation via magnetic separation, thereby reducing operational costs, recycling efforts, and environmental risks [24,39]. Recent studies have demonstrated that magnetic composites such as magnetic granular activated carbon (MGAC) and nitrogen-doped magnetic activated carbon (MNAC) not only enhance hydrogen production by over 60% but also allow for more than 90% recovery and reuse across multiple fermentation cycles [40,41]. Additionally, the use of bimetallic ferrites like $NiFe_2O_4$ and $CoFe_2O_4$ has shown improved electron transfer capabilities and system stability under both mesophilic and thermophilic conditions, further underscoring the practical advantages of MNPs in sustainable biohydrogen processes [42].

Despite a growing number of studies reporting promising phenomenological improvements, a considerable gap remains between the observed performance enhancements and a deep mechanistic understanding. A significant portion of current research remains focused on synthesizing novel magnetic materials and reporting empirical hydrogen production data, which are often limited to correlative analyses. The fundamental mechanisms—how and why MNPs enhance hydrogen production—are often inferred, rather than conclusively established. Critical knowledge gaps persist, particularly regarding the structure–activity relationships between MNPs' physicochemical properties (e.g., size, morphology, crystal phase, surface charge) and their biological effects [13,43]. Upstream regulatory signaling pathways through which MNPs or their ions trigger transcriptional upregulation of key genes (e.g., hydrogenases and metabolic enzymes) remain largely unexplored [44]. Furthermore, the unique "magnetobiological effects" of MNPs and their potential synergy with chemical influences are often overlooked, as most studies treat MNPs merely as conventional chemical additives [45,46]. This fragmented and superficial mechanistic insight impedes the rational design of next-generation MNPs and the strategic optimization of MNP–microbe hybrid systems for scalable biohydrogen production.

Moreover, the mechanisms through which MNPs enhance hydrogen production vary substantially between pure cultures and mixed cultures. Existing reviews tend to lump different systems together without systematically unraveling the distinct biological interactions involved. For instance, in well-defined pure cultures, the uniform microbial background allows for a detailed investigation into micro-scale interactions between MNPs and cellular or enzymatic components—such as direct regulation of hydrogenase activity [39,47], serving as electron carriers [48,49], and redirecting central carbon metabolic pathways [48,50]. In contrast, mixed cultures with more realistic engineering applications consist of multiple microorganisms, including hydrolytic bacteria, acid-producing bacteria, hydrogen-producing bacteria, and even hydrogen-consuming bacteria, collectively forming a miniature ecosystem. Within such a setting, MNPs act not merely as interaction partners with individual cells, but as a key environmental factor influencing the entire community. Their enhancement mechanisms are mainly reflected in the regulation of the microbial community structure [51,52] and the stabilization of hydrogen-producing populations [51,53]. Therefore, in mixed culture systems, the role of MNPs operates more at the ecosystem level, with their promoting effects arising from the synergy of multiple ecological mechanisms.

This review, therefore, aims to systematically summarize relevant research progress, deeply explore the enhancement mechanisms of MNPs with a separate focus on pure culture systems and mixed culture systems, and integrate the latest research findings to provide a clear theoretical framework and directional guidance for subsequent research and application. To ensure comprehensive and systematic analysis, the review was conducted through a structured methodological approach. The relevant literature was identified by searching major scientific databases, including Web of Science, Scopus, PubMed, and Google Scholar, using keyword combinations related to magnetic nanoparticles, dark fermentation, biohydrogen production, electron transfer, and microbial metabolism. Articles published between 2005 and 2024 were screened, with a focus on experimental studies that reported mechanistic insights into MNP-enhanced hydrogen production in either pure or mixed cultures. Data from selected studies were extracted and synthesized thematically to compare mechanisms across different microbial systems, summarize key performance outcomes in tabular form, and illustrate proposed mechanisms in schematic diagrams [54]. This methodological framework ensures that the review provides an evidence-based and integrated perspective on the multi-level roles of MNPs, while also identifying critical knowledge gaps and the future research directions needed to advance the field toward scalable and sustainable biohydrogen production.

2. Enhancement Effects and Mechanisms of MNPs in Pure Culture Hydrogen Production Systems

In pure culture fermentation systems, the enhancement effects of MNPs on hydrogen production processes and underlying micro-mechanisms could be more directly revealed due to single microbial species and clear metabolic backgrounds.

2.1. Comprehensive Improvement of Hydrogen Production Efficiency and Kinetic Parameters

The addition of MNPs most directly and significantly enhances ultimate hydrogen production, maximum hydrogen production rate, and shortens the lag phase. A summary of the key performance enhancements is provided in Table 1. For instance, Nath et al. [55] reported that adding 100 mg/L of zero-valent iron nanoparticles (Fe(0) NPs) to a glucose fermentation system with *E. cloacae* DH-89 increased the hydrogen yield from approximately 0.95 to 1.9 mol H_2/mol glucose, achieving a 100% increase. This yield doubling stems not only from increased microbial numbers but also from the enhanced hydrogen production metabolic activity per cell after nanomaterial intervention.

Beckers et al. [43] demonstrated that encapsulating ultrafine Fe_2O_3 NPs (~2–3 nm) in porous silica could increase batch hydrogen production of *butyricum* by 38%, even at very low concentrations (10^{-6} mol/L), highlighting the critical impact of nanomaterial size effects and dispersion state on catalytic efficiency. Ramprakash et al. [44] found that adding 30 mg/L of Fe_3O_4 NPs to *Enterobacter aerogenes* using acid-ultrasound combined pretreated garden waste hydrolysate increased the hydrogen yield by 32%, raised the maximum hydrogen production rate to 1.3 mL/h, and significantly shortened the lag phase.

Table 1. Research progress on enhancing dark fermentative hydrogen production performance in pure cultures using MNPs.

Nanomaterial	Strain	Substrate	Optimal Concentration	Hydrogen Production Enhancement Effect	Reference
Fe(0) NPs	*Enterobacter cloacae* DH-89	Glucose	100 mg/L	1.9 mol H_2/mol glucose, 100% increase	[55]
Fe^{2+} ions	*Ethanoligenens harbinense*	Glucose	100 mg/L	1.97 mol H_2/mol glucose, 13% increase	[56]
Fe_3O_4 NPs	*Enterobacter aerogenes*	Garden waste hydrolysate	30 mg/L	98 mL/g GW, 32% increase	[44]
γ-Fe_2O_3 NPs	*Clostridium acetobutylicum* NCIM 2337	Glucose	175 mg/L	2.33 mol H_2/mol glucose, 34% increase	[45]
Ferrihydrite nanorods	*Clostridium pasteurianum* DSM525	Glucose	100 mg/L	3.55 mol H_2/mol glucose, 15.6% increase	[47]
NiO NPs	*Bacillus anthracis* PUNAJAN 1	POME	1.5 mg/L	0.563 L H_2/g-COD, 51% increase	[46]
CoO NPs	*Bacillus anthracis* PUNAJAN 1	POME	1.0 mg/L	0.487 L H_2/g-COD, 67% increase	[46]
CoO NPs	*Clostridium beijerinckii* DSM 791	Rice mill wastewater	1.5 mg/L	195.7 mL H_2/L, 90% increase	[48]
NiO NPs	*Clostridium beijerinckii* DSM 791	Rice mill wastewater	1.5 mg/L	214.9 mL H_2/L, 109% increase	[48]
LPBC	*Clostridium pasteurianum*	Glucose	200 ppm	1.76 mol H_2/mol glucose, 57% increase	[50]

Recent studies further confirm significant enhancement effects of different MNPs in pure culture systems. Mohanraj et al. [45] used green-synthesized γ-Fe_2O_3 NPs (175 mg/L) to increase the hydrogen yield of *Clostridium acetobutylicum* NCIM 2337 from 1.74 to 2.33 mol H_2/mol glucose: a 34% increase. Mishra et al. [46] studied NiO and CoO NPs' roles in hydrogen production from palm oil mill effluent (POME) fermentation by *Bacillus anthracis* PUNAJAN 1, finding that 1.5 mg/L NiO NPs increased the hydrogen yield by 1.51-fold (0.563 L H_2/g-COD) and 1.0 mg/L CoO NPs by 1.67-fold (0.487 L H_2/g-COD). Zhang et al. [47] introduced ferrihydrite nanorods (100 mg/L), increasing the hydrogen yield of *C. pasteurianum* DSM525 by 15.6% to 3.55 mol H_2/mol glucose and significantly improving the glucose conversion efficiency (100% vs. 68% control).

Notably, Rambabu et al. [48] demonstrated that NiO and CoO NPs (~26 nm and ~50 nm, respectively) significantly enhanced biohydrogen production from rice mill wastewater (RMWW) using *Clostridium beijerinckii* DSM 791. At the optimal concentration of 1.5 mg/L, NiO NPs increased the cumulative hydrogen yield to 214.9 mL H_2/L (2.09-fold increase), while CoO NPs reached 195.7 mL H_2/L (1.9-fold increase). Gompertz model effectively described fermentation kinetics, showing a shortened lag phase and increased maximum hydrogen production rate. Zhang et al. [49] reported that magnetite and ferrihydrite enhanced the acetate production pathway and increased the total extracellular electron output efficiency by up to 53.5%, indicating a metabolic shift towards more efficient electron utilization.

Nanobiocatalysts with highly ordered atomic structures exhibit exceptional capabilities. Yang et al. [50] developed an $L1_0$-FePt@PPy nanobiocatalyst (LPBC) that increased the hydrogen production rate by ~103% and the yield by 57% (1.76 mol H_2/mol glucose) in *Clostridium pasteurianum*. The LPBC/CH system optimized electron flow between NADH oxidation and [FeFe]-hydrogenase, maintaining stable performance over 30 days of cyclic use, demonstrating excellent long-term operational potential.

2.2. Promotion of Cell Growth and Biomass Accumulation

Appropriate amounts of MNPs could serve as trace element sources required for microbial growth, promoting cell proliferation. Tychounian et al. [57] found that adding metal ions, such as Ni^{2+} and Fe^{2+}, to a culture medium could nearly double the microbial biomass concentration. When added in nanoparticle form, their slow-release characteristics could more sustainably maintain effective metal ion concentrations in the medium, providing stable support for the microbial exponential growth phase, thereby establishing a solid biomass foundation for high hydrogen production.

Zhang et al. [47] confirmed that ferrihydrite nanorods significantly promoted *Clostridium pasteurianum* cell growth, with biomass positively correlating with nanomaterial concentration, which was consistent with the maximum cumulative hydrogen production trend. Rambabu et al. [48] reported that 1.5 mg/L NiO and CoO NPs increased the *Clostridium beijerinckii* biomass concentrations to 3.93 µg/L and 3.51 µg/L, respectively, approximately two-fold higher than control. This observation proved beneficial for the roles of NPs inclusion for microbial growth and multiplication.

3. Mechanisms of MNPs for Enhancing Hydrogen Production in Pure Cultures

Building upon the observed significant enhancement phenomena, an in-depth exploration of intrinsic molecular and cellular level mechanisms is crucial for the rational design and application of MNPs. Their enhancement mechanisms primarily stem from three core aspects.

3.1. *Serving as Enzyme Cofactor Reservoirs and Regulating Hydrogenase Activity*

Hydrogenase is the core enzyme catalyzing proton reduction to hydrogen in dark fermentative microorganisms, and its active center assembly and function (e.g., H-cluster of [FeFe]-hydrogenase, heteronuclear active site of [NiFe]-hydrogenase) highly depend on metal ions such as Fe and Ni [58]. MNPs could act as sustained and controllable release sources of these metal ions in fermentation broth.

Regarding the cofactor supply, Fe(0) NPs and iron oxide NPs slowly corrode and release Fe^{2+} under anaerobic conditions; these ions could be absorbed by microorganisms and directly used for iron–sulfur cluster synthesis (e.g., ferredoxin) and [FeFe]-hydrogenase active centers. Mohanraj et al. [45] pointed out that FeNPs enhance hydrogenase activity by increasing ferredoxin activity and promoting electron transfer. Nath et al. [54] further emphasized that iron is a key component of ferredoxin and hydrogenase, with FeNPs' addition significantly enhancing hydrogenase activity.

Beyond merely supplying cofactors, nanoparticles could regulate key enzyme synthesis at a transcriptional level. Zhang et al. [59] found that NiO NPs' addition significantly upregulated genes encoding hydrogenase and formate hydrogenase in *Klebsiella* sp. WL1316, resulting in a remarkable 623% increase in total hydrogenase activity. Transcriptomic analysis by Zhang et al. [47] demonstrated that ferrihydrite nanorods significantly upregulated hydrogenase-related genes (e.g., HypC/HybG/HupF family chaperones and [Ni/Fe] hydrogenase large subunits) in *Clostridium pasteurianum*, concurrently boosting hydrogenase specific activity by 1.3-fold, indicating dual regulation at the transcriptional and enzymatic activity levels.

Mishra et al. [46] discovered that NiO NPs, as [Ni-Fe] hydrogenase active site components, directly participate in redox processes, whereas CoO NPs, as cobaltin coenzyme constituents, regulate multiple metabolic enzyme activities. This indicates that MNPs act as "metabolic signal triggers," deeply reconstructing the strain hydrogen production metabolic network at a genetic level. Regarding the transcriptional activation effect of nanobiocatalysts, Yang et al. [50] found that LPBC not only significantly upregulates [FeFe]-hydrogenase (CLPA-c33960) and [NiFeSe]-hydrogenase gene expression in *Clostridium pasteurianum* but also activates relevant metabolic pathways, such as glycolysis and acetate production, synergistically enhancing hydrogen production through multiple pathways.

3.2. Acting as Electron Carriers and Optimizing Electron Transfer Pathways

The essence of substrate decomposition to H_2 generation is electron transfer and flow. MNPs, especially those with good conductivity like Fe_3O_4 and zero-valent iron, could greatly optimize this process.

Regarding extracellular electron transfer, these nanoparticles could act as solid-state electron mediators, accepting electrons transferred from reductive carriers (e.g., reduced ferredoxin) and directly transferring them to hydrogenase or solution protons, thereby "short-circuiting" potentially less efficient intracellular electron transfer chains and accelerating the H_2 production rate [60]. Zhang et al. [47] found that ferrihydrite nanorods were reduced to Fe(II) during fermentation, providing both iron sources required for hydrogenase synthesis and acting as an electron mediator to promote electron transfer.

Concerning direct interspecies electron transfer (DIET) in pure cultures, although species are single, synergy between the cells may still exist. Studies show that after adding Fe_3O_4 NPs, formation of nanomaterial-mediated conductive networks or enhanced "bacterial nanowires" conductivity among cells could be promoted [61]. This structure facilitates direct electron exchange between cells, enabling more balanced and efficient electron distribution and utilization within the population, alleviating metabolic inhibition caused by local electron excess. Ramprakash et al. [44] further proposed that Fe_3O_4 NPs could act as electron mediators, promoting electron transfer between *E. aerogenes* cells and optimizing NADH/NAD^+ balance, thereby increasing hydrogen yield.

In targeted electron flow regulation, LPBC constructs efficient electron transfer shortcuts by catalyzing NADH oxidation to generate NAD^+ and directly transferring the released electrons to [FeFe]-hydrogenase. Molecular dynamics simulations indicated that LPBC could bind to the [4Fe-4S] ferredoxin domain of [FeFe]-hydrogenase, shortening the electron transfer distance and significantly enhancing hydrogenase activity [50]. Mishra et al. [46] noted that NiO and CoO NPs' high conductivity enables them as electron conductors, accelerating electron transfer from reductive carriers to hydrogenase or protons, thus increasing the hydrogen production rate.

3.3. Regulation of Central Carbon Metabolism and Energy Metabolism

MNPs could influence the strain's central metabolic network, guiding carbon flow and reducing power towards hydrogen production pathways. This regulatory effect is phenotypically evidenced by a significant shift in the distribution of fermentation end-products. The addition of MNPs, such as Fe(0) NPs or Fe_3O_4, consistently results in a marked increase in acetate concentration, coupled with a suppression of non-hydrogenic or low-hydrogen-yield products like ethanol, propionate, and lactate. This shift visually reflects a fundamental reprogramming of the strain's metabolic network, channeling more substrate carbon and reducing equivalents into the high-hydrogen-yield acetate-type fermentation pathway, thereby intrinsically improving the substrate-to-hydrogen conversion efficiency.

Table 2 summarizes key cases where MNPs redirect central carbon metabolism from low-hydrogen-yield pathways (e.g., ethanol, lactate production) toward high-yield acetate-type fermentation in various pure cultures. This metabolic reprogramming not only enhances the hydrogen yield but also improves the overall metabolic efficiency and system stability by optimizing electron flow and energy distribution. For instance, Ramprakash et al. [44] confirmed that Fe_3O_4 NPs addition significantly increased the acetate proportion in end products to over 60% while limiting ethanol production: a change that positively correlated with the observed increase in hydrogen yield. Mohanraj et al. [45] further noted that FeNPs' addition shifted the fermentation profile of *Clostridium acetobutylicum* from a butyrate type toward an acetate/butyrate mixed type, which is more favorable for hydrogen production. Similar metabolic steering was observed with NiO and CoO NPs, which favored acetate and butyrate pathways, as indicated by increased ethanol-to-acetate (E/A) and butyrate-to-acetate (B/A) ratios at optimal concentrations [48]. Zhang et al. [49] provided a clear example of this electron and carbon rerouting, observing that both magnetite and ferrihydrite enhanced the acetate production pathway while lowering butyrate production in *Clostridium bifermentans*, thereby redirecting the carbon and electron fluxes to support higher electron output efficiency. The underlying reason for this metabolic shift lies in the ability of MNPs or their released ions to influence key enzymatic nodes and cellular redox states. Nanoparticles may inhibit key enzymes in competitive pathways, such as alcohol dehydrogenase, or alter the intracellular $NADH/NAD^+$ ratio. To regenerate oxidized coenzyme NAD^+ and sustain continuous glycolysis, the strain is compelled to dispose of excess reducing power, primarily through the hydrogen production pathway, thereby indirectly but powerfully driving hydrogen generation [62]. This is corroborated by findings that CoO NPs, through participation in cobalt–porphyrin protein reactions, could accelerate the electron carrier capacity and optimize the energy distribution of carbon metabolic flow [46].

Table 2. Summary of metabolic shifts induced by MNPs in pure culture systems.

MNPs Type	Strain	Substrate	Main Changes in Metabolic Products (vs. Control)	Metabolic Shift Outcomes	Reference
Fe_3O_4 NPs	*Enterobacter aerogenes*	Garden waste	Acetate proportion >60%; ethanol significantly decreased	Redirected carbon flux toward acetate-type fermentation (higher H_2 yield), suppressing ethanol pathway, resulting in a 32% increase in H_2 yield.	[44]
FeNPs	*Clostridium acetobutylicum*	Glucose	Shift from butyrate type to acetate/butyrate mixed type	Optimized carbon flow distribution and enhanced electron utilization efficiency, leading to a 34% increase in H_2 yield.	[45]
NiO/CoO NPs	*Clostridium beijerinckii*	Rice mill wastewater	Increased acetate and butyrate ratios (elevated E/A and B/A ratios)	Promoted acetate and butyrate pathways, enhancing reducing power release, which increased H_2 yield by 2.09-fold (NiO).	[48]
Magnetite/Ferrihydrite	*Clostridium bifermentans*	Glucose	Enhanced acetate pathway; decreased butyrate production	Improved total extracellular electron output efficiency by up to 53.5%, optimizing carbon and electron fluxes toward H_2 production.	[49]
Fe(0) NPs	*Enterobacter cloacae*	Glucose	Acetate accumulation; reduced ethanol, propionate, lactate	Redirected carbon flow and reduced equivalents toward H_2-producing pathways, doubling the H_2 yield (100% increase).	[54]
LPBC Nanobiocatalyst	*Clostridium pasteurianum*	Glucose	Enhanced acetate pathway; upregulation of glycolysis and acetate-genesis genes	Synergistically regulated metabolism and electron transfer, achieving a 57% increase in H_2 yield and a ~103% increase in production rate.	[50]

Furthermore, this metabolic redirection confers a significant energy advantage. By promoting the more efficient acetate pathway, which yields a net gain of ATP, over pathways like butyrate fermentation, and by optimizing electron transfer processes to minimize energy loss, MNPs help the strain to establish a metabolic state with a higher energy charge. This energized state not only benefits microbial growth and biomass accumulation but also provides the necessary energy foundation for sustained and efficient hydrogen production.

3.4. *Overview of the Integrated Enhancement Mechanisms Induced by MNPs in Pure Culture Systems*

In summary, the enhancement of dark fermentative hydrogen production by MNPs in pure cultures arises from the synergistic interplay of three core mechanisms operating at molecular and cellular scales. These include: (1) serving as sustained-release reservoirs of essential metal cofactors (e.g., Fe, Ni) to facilitate hydrogenase biosynthesis and activation; (2) functioning as efficient electron carriers that create conductive shortcuts for intra- and intercellular electron transfer, directly channeling electrons toward proton reduction; and (3) exerting systematic reprogramming of central carbon and energy metabolism, redirecting the metabolic flux from low-hydrogen-yield pathways (e.g., ethanol, lactate) toward high-yield routes (e.g., acetate), while optimizing cellular energy status (Figure 1). A holistic understanding of these integrated mechanisms provides a critical foundation for the rational design of next-generation MNPs tailored to specific biohydrogen-producing strains.

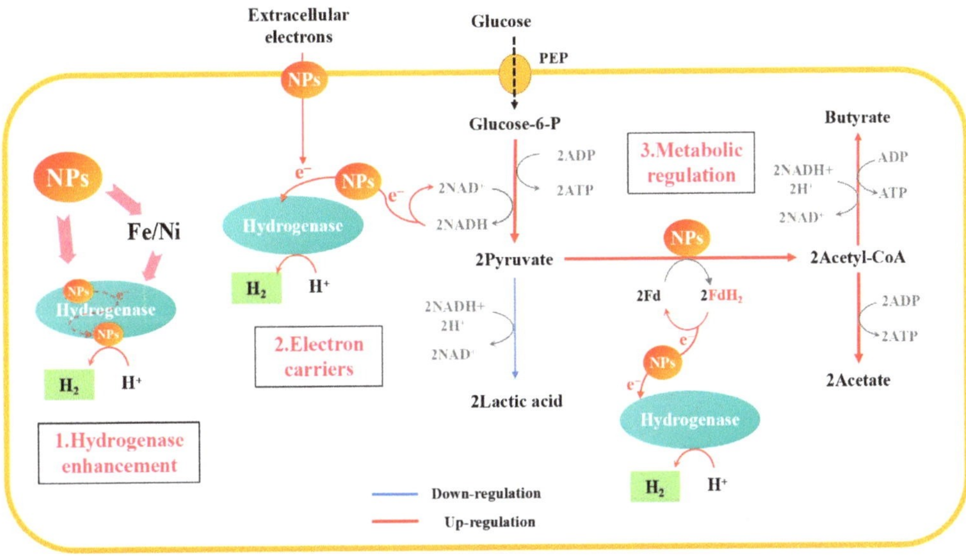

Figure 1. The mechanism by which MNPs enhance hydrogen production in pure bacterial cultures.

4. Current Limitations in Mechanistic Understanding and Future Research Directions

Although existing studies have preliminarily revealed the enhancement mechanisms of MNPs in pure culture systems from perspectives such as enzyme activity, electron transfer, and metabolic regulation, significant knowledge gaps remain in understanding many key aspects, creating bottlenecks in moving from phenomenological observation to mechanistic elucidation [45,48].

4.1. The Precise "Structure–Activity Relationship" Between Material Physicochemical Properties and Biological Effects Remains Unclear

Core parameters of MNPs—such as size, morphology, crystal phase, surface charge, and magnetic strength—and how they specifically regulate interactions with cell membranes and enzyme active sites are currently mostly limited to correlative analyses [35,36].

Size effects: Smaller MNPs generally exhibit higher surface-area-to-volume ratios, potentially enhancing interactions with microbial cells and enzymes. For instance, ultrafine Fe_2O_3 NPs (~2–3 nm) encapsulated in porous silica increased hydrogen production by *butyricum* by 38%, even at extremely low concentrations (10^{-6} mol/L) [43]. In contrast, larger particles may serve as more stable scaffolds for microbial attachment or electron transfer networks. However, a systematic comparison of size-dependent effects across a defined range (e.g., 5 nm vs. 50 nm vs. 100 nm) for the same MNP composition is largely absent from the literature, making it difficult to establish optimal size thresholds for biohydrogen enhancement.

Shape effects: Particle morphology influences the interfacial contact area and orientation with biological surfaces. Rod-shaped particles, such as ferrihydrite nanorods [47], may present different facet reactivity and alignment compared to spherical Fe_3O_4 NPs [44]. This could affect the efficiency of electron transfer, metal ion release kinetics, and physical interactions with cell membranes or extracellular polymeric substances. Nevertheless, direct comparative studies controlling for size and surface chemistry while varying only in shape are scarce, leaving shape–activity correlations largely speculative.

Surface functionalization and coatings: Modifying the MNP surface with organic or inorganic coatings can drastically alter biocompatibility, dispersion stability, surface charge, and functionality.

Carbon-based composites: Coating or embedding MNPs within carbon matrices, as in MGAC [40] or MNAC [41], significantly enhanced conductivity, provided a high surface area for microbial immobilization, and introduced additional surface functional groups (e.g., -COOH, -OH, pyridinic-N), which were shown to improve electron transfer efficiency.

Conductive polymer coatings: The LPBC utilized a polypyrrole (PPy) coating to create a conductive interface that optimized the electron flow between NADH oxidation and the [FeFe]-hydrogenase active site, leading to a ~103% increase in the hydrogen production rate [50].

Biological coatings: "Green-synthesized" MNPs, using plant extracts, may carry residual biomolecular coatings (e.g., polyphenols, flavonoids) that could influence initial microbial adhesion, reduce cytotoxicity, or provide supplemental carbon sources [45,63].

In summary, while preliminary evidence suggests that tuning MNP size, shape, and surface chemistry could optimize biohydrogen production, the field lacks systematic, controlled studies that isolate these variables [64,65]. Future work must employ well-defined MNP libraries to establish the quantitative relationships between these physicochemical descriptors and biological outcomes, enabling the rational design of "tailored" MNPs for specific biohydrogen applications.

4.2. The Specific Pathways and Relative Efficiency of MNPs Acting as "Electron Bridges" Remain Ambiguous

While MNPs are widely regarded as promoters of electron transfer, the underlying microscopic mechanism remains ambiguous, with multiple plausible pathways that are challenging to differentiate experimentally. Currently, most studies remain at a qualitative or semi-quantitative level [66,67]. A key research gap lies in the absence of standardized in situ methodologies that are capable of dynamically quantifying the proportion of electron flux mediated by MNPs, relative to the total electron flow in active fermentation systems. More integrated and advanced electrochemical and spectroscopic techniques are required to quantitatively resolve this process. For instance, chronoamperometry (CA) and linear sweep voltammetry (LSV) can be employed to measure steady-state and potential-dependent electron fluxes between MNPs and microbial cells or purified enzymes, providing direct electron transfer rate data [68,69]. Electrochemical impedance spectroscopy (EIS) is particularly suitable for characterizing the charge transfer resistance and capacitance at the MNP—microbe interface, revealing how MNPs lower the energy barrier for electron exchange [70,71]. Cyclic voltammetry (CV) can help determine the redox potentials of MNPs and key biological cofactors (e.g., ferredoxins, cytochromes), suggesting possible redox-mediated pathways [72,73]. Spectroelectrochemical approaches, such as UV–Vis spectroscopy coupled with electrochemical control, enable in situ monitoring of valence state changes in MNPs during fermentation, correlating nanoparticle redox cycling with hydrogen evolution [74]. Furthermore, conductive atomic force microscopy

(C-AFM) and scanning electrochemical microscopy (SECM) have emerged as powerful tools for mapping local conductivity and electron transfer activity at nanoscale resolution, offering visual evidence of conductive networks formed between MNPs and microbial aggregates [75,76].

Nevertheless, fundamental questions persist regarding the precise mode of MNP-mediated electron transfer: are electrons transferred via direct "point-to-point" contact between MNPs and the cell surface? Do MNPs act as reversible redox mediators, shuttling between cells and the solution? Or do they indirectly regulate the electron transfer efficiency of membrane-bound hydrogenases or transporter proteins by influencing their conformation [48]? More importantly, within the complex intracellular electron transfer network (involving multiple pathways such as ferredoxin, NADH, and flavoproteins), the proportion of electron flux mediated by MNPs and how it synergizes or competes with the natural electron transport chain thermodynamically and kinetically remain speculative, as current technical means struggle to track and quantify this microscopic, rapid interfacial process in real time and in situ [44].

4.3. The Upstream Regulatory Signaling Pathways of MNPs at the Genetic Level Are Largely Unknown

The observed upregulation of hydrogenase and related metabolic genes is an outcome, but what is the initiating signal? How are nanoparticles or their released ions sensed by the cell as a specific "signal"? Potential sensing mechanisms include the following: localized reactive oxygen species (ROS) bursts, induced by MNPs being perceived as a stress signal; specific interactions of MNPs with membrane receptors; or dissolved metal ions (e.g., Fe^{2+}/Ni^{2+}) being captured by intracellular metal-sensing proteins (e.g., Fur, NikR) [59]. However, the complete signal transduction cascade—from these initial sensing events to the final transcriptional activation in the nucleus—involving specific second messengers, phosphorylation cascades, or transcription factors, remains a "black box". This prevents active intervention or optimization of this signaling pathway to maximize its positive regulatory effects.

Finally, the unique magnetobiological effects of MNPs and their synergistic mechanisms are severely overlooked. Most studies treat MNPs as ordinary chemical additives, completely ignoring the potential unique biological effects arising from their magnetism itself [48,63]. The weak magnetic fields generated by MNPs, or enhanced under an externally applied magnetic field, might influence transmembrane transport of charged ions via Lorentz forces, alter the conformation and vibrational modes of membrane proteins via magneto-mechanical effects, or modulate redox processes by affecting radical pair reactions. These "non-chemical" physical field effects could synergize with the chemical actions of MNPs to collectively influence microbial metabolic activity and the hydrogen production capacity. However, due to the lack of carefully designed control experiments (e.g., using non-magnetic but chemically similar nanoparticles for comparison, or applying equivalent magnetic fields in the absence of MNPs) this potential unique magnetobiological contribution is entirely absent from current enhancement mechanism models, leading to a fundamentally incomplete understanding of MNPs' action mechanisms [48].

To address the above limitations, future research urgently needs to move towards interdisciplinary integration and technological innovation. In material characterization, advanced techniques such as synchrotron radiation X-ray absorption spectroscopy (XAS), spherical aberration-corrected transmission electron microscopy (TEM), and magnetic force microscopy (MFM) should be actively employed to analyze the chemical state, structure, and spatial distribution of MNPs within bacterial cells at the atomic scale and in real space [35,47]. In process probing, developing in situ electrochemical testing platforms for microbial systems, combined with spectroscopic methods, is needed to monitor interfacial electron flow in real time [60]. In biological mechanism exploration, integrating transcriptomics, proteomics, and metabolomics, and utilizing gene-knockout strains, will allow for precise dissection of the complete pathway from signal perception to metabolic remodeling [50,59]. Ultimately, through multi-physics coupling models, the physical properties (e.g., magnetism) and chemical properties (composition, valence state) of materials should be correlated with biological responses to construct a more comprehensive theoretical framework of "MNPs–microbe" interactions that transcends the current chemical perspective, providing a solid theoretical foundation for the precise design and optimization of efficient and stable bio-hybrid hydrogen production systems [50].

5. Enhancement Effects of MNPs in Mixed Cultures Hydrogen Production Systems

Mixed culture systems better approximate complex environments of practical applications, containing hydrolytic bacteria, acidogenic bacteria, hydrogen-producing bacteria, and potential hydrogen-consuming bacteria. In these systems, MNPs' enhancement effects exhibit macro-ecological effects that are different from pure culture systems, and novel magnetic composite material applications are increasingly widespread.

5.1. Significant Improvement of Hydrogen Production Performance from Complex Substrates

When treating actual wastes, such as distillery wastewater, food waste, and algal biomass, MNPs demonstrate powerful enhancement capabilities [77]. A summary of key performance enhancements is provided in Table 3. For instance, Gadhe et al. [26] reported that when treating complex distillery wastewater, 200 mg/L Fe_2O_3 NPs addition increased the hydrogen production rate by 45%, reaching 8.0 mmol H_2/g COD. More strikingly, Yin and Wang [64] found that combined Fe(0) and Ni(0) NPs use increased macroalgae hydrogen production by 550%. Abdul Aziz et al. [65] used green-synthesized Fe_3O_4 NPs (~75 nm) to treat glucose substrate. At 200 mg Fe/L concentration, hydrogen production increased by 72.33%, the hydrogen production rate increased to 3.30 mL/h, and the lag phase shortened by 7.12 h. The study also pointed out that the high temperature (55 °C) and NPs coupling effect further enhanced the system's metabolic activity. This indicates that MNPs effectiveness is not only for simple sugars but also real substrate systems with complex compositions and potential inhibitors.

In recent years, researchers developed various magnetic composite materials to further improve hydrogen production efficiency and material recyclability. MGAC embeds MNPs (e.g., $NiFe_3O_4$) into granular activated carbon (GAC), combining excellent GAC carrier properties with MNPs' catalytic and magnetic response properties. Jamaludin et al. [40] optimized MGAC preparation conditions using response surface methodology (Ni:Fe = 0.53, MNP:GAC = 0.02, using coconut shell-derived GAC-O as a carrier). The hydrogen production rate in the thermophilic dark fermentation system reached 20.33 mL H_2/L·h: 63.99% higher than non-magnetic GAC. MGAC's high conductivity (95.64% higher than non-magnetic GAC) effectively promoted electron transfer between microorganisms.

Magnetic nitrogen-doped activated carbon (MNAC), prepared by Tian et al. [41] at 600 mg/L concentration, achieved 304.8 mL/g glucose hydrogen production, 51.94% increase compared to control. Nitrogen doping optimized the material pore structure and electron transport capacity, while magnetic components (FeO, Fe_3O_4) facilitated recovery and provided the Fe element required for hydrogen production.

Multi-element ferrite nanoparticles such as $MgFe_2O_4$ and $Ca_{0.5}Mg_{0.5}Fe_2O_4$ also show good application prospects. Wang et al. [64] compared and found that 100 mg/L $MgFe_2O_4$ NPs increased hydrogen production by 35.2% (183.6 mL/g glucose), outperforming $Ca_{0.5}Mg_{0.5}Fe_2O_4$ NPs, which achieved a 26.6% increase (171.9 mL/g glucose) at 400 mg/L.

Both NPs promoted butyrate-type fermentation and effectively alleviated humic acid hydrogen production inhibition.

Table 3. Research progress on enhancing dark fermentative hydrogen production performance in mixed microbial systems, using MNPs.

Nanomaterial	Consortium/Inoculum	Substrate	Optimal Concentration	Hydrogen Production Enhancement Effect	Reference
Fe(0) NPs	Anaerobic sludge	Grass biomass	400 mg/L	64.8 mL/g dry grass, 74% increase	[52]
Fe(0) + Ni(0) NPs	Anaerobic sludge	Macroalgae	200 mg/L	20.25 mL/g VS, 550% increase	[64]
Fe^{2+} ions	Anaerobic sludge	Glucose	350 mg/L	311 mL/g glucose, 59% increase	[78]
Fe_2O_3 NPs	Anaerobic sludge	Distillery wastewater	200 mg/L	8.0 mmol/g COD, 45% increase	[26]
Fe_3O_4 NPs	Anaerobic sludge	Sugarcane bagasse	200 mg/L	1.21 mol H_2/mol glucose, 41% increase	[37]
Fe_3O_4 NPs	Thermophilic mixed culture	Glucose	200 mg Fe/L	0.79 mol H_2/mol glucose, 72.33% increase	[65]
Ferrihydrite	Heat-shocked marine sediment consortium	Glucose	100 mg/L	1.98 mol H_2/mol glucose; induced H_2 production exclusively by Clostridium	[79]
Ni(0) NPs	Anaerobic sludge	Glucose	5.7 mg/L	250 mL/g VS, 0.9% increase	[80]
NiO NPs	Anaerobic sludge	Glucose	25 mg/L	98 mL, 55.6% increase	[80]
NiO NPs	Anaerobic sludge	Dairy wastewater	10 mg/L	16% increase	[26]
NiO + Fe_2O_3 NPs	Anaerobic sludge	Distillery wastewater	5+200 mg/L	8.83 mmol/g COD, 62% increase	[80]
Co NPs	Activated sludge	Waste-activated sludge	7 mg/g VS	5.40 mL H_2/g VS, >200% increase	[81]
$NiCo_2O_4$ NPs	Anaerobic sludge	Glucose	400 mg/L	259.67 mL/g glucose, 34% increase	[82]
$CoFe_2O_4$ NPs	Anaerobic sludge	Glucose	0.4 g/L	205.24 mL/g glucose, 31.8% increase	[42]
$MnFe_2O_4$ NPs	Anaerobic sludge	Glucose	400 mg/L	272.7 mL/g glucose, 40.1% increase	[83]
$LaFeO_3$ NPs	Mixed culture	Glucose	100 mg/L	289.8 mL/g glucose, 47.6% increase	[84]
MGAC	Thermophilic mixed culture	Glucose	0.02 MNP:GAC	20.33 mL H_2/L·h, 63.99% increase	[40]
MNAC	Mixed culture	Glucose	600 mg/L	304.8 mL/g glucose, 51.94% increase	[41]
$MgFe_2O_4$ NPs	Mixed culture	Glucose	100 mg/L	183.6 mL/g glucose, 35.2% increase	[77]
$Ca_{0.5}Mg_{0.5}Fe_2O_4$ NPs	Mixed culture	Glucose	400 mg/L	171.9 mL/g glucose, 26.6% increase	[77]

5.2. Enhanced System Stability and Process Continuity

In continuous-flow reactors with mixed cultures, MNPs' addition helps maintain long-term system stability. They create a better growth environment for strict anaerobic hydrogen-producing bacteria by reducing the redox potential, while simultaneously inhibiting aerobic and facultative anaerobic contaminants [85]. Furthermore, they play positive roles in mitigating the system acidification caused by organic acid accumulation and adsorbing/removing certain toxic substances (e.g., heavy metals), thereby ensuring a stable continuous dark fermentation process within the wider operating window. Composite materials like MGAC and MNAC, due to porous structures, could better immobilize microorganisms, form stable biofilms, and extend microbial retention time, further enhancing system stability [40].

Additionally, rapid accumulation of organic acids during efficient hydrogen production could lead to sharp pH drops, inhibiting microbial activity. Certain MNPs, particularly zero-valent iron, consume protons (H^+) during their corrosion or reduction processes, thereby providing pH buffering capacity. This helps stabilize the fermentation system against acid inhibition, maintaining optimal conditions for hydrogen-producing bacteria such as *Clostridium* [47]. The combined effects of heavy metal adsorption and acid mitigation significantly enhance the overall stability and operational window of mixed-culture dark fermentation systems. The pH buffering capacity of certain MNPs contributes significantly to system stability. For instance, ferrihydrite reduction consumes protons, which helps counteract acidification due to volatile fatty acid (VFA) accumulation during fermentation [79]. This stabilization of pH values creates a more favorable environment for hydrogen-producing bacteria and supports prolonged hydrogen production.

6. Mechanisms of MNPs for Enhancing Hydrogen Production in Mixed Cultures

In mixed culture systems, MNPs' enhancement mechanisms are more complex and subtle, which is mainly reflected in ecological-level regulation of interspecies relationships and system functions. MNPs exert a selective pressure on mixed cultures, shaping the ecosystem through intrinsic mechanisms including selective growth promotion and the inhibition of hydrogen consumption pathways.

6.1. Selective Enrichment of Hydrogen-Producing Microbiota via Ecological Regulation

The introduction of MNPs into mixed-culture systems initiates a profound ecological succession that strategically reshapes the microbial community toward a structure and function that are optimized for hydrogen production. Robust evidence from high-throughput sequencing consistently reveals that MNPs supplementation, particularly with iron- and nickel-based materials, selectively enriches canonical hydrogen-producing bacteria (HPB) such as *Clostridium* while suppressing non-hydrogenic or hydrogen-consuming competitors [65]. This directed succession is unequivocally linked to enhanced hydrogen yields. For instance, the addition of $LaFeO_3$ NPs elevated the relative abundance of the key HPB *Clostridium sensu stricto 1* from 39.3% to 53.5%, correlating with a 47.6% increase in hydrogen production [84]. A more pronounced structural shift was driven by MNAC, which boosted *Clostridium sensu stricto 1* to 64.92% of the community and achieved a 51.94% increase in hydrogen yield [41]. A particularly compelling demonstration of this principle was shown in a heat-shocked marine sediment consortium, where the microbial community bifurcated, based on the presence of ferrihydrite. In its absence, the consortium was dominated by *Bacillus* (99.5%), performing non-hydrogenic lactic acid fermentation. In stark contrast, ferrihydrite enrichment led to a *Clostridium*-dominated community (57.3%) that performed high-yield butyric/acetic acid fermentation, fundamentally redirecting the metabolic output of the system [79]. Concomitant with this community restructuring is a consistent optimization of metabolic pathways, with MNPs-amended systems typically shifting toward high-hydrogen-yield butyrate-type fermentation.

The observed phenomenological shifts are driven by multifaceted mechanisms through which MNPs modify the ecological niche and act as selective agents. A primary mechanism involves modification of the microenvironmental conditions. Materials such as zero-valent iron (Fe(0)) and magnetite (Fe_3O_4) consume protons and lower the system oxidation-reduction potential (ORP) during their corrosion or reduction, thereby fostering a stricter anaerobic environment that favors obligate anaerobes like *Clostridium* while inhibiting facultative competitors [85]. Beyond these broad physicochemical changes, the specific release of metal ions from MNPs provides a more targeted selective pressure. Hydrogen-producing bacteria often possess superior uptake and tolerance mechanisms for essential elements like Fe and Ni. For example, the gradual release of La^{2+} and Fe ions from $LaFeO_3$ NPs has been proposed to enhance transmembrane transport, selectively promoting the growth and activity of HPB, as evidenced by a corresponding increase in the gene copy number of [Fe-Fe] hydrogenase [84].

A deeper, ecological filtering mechanism arises from the ability of certain MNPs to function as terminal electron acceptors. This process selectively enriches microorganisms that are capable of dissimilatory metal reduction: a metabolism that is intrinsically linked to hydrogen production in many fermentative bacteria [86]. The case of ferrihydrite provides a clear paradigm: its reduction acted as a powerful ecological filter that exclusively enriched the hydrogen-producing *Clostridium* species from a mixed inoculum, as these organisms could utilize Fe(III) as an electron sink [79]. This electron sink effect not only provides HPB with a competitive advantage by offering an additional respiratory pathway but also strategically diverts the electron flow away from the metabolic routes that generate low-hydrogen-yield products like lactate or ethanol.

In summary, MNPs function as active ecological engineers, rather than passive additives. Through an integrated strategy encompassing niche modification and electron sink effects, they selectively reshape the microbial community into a functionally robust ecosystem that is primed for efficient and stable hydrogen production.

6.2. *Construction of DIET Networks to Activate Collective Microbial Activity*

In mixed microbial fermentation systems, interspecies electron transfer is a key factor influencing microbial activity. Traditional interspecies hydrogen transfer (IHT) relies on hydrogen molecules as electron carriers, exhibiting low efficiency and susceptibility to hydrogen partial pressure inhibition. MNPs, particularly those with excellent conductivity (e.g., Fe_3O_4, MGAC, MNAC), could serve as "electron bridges" to facilitate DIET.

The mechanism by which MNPs promote DIET is mainly reflected in the construction of efficient material–microbe interfaces, which could be systematically described as follows:

Physical contact: The high specific surface area of nanoparticles provides ample contact sites with microbial cells, and their morphology and size influence the effective contact area with microorganisms.

Microenvironment regulation: Nanoparticles influence the local pH, ion concentration, and redox status near the interface through their surface characteristics or reactions (e.g., Fe^{2+}/Fe^{2+} cycle), creating a more favorable microenvironment for electron transfer while potentially providing physical protection. Such extracellular electron transfer occurs both intra- and inter-species.

Metabolic response: The enhanced electron flux through DIET directly stimulates the metabolic activity of the microbial consortium, enabling faster decomposition and utilization of fermentation substrates. This accelerated substrate metabolism provides more abundant reducing power (e.g., NADH) and intermediates (e.g., pyruvate) for the hydrogen production process. Concurrently, it drives carbon flow toward hydrogen-efficient fermentation pathways (e.g., acetate–butyrate type) rather than low-hydrogen-yield pathways (e.g., ethanol or lactate production), thereby fundamentally increasing the system's hydrogen production rate.

For example, Fe_3O_4 NPs act as "public nano-wires", facilitating a direct electron exchange between syntrophic partners such as Clostridium and Syntrophomonas [66]. This electrical connection not only accelerates the degradation of organic acids (e.g., butyrate, propionate) but also helps maintain low hydrogen partial pressure, sustaining favorable thermodynamic conditions for continuous hydrogen production.

Composite materials like MGAC and MNAC further optimize DIET efficiency. MGAC combines high surface area and excellent conductivity with embedded magnetic nanoparticles (e.g., $NiFe_3O_4$), significantly enhancing electron transfer between microorganisms [40]. Similarly, nitrogen doping in MNAC improves its electronic and pore structure, strengthening its role as an electron mediator and promoting synergistic microbial metabolism [41]. Additionally, MNPs contribute to the formation of conductive biofilms or microbe–

nanomaterial aggregates, which further stabilize the DIET network. For instance, Fe_3O_4 NPs could integrate with extracellular polymeric substances to form a three-dimensional conductive matrix that enhances the electron transfer efficiency [87].

In summary, by establishing robust DIET networks, MNPs fundamentally enhance electron flow and metabolic cooperation in mixed cultures. This "electron-bridging" effect activates collective microbial activity, accelerates substrate conversion, and directs the metabolic flux toward hydrogen production, thereby significantly improving both the rate and stability of hydrogen generation.

6.3. Enhancement of Overall System Functional Robustness

Mixed cultures face environmental fluctuation and toxin shock challenges in practical applications. MNPs enhance system resistance and stability in various ways:

Toxin adsorption and passivation: MNPs and their composites (e.g., MGAC, MNAC, biochar-supported Fe(0) NPs) possess high specific surface areas, enabling the effective adsorption and immobilization of inhibitors such as heavy metals (e.g., Pb, Cr, Cd) and organic toxins. This reduces their bioavailability and creates a cleaner microenvironment for functional microbiota [88,89]. Furthermore, by attaching to bacterial surfaces, MNPs could stimulate the secretion of extracellular polymeric substances (EPS) and accelerate biofilm formation. This self-reinforcing process not only provides additional physical protection for the microbial community against environmental toxins but also enhances bacterial tolerance to adverse conditions by creating a stable, aggregated micro-ecosystem [90]. For instance, Fe(0) NPs and Fe_3O_4 have been shown to stabilize heavy metals in sludge via adsorption or Fe–Mn interactions, alleviating metal stress and maintaining microbial activity [37]. Wang et al. [77] further demonstrated that $MgFe_2O_4$ and $Ca_{0.5}Mg_{0.5}Fe_2O_4$ NPs effectively counteract the inhibitory effect of humic acid on hydrogen production.

Mitigation of acid inhibition: During efficient hydrogen production, rapid organic acid accumulation may lead to a sharp pH drop. Some MNPs' corrosion processes consume protons, providing certain pH buffering capacity. Furthermore, the DIET networks they promote could consume organic acids faster, thereby alleviating the system's acid inhibition. Ferrihydrite reduction was identified as a mechanism that protects against system acidification, due to organic acid accumulation. The consumption of protons during the reduction of Fe(III) to Fe(II) helped stabilize the pH, preventing the sharp drop that could inhibit hydrogen-producing bacteria. This buffering action, combined with the selective enrichment of robust hydrogen producers like *Clostridium*, significantly contributed to the system's functional robustness and elevated carbon and electron conversion efficiencies [79].

Enhanced thermal stability: Abdul Aziz et al. [65] pointed out that green-synthesized Fe_3O_4 NPs maintained high catalytic activity and biocompatibility under high-temperature (55 °C) conditions, significantly improving the system's thermal stability and hydrogen production persistence. MGAC's successful application in thermophilic dark fermentation systems also confirms this [40].

6.4. Overview of the Integrated Enhancement Mechanisms Induced by MNPs in Mixed Culture Systems

In conclusion, the enhancement of hydrogen production by MNPs in mixed cultures is achieved through a synergistic combination of ecological and engineering mechanisms. These include the following: (1) selective enrichment of hydrogen-producing microbiota via niche modification and electron sink effects; (2) enhanced metabolic activity through the construction of DIET networks that act as conductive electron bridges, optimizing interspecies electron flow and directing carbon metabolism toward high-yield pathways; and (3) enhanced system robustness via toxin adsorption, pH buffering, and thermal stabilization, which collectively maintain functional stability under variable or stressful conditions (Figure 2). A holistic understanding of this integrated ecological strategy is essential for the rational design of MNP-based interventions to optimize complex microbial communities in practical biohydrogen production systems.

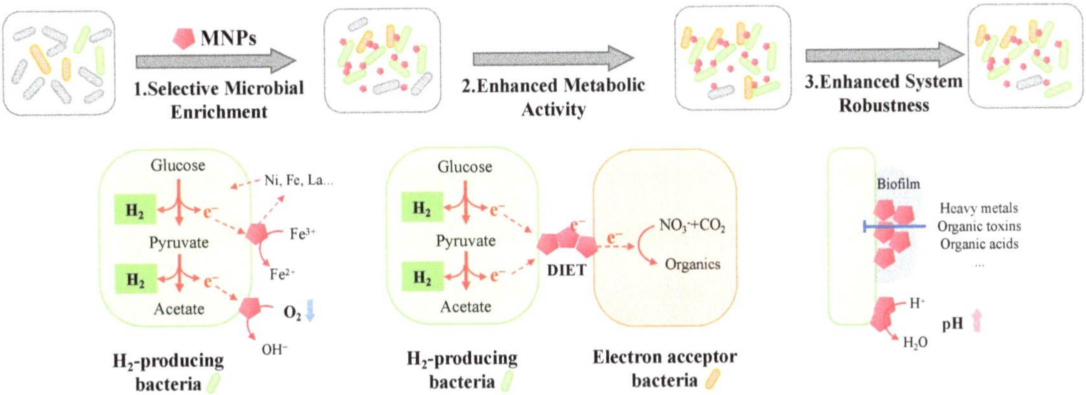

Figure 2. The mechanism by which MNPs enhance hydrogen production in mixed microbial systems.

7. Research Challenges and Perspectives on the Mechanisms of MNPs in Mixed Culture Systems

In mixed microbial community systems, the synergistic mechanisms of MNPs are far more complex than in pure culture systems. Their ecological-level regulatory properties also imply that current research faces more urgent bottlenecks and blind spots in elucidating these mechanisms.

7.1. Contradiction Between Selectivity and Universality in Niche Regulation

Current studies generally observe that MNPs could selectively enrich hydrogen-producing bacteria (HPB) like Clostridium, but the intrinsic "selectivity" rules remain unclear [41,65]. First, the universality of this enrichment effect is questionable. Different inoculum sources (e.g., anaerobic sludge, compost, river sediment) harbor vastly different initial microbial community structures. Will the directional enrichment effect demonstrated by MNPs in one inoculum remain valid in a system with a different ecological background? Second, the current understanding of "selectivity" is mostly based on macroscopic statistics of relative abundance changes, overlooking heterogeneity at the individual microbial level. Do MNPs favor the growth of strains with specific surface proteins or membrane permeability, or do they indirectly eliminate competitors by altering environmental factors? The molecular basis of this selectivity has not been revealed [41,84]. More importantly, when MNPs suppress hydrogen-consuming bacteria (e.g., homoacetogens and hydrogenotrophic methanogens), is this inhibition specific or a non-specific broad-spectrum toxic effect? If the latter, there is a risk of simultaneously inhibiting beneficial hydrolytic and acidogenic bacteria, which could be detrimental to the overall system function [79,85]. Therefore, there is an urgent need to establish more refined controlled experiments combined with single-cell technologies to clarify the specific targets and microscopic processes of MNPs in "fostering the superior and eliminating the inferior" within complex consortia.

7.2. Dilemmas in Evidencing and Quantifying DIET Networks

Although conductive MNPs (e.g., Fe_3O_4, MGAC) are considered to act as "electron bridges" promoting DIET, the current evidence chain remains incomplete [71,86]. First, most studies attribute this effect to the high conductivity of the materials, but the direct causal link between increased conductivity and enhanced hydrogen production performance has not been strictly verified. In complex mixed consortia, electron donor and acceptor bacteria could connect via multiple pathways (e.g., pili, extracellular polymeric substances). Are MNPs serving as the primary electron conduit, or merely enhancing the efficiency or stability of existing conductive networks [72,86]? Second, there is a severe lack of quantitative assessment of the DIET flux. What proportion of the total electron flow in a mixed consortium is contributed by the MNP-mediated DIET pathway? How much more efficient is it compared to traditional interspecies hydrogen transfer (IHT)? These key parameters cannot be obtained from macroscopic gas production data alone [79]. Future work needs to develop in situ electrochemical monitoring techniques and combine them with metabolic flux analysis, using isotopic labels (e.g., ^{13}C or ^{15}N) to accurately quantify the DIET flux and thus assess its true ecological and metabolic contribution [79].

7.3. Maintenance Mechanisms of System Functional Robustness and Long-Term Risks

While MNPs have been shown to enhance system stability against acid inhibition and toxin shocks, the deeper maintenance mechanisms underlying this "robust" state are poorly studied [77,79]. For example, the corrosion of MNPs (e.g., Fe(0) NPs) consumes protons, providing pH buffering capacity, but how does this chemical buffering synergize with the physiological buffering capacity generated by the self-regulation of the microbial community [85]? During long-term operation, as MNPs are consumed and this chemical buffering capacity decays, could the microbial community achieve a smooth functional transition? Furthermore, the adsorption of toxins by MNPs has been proven to be beneficial, but this adsorption capacity is finite. Could saturation lead to a sudden functional failure [77,88]? More importantly, is the MNP-induced, HPB-dominated community structure, while efficient in the short term, posing a long-term ecological risk due to its lower species diversity [41,79]? Would such a system be more prone to collapse when facing novel shock loads (e.g., antibiotics) that MNPs cannot mitigate? These questions concerning long-term system resilience and succession patterns cannot be answered by current short-term batch experiments [65,77].

7.4. Systematic Neglect of Magnetic Field Effects in Complex Systems

Similarly to pure culture systems, the magnetobiological effects of MNPs in mixed systems are almost entirely submerged in their chemical effects [48,63]. Mixed consortia may contain various microorganisms with different responses to magnetic fields. Weak magnetic fields might differentially affect the growth rates, cell aggregation behaviors, and gene expressions of different species [63]. For instance, could magnetic fields indirectly alter the niche structure of the entire community by influencing magnetotactic bacteria or other magneto-sensitive microorganisms? When constructing conductive aggregates centered around MNPs, do magnetic forces influence the spatial arrangement between microbes and materials, thereby optimizing the electron transfer paths [61]? Currently, due to the lack of systematic studies under strictly controlled conditions (e.g., using inert magnetic microspheres instead of active MNPs), we cannot assess the independent role of this physical field in complex ecological interactions, undoubtedly leading to a biased understanding of the overall mechanism of MNPs action [48,63].

In summary, future research on the mechanisms of MNPs in mixed systems must adopt a more systematic ecological perspective and employ more advanced in situ analytical techniques. It is necessary to resolve the dynamic network of community functional succession via metagenomics/metatranscriptomics, visualize the transport of elements within the consortium by using techniques like nano-secondary ion mass spectrometry (NanoSIMS), and design sophisticated long-term continuous experiments to evaluate system stability and resilience [79,84]. Only by treating MNPs as an active "abiotic factor" within the mixed microbial ecosystem and coupling their chemical and physical (magnetic) properties with the ecological and evolutionary dynamics of microbes could we fundamentally reveal the deep-seated mechanisms by which they enhance dark fermentative hydrogen production, thereby promoting the transition of this technology from the laboratory to the engineering application [65].

8. Challenges and Prospects

Although MNPs show great potential in enhancing dark fermentative hydrogen production, their transition from laboratory to large-scale industrial application still faces many challenges, indicating future research directions.

In-depth analysis of multi-dimensional action mechanisms: The electron transfer pathways, kinetics, and regulatory factors at the nanomaterial–microbe interface remain poorly understood, representing a critical knowledge gap in optimizing bio-nano interactions. To address this, interdisciplinary research integrating electrochemical methods (e.g., cyclic voltammetry, electrochemical impedance spectroscopy) [91,92], advanced characterization techniques (e.g., synchrotron radiation-based spectroscopy, cryo-electron microscopy) [93,94], and multi-omics approaches (metagenomics, transcriptomics, and proteomics) is essential. Such integrated methodologies could systematically elucidate the structural, electronic, and dynamic interactions between nanoparticles and microbial entities—down to the atomic and molecular scales.

Additionally, MNPs uniqueness compared to other nanoparticles lies in "magnetism". Future needs include designing precise control experiments to specifically study the direct effects of nanoparticles' own magnetic field and behavior under external magnetic fields on microbial metabolism, enzyme conformation, biofilm formation, etc., providing new ideas for designing magnetically regulated hydrogen production systems [48,63]. To address this, advanced analytical techniques that are capable of probing magnetobiological interactions are indispensable. Notably, in situ monitoring of magnetically influenced processes is crucial [95]. For instance, magneto-electrochemical setups, integrating electrochemical workstations with programmable electromagnets, enables a real-time measurement of electron transfer rates via the constant current method, or real-time determination of charge transfer resistance and other parameters through electrochemical impedance spectroscopy under applied magnetic fields [96,97]. Furthermore, synchrotron-based techniques like X-ray magnetic circular dichroism (XMCD) could provide element-specific information on the valence and magnetic state of metal ions within MNPs or even in microbial metalloenzymes in operando, thereby directly linking magnetic states to redox activity [98,99]. Concurrently, analyzing the biological response under magnetic stimuli is essential. Comparative multi-omics analyses—including transcriptomics and proteomics—of microbial systems exposed to MNPs versus non-magnetic analogs or to external magnetic fields alone could reveal unique genetic and metabolic signatures that are specifically triggered by magnetic forces [100,101]. Advanced imaging techniques, such as cryo-electron tomography, could, in principle, visualize the structural organization of cells and biofilms in the presence of MNPs, potentially revealing magnetically induced alignment or clustering phenomena [102,103].

Building on such foundational insights, predictive mechanism models should be developed to describe how MNPs enhance microbial hydrogen production. These models must account for system variability by examining both defined pure-culture systems and complex mixed-microbiome environments. By comparing and reconciling findings across both systems, a unified "pure-culture–mixed-culture hybrid model" could be constructed.

This integrative framework will enable the holistic prediction, control, and optimization of the system's performance—bridging fundamental nanomaterial properties with the microbiome function to advance scalable and efficient bioenergy applications.

Engineering upgrade from batch experiments to continuous systems: Current research is mostly limited to batch experiments, differing from continuous industrial operation needs. Future efforts should focus on verifying nanoparticles' long-term enhancement effects and stability in continuous-flow reactors (e.g., UASB, EGSB) and optimizing online dosing, dispersion, and recovery strategies to solve potential clogging and wear problems [13]. Additionally, emphasis should be placed on exploring the application of magnetic nanoparticles for treating more complex organic wastes, such as lignocellulosic biomass and municipal organic waste. Simultaneously, future studies should actively explore synergistic effects with other enhancement strategies (e.g., pretreatment, key microbiota enrichment) to achieve hydrogen production efficiency multiplication through combined processes.

Construction of cost-effectiveness and recycling systems: Although MNPs are easy to recover via magnetic separation, the key advantages for realizing circular economy, recovery efficiency, retained catalytic activity after cyclic use, and entire process cost-effectiveness still need systematic evaluation. Developing low-cost, high-stability MNPs (e.g., synthesized using industrial by-products) and designing integrated magnetic separation reactors are key future directions. Ramprakash et al. [44] showed that Fe_3O_4 NPs maintained over 90% relative hydrogen production activity after four consecutive hydrogen production cycles, demonstrating good reusability. MNAC and MGAC's magnetic separation properties also facilitated recovery and reuse [40,41]. LPBC maintained high catalytic activity after 30 days of cycling, with a magnetic recovery rate of >95%, demonstrating excellent long-term operation and economic potential [50].

However, implementing magnetic separation in continuous systems presents three major challenges [104,105]:

MNP retention and separation in continuous flow: Designing integrated reactors to prevent MNP washout while enabling efficient, continuous recovery is non-trivial. Inadequate separation increases costs and poses environmental risks from metal residues in the effluent [13]. An effective design is therefore critical to prevent metal carry-over [13], and dedicated engineering is needed to adapt this batch-proven advantage [13,106].

Long-term stability and recyclability: Ensuring MNP stability and activity over extended continuous operation is paramount. Surface coatings can enhance dispersibility and durability [107,108], but long-term exposure to fermentation media may lead to deactivation or toxicity. Notably, current evidence is predominantly from batch studies [105,106], with a significant gap in long-term, continuous-system performance data [13]. Future work must focus on synthesizing cost-effective, stable MNPs (e.g., via green methods [109]) and rigorously testing their reusability in continuous reactors [106].

System integration and economics: Developing economically viable processes hinges on an integrated reactor design that couples fermentation with efficient in-line MNP separation and recycling [13]. The overall cost–benefit depends on MNP synthesis cost, operational lifespan, and separation energy. Advancing technology necessitates moving beyond batch experiments to pilot-scale, continuous validation to obtain scalable engineering data [13,106].

Addressing these challenges will require collaborative efforts across materials science, microbiology, and engineering to realize the full potential of MNPs in sustainable bioenergy systems.

9. Conclusions

Magnetic nanoparticles (MNPs) significantly enhance dark fermentative hydrogen production through multi-level mechanisms. In pure culture systems, MNPs function primarily at the molecular and cellular levels by serving as sustained-release sources of essential enzyme cofactors, acting as efficient electron carriers, and redirecting the central carbon metabolism toward high-yield acetate-type fermentation pathways. In mixed culture systems that better represent practical applications, MNPs operate at ecological levels by modifying microenvironmental niches to selectively enrich hydrogen-producing consortia, constructing conductive networks for DIET, and enhancing system robustness through toxin adsorption and pH buffering capabilities.

Despite substantial progress in understanding MNPs' enhancement phenomena, critical knowledge gaps persist in mechanistic elucidations. The structure–activity relationships between MNPs' physicochemical properties and biological effects remain ambiguous, while the specific pathways and efficiency of MNP-mediated electron transfer lack quantitative characterization. At genetic levels, the signaling mechanisms through which MNPs are perceived by cells and trigger transcriptional responses constitute a "black box" in current understanding. Furthermore, the unique magnetobiological effects of MNPs are frequently overlooked in mechanistic models, without effective differentiation from or integration with chemical effects.

Future research should prioritize interdisciplinary integration of electrochemical monitoring, multi-omics analyses, and advanced characterization techniques to deepen the mechanistic understanding of nanomaterial–microbe interactions. Such fundamental insights should guide the rational design of smarter, highly biocompatible MNPs composites with tailored functionalities. Ultimately, transitioning from batch experiments to continuous-flow systems will be crucial for enhancing long-term operational stability, recyclability, and ecological safety assessment. These coordinated efforts will advance MNP–microorganism hybrid systems toward efficient, stable, economical, and sustainable biohydrogen production, thereby accelerating the industrialization of this promising bioenergy technology.

Author Contributions: J.Y.: Investigation, Visualization, Writing—original draft; Z.X.: Conceptualization, Funding acquisition, Supervision, Writing—review and editing. All authors have read and agreed to the published version of the manuscript.

Funding: This work was financially supported by the National Natural Science Foundation of China (Grant No. 42377134), the Guangdong Foundation for Program of Science and Technology Research (Grant No. 2023B1212060044), the Young Talent Project of GDAS (2024GDASQNRC-0204), and the GDAS' Project of Science and Technology Development (2022GDASZH-2022010102 and 2024GDASZH-2024010102).

Data Availability Statement: Data will be made available on request.

Conflicts of Interest: The authors declare that they have no known competing financial interests or personal relationships that could have appeared to influence the work reported in this paper.

References

1. Dincer, I.; Acar, C. Review and evaluation of hydrogen production methods for better sustainability. *Int. J. Hydrogen Energy* **2015**, *40*, 11094–11111. [CrossRef]
2. Show, K.Y.; Yan, Y.G.; Lee, D.J. Biohydrogen production: Status and perspectives. In *Biofuels: Alternative Feedstocks and Conversion Processes for the Production of Liquid and Gaseous Biofuels*, 2nd ed.; Academic Press: Nanjing, China, 2019; pp. 693–713.
3. Soares, J.F.; Confortin, T.C.; Todero, I.; Mayer, F.D.; Mazutti, M.A. Dark fermentative biohydrogen production from lignocellulosic biomass: Technological challenges and future prospects. *Renew. Sustain. Energy Rev.* **2020**, *117*, 109484. [CrossRef]

4. Ghimire, A.; Frunzo, L.; Pirozzi, F.; Trably, E.; Escudié, R.; Lens, P.N.L.; Esposito, G. A Review on Dark Fermentative Biohydrogen Production from Organic Biomass: Process Parameters and Use of By-Products. *Appl. Energy* **2015**, *144*, 73–95. [CrossRef]
5. Lee, D.J.; Show, K.Y.; Su, A. Dark Fermentation on Biohydrogen Production: Pure Culture. *Bioresour. Technol.* **2011**, *102*, 8393–8402. [CrossRef] [PubMed]
6. Rao, R.; Basak, N. Optimization and Modelling of Dark Fermentative Hydrogen Production from Cheese Whey by Enterobacter aerogenes 2822. *Int. J. Hydrogen Energy* **2021**, *46*, 1777–1800. [CrossRef]
7. Pachapur, V.L.; Kutty, P.; Pachapur, P.; Brar, S.K.; Le Bihan, Y.; Galvez-Cloutier, R.; Buelna, G. Seed Pretreatment for Increased Hydrogen Production Using Mixed-Culture Systems with Advantages over Pure-Culture Systems. *Energies* **2019**, *12*, 530. [CrossRef]
8. Sinha, P.; Roy, S.; Das, D. Genomic and Proteomic Approaches for Dark Fermentative Biohydrogen Production. *Renew. Sustain. Energy Rev.* **2016**, *56*, 1308–1321. [CrossRef]
9. Chandrasekhar, K.; Mohan, S.V. Bio-Electrohydrolysis as a Pretreatment Strategy to Catabolize Complex Food Waste in Closed Circuitry: Function of Electron Flux to Enhance Acidogenic Bio-Hydrogen Production. *Int. J. Hydrogen Energy* **2014**, *39*, 11411–11422. [CrossRef]
10. Yang, G.; Wang, J. Changes in Microbial Community Structure During Dark Fermentative Hydrogen Production. *Int. J. Hydrogen Energy* **2019**, *44*, 25541–25550. [CrossRef]
11. Pakarinen, O.; Lehtomäki, A.; Rintala, J. Batch Dark Fermentative Hydrogen Production from Grass Silage: The Effect of Inoculum, pH, Temperature and VS Ratio. *Int. J. Hydrogen Energy* **2008**, *33*, 594–601. [CrossRef]
12. Alexandropoulou, M.; Antonopoulou, G.; Lyberatos, G. Modeling of continuous dark fermentative hydrogen production in an anaerobic up-flow column bioreactor. *Chemosphere* **2022**, *293*, 133527. [CrossRef] [PubMed]
13. Hidalgo, D.; Martin-Marroquin, J.M.; Corona, F. The role of magnetic nanoparticles in dark fermentation. *Biomass Convers. Biorefinery* **2023**, *13*, 16299–16320. [CrossRef]
14. Patel, S.K.; Das, D.; Kim, S.C.; Cho, B.-K.; Kalia, V.C.; Lee, J.-K. Integrating strategies for sustainable conversion of waste biomass into dark-fermentative hydrogen and value-added products. *Renew. Sustain. Energy Rev.* **2021**, *150*, 111491. [CrossRef]
15. Bann, J.R.; Kavitha, S.; Kannah, R.Y.; Bhosale, R.R.; Kumar, G. Industrial wastewater to biohydrogen: Possibilities towards successful biorefinery route. *Bioresour. Technol.* **2020**, *298*, 122378. [CrossRef]
16. Mishra, P.; Krishnan, S.; Rana, S.; Singh, L.; Sakinah, M.; Ab Wahid, Z. Outlook of fermentative hydrogen production techniques: An overview of dark, photo and integrated dark-photo fermentative approach to biomass. *Energy Strategy Rev.* **2019**, *24*, 27–37. [CrossRef]
17. Sivagurunathan, P.; Kumar, G.; Bakonyi, P.; Kim, S.-H.; Kobayashi, T.; Xu, K.Q.; Lakner, G.; Tóth, G.; Nemestóthy, N.; Bélafi-Bakó, K. A critical review on issues and overcoming strategies for the enhancement of dark fermentative hydrogen production in continuous systems. *Int. J. Hydrogen Energy* **2016**, *41*, 3820–3836. [CrossRef]
18. Bundhoo, M.Z.; Mohee, R. Inhibition of dark fermentative bio-hydrogen production: A review. *Int. J. Hydrogen Energy* **2016**, *41*, 6713–6733. [CrossRef]
19. Wang, S.; Zhang, T.; Bao, M.; Su, H.; Xu, P. Microbial production of hydrogen by mixed culture technologies: A review. *Biotechnol. J.* **2020**, *15*, 1900297. [CrossRef]
20. Cao, Y.; Liu, H.; Liu, W.; Guo, J.; Xian, M. Debottlenecking the biological hydrogen production pathway of dark fermentation: Insight into the impact of strain improvement. *Microb. Cell Fact.* **2022**, *21*, 1–16. [CrossRef] [PubMed]
21. Birrell, J.A.; Rodríguez-Maciá, P.; Reijerse, E.J.; Martini, M.A.; Lubitz, W. The catalytic cycle of [FeFe] hydrogenase: A tale of two sites. *Coord. Chem. Rev.* **2021**, *449*, 214191. [CrossRef]
22. Pugazhendhi, A.; Shobana, S.; Nguyen, D.D.; Banu, J.R.; Sivagurunathan, P.; Chang, S.W.; Ponnusamy, V.K.; Kumar, G. Application of nanotechnology (nanoparticles) in dark fermentative hydrogen production. *Int. J. Hydrogen Energy* **2019**, *44*, 1431–1440. [CrossRef]
23. Maroušek, J. Nanoparticles can change (bio)hydrogen competitiveness. *Fuel* **2022**, *328*, 125318. [CrossRef]
24. Shanmugam, S.; Hari, A.; Pandey, A.; Mathimani, T.; Felix, L.; Pugazhendhi, A. Comprehensive review on the application of inorganic and organic nanoparticles for enhancing biohydrogen production. *Fuel* **2020**, *270*, 117453. [CrossRef]
25. Elreedy, A.; Fujii, M.; Koyama, M.; Nakasaki, K.; Tawfik, A. Enhanced fermentative hydrogen production from industrial wastewater using mixed culture bacteria incorporated with iron, nickel, and zinc-based nanoparticles. *Water Res.* **2019**, *151*, 349–361. [CrossRef] [PubMed]
26. Gadhe, A.; Sonawane, S.S.; Varma, M.N. Enhancement effect of hematite and nickel nanoparticles on biohydrogen production from dairy wastewater. *Int. J. Hydrogen Energy* **2015**, *40*, 4502–4511. [CrossRef]
27. Khan, M.M.; Lee, J.; Cho, M.H. Electrochemically active biofilm mediated bio-hydrogen production catalyzed by positively charged gold nanoparticles. *Int. J. Hydrogen Energy* **2013**, *38*, 5243–5250. [CrossRef]

28. Ganachari, S.V.; Patil, V.B.; Bhat, R.R.; Nadaf, H.A.; Bali, G. Hybrid nano and microbial consortium technologies to harvest biofuel (biomethane) from organic and agri waste. In *Microbial Resource Technologies for Sustainable Development*; Elsevier: Karnataka, India, 2022; pp. 369–393.
29. Lin, R.; Cheng, J.; Ding, L.; Song, W.; Liu, M.; Zhou, J.; Cen, K. Enhanced dark hydrogen fermentation by addition of ferric oxide nanoparticles using Enterobacter aerogenes. *Bioresour. Technol.* **2016**, *207*, 213–219. [CrossRef]
30. Mohanraj, S.; Anbalagan, K.; Kodhaiyolii, S.; Pugalenthi, V. Comparative evaluation of fermentative hydrogen production using Enterobacter cloacae and mixed culture: Effect of Pd(II) ion and phytogenic palladium nanoparticles. *J. Biotechnol.* **2014**, *192*, 87–95. [CrossRef]
31. Sekoai, P.T.; Ouma, C.N.M.; du Preez, S.P.; Modisha, P.; Engelbrecht, N.; Bessarabov, D.G.; Ghimire, A. Application of nanoparticles in biofuels: An overview. *Fuel* **2019**, *237*, 380–397. [CrossRef]
32. Bossi, S.; Rajamohan, N. Nanotechnology approach for enhancement in biohydrogen production—Review on applications of nanocatalyst and life cycle assessment. *Fuel* **2022**, *323*, 124351. [CrossRef]
33. Li, J.; Li, C.; Zhao, L.; Pan, X.; Cai, G.; Zhu, G. The application status, development and future trend of nano-iron materials in anaerobic digestion system. *Chemosphere* **2021**, *269*, 129389. [CrossRef]
34. Ameen, F.; Alsamhary, K.; Alabdullatif, J.A.; Alnadhari, S. A review on metal-based nanoparticles and their toxicity to beneficial soil bacteria and fungi. *Ecotoxicol. Environ. Saf.* **2021**, *213*, 112027. [CrossRef]
35. Reddy, L.H.; Arias, J.L.; Nicolas, J.; Couvreur, P. Magnetic nanoparticles: Design and characterization, toxicity and biocompatibility, pharmaceutical and biomedical applications. *Chem. Rev.* **2012**, *112*, 5818–5878. [CrossRef]
36. Ali, A.; Shah, T.; Ullah, R.; Zhou, P.; Guo, M.; Ovais, M.; Tan, Z.; Rui, Y.K. Review on recent progress in magnetic nanoparticles: Synthesis, characterization, and diverse applications. *Front. Chem.* **2021**, *9*, 629054. [CrossRef]
37. Reddy, K.; Nasr, M.; Kumari, S.; Kumar, S.; Gupta, S.K.; Enitan, A.M.; Bux, F. Biohydrogen production from sugarcane bagasse hydrolysate: Effects of pH, S/X, Fe^{2+}, and magnetite nanoparticles. *Environ. Sci. Pollut. Res.* **2017**, *24*, 8790–8804. [CrossRef] [PubMed]
38. Taherdanak, M.; Zilouei, H.; Karimi, K. The effects of Fe^0 and Ni^0 nanoparticles versus Fe^{2+} and Ni^{2+} ions on dark hydrogen fermentation. *Int. J. Hydrogen Energy* **2016**, *41*, 167–173. [CrossRef]
39. Kumar, G.; Mathimani, T.; Rene, E.R.; Pugazhendhi, A. Application of nanotechnology in dark fermentation for enhanced biohydrogen production using inorganic nanoparticles. *Int. J. Hydrogen Energy* **2019**, *44*, 13106–13113. [CrossRef]
40. Jamaludin, N.F.M.; Jamali, N.S.; Abdullah, L.C.; Idrus, S.; Engliman, N.S.; Abdul, P.M. Biohydrogen production with utilisation of magnetite nanoparticles embedded in granular activated carbon from coconut shell. *Int. J. Hydrogen Energy* **2023**, *48*, 11695–11708. [CrossRef]
41. Tian, K.; Zhang, J.; Zhou, C.; Yang, M.; Zhang, X.; Yan, X.; Zang, L. Magnetic nitrogen-doped activated carbon improved biohydrogen production. *Environ. Sci. Pollut. Res.* **2023**, *30*, 87215–87227. [CrossRef]
42. Zhang, J.; Li, W.; Yang, J.; Li, Z.; Zhang, J.; Zhao, W.; Zang, L. Cobalt ferrate nanoparticles improved dark fermentation for hydrogen evolution. *J. Clean. Prod.* **2021**, *316*, 128275. [CrossRef]
43. Beckers, L.; Hiligsmann, S.; Lambert, S.D.; Heinrichs, B.; Thonart, P. Improving effect of metal and oxide nanoparticles encapsulated in porous silica on fermentative biohydrogen production by Clostridium butyricum. *Bioresour. Technol.* **2013**, *133*, 109–117. [CrossRef]
44. Ramprakash, B.; Incharoensakdi, A. Supplementation of magnetic nanoparticles for enhancement of dark fermentative hydrogen production from pretreated garden wastes using Enterobacter aerogenes. *Fuel* **2023**, *342*, 127857. [CrossRef]
45. Mohanraj, S.; Kodhaiyolli, S.; Rengasamy, M.; Pugalenthi, V. Green synthesized iron oxide nanoparticles effect on fermentative hydrogen production by Clostridium acetobutylicum. *Appl. Biochem. Biotechnol.* **2014**, *173*, 318–331. [CrossRef] [PubMed]
46. Mishra, P.; Thakur, S.; Mahapatra, D.M.; Wahid, Z.A.; Liu, H.; Singh, L. Impacts of nano-metal oxides on hydrogen production in anaerobic digestion of palm oil mill effluent—A novel approach. *Int. J. Hydrogen Energy* **2018**, *43*, 2666–2676. [CrossRef]
47. Zhang, Y.; Xiao, L.; Wang, S.; Liu, F. Stimulation of ferrihydrite nanorods on fermentative hydrogen production by Clostridium pasteurianum. *Bioresour. Technol.* **2019**, *283*, 308–315. [CrossRef]
48. Rambabu, K.; Bharath, G.; Thanigaivelan, A.; Das, D.B.; Show, P.L.; Banat, F. Augmented biohydrogen production from rice mill wastewater through nano-metal oxides assisted dark fermentation. *Bioresour. Technol.* **2021**, *319*, 124243. [CrossRef]
49. Zhang, Y.; Liu, F.; Xu, H.; Xiao, L. Extraction of electrons by magnetite and ferrihydrite from hydrogen-producing Clostridium bifermentans by strengthening the acetate production pathway. *Sci. China Technol. Sci.* **2019**, *62*, 1719–1725. [CrossRef]
50. Yang, L.; Dong, Y.; Zhao, D.; Li, X.; He, J.; Lai, T.P.; Zhou, E.; Ueki, T.; Li, Y.; Meng, X.; et al. A nanobiocatalyst-driven hybrid system for efficient and sustainable hydrogen production via electron flow optimization. *Adv. Mater.* 2025; *in press*.
51. Tychounian, K.; Poladyan, A.; Tychounian, A. Enhancement of Escherichia coli bacterial biomass and hydrogen production by some heavy metal ions and their mixtures during glycerol vs glucose fermentation at a relatively wide range of pH. *Int. J. Hydrogen Energy* **2017**, *42*, 6590–6597. [CrossRef]

52. Yang, G.; Wang, J. Improving mechanisms of biohydrogen production from grass using zero-valent iron nanoparticles. *Bioresour. Technol.* **2018**, *266*, 413–420. [CrossRef]
53. Yin, Y.; Wang, J. Enhanced biohydrogen production from macroalgae by zero-valent iron nanoparticles: Insights into microbial and metabolites distribution. *Bioresour. Technol.* **2019**, *282*, 110–117. [CrossRef]
54. Walke, V.; Das, S.; Mittal, A.; Agrawal, A. Tumor Educated Platelets as a Biomarker for Diagnosis of Lung cancer: A Systematic Review. *Asian Pac. J. Cancer Prev.* **2024**, *25*, 1911. [CrossRef]
55. Nath, D.; Manhar, A.K.; Gupta, K.; Saikia, D.; Das, S.K.; Mandal, M. Phytosynthesized iron nanoparticles: Effects on fermentative hydrogen production by Enterobacter cloacae DH-89. *Bull. Mater. Sci.* **2015**, *38*, 1533–1538. [CrossRef]
56. Zhao, X.; Xing, D.; Qi, N.; Zhao, Y.; Hu, X.; Ren, N. Deeply mechanism analysis of hydrogen production enhancement of Ethanoligenens harbinense by Fe^{2+} and Mg^{2+}: Monitoring at growth and transcription levels. *Int. J. Hydrogen Energy* **2017**, *42*, 19695–19700. [CrossRef]
57. Lubitz, W.; Ogata, H.; Rüdiger, O.; Reijerse, E. Hydrogenases. *Chem. Rev.* **2014**, *114*, 4081–4148. [CrossRef]
58. Zhang, Q.; Xu, S.; Li, Y.; Ding, P.; Zhang, Y.; Zhao, P. Green-synthesized nickel oxide nanoparticles enhances biohydrogen production of Klebsiella sp. WL1316 using lignocellulosic hydrolysate and its regulatory mechanism. *Fuel* **2021**, *305*, 121543. [CrossRef]
59. Jiang, X.; Hu, J.; Lieber, A.M.; Jackan, C.S.; Biffinger, J.C.; Fitzgerald, L.A.; Ringeisen, B.R.; Lieber, C.M. Nanoparticle facilitated extracellular electron transfer in microbial fuel cells. *Nano Lett.* **2014**, *14*, 6737–6742. [CrossRef]
60. El-Naggar, M.Y.; Wanger, G.; Leung, K.M.; Yuzvinsky, T.D.; Southam, G.; Yang, J.; Lau, W.M.; Nealson, K.H.; Tiedje, G.J.M. Electrical transport along bacterial nanowires from Shewanella oneidensis MR-1. *Proc. Natl. Acad. Sci. USA* **2010**, *107*, 18127–18131. [CrossRef] [PubMed]
61. Zhao, W.; Zhang, Y.; Du, B.; Wei, D.; Wei, Q.; Zhao, Y. Enhancement effect of silver nanoparticles on fermentative biohydrogen production using mixed bacteria. *Bioresour. Technol.* **2013**, *142*, 240–245. [CrossRef]
62. Khokhlova, G.; Abashina, T.; Belova, N.; Panchelyuga, V.; Petrov, A.; Abreu, F.; Vainshtein, M. Effects of combined magnetic fields on bacteria Rhodospirillium rubrum VEM-B-1621. *Bioelectromagnetics* **2021**, *42*, 485–490.
63. Abdul Aziz, A.H.; Engliman, N.S.; Mansor, M.F.; Abdul, P.M.; Arisht, S.N.; Jamali, N.S.; Tiang, M.F. Synergistic enhancement of biohydrogen production by supplementing with green synthesized magnetic iron nanoparticles using thermophilic mixed bacteria culture. *Int. J. Hydrogen Energy* **2022**, *47*, 40683–40695. [CrossRef]
64. Wang, R.; Zhang, H.; Zhang, J.; Zhou, C.; Zhang, X.; Yan, X.; Yu, F.; Zhang, J. Comparison of calcium magnesium ferrite nanoparticles for boosting biohydrogen production. *Bioresour. Technol.* **2024**, *395*, 130410. [CrossRef] [PubMed]
65. Wang, J.; Wan, W. Effect of Fe^{2+} concentration on fermentative hydrogen production by mixed cultures. *Int. J. Hydrogen Energy* **2008**, *33*, 1215–1220. [CrossRef]
66. Cheng, J.; Li, H.; Ding, L.; Zhou, J.; Song, W.; Li, Y.Y.; Lin, R. Improving Hydrogen and Methane Co-Generation in Cascading Dark Fermentation and Anaerobic Digestion: The Effect of Magnetite Nanoparticles on Microbial Electron Transfer and Syntrophism. *Chem. Eng. J.* **2020**, *397*, 125394. [CrossRef]
67. Thakur, A.; Sharma, S.; Khajuria, T.; Chib, M.; Bangotra, R.; Kapoor, N.; Mahajan, R.; Bajaj, B.K. Nanobiocatalysts for Efficient Conversion of Microwave Aided Ionic Liquid Pretreated Rice Straw Biomass to Biofuel. *Biomass Convers. Biorefin.* **2025**, *10*, 15. [CrossRef]
68. Chandra, S.; Arora, K.; Bahadur, D. Impedimetric Biosensor Based on Magnetic Nanoparticles for Electrochemical Detection of Dopamine. *Mater. Sci. Eng. B* **2012**, *177*, 1531–1537. [CrossRef]
69. Saeed, M.; Marwani, H.M.; Rabbee, M.F.; Alfaifi, S.Y.; Khan, M.M.R.; Rahman, M.M. Fabrication of 4-Methoxyphenol Chemical Sensor with Graphene Oxide Conjugated Binary Co3O4/Y2O3 Nanocomposite by Linear Sweep Voltammetry. *Microchem. J.* **2024**, *207*, 112231. [CrossRef]
70. Liang, R.P.; Yao, G.H.; Fan, L.X.; Qiu, J.D. Magnetic Fe3O4@Au Composite-Enhanced Surface Plasmon Resonance for Ultrasensitive Detection of Magnetic Nanoparticle-Enriched α-Fetoprotein. *Anal. Chim. Acta* **2012**, *737*, 22–28. [CrossRef]
71. Zamfir, L.G.; Geana, I.; Bourigua, S.; Rotariu, L.; Bala, C.; Errachid, A.; Jaffrezic-Renault, N. Highly Sensitive Label-Free Immunosensor for Ochratoxin A Based on Functionalized Magnetic Nanoparticles and EIS/SPR Detection. *Sens. Actuators B Chem.* **2011**, *159*, 178–184. [CrossRef]
72. Martin, M.; Carballo, P.A.S.; Campuzano, S.; Villalonga, R.; Gonzalez-Mora, J.L. Amperometric Magnetobiosensors Using Poly(dopamine)-Modified Fe3O4 Magnetic Nanoparticles for the Detection of Phenolic Compounds. *Anal. Methods* **2015**, *7*, 8801–8808. [CrossRef]
73. González-Gutiérrez, A.G.; Quionez-López, R.R.; Cano, M.E.; Quintero, L.H.; Casillas, N. Magnetic and Electrochemical Characterization of Magnetite Nanoparticles Modified with Tetrahydroxyquinone. *Appl. Nanosci.* **2024**, *14*, 1103–1113. [CrossRef]
74. Arif, M.; Raza, H.; Akhter, T. UV-Vis Spectroscopy in the Characterization and Applications of Smart Microgels and Metal Nanoparticle-Decorated Smart Microgels: A Critical Review. *RSC Adv.* **2024**, *14*, 38120–38134. [CrossRef]

75. Dinelli, F.; Modestino, M.; Galluzzi, A.; Posati, T.; Seri, M.; Zamboni, R.; Sotgiu, G.; Corticelli, F.; Polichetti, M. Magnetic Behaviour of Iron Oxide/Dextran Nanoparticles in a Keratin Matrix. *Appl. Sci.* **2024**, *14*, 1106. [CrossRef]
76. Tel-Vered, R.; Bard, A.J. Generation and Detection of Single Metal Nanoparticles Using Scanning Electrochemical Microscopy Techniques. *J. Phys. Chem. B* **2006**, *110*, 25279. [CrossRef]
77. Zhang, Y.; Xiao, L.; Hao, Q.; Li, X.; Liu, F. Ferrihydrite reduction exclusively stimulated hydrogen production by Clostridium with community metabolic pathway bifurcation. *ACS Sustain. Chem. Eng.* **2020**, *8*, 7574–7580. [CrossRef]
78. Gadhe, A.; Sonawane, S.S.; Varma, M.N. Influence of nickel and hematite nanoparticle powder on the production of biohydrogen from complex distillery wastewater in batch fermentation. *Int. J. Hydrogen Energy* **2015**, *40*, 10734–10743. [CrossRef]
79. Córdova-Lizama, A.; Carrera-Figuetiras, C.; Palacios, A.; Castro-Oliviera, P.M.; Ruiz-Espinoza, J. Improving hydrogen production from the anaerobic digestion of waste activated sludge: Effects of cobalt and iron zero valent nanoparticles. *Int. J. Hydrogen Energy* **2022**, *47*, 30074–30084. [CrossRef]
80. Li, Z.; Wang, J.; Tian, K.; Zhou, C.; Pei, Y.; Zhang, J.; Zang, L. Nickel-cobalt oxide nanoparticle-induced biohydrogen production. *ACS Omega* **2022**, *7*, 41594–41605. [CrossRef]
81. Sun, H.; Shen, J.; Hu, M.; Zhang, J.; Cai, Z.; Zang, L.; Ji, D. Manganese ferrite nanoparticles enhanced biohydrogen production from mesophilic and thermophilic dark fermentation. *Energy Rep.* **2021**, *7*, 6234–6245. [CrossRef]
82. Yang, J.; Zhang, H.; Liu, H.; Zhang, J.; Pei, Y.; Zang, L. Unraveling the roles of lanthanum-iron oxide nanoparticles in biohydrogen production. *Bioresour. Technol.* **2022**, *351*, 127027. [CrossRef] [PubMed]
83. Feng, Y.; Zhang, Y.; Quan, X.; Chen, S. Enhanced anaerobic digestion of waste activated sludge digestion by the addition of zero valent iron. *Water Res.* **2014**, *52*, 242–250. [CrossRef]
84. Kato, S.; Hashimoto, K.; Watanabe, K. Methanogenesis facilitated by electric syntrophy via (semi)conductive iron oxide minerals. *Environ. Microbiol.* **2012**, *14*, 1646–1654. [CrossRef]
85. Sinharoy, A.; Pakshirajan, K. A novel application of biologically synthesized nanoparticles for enhanced biohydrogen production and carbon monoxide bioconversion. *Renew. Energy* **2020**, *147*, 864–873. [CrossRef]
86. Marcelo, L.R.; de Gois, J.S.; da Silva, A.A.; Cesar, D.V. Synthesis of iron-based magnetic nanocomposites and applications in adsorption processes for water treatment: A review. *Environ. Chem. Lett.* **2021**, *19*, 1229–1274. [CrossRef]
87. Soylak, M.; Ozalp, O.; Uzcan, F. Magnetic nanomaterials for the removal, separation and preconcentration of organic and inorganic pollutants at trace levels and their practical applications: A review. *Trends Environ. Anal. Chem.* **2021**, *29*, e00109. [CrossRef]
88. Mercier, A.; Bertaux, J.; Lesobre, J.; Gravouil, K.; Verdon, J.; Imbert, C.; Valette, E.; Héchard, Y. Characterization of biofilm formation in natural water subjected to low-frequency electromagnetic fields. *Biofouling* **2016**, *32*, 287–299. [CrossRef] [PubMed]
89. Xu, Y.; Hou, M.; Ruan, J.; Qu, M.; Sun, H.; Xu, J.; Zhou, S. Effect of magnetic field on surface properties of Bacillus cereus CrA and its extracellular polymeric substances (EPS). *J. Adhes. Sci. Technol.* **2014**, *28*, 2196–2208. [CrossRef]
90. Suanon, F.; Sun, Q.; Mama, D.; Li, J.; Dimon, B.; Yu, C.-P. Effect of nanoscale zero-valent iron and magnetite (Fe_3O_4) on the fate of metals during anaerobic digestion of sludge. *Water Res.* **2016**, *88*, 897–903. [CrossRef]
91. Premaratne, G.; Bhandari, S.; Walgama, C.; Chikkaveeraiah, B.V.; Jin, A.; Krishnan, S. Size-Dependent Electrochemical and Morphological Properties of Magnetite Nanoparticles Adsorbed on Electrodes. *ACS Meas. Sci. Au* **2025**, *5*, 325–331. [CrossRef]
92. Shahzad, U.; Marwani, H.M.; Rabbee, M.F.; Alfaifi, S.Y.; Alzahrani, K.A.; Khan, M.M.R.; Rahman, M.M. Efficient Sensitive Detection of Nitrite with Binary Y/Fe-Modified Multiwalled Carbon Nanotube Nanocomposite by Electrochemical Approaches. *Mater. Chem. Phys.* **2024**, *328*, 130000. [CrossRef]
93. Liu, C.H.; Chuang, Y.H.; Chen, T.Y.; Tian, Y.; Hui, L.; Wang, M.K.; Wei, Z. Mechanism of Arsenic Adsorption on Magnetite Nanoparticles from Water: Thermodynamic and Spectroscopic Studies. *Environ. Sci. Technol.* **2015**, *49*, 7726–7734. [CrossRef]
94. Amardeep, B.; Agrawal, A.K.; Balwant, S.; Sanjeev, G.; Navdeep, G. Surface Plasmon Band Tailoring of Plasmonic Nanostructure Under the Effect of Water Radiolysis by Synchrotron Radiation. *J. Synchrotron Radiat.* **2017**, *24*, 1209–1217. [CrossRef]
95. Agiotis, L.; Theodorakos, I.; Samothrakitis, S.; Papazoglou, S.; Zergioti, I.; Raptis, Y.S. Magnetic Manipulation of Superparamagnetic Nanoparticles in a Microfluidic System for Drug Delivery Applications. *J. Magn. Magn. Mater.* **2016**, *401*, 956–964. [CrossRef]
96. Hou, X.; Yan, X.; Chen, W.; Xu, Y.; Li, C. Magneto-Acousto-Electrical Tomography of Magnetic Nanoparticles with Electrode Detection. *IEEE Trans. Instrum. Meas.* **2025**, *74*, 4512612. [CrossRef]
97. Zhao, F.; Li, Y.; Liu, G. Magnetic Actuation for Magnetic Nanoparticles Under Magneto-Thermo-Acoustic Imaging. *AIP Adv.* **2025**, *15*, 025222. [CrossRef]
98. Dawn, R.; Kiran, C.; Weng, W.T.; Aabdin, Z.; Zzaman, M.; Faizal, F.; Panatarani, C.; Joni, I.M.; Chandra, G.; Amemiya, K. Effect of Spin and Orbital Selectivity of DNA Induced SiO2 Coated Fe3O4 Nanoparticles: X-Ray Magnetic Circular Dichroism Study. *Colloids Surf. A* **2025**, *709*, 136111. [CrossRef]
99. Mirzajani, F.; Rostamzadeh, A.; Tahmasian, Z.; Obaid, H.S.A.; Motevalli, S.M. Effects of MRI Magnetic Iron Oxide Nanoparticles on the Structural and Enzymatic Properties of Liver-Related Enzymes. *Micro Nano Syst. Lett.* **2024**, *12*, 13. [CrossRef]

100. Shin, T.H.; Lee, G. Reduced Lysosomal Activity and Increased Amyloid Beta Accumulation in Silica-Coated Magnetic Nanoparticles-Treated Microglia. *Arch. Toxicol.* **2024**, *98*, 14. [CrossRef]
101. Xi, P.; Liu, S.; Tang, J.; Wang, X.; Liu, Y.; Wang, X.; Hu, S.; Wang, K.; Li, W.; Cai, Z. Single-Cell Transcriptomics Reveals Ferrimagnetic Vortex Iron Oxide Nanoring-Mediated Mild Magnetic Hyperthermia Exerts Antitumor Effects by Alleviating Macrophage Suppression in Breast Cancer. *Biomed. Pharmacother.* **2024**, *170*, 115954. [CrossRef]
102. Roussel, T.; Ferry, D.; Kosta, A.; Miele, D.; Sandri, G.; Tansi, F.L.; Steiniger, F.; Southern, P.; Pankhurst, Q.A.; Peng, L. Insight into the Internal Structure of High-Performance Multicore Magnetic Nanoparticles Used in Cancer Thermotherapy. *ACS Mater. Au* **2024**, *4*, 489–499. [CrossRef]
103. Poplawsky, J.D.; Zand, F.; Monai, M.; Weckhuysen, B.M. High-Pressure Resin Embedment of Mesoporous Silica Supported Nanoparticles for High-Quality Atom Probe Tomography Analysis. *Microsc. Microanal.* **2024**, *30*, 93–94. [CrossRef]
104. Kwon, D.; Lee, B.; Sung, S. Design and Performance Validation of Tightly Coupled Magneto-Inertial Navigation System for Robust Outdoor Application. *IEEE Access* **2024**, *12*, 142215–142226. [CrossRef]
105. Liu, S.; Yu, B.; Wang, S.; Shen, Y.; Cong, H. Preparation, Surface Functionalization and Application of Fe3O4 Magnetic Nanoparticles. *Adv. Colloid Interface Sci.* **2020**, *281*, 102165. [CrossRef] [PubMed]
106. Wu, Y.; Xia, C.; Cao, J.; AL Garalleh, H.; Garaleh, M.; Khouj, M.; Pugazhendhi, A. A Review on Current Scenario of Nanocatalysts in Biofuel Production and Potential of Organic and Inorganic Nanoparticles in Biohydrogen Production. *Fuel* **2023**, *338*, 127216. [CrossRef]
107. Biehl, P.; Von der Lühe, M.; Dutz, S.; Schacher, F.H. Synthesis, Characterization, and Applications of Magnetic Nanoparticles Featuring Polyzwitterionic Coatings. *Polymers* **2018**, *10*, 91. [CrossRef]
108. Nel, A.E.; Mälder, L.; Velegol, D.; Xia, T.; Hoek, E.M.; Somasundaran, P.; Klaessig, F.; Castranova, V.; Thompson, M. Understanding Biophysicochemical Interactions at the Nano-Bio Interface. *Nat. Mater.* **2009**, *8*, 543–557. [CrossRef]
109. Kaur, L.; Sidhu, A.K. Green Synthesis: An Eco-friendly Route for the Synthesis of Iron Oxide Nanoparticles. *Front. Nanotechnol.* **2021**, *3*, 655062. [CrossRef]

hydrogen

MDPI

Review

Proton Exchange Membrane Fuel Cells for Aircraft Applications: A Comprehensive Review of Key Challenges and Development Trends

Xinfeng Zhang [1,2,*], Han Yue [1,2], Hui Zheng [1,2], Lixing Tan [1], Zhiming Zhang [3] and Feng Li [4]

1 School of Information and Electrical Engineering, Hangzhou City University, Hangzhou 310015, China; 2023028085801040@ecjtu.edu.cn (H.Y.)
2 School of Electrical and Automation Engineering, East China Jiaotong University, Nanchang 330013, China
3 College of Automotive Studies, Tongji University, Shanghai 200092, China
4 Department of Electrical and Electronic Engineering, Hong Kong Polytechnic University, Hong Kong 999077, China
* Correspondence: zxf@hzcu.edu.cn; Tel.: +86-13817829495

Abstract

Hydrogen energy is a pivotal alternative to lithium-ion batteries for low-altitude aircraft, offering a pathway to sustainable aviation with its zero emissions and high energy density. Nevertheless, its broader application is hindered by challenges in storage, safety, and performance under extreme conditions such as low pressure and low temperature at high altitudes. This paper systematically evaluates various hydrogen power technologies—including water-cooled and air-cooled proton exchange membrane fuel cells (PEMFCs) as well as hydrogen turbines—highlighting their respective advantages, limitations, and suitability for different aircraft types. Among these, water-cooled PEMFCs are identified as the most viable option for manned low-altitude aircraft due to their balanced performance in power density and startup capability. In contrast, air-cooled PEMFCs demonstrate distinct cost-effectiveness for lightweight drones, while hydrogen turbines show promise for long-range regional transport. Furthermore, we analyze current progress in integrating PEMFCs into aircraft platforms and discuss persistent challenges in system compatibility and environmental adaptation. Finally, potential future development directions for PEMFC applications in low-altitude aviation are outlined.

Keywords: PEMFC; hydrogen propulsion system; low-altitude aircraft; aerospace domain

1. Introduction

The increasing attention to energy conservation and environmental issues is driving both aviation companies and consumers to favor more environmentally friendly products [1]. Despite long-term investments in technological innovation by aircraft and engine manufacturers, the rapid growth of air traffic has still led to a significant increase in carbon emissions. A recent study by the International Council on Clean Transportation (ICCT) reported a 32% increase in CO_2 emissions from airlines between 2013 and 2018. This growth, with a compound annual growth rate (CAGR) of 5.7%, surpassed the International Civil Aviation Organization (ICAO)'s estimated average of 3.3% by 70% [2]. The European Commission believes that Global emissions from international aviation are expected to increase by approximately 70% compared to 2005 levels, and the ICAO predicts that this figure will further grow by 300–700% by 2050 [3].

https://doi.org/10.3390/hydrogen6040116

Currently, most low-altitude electric aircraft are powered by electrical energy, with lithium batteries serving as the primary energy source [4]. The energy density of lithium batteries reaches 230–260 Wh/kg [5], and the latest solid-state lithium batteries can even achieve energy densities of 350–400 Wh/kg [6]. However, as the demand for large-payload and long-endurance industrial drones increases, traditional lithium batteries are no longer sufficient to meet their power requirements [7]. The integration of lithium batteries with fuel cells forms a hybrid energy system designed to mitigate the inherent limitations of fuel cells, particularly their insufficient dynamic response. By implementing optimized energy management strategies, such a system achieves optimal power distribution between the fuel cell and the lithium battery. This coordinated approach not only minimizes hydrogen consumption and extends the operational endurance of the system but also contributes to prolonging the service life of the lithium battery by mitigating stressful operating conditions [8]. For unmanned aerial vehicles (UAVs) powered by such hydrogen-electric hybrid propulsion systems, the carried hydrogen capacity is a primary determinant of the vehicle's endurance [9].

The five major hydrogen power technologies are divided into proton exchange membrane fuel cells (air-cooled/water-cooled), alkaline fuel cells (AFC), hydrogen turbine generators, and hydrogen piston internal combustion engines [10]. Among them, PEMFCs have a relatively short history, but they are favored for their simplicity, light weight, high efficiency, high specific power density, low waste, typical operating temperature and pressure, technological maturity, and low cost [11]. Given the advantages of PEMFCs, NASA has proposed fuel cell-powered propulsion technology in an attempt to discover a low-cost, safe, environmentally compatible, and silent aircraft for the "21st century" [12]. The U.S. Navy's "Ion Tiger" drone achieved a 26 h flight using PEMFC [13], and Airbus and Boeing have successfully tested PEMFC based auxiliary power units (APUs) on the A320 and B737 platforms, demonstrating a 15% reduction in fuel consumption [14]. Despite these advancements, challenges remain. PEMFCs face limitations in power density (0.5–1.0 kW/kg (air-cooled) and 1.2–2.0 kW/kg (water-cooled)), hydrogen storage efficiency, and operational safety under extreme conditions. High costs, the immaturity of hydrogen infrastructure and the limited volumes of green or low-carbon hydrogen currently available relative to demand are further hindering commercialization. Moving forward, the primary research directions for aircraft are hydrogen turbine power and high-power water-cooled fuel cell stacks as power sources, with liquid hydrogen as the main energy storage method. The key breakthroughs will focus on increasing stack power density, wide-temperature and variable-altitude environmental adaptability, onboard hydrogen-electric hybrid propulsion, and energy management.

This paper systematically reviews the application and technological progress of Proton Exchange Membrane Fuel Cells (PEMFCs) in the aviation sector. It provides a comparative analysis of various hydrogen power technologies, evaluating their respective advantages and disadvantages, system efficiency, operating temperature ranges, energy and power density, and safety considerations. The feasibility of different hydrogen storage methods—including high-pressure gaseous hydrogen and cryogenic liquid hydrogen—for aviation scenarios is also critically assessed, focusing on their volumetric and gravimetric energy density, technical maturity, and adaptation to the stringent weight and space constraints of aircraft. Subsequently, the research and development advancements of hydrogen power systems within aviation are analyzed. This analysis highlights the core applications of fuel cells across diverse aircraft platforms, encompassing large-scale manned aircraft, regional aircraft, and unmanned aerial vehicles (UAVs). The current research status and emerging development trends of fuel cell technology in aviation are examined, with particular emphasis on the application of PEMFCs as primary power sources, auxiliary power units (APUs), and within multifunctional fuel cell (MFC) systems that co-produce usable water and deoxygenated air. Finally, the paper summarizes the application status, identifies key future development trends, and outlines the primary challenges—such as durability under aerial conditions, thermal management, and system integration—facing the deployment of PEMFCs in low-altitude aircraft. The core purpose of applying hydrogen power in aircraft is to propel the aviation industry towards deep decarbonization and sustainable development. It aims to fundamentally address the high carbon emissions associated with traditional aviation fuels by introducing a zero-carbon fuel, while maintaining or even enhancing flight performance.

2. The Advantages and Disadvantages of Various Hydrogen Power Technologies and Their Feasibility for Applications in the Aviation Field

In the wave of the energy revolution for low-altitude aircraft, hydrogen has emerged as a key alternative to lithium batteries due to its zero emissions and high energy density [15]. However, the five major hydrogen power technology pathways—proton exchange membrane fuel cells (air-cooled/water-cooled), AFC, hydrogen turbine generators, and hydrogen piston internal combustion engines—each have their own strengths and weaknesses. The selection of these technologies must be based on the payload capacity, range requirements, safety, and environmental adaptability of the aircraft. In this context, this paper summarizes and compares the advantages and disadvantages, efficiency, operating temperature, energy/power density, and safety of various hydrogen power technologies, as shown in Table 1.

Table 1. Advantages, disadvantages and parameters of various hydrogen power technologies.

Type	PEMFC (Air-Cooled)	PEMFC (Water-Cooled)	AFC	Hydrogen Turbine Generator	Hydrogen Piston Internal Combustion Engine
Fuel	High-purity H_2	H_2 (purity $\geq$ 99.97%)	Pure H_2 + pure O_2	Hydrogen/kerosene blend	H_2 (methane blending)
Electrolyte/Structure	Perfluorosulfonic acid membrane	Thin-layer Perfluorosulfonic acid membrane	30% KOH solution	Combustion chamber + turbine assembly + generator	Internal combustion engine + generator
Operating Temperature	45–75 °C	60–85 °C	60–120 °C	Turbine inlet temperature > 1000 °C	In-cylinder combustion temperature > 200 °C
Single Module Power Range	0.3–5 kW	5 kW–1 MW	0.3–20 kW	2–30 MW	250–450 kW
Efficiency	40–50%	50–60%	60–70%	35–45%	25–40%
Power Density	0.5–1.0 kW/kg	1.2–2.0 kW/kg	0.3–0.8 kW/kg	3–5 kW/kg	0.8–1.5 kW/kg
Low-Temperature (−30 °C~−20 °C) Start-up Time	30~60 s	30 s~7 min	The electrolyte exhibits high activity at low temperatures.	Preheating time of 5–15 min is required.	A wide flammability limit facilitates cold-start times of less than 10 s.
Room-Temperature Start-up Time	<30 s	<30 s	<15 s	2~5 min	<1 s
Fault Tolerance	Moderate (dependent on ambient temperature)	High (requires precise thermal management)	Very high (tolerant to impurities)	Low (requires flashback prevention)	Moderate (requires pre-ignition prevention)
References	[16–18]	[19,20]	[18,21,22]	[23–25]	[26,27]

Based on the data, advantages and disadvantages, and safety considerations provided in Table 1, a comprehensive analysis of the applicability of the five types of hydrogen power systems for low-altitude aircraft is conducted.

Air-cooled PEMFCs are considered particularly suitable for small unmanned aerial vehicles (UAVs) due to their simplicity, lightweight, and high reliability [16]. Air-cooled PEMFCs eliminate the need for liquid cooling pipelines and radiators, reducing system weight by 15–20% and significantly enhancing payload capacity. Additionally, the relatively simple structure of air-cooled PEMFCs reduces system complexity and potential failure points, thereby improving system reliability and maintenance convenience. These fuel cells exhibit excellent low-temperature start-up performance, enabling rapid start-up and stable operation in extremely cold environments [17]. Although the heat dissipation efficiency of air-cooled PEMFCs may be lower than that of water-cooled systems, they can effectively manage heat within the power demand range of small UAVs, ensuring stable cell output. However, several challenges remain, including the risk of hydrogen leakage leading to combustion and explosion, catalyst poisoning by air impurities (e.g., CO_2), limited heat dissipation capacity, poor altitude adaptability, and low upper limit of power density [18].

Water-cooled PEMFCs, equipped with an efficient cooling system, can effectively manage the heat generated during high-power operation, ensuring stable cell performance across a wide temperature range [19]. The integrated heater in the water-cooled system enables rapid temperature increase, facilitating quick start-up in low-temperature conditions. Moreover, the high-power density and good environmental adaptability of water-cooled PEMFCs make them well-suited to meet the operational demands of aircraft in complex environments [20]. As a result, water-cooled PEMFCs are irreplaceable in medium and large cargo UAVs, manned electric Vertical Take-Off and Landing (eVTOL), and high-altitude, long-endurance platforms, due to their high-power density, uniform temperature distribution, and environmental adaptability.

AFCs exhibit high energy conversion efficiency, effectively utilizing the energy of the fuel. They also demonstrate excellent low-temperature start-up performance, enabling rapid and stable operation at extremely low temperatures. As one of the earliest fuel cell technologies to achieve practical application [21], AFCs are primarily used in aerospace vehicles and UAVs, where high energy density and reliability are crucial. Despite their high fuel purity requirements and sensitivity to CO_2, which limit their application in some civilian scenarios, AFCs can potentially expand their scope through technological advancements, such as the development of new catalysts and electrolyte materials, as well as optimized system design [22].

Hydrogen turbine generators feature extremely high-power density, meeting the large power demands of large aircraft during takeoff, climb, and cruise phases. Their high energy conversion efficiency significantly reduces fuel consumption, enhancing aircraft economy [23]. Additionally, the operational characteristics of hydrogen turbine generators are similar to those of traditional aero-engines, facilitating integration into existing aero-power system architectures and reducing the difficulty of technological transition [24]. Their scalable design allows flexible adaptation to aircraft of different sizes and payload requirements. However, challenges remain, including noise levels exceeding 100 dB, hydrogen embrittlement risks in turbine blades, the need for high-temperature materials, and cavitation issues in liquid hydrogen pumping [25].

Hydrogen piston internal combustion engines (ICEs) offer rapid start-up and powerful output without preheating in low-temperature conditions. They inherit the mature technology and reliability of traditional ICEs, making them suitable for small and medium low-altitude aircraft [26]. However, the storage and transportation of hydrogen fuel require specialized containers and infrastructure, increasing system complexity and weight [27]. Moreover, issues such as increased dead weight from mechanical transmission components, low power density, flashback risk requiring lean-burn technology, high hydrogen diffusion rate leading to accumulation, and high compression ratio-induced knocking need to be addressed [24]. Despite these challenges, the application prospects of hydrogen piston ICEs in low-altitude aircraft remain promising as hydrogen fuel infrastructure improves and related technologies continue to advance.

In summary, air-cooled PEMFCs offer the best weight reduction and are suitable for small and medium UAVs but require high performance in low-temperature start-up and heat management. Water-cooled PEMFCs provide the best overall energy efficiency and low-temperature start-up performance and are suitable for eVTOLs requiring long endurance and high reliability. AFCs are mainly applied in aerospace vehicles and UAVs with high energy density and reliability requirements. Hydrogen turbine generators are suitable for large aircraft due to their high power and low-temperature adaptability but need to address noise and system complexity issues. Hydrogen piston ICEs are suitable for small and medium low-altitude aircraft but currently face multiple challenges.

3. Current Status of PEMFC Research in Low-Altitude Aircraft

3.1. Commercial Aircraft

Significant breakthroughs have been made in the research and development of hydrogen-powered systems in the aviation field. The application process of PEMFC in aviation is shown in Figure 1.

PEMFC has become an important clean energy technology in the aviation field, with advantages such as high efficiency, zero emissions, and low noise [28]. Power systems primarily based on high-power water-cooled reactors, with liquid hydrogen as the main energy storage system, have been launched or announced in various countries. Recent specific research developments include the following: In September 2023, CAJU (Clean Aviation Joint Undertaking) determined that 86 million euros of EU funding would be specifically allocated to the hydrogen-powered aircraft sector in the second batch of projects, encompassing three projects: the hydrogen propulsion technology research (Trophy) project led by Safran Group [29];The megawatt-class hydrogen fuel cell power system (FAME) project led by Airbus; and the hydrogen-electric zero-emission propulsion system (HEROPS) project led by MTU. In January 2023, U.S.-based Zero Avia successfully test-flew a 19-seat Dornier aircraft. The twin-engine aircraft was modified with Zero Avia's hydrogen-electric power system ZA600 installed on the left wing, while the right wing housed a Honeywell TPE-331 turboprop engine. The hydrogen-electric power train includes two fuel cell stacks, hydrogen storage tanks, and a fuel cell power generation system installed within the cabin [30]. In June 2023, Airbus announced that it had conducted a full-power operation test of a 1.2 MW fuel cell system in its electric aircraft system test chamber. The test was achieved by coupling multiple power channels to drive a single propeller. In July 2023, the U.S. National Aeronautics and Space Administration's Efficient Aircraft Electrical Technology Center advanced the development of a zero-emission concept passenger aircraft. The concept aircraft uses liquid hydrogen as both fuel and coolant to achieve superconducting performance, to enhance efficiency and reduce weight. To effectively integrate numerous subsystems such as liquid hydrogen tanks, the fuselage, and fans, the plan is to first conduct scaled-down model test flights using gaseous hydrogen fuel. The second phase will involve testing sub components such as a 300 kW cryogenic cooling motor, a 300 kW zero-loss superconducting cable, and a 300 L liquid hydrogen tank. Among these, the weight index (GI) of the liquid hydrogen tank is planned to reach 60% (current liquid hydrogen tanks are less than 30%) [31]. In September 2023, Germany's H2Fly successfully completed the world's first test flight of the HY4, a manned electric aircraft powered entirely by liquid hydrogen. The test results showed that the use of liquid hydrogen doubled the maximum range of the HY4 aircraft to 1500 km [32].

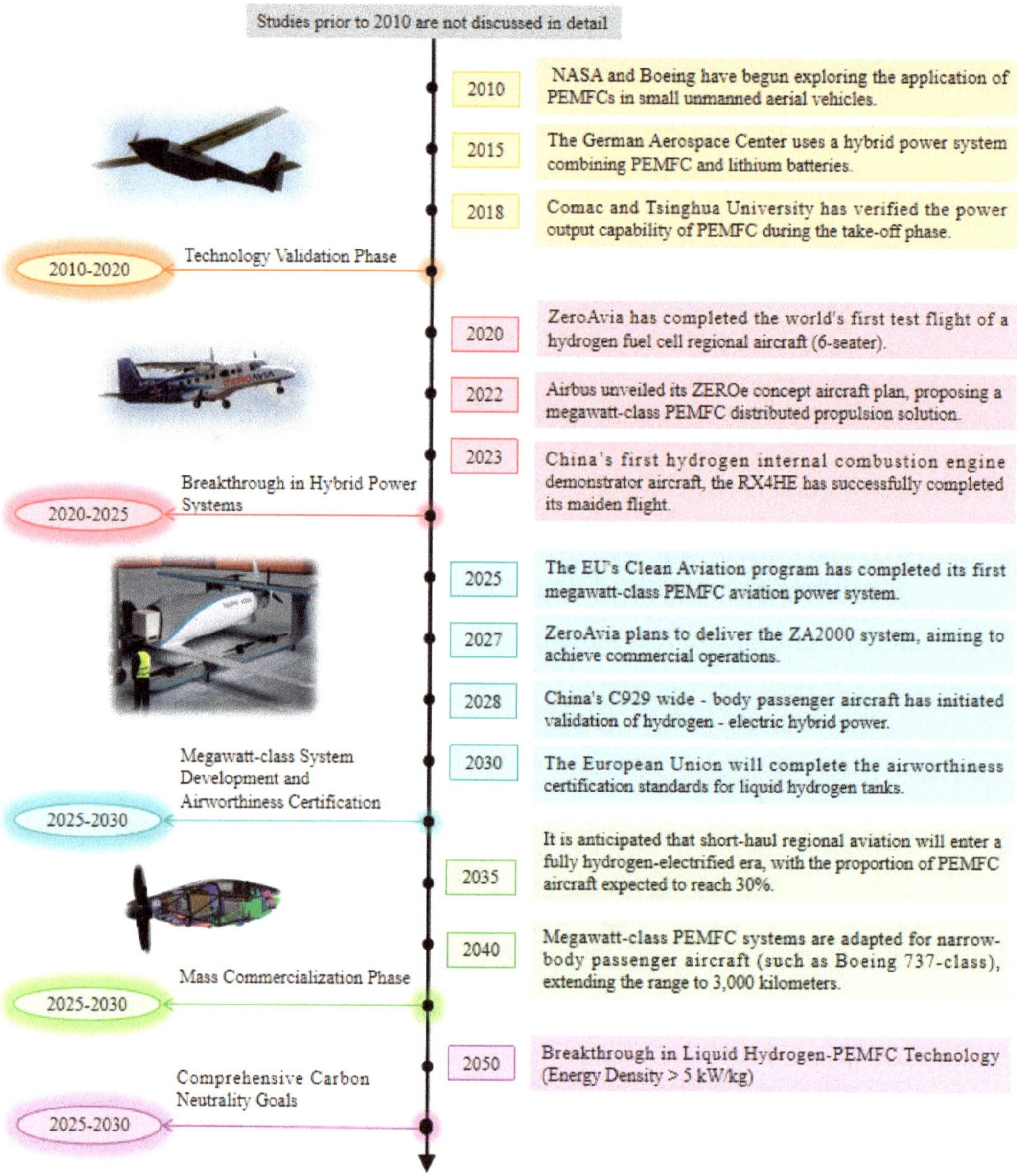

Figure 1. Application of PEMFC in aviation [28–32].

3.2. Electric Vertical Takeoff and Landing Aircraft

Low-altitude aircraft have more advantages, and there is greater demand for onboard power sources with a wide temperature range and long flight times [33]. In hydrogen-electric hybrid power systems with different architectures, the characteristics at the consumption end (i.e., power demand) and the supply end (i.e., power provision) differ significantly. These differences primarily arise from the varying power demands placed on the lithium battery and the fuel cell during different operational phases (e.g., startup, acceleration, cruise, and descent), which in turn dictate the distinct output responses and operational strategies of the supply side components. For instance, during high-power phases such as takeoff and climb, the lithium battery typically provides peak power supplementation, whereas the fuel cell operates at a stable rated power output during cruise phases to maintain high efficiency, as shown in Table 2. In the low-altitude economy and UAV application sectors, a highly dynamic and innovative integrated economic model is formed by 85% UAVs and 15% general aviation operations at altitudes below 1000 m (with special cases extending up to 3000 m). In this process, the extensive application areas and operational scenarios of low-altitude industrial UAVs, such as passenger and cargo transport, agricultural and forestry pest control, security monitoring, geographic surveying, pipeline inspection, and emergency firefighting, can continuously generate economic benefits for this sector [34]. According to statistics, the global hydrogen fuel cell drone market size reached US$0.27 billion in 2023 and is expected to increase to US$1.236 billion by 2030, with a compound annual growth rate of 76.3%. In 2024, the global fuel cell installation volume is expected to grow by 8.6% year-on-year [35].

Table 2. Consumption-side and supply side characteristics of hydrogen-electric hybrid vehicles with different architectures.

Architecture	Multi-Rotor	Composite Wing	Tiltrotor
Consumption characteristics of hydrogen-electric hybrid vehicles	The vertical takeoff and landing phase relies on multiple rotors working simultaneously, requiring high instantaneous power demand. Hydrogen fuel cells must work in conjunction with lithium batteries to provide peak power support.	Vertical takeoff and landing is powered by independent rotors, while cruise flight is driven by fixed wings and horizontal propellers, allowing hydrogen fuel cells to focus on supplying power during cruise flight.	Tilting rotors or wings require dynamic adjustment of power output direction, placing extremely high demands on fuel cell power density and dynamic response.
Supply side characteristics of hydrogen-electric hybrid power	Hydrogen fuel cells are mainly used to supplement range, with lithium batteries still being the primary source of power.	Focus on the synergy between hydrogen energy and batteries to balance load capacity and range.	High-energy-density liquid hydrogen storage systems and high-power fuel cells are required to meet the long-range requirements of the cruise phase.

The development pace of hydrogen-powered low-altitude aircraft has progressed more slowly compared to other hydrogen-powered flight vehicles. Nevertheless, Proton Exchange Membrane Fuel Cell (PEMFC) technology has already established applications within low-altitude aircraft platforms. The timeline of PEMFC applications in low-altitude aircraft is shown in Figure 2 [11,36–49]. The earliest recorded application of PEMFC in low-altitude aircraft was in 2003 with the 38-cm-wingspan Hornet demonstrator, which proved that fuel cells can achieve aerial flight even at a small scale [11]. Although the research and development of fuel cell-powered drones can be traced back to the 1950s, manned low-altitude aircraft still require further technological advancements. Many aspects of fuel-cell-based propulsion systems need to be improved, such as durability, reliability, and fuel efficiency. The technological accumulation at this stage laid the foundation for the breakthroughs in hybrid architectures and high-power-density stacks after 2006. Although there have been significant breakthroughs in material innovation, hybrid architectures, and thermal management technologies, PEMFC still faces core bottlenecks in energy density, dynamic response, and environmental adaptability in the field of low-altitude aircraft [48].

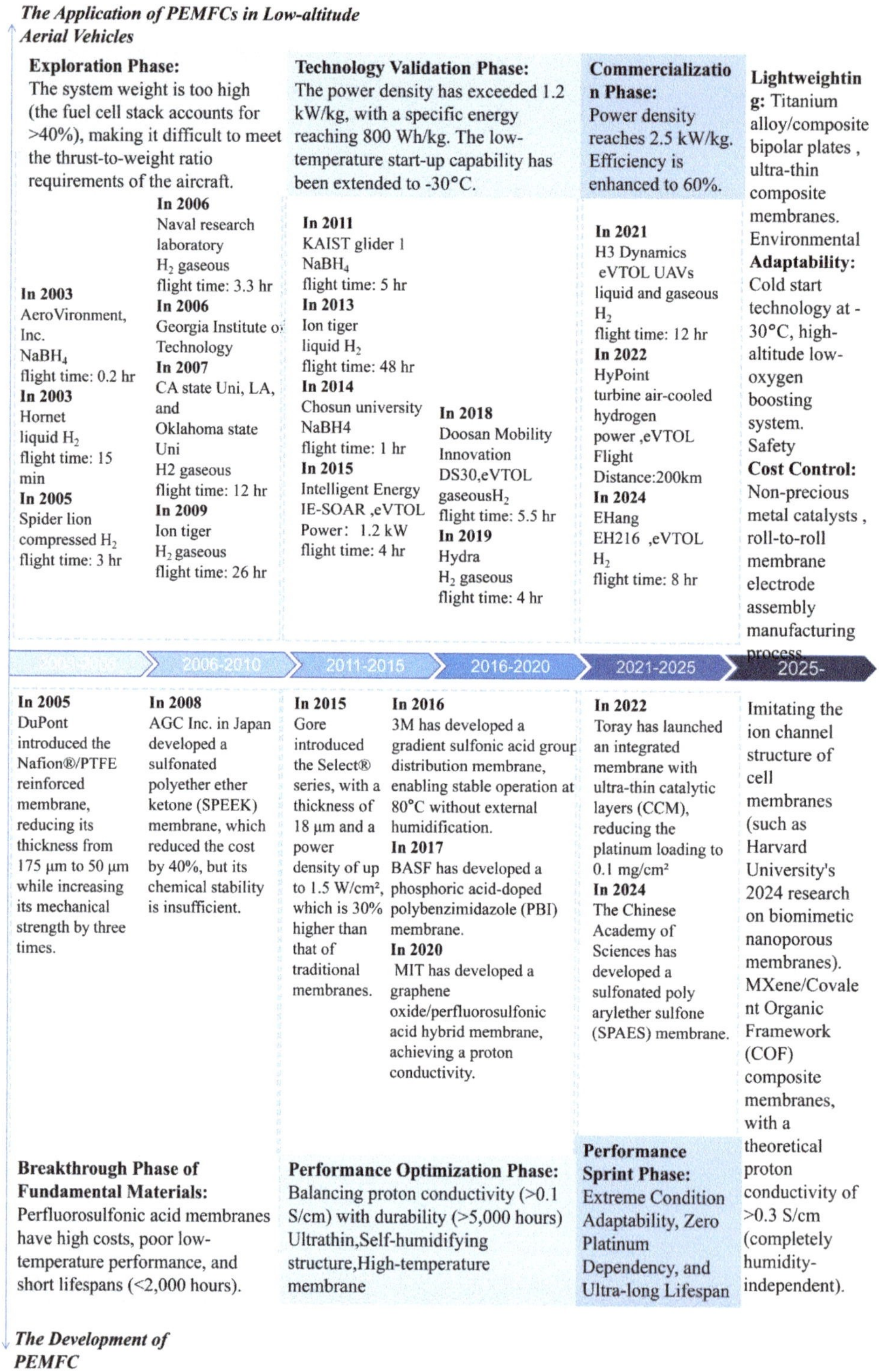

Figure 2. Application of PEMFC in Low-Altitude Aircraft and Development of PEMFC.

Between 2011 and 2020, with the accelerated development of hydrogen energy technology and the increasing demand for carbon reduction in aviation, the advantages of PEMFC in long-endurance, low-noise, and zero-emission scenarios were further highlighted. Zero Avia's HyFlyer completed the world's first test flight of a hydrogen-fuel-cell-powered commercial aircraft, with a range of 800 km. It used 70 MPa Type IV hydrogen storage tanks and 3D flow-field stacks [50]. Alaka'I Skai eVTOL, the first hydrogen fuel cell-powered manned eVTOL, has an endurance of 4 h, equipped with self-healing membranes and two-phase cooling systems. Airworthiness certification is driving the establishment of industry standards. However, it still faces core challenges such as system complexity, cost, and airworthiness certification [51].

Between 2021 and 2025, the application of PEMFC in low-altitude aircraft has made great progress. BASF's Avantin™ coolant additive, which imitates the antifreeze protein of Arctic fish, can achieve non-freezing at −50 °C and reduce the energy consumption of cold start by 70%. In addition, Ballard's 3D-printed gradient porous diffusion layer can increase the drainage rate by three times under high humidity conditions [15]. H3 Dynamics' HYDRA-12 uses non-platinum catalysts (Fe-N-C), reducing costs by 40%, which has led to rapid progress in low-altitude aircraft [52]. EHang's EH216-H2, the world's first hydrogen-powered manned drone certified by the civil aviation authority. It is equipped with a 5 kW PEMFC system and has an endurance of 8 h. Titanium alloy bipolar plates (thickness 0.6 mm) and a multi-level hydrogen safety monitoring system are used, with a leakage response time of less than 50 ms [53,54].

Many companies around the world have already invested in the research and development of hybrid-electric eVTOLs. For example, companies such as Lilium in Germany, Joby Aviation in the United States, and Zero Tech in China are all actively exploring the application of hybrid-electric technologies [55]. To overcome the range limitations inherent in all-electric propulsion systems, the hybrid-electric propulsion system (HEPS) has emerged as a pivotal research focus. This system demonstrates notable advantages in power density, with reported overall efficiency ranging from 42% to 48% as shown in Figure 3. Concurrently, advancements are being pursued not only in the core energy conversion technologies but also through the integration of innovative approaches across the entire design and manufacturing life-cycle. HEPS can extend the range of eVTOLs to over 600 km (a 120% increase compared to all electric systems), but at the cost of an additional 18–22% in system weight. HyPoint, in collaboration with NASA, has developed a high-altitude drone using a turbocharged PEMFC system with a power-density of 2.2 kW/kg (1.5 times that of traditional systems) that can maintain 95% performance at an altitude of 5000 m. The use of high temperature membrane electrodes (capable of withstanding 120 °C) has solved the heat dissipation problem at high altitudes, and the system has passed a 1000 h durability test [55]. Airbus 's Zephyr H3 solar hydrogen hybrid drone charges with solar energy during the day and is powered by hydrogen fuel cells at night, with the goal of achieving continuous flight for 30 days. It uses ultra-thin composite membranes (10 µm) and flexible hydrogen storage bags, achieving a specific energy of 2000 Wh/kg [56]. However, hybrid-electric eVTOLs do indeed require an increase in system weight and greater operational complexity during the design phase [9]. In the future, it is necessary to promote technological development through energy management optimization, algorithm light weighting, and policy coordination.

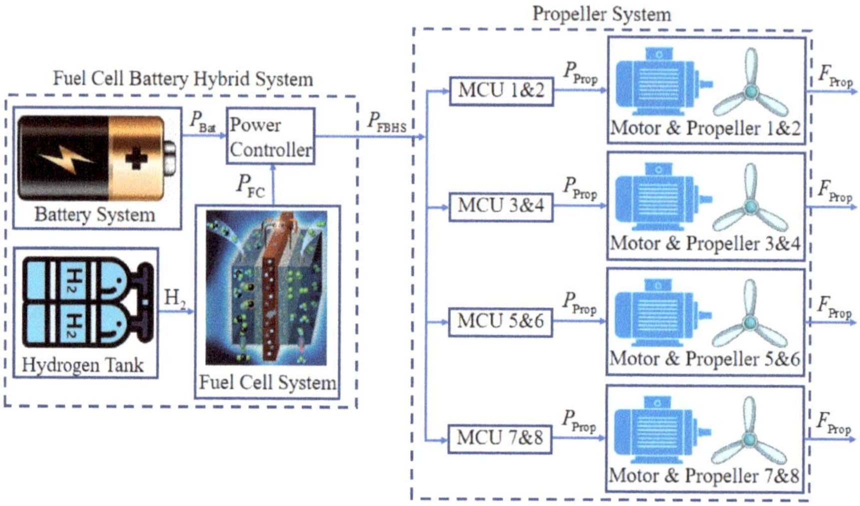

Figure 3. Hybrid Electric Propulsion System.

4. Challenges in the Technological Development of PEMFCs for Low-Altitude Aircraft

4.1. Hydrogen Storage and Spatial Constraints

In the application of PEMFCs to low-altitude aircraft, the core challenge of hydrogen storage systems lies in balancing gravimetric and volumetric hydrogen storage densities. Currently, the main hydrogen storage technologies include compressed gaseous hydrogen, liquid hydrogen, and solid-state hydrogen storage [57].

Compressed gaseous hydrogen storage achieves hydrogen storage by compressing hydrogen gas up to 700 bar, with a gravimetric density of approximately 5–7%. However, this method suffers from low volumetric efficiency, high energy consumption for liquefaction, evaporation losses, costly cryogenic insulation technologies, and complex tank structures required to withstand high pressures or ultra-low temperatures [58].

Significant progress has also been made in solid-state hydrogen storage materials. Metal hydrides and nanostructured carbon nanotube/graphene composites have improved hydrogen storage capacity to 3–5% through nano-structuring and alloying. Nevertheless, these materials still exhibit poor cycling stability, gravimetric densities below the practical threshold of 5%, and require elevated temperatures for hydrogen release [59].

4.2. Thermal Management and System Integration Optimization

PEMFCs operate at relatively low temperatures, necessitating precise control of both reaction humidity and temperature. During high-altitude flight operations, low ambient temperatures may reduce fuel cell efficiency, while the design of the thermal management system must remain compatible with the aircraft's aerodynamic layout. A key technical challenge lies in redesigning the air intake system to accommodate required oxygen flow rates, as conventional engine cooling strategies cannot be directly applied to PEMFC systems [60].

(1) Integration and Optimization of Liquid Cooling Systems: For high-power PEMFC applications, liquid cooling technology achieves stable stack temperature control—within a temperature difference of less than 5 °C—through coordinated regulation of circulating pumps and radiators [61]. This approach mitigates the drawbacks of traditional high-pressure hydrogen storage systems, such as large volume and uneven heat dissipation.

Additionally, flow-following control strategies help decouple the relationship between temperature and thermal gradients [62]. However, multi-stack PEMFC configurations require either series or parallel cooling loops. Series arrangements increase coolant temperature gradients, while parallel configurations result in significant pressure drops, making it difficult to balance heat dissipation efficiency and energy consumption. Moreover, inherited designs often exhibit poor spatial compatibility with low-altitude aircraft, potentially reducing passenger capacity [63].

(2) Phase-Change Cooling, Nanofluid Technology, and Multi-Heat Source Coordinated Management: Phase-change cooling enhances heat dissipation efficiency by utilizing the latent heat of water. Experimental results show that injecting liquid water into the cathode flow channel can limit temperature fluctuations to within 1.2 °C. Furthermore, nanofluids demonstrate a 20–23% improvement in heat transfer capability compared to conventional coolants, addressing the conflict between high heat dissipation demands and increased flow resistance in high-power scenarios. However, the high preparation cost of nanofluid coolants remains a concern, particularly due to unresolved issues such as the dispersion stability of Al_2O_3 nanoparticles [64]. For hydrogen-electric hybrid power systems, model simplification errors and insufficient parameter self-adjustment capabilities necessitate the integration of digital twin technology and reinforcement learning to achieve real-time dynamic optimization [65].

4.3. Cost and Durability Issues in Aviation Applications

Regarding cost issues, platinum catalysts and proton exchange membrane (PEM) materials are expensive, accounting for over 40% of the total cost of fuel cells.

Regarding durability, a multi-condition lifespan prediction model has been developed based on machine learning and digital twin technologies. This model enables dynamic assessment of the remaining useful life (RUL) of PEMFCs by combining historical data to predict performance degradation trends. By dynamically adjusting voltage, load, and cooling strategies, the degradation rate can be reduced by more than 30%, effectively mitigating degradation under low-load operating conditions [66]. Additionally, to address the chemical degradation of proton exchange membranes, doping with transition metals has been employed to reduce the hydrogenation enthalpy, thereby extending membrane lifespan and suppressing carbon corrosion and platinum dissolution [67].

However, when multiple PEMFC stacks operate in parallel, uneven distribution of coolant flow can lead to local temperature gradients exceeding 10 °C, accelerating membrane electrode degradation. The core durability challenges of PEMFCs in low-altitude aircraft arise from the coupled effects of dynamic operating conditions, vibration environments, conflicts in water and thermal management, spatial constraints, and contamination-induced degradation.

4.4. Performance of Hybrid Power Systems

(1) Dynamic Response: The three-dimensional non-isothermal model based on digital twin technology has further improved the accuracy of temperature field prediction. Moreover, intelligent control strategies have significantly enhanced the dynamic response capability of PEMFCs under sudden load changes [68]. However, under frequent load variations, voltage undershoot in PEMFCs may still reduce stack lifespan, and existing control strategies struggle to fully eliminate the impact of voltage fluctuations on durability. Additionally, the uniformity of individual cell voltages deteriorates during high-power operation, and there is a lack of real-time monitoring technology for high-precision three-dimensional temperature fields [69].

(2) Hybrid Power Requirements: The relatively slow dynamic response of PEMFCs makes it difficult to meet the high-power demands of aircraft during takeoff, climb, and other high-load scenarios. Therefore, hybrid systems combining PEMFCs with lithium batteries or supercapacitors are necessary. NASA's proposed SOFC/GT hybrid system increases power generation efficiency to 70% by recovering waste heat from high-temperature fuel cells and gas turbines, enabling cascaded energy utilization [70]. However, power switching between fuel cells and batteries presents voltage matching challenges, and transient response delays may trigger protective shutdowns of the system. Additionally, uneven coolant distribution during multi-stack parallel operation can lead to local temperature gradients. Balancing lightweight requirements with heat dissipation efficiency remains a challenge [71].

Hydrogen-electric hybrid algorithms have enabled breakthroughs in range, energy efficiency, and airworthiness certification in the eVTOL sector. Lilium's hydrogen-electric eVTOL employs a distributed hydrogen fuel cell design combined with dynamic power allocation algorithms. During vertical takeoff and landing phases, up to 90% of auxiliary power is supplied by batteries, while in cruise flight, the system switches to hydrogen fuel cell dominance, extending the range to 600 km [72]. Joby Aviation's S4 model, powered by liquid hydrogen and enhanced by reinforcement learning algorithms, has achieved a verified flight range of 841 km [73]. However, key bottlenecks remain in hybrid power algorithms, particularly in power distribution strategies, algorithm real-time performance, and system integration.

4.5. Environmental Adaptability

(1) Low-Temperature Environmental Adaptability: In low-temperature environments (<-20 °C) or high-altitude, low-oxygen conditions, changes in the elastic modulus of sealing materials lead to reduced interfacial contact stress, resulting in hydrogen leakage rate prediction errors exceeding 15%. Moreover, the absence of dynamic thermodynamic coupling models and the reliance on laboratory-based steady-state experimental data limit the ability to accurately simulate the transient operating conditions encountered during actual eVTOL flight missions [74].

(2) Low-Temperature Startup Performance: Traditional constant-current or constant-voltage startup strategies are prone to ice blockage and performance degradation in extremely cold environments. Hybrid control strategies address the limitations of single-mode control by dynamically balancing heat generation and ice formation, thereby improving startup success rates [75]. However, at temperatures below -40 °C, existing heating systems require more than 10 min to reach operational temperature, which fails to meet the rapid response requirements of eVTOL emergency takeoff and landing scenarios. Additionally, under high-altitude, low-oxygen conditions, reduced air density decreases oxygen transport efficiency. Existing flow field designs are susceptible to localized flooding at low atmospheric pressures, exacerbating the risk of ice blockage. Furthermore, during parallel operation of multiple fuel cell stacks, local temperature gradients exceeding 10 °C can accelerate membrane electrode degradation [76].

(3) Vibration and Impact Resistance: Conventional flow field structures are prone to reactant maldistribution and flooding under vibration conditions. However, through multi-physics coupling analysis and optimization of flow field structures and gas diffusion layers, the structural stability of PEMFCs under dynamic environments has been significantly improved. Moreover, the use of high-strength ceramic composite materials and metal-based alloy electrodes, combined with 3D printing techniques to optimize structural porosity, has enhanced the mechanical fatigue resistance of electrodes by 30% in vibration environments [77]. Nevertheless, in high-frequency vibration scenarios typical of eVTOL

operations, prediction errors for material fatigue life still exceed 15%, and there is a lack of dynamic thermo-mechanical coupling models to simulate microcrack propagation paths. Additionally, under the combined effects of vibration and impact, existing control strategies exhibit limited adaptability to multiple disturbances, and voltage recovery time remains a critical area for further optimization [78].

5. Development Trends of PEMFC Applications in Low-Altitude Aircraft

5.1. General Development Trends

(1) Material Innovation: Significant efforts are focused on material innovation to reduce costs and enhance efficiency. This includes the development of ultra-thin composite proton exchange membranes, such as perfluorosulfonic acid (PFSA) composite membranes, and, in response to environmental concerns related to per- and polyfluoroalkyl substances (PFAS), the investigation of alternatives to fluorinated membranes, and the exploration of non-precious metal catalysts, like Fe-N-C, to replace expensive platinum-group metals [79].

(2) Breakthroughs in Hydrogen Storage Technology: Breakthroughs in hydrogen storage technology are crucial for practical applications. The aim is to advance liquid hydrogen storage (e.g., Hylium's LH60 system) and chemical hydrogen storage (e.g., hydrogen generation from $NaBH_4$ hydrolysis) to achieve gravimetric hydrogen densities exceeding 8.5 wt%.

(3) Enhancing Fuel Cell Stack Power Density: To improve the effective payload or flight time of aerial vehicles, it is essential to enhance fuel cell stack power density. This will be achieved through lightweight design, increasing the power-to-weight ratio of hydrogen propulsion systems [80].

(4) Wide Temperature Range and Varying Altitude Environmental Adaptability: Adapting to wide temperature ranges and varying altitudes is a significant challenge. Under standard atmospheric conditions, temperature typically decreases by about 0.65 °C for every 100 m increase in altitude (standard lapse rate), while atmospheric pressure drops by approximately 1 kPa. Consequently, within low-altitude airspace (below 3000 m), temperatures can vary by up to 18 °C, and pressure by 30%. The extreme cold and low-pressure environments at high altitudes severely impact the power and efficiency of hydrogen propulsion system.

5.2. Hydrogen Fuel Cell as Main Power Supply

As main power supply, the hydrogen fuel cell system play a very important role in electrical propulsion, requiring no additional external power supply.

System Integration Optimization: System integration optimization focuses on enhancing thermal management efficiency and achieving lightweight designs through advanced techniques. This includes the implementation of micro-channel cooling and self-adaptive humidification technologies.

5.3. Hydrogen-Electric Hybrid Power System

The hydrogen-electric hybrid power system integrates hydrogen fuel cells with lithium batteries or other energy sources, creating a complementary architecture.

(1) Smart Energy Management Algorithms: Smart energy management algorithms are being developed to dynamically allocate energy based on the specific flight phase.

(2) Modular Design: Modular design focuses on creating standardized hydrogen-electric hybrid power modules. This approach allows for greater flexibility and adaptability, as these per-designed modules can be easily integrated into various aircraft types.

5.4. Hydrogen Fuel Cell as Auxiliary Power

The hydrogen fuel cell acts as an auxiliary power source, primarily dedicated to specific functions such as range extension or providing emergency backup power.

(1) Miniaturized Fuel Cells: Miniaturized fuel cells focus on creating compact, air-cooled fuel cells.

(2) Multi-Scenario Integration: Multi-scenario integration involves combining fuel cells with other energy sources like solar power and hydrogen turbine engines to form multi-energy complementary systems.

6. Conclusions

This paper systematically reviews the advantages, disadvantages, performance characteristics, and feasibility of various hydrogen power technologies for aviation applications. Subsequently, it synthesizes the research and development progress of fuel cells within the aviation sector. A critical analysis of the current application status, future development trends, and challenges of Proton Exchange Membrane Fuel Cells (PEMFCs) in aircraft is presented. The primary challenges identified include hydrogen storage and space limitations, thermal management, cost reduction, and enhancing performance, reliability, and environmental adaptability. Based on the power supply configurations of fuel cells in low-altitude aircraft, the systems are categorized into three distinct types: pure hydrogen fuel cell power systems, hydrogen-electric hybrid power systems, and hydrogen-electric auxiliary power systems. Common development priorities across these systems focus on improving stack power density, enhancing adaptability to wide temperature ranges and variable altitude environments, and advancing hydrogen storage technologies. The principal differences lie in their core optimization strategies: pure hydrogen fuel cell systems emphasize system integration optimization; hydrogen-electric hybrid systems prioritize sophisticated on-board power and energy management strategies; while hydrogen-electric auxiliary systems tend toward the miniaturization of fuel cell components.

Author Contributions: H.Y. and X.Z.; methodology, H.Y.; formal analysis, X.Z., H.Y., H.Z. and L.T.; investigation, Z.Z.; data curation, F.L.; writing—original draft preparation, H.Y.; writing—review and editing, H.Z. and L.T.; supervision, X.Z.; project administration, H.Y.; funding acquisition, X.Z. All authors have read and agreed to the published version of the manuscript.

Funding: This research was funded by the Zhejiang Provincial Emergency Department Project, grant number 2024YJ023 and This research was funded by Key R&D Program of. the Ministry of Science and Technology, grant number 2023YFB3209805.

Data Availability Statement: The original contributions presented in this study are included in the article. Further inquiries can be directed to the corresponding author.

Conflicts of Interest: The authors declare no conflicts of interest.

Abbreviations

The following abbreviations are used in this manuscript:

eVTOL	electric Vertical Take-Off and Landing
PEMFCs	Proton Exchange Membrane Fuel Cells
PEM	Proton Exchange Membrane
ICCT	International Council on Clean Transportation
CAGR	The compound annual growth rate
ICAO	International Civil Aviation Organization
AFC	Alkaline Fuel Cells
APUs	Auxiliary Power Units

UAVs	Unmanned Aerial Vehicles
ICE	Internal Combustion Engines
CAJU	Clean Aviation Joint Undertaking
HEROPS	Hydrogen-Electric Zero-Emission Propulsion System
HEPS	Hybrid-Electric Propulsion System
RUL	Remaining Useful Life
GI	weight index

References

1. Bahari, M.; Rostami, M.; Entezari, A.; Ghahremani, S.; Etminan, M. Performance evaluation and multi-objective optimization of a novel UAV propulsion system based on PEM fuel cell. *Fuel* **2022**, *311*, 122554. [CrossRef]
2. Zheng, B. CO_2 Emissions from Commercial Aviation: 2013, 2018, and 2019 Analysis, International Council on Clean Transportation. United States of America. 2020. Available online: https://coilink.org/20.500.12592/62jb1j (accessed on 5 June 2025).
3. Bergero, C.; Gosnell, G.; Gielen, D.; Kang, S.; Bazilian, M.; Davis, S.J. Pathways to net-zero emissions from aviation. *Nat. Sustain.* **2023**, *6*, 404–414. [CrossRef]
4. Dixit, M.; Bisht, A.; Witherspoon, B.; Essehli, R.; Amin, R.; Duncan, A.; Hines, J.; Kweon, C.-B.M.; Belharouak, I. Battery Electrolyte Design for Electric Vertical Takeoff and Landing (eVTOL) Platforms. *Adv. Energy Mater.* **2024**, *14*, 2400772. [CrossRef]
5. Muralidharan, N.; Self, E.C.; Dixit, M.; Du, Z.; Essehli, R.; Amin, R.; Nanda, J.; Belharouak, I. Next generation cobalt free cathodes a prospective solution to the battery industry's cobalt problem. *Transit. Met. Oxides Electrochem. Energy Storage* **2022**, *12*, 33–53.
6. Toghyani, S.; Cistjakov, W.; Baakes, F.; Krewer, U. Conceptual design of solid-state Li-battery for urban air mobility. *J. Electrochem. Soc.* **2023**, *170*, 103510. [CrossRef]
7. Ma, X.; Ding, S. *Overview and Research on Airworthiness and Safety of Electrical Propulsion and Battery Technologies in eVTOL*; SAE Technical Paper; SAE International: Warrendale, PA, USA, 2023. [CrossRef]
8. Park, J.; Lee, D.; Lim, D.; Yee, K.J. A refined sizing method of fuel cell-battery hybrid system for eVTOL aircraft. *Appl. Energy* **2022**, *328*, 120160. [CrossRef]
9. An, J.H.; Kwon, D.Y.; Jeon, K.S.; Tyan, M.; Lee, J.W. Advanced sizing methodology for a multi-mode EVTOL UAV powered by a hydrogen fuel cell and battery. *Aerospace* **2022**, *9*, 71. [CrossRef]
10. Adler, E.J.; Martins, J.R.R.A. Hydrogen-powered aircraft: Fundamental concepts, key technologies, and environmental impacts. *Prog. Aerosp. Sci.* **2023**, *141*, 100922. [CrossRef]
11. Gong, A.; Verstraete, D. Fuel cell propulsion in small fixed-wing unmanned aerial vehicles: Current status and research needs. *Int. J. Hydrogen Energy* **2017**, *42*, 21311–21333. [CrossRef]
12. Seidel, J.A.; Sehra, A.K.; Colantonio, R.O. NASA aeropropulsion research: Looking forward. In Proceedings of the Fifteenth International Symposium on Airbreathing Engines, Bangalore, India, 2–7 September 2001.
13. Timlon, J. *ZeroAvia: Building a Green Business at Scale to Accelerate Sustainable Aviation*; SAGE Publications: Thousand Oaks, CA, USA, 2025. [CrossRef]
14. Lo Verde, A. Competition Dynamics and Market Power: Analysis of the Duopoly Between Airbus and Boeing in the Civil Aircraft Manufacturing Industry. Master's Thesis, Politecnico di Torino, Torino, Italia, 2024.
15. Qasem, N.A.A.; Abdulrahman, G.A.Q. A recent comprehensive review of fuel cells: History, types, and applications. *Int. J. Energy Res.* **2024**, *2024*, 7271748. [CrossRef]
16. Zhijing, G.; Shufan, H.; Lihong, Y.; Jun, C. Heat Transfer Performance of Flat Heat Pipes for Proton Exchange Membrane Fuel Cells. *Chem. Eng.* **2024**, *52*, 41–46. (In Chinese)
17. Luo, L.; Jian, Q.; Huang, B.; Huang, Z.; Zhao, J.; Cao, S. Experimental study on temperature characteristics of an air-cooled proton exchange membrane fuel cell stack. *Renew. Energy* **2019**, *143*, 1067–1078. [CrossRef]
18. Zhao, C.; Wang, F.; Wu, X. Analysis and review on air-cooled open cathode proton exchange membrane fuel cells: Bibliometric, environmental adaptation and prospect. *Renew. Sustain. Energy Rev.* **2024**, *197*, 114408. [CrossRef]
19. Li, Q.; Liu, Z.; Sun, Y.; Yang, S.; Deng, C.W. A review on temperature control of proton exchange membrane fuel cells. *Processes* **2021**, *9*, 235. [CrossRef]
20. Zhu, K.Q.; Ding, Q.; Xu, J.H.; Yang, C.; Zhang, J.; Zhang, Y.; Huang, T.M.; Wan, Z.-M.; Wang, X.-D. Dynamic performance for a kW-grade air-cooled proton exchange membrane fuel cell stack. *Int. J. Hydrogen Energy* **2022**, *47*, 35398–35411. [CrossRef]
21. Hamada, A.T.; Orhan, M.F.; Kannan, A.M. Alkaline fuel cells: Status and prospects. *Energy Rep.* **2023**, *9*, 6396–6418. [CrossRef]
22. Azzam, M.; Qaq, Z.; Orhan, M. Design and analysis of an alkaline fuel cell. *J. Therm. Eng.* **2023**, *9*, 138–160. [CrossRef]
23. Marin, G.E.; Osipov, B.M.; Titov, A.V.; Akhmetshin, A.R. Gas turbine operating as part of a thermal power plant with hydrogen storages. *Int. J. Hydrogen Energy* **2023**, *48*, 33393–33400. [CrossRef]

24. Boretti, A. Towards hydrogen gas turbine engines aviation: A review of production, infrastructure, storage, aircraft design and combustion technologies. *Int. J. Hydrogen Energy* **2024**, *88*, 279–288. [CrossRef]
25. Sarakhanova, R.Y.; Kharitonov, S.A.; Zharkov, M.A.; Udovichenko, A.V. Starter-generator system of a hydrogen-fueled aircraft. *Int. J. Hydrogen Energy* **2025**, *99*, 632–639. [CrossRef]
26. Gürbüz, H. Optimization of combustion and performance parameters by intake-charge conditions in a small-scale air-cooled hydrogen fuelled SI engine suitable for use in piston-prop aircraft. *Aircr. Eng. Aerosp. Technol.* **2021**, *93*, 448–456. [CrossRef]
27. Depcik, C.; Cassady, T.; Collicott, B.; Burugupally, S.P.; Li, X.; Alam, S.S.; Arandia, J.R.; Hobeck, J. Comparison of lithium ion Batteries, hydrogen fueled combustion Engines, and a hydrogen fuel cell in powering a small Unmanned Aerial Vehicle. *Energy Convers. Manag.* **2020**, *207*, 112514. [CrossRef]
28. Agyekum, E.B.; Ampah, J.D.; Wilberforce, T.; Afrane, S.; Nutakor, C. Research progress, trends, and current state of development on PEMFC-new insights from a bibliometric analysis and characteristics of two decades of research output. *Membranes* **2022**, *12*, 1103. [CrossRef] [PubMed]
29. de Boer, P.C.; Wit, A.; van Benthem, R.C. *Development of a Liquid Hydrogen-Based Fuel Cell System for the HYDRA-2 Drone;* AIAA SciTech 2022 Forum; Netherlands Aerospace Centre NLR: Amsterdam, The Netherlands, 2022; p. 443.
30. Sparano, M.; Sorrentino, M.; Troiano, G.; Cerino, G.; Piscopo, G.; Basaglia, M.; Pianese, C. The future technological potential of hydrogen fuel cell systems for aviation and preliminary co-design of a hybrid regional aircraft powertrain through a mathematical tool. *Energy Convers. Manag.* **2023**, *281*, 116822. [CrossRef]
31. Wu, J.; Ji, Y. Analysis and Prospects of Key Technologies for Hydrogen-Electric Regional Aircraft. In Proceedings of the 6th China Aeronautical Science and Technology Conference, Wuzhen, China, 26–27 September 2023; Springer: Singapore, 2023; pp. 587–595.
32. Mukhopadhaya, J. *Performance Analysis of Fuel Cell Retrofit Aircraft;* ICCT White Paper; ICCT: Washington, DC, USA, 2023.
33. Thipphavong, D.P.; Apaza, R.; Barmore, B.; Battiste, V.; Belcastro, C.; Burian, B.; Dao, A.-Q.V.; Feary, M.; Go, S.; Goodrich, K.H.; et al. Urban air mobility airspace integration concepts and considerations. In Proceedings of the 2018 Aviation Technology, Integration, and Operations Conference, Atlanta, GA, USA, 25–29 June 2018; p. 3676.
34. National Academies of Sciences and Medicine, Engineering. *Advancing Aerial Mobility: A National Blueprint;* The National Academies Press: Washington, DC, USA, 2020; p. 82.
35. Joby Aviation Generates First Revenue, Takes Key Step towards Certifying Aircraft. Available online: https://www.jobyaviation.com/news/joby-aviation-generates-first-revenue-takes-key-step-towards-certifying-aircraft/ (accessed on 9 February 2021).
36. Baroutaji, A.; Wilberforce, T.; Ramadan, M.; Olabi, A.G. Comprehensive investigation on hydrogen and fuel cell technology in the aviation and aerospace sectors. *Renew. Sustain. Energy Rev.* **2019**, *106*, 31–40. [CrossRef]
37. Mobariz, K.N.; Youssef, A.M.; Abdel-Rahman, M. Long endurance hybrid fuel cell-battery powered UAV. *World J. Model. Simul.* **2015**, *11*, 69–80.
38. Özbek, E.; Yalin, G.; Ekici, S.; Karakoc, H.T. Evaluation of design methodology, limitations, and iterations of a hydrogen fuelled hybrid fuel cell mini UAV. *Energy* **2020**, *213*, 118757. [CrossRef]
39. Swider-Lyons, K.; Stroman, R.O.; Schuette, M.; Mackrell, J.; Page, G.; Rodgers, J.A. Hydrogen fule cell propulsion for long endurance small UVAs. In Proceedings of the AIAA Centennial of Naval Aviation Forum "100 Years of Achievement and Progress", Virginia Beach, VA, USA, 21–22 September 2011; p. 6975.
40. Velev, O. Summary of fuel cell projects: AeroVironment 1997–2007. In *National Hydrogen Association Fall 2007;* Topical Forum Columbia: New York, NY, USA, 2007.
41. McConnell, V.P. Military UAVs claiming the skies with fuel cell power. *Fuel Cells Bull.* **2007**, *2007*, 12–15. [CrossRef]
42. Stroman, R.; Kellogg, J.C.; Swider-Lyons, K. Testing of a PEM fuel cell system for small UAV propulsion. *Power* **2000**, *60*. Available online: https://www.researchgate.net/publication/264543086_Testing_of_a_PEM_fuel_cell_system_for_small_UAV_propulsion (accessed on 9 May 2021).
43. Bradley, T.H.; Moffitt, B.A.; Mavris, D.N.; Parekh, D.E. Development and experimental characterization of a fuel cell powered aircraft. *J. Power Sources* **2007**, *171*, 793–801. [CrossRef]
44. Spencer, K.M.; Martin, C.A. *Investigation of Potential Fuel Cell Use in Aircraft, Institute for Defense Analyses;* Institute for Defense Analyses: Alexandria, VA, USA, 2013.
45. Kim, K.; Kim, T.; Lee, K.; Kwon, S. Fuel cell system with sodium borohydride as hydrogen source for unmanned aerial vehicles. *J. Power Sources* **2011**, *196*, 9069–9075. [CrossRef]
46. Seo, J.E.; Kim, Y.; Kim, Y.; Kim, K.; Lee, J.H.; Lee, D.H.; Kim, Y.; Shin, S.J.; Kim, D.M.; Kim, S.Y.; et al. Portable ammonia-borane based H2 power-pack for unmanned aerial vehicles. *J. Power Sources* **2014**, *254*, 329–337. [CrossRef]
47. Gundlach, J.; Foch, R.J. *Unmanned Aircraft Systems Innovation at the Naval Research Laboratory, Aerospace Research Central;* AIAA: Reston, VA, USA, 2014. [CrossRef]
48. Kim, T. NaBH4 (sodium borohydride) hydrogen generator with a volume-exchange fuel tank for small unmanned aerial vehicles powered by a PEM (proton exchange membrane) fuel cell. *Energy* **2014**, *69*, 721–727. [CrossRef]

49. H3 Dynamic. Dreaming Big the Future of Aviation is Hydrogen-Electric, Autonomous, and Digital. 2023. Available online: https://www.h3dynamics.com/ (accessed on 11 May 2025).
50. Dziwulski, E.; Ribeiro, W.F.; Basilio, F.L. Avançando Para o Azul: Viabilidade e Impactos Dos Aviões Zeroe da Airbus Movidos a Hidrogênio: Moving Towards the Blue: Feasibility and Impacts of Airbus Zeroe Hydrogen-Powered Aircraft. *Rev. Bras. Aviação Civ. Ciências Aeronáuticas* **2025**, *5*, 27–56.
51. Swaminathan, N.; Reddy, S.R.P.; RajaShekara, K.; Haran, K.S. Flying cars and eVTOLs—Technology advancements, powertrain architectures, and design. *IEEE Trans. Transp. Electrif.* **2022**, *8*, 4105–4117. [CrossRef]
52. Mus, J.; Mylle, S.; Schotte, S.; Fevery, S.; Latré, S.K.; Buysschaert, F. CFD Modelling and Simulation of PEMFCs in STAR-CCM+. In Proceedings of the 2022 11th International Conference on Renewable Energy Research and Application (ICRERA), Istanbul, Turkey, 18–21 September 2022; IEEE: Piscataway, NJ, USA, 2022; pp. 260–267.
53. Scanff, R.; Cauville, B.; Néron, D.; Ladevèze, P. Reduced-order models for aeronautics based on the LATIN-PGD method within non-linear industrial Simcenter Samcef software. In Proceedings of the 9th European Conference for Aeronautics and Space Sciences (EUCASS-3AF 2022), Lille, France, 27 June–1 July 2022. [CrossRef]
54. Kwon, K. Hydrogen: Coming to an aircraft near you. *Aerosp. Am.* **2022**, *60*, 22.
55. Gao, Y.; Jausseme, C.; Huang, Z.; Yang, T. Hydrogen-Powered Aircraft: Hydrogen–electric hybrid propulsion for aviation. *IEEE Electrif. Mag.* **2022**, *10*, 17–26. [CrossRef]
56. Qasem, N.A.A. A recent overview of proton exchange membrane fuel cells: Fundamentals, applications, and advances. *Appl. Therm. Eng.* **2024**, *252*, 123746. [CrossRef]
57. Preuster, P.; Alekseev, A.; Wasserscheid, P. Hydrogen storage technologies for future energy systems. *Annu. Rev. Chem. Biomol. Eng.* **2017**, *8*, 445–471. [CrossRef]
58. Li, H.; Cao, X.; Liu, Y.; Shao, Y.; Nan, Z.; Teng, L.; Peng, W.; Bian, J. Safety of hydrogen storage and transportation: An overview on mechanisms, techniques, and challenges. *Energy Rep.* **2022**, *8*, 6258–6269. [CrossRef]
59. Massaro, M.C.; Biga, R.; Kolisnichenko, A.; Marocco, P.; Monteverde, A.H.A.; Santarelli, M. Potential and technical challenges of on-board hydrogen storage technologies coupled with fuel cell systems for aircraft electrification. *J. Power Sources* **2023**, *555*, 232397. [CrossRef]
60. He, L.; Yang, Y.; Zhang, Y.; Li, P.; Xin, Y. A review of thermal management of proton exchange membrane fuel cell systems. *J. Renew. Sustain. Energy* **2023**, *15*, 012703. [CrossRef]
61. Xia, Q.; Zhang, T.; Sun, Z.; Gao, Y. Design and optimization of thermal strategy to improve the thermal management of proton exchange membrane fuel cells. *Appl. Therm. Eng.* **2023**, *222*, 119880. [CrossRef]
62. Deng, B.; Zhang, X.; Yin, C.; Luo, Y.Q.; Tang, H. Improving a fuel cell system's thermal management by optimizing thermal control with the particle swarm optimization algorithm and an artificial neural network. *Appl. Sci.* **2023**, *13*, 12895. [CrossRef]
63. Yang, L.; Nik-Ghazali, N.-N.; Ali, M.A.H.; Chong, W.T.; Yang, Z.; Liu, H. A review on thermal management in proton exchange membrane fuel cells: Temperature distribution and control. *Renew. Sustain. Energy Rev.* **2023**, *187*, 113737. [CrossRef]
64. Madheswaran, D.K.; Vengatesan, S.; Varuvel, E.G.; Praveenkumar, T.; Jegadheeswaran, S.; Pugazhendhi, A.; Arulmozhivarman, J. Nanofluids as a coolant for polymer electrolyte membrane fuel cells: Recent trends, challenges, and future perspectives. *J. Clean. Prod.* **2023**, *424*, 138763. [CrossRef]
65. Zhou, S.; Chen, C.; Lei, F. A Review of Thermal Management for Proton Exchange Membrane Fuel Cell Power Systems. *Automot. Dig.* **2023**, *15*, 02-271. (In Chinese)
66. Dirkes, S.; Leidig, J.; Fisch, P.; Pischinger, S. Prescriptive Lifetime Management for PEM fuel cell systems in transportation applications, Part I: State of the art and conceptual design. *Energy Convers. Manag.* **2023**, *277*, 116598. [CrossRef]
67. Ishimoto, Y.; Wulf, C.; Schonhoff, A.; Kuckshinrichs, W. Life cycle costing approaches of fuel cell and hydrogen systems: A literature review. *Int. J. Hydrogen Energy* **2024**, *54*, 361–374. [CrossRef]
68. Jiang, D.; Long, Y.; Fu, P.; Guo, C.; Tang, Y.; Huang, H. A novel multi-stack fuel cell hybrid system energy management strategy for improving the fuel cell durability of the hydrogen electric Multiple Units. *Int. J. Green Energy* **2024**, *21*, 1766–1775. [CrossRef]
69. Salem, K.A.; Palaia, G.; Quarta, A.A. Review of hybrid-electric aircraft technologies and designs: Critical analysis and novel solutions. *Prog. Aerosp. Sci.* **2023**, *141*, 100924. [CrossRef]
70. Hince, S.B.; Ansell, P.J. Hybrid H2 Powertrain Characterization for Aircraft Design and Integration. In Proceedings of the AIAA Aviation Forum and Ascend 2024, Las Vegas, NV, USA, 29 July–2 August 2024; p. 3965.
71. Tian, W.; Zhang, X.; Yang, D. Double-layer fuzzy adaptive NMPC coordinated control method of energy management and trajectory tracking for hybrid electric fixed wing UAVs. *Int. J. Hydrogen Energy* **2022**, *47*, 39239–39254. [CrossRef]
72. Marzouk, O. Edelweiss Applied Science and Technology. *Edelweiss Appl. Sci. Technol.* **2025**, *9*, 413–442. [CrossRef]
73. Oladosu, T.L.; Pasupuleti, J.; Kiong, T.S.; Koh, S.P.J.; Yusaf, T. Energy management strategies, control systems, and artificial intelligence-based algorithms development for hydrogen fuel cell-powered vehicles: A review. *Int. J. Hydrogen Energy* **2024**, *61*, 1380–1404. [CrossRef]

74. Shi, D.; Cai, L.; Zhang, C.; Chen, D.; Pan, Z.; Kang, Z.; Liu, Y.; Zhang, J. Fabrication methods, structure design and durability analysis of advanced sealing materials in proton exchange membrane fuel cells. *Chem. Eng. J.* **2023**, *454*, 139995. [CrossRef]
75. Cardone, L.M.; Petrone, G.; De Rosa, S.; Franco, F.; Greco, C.S. Review of the recent developments about the hybrid propelled aircraft. *Aerotec. Missili Spaz.* **2024**, *103*, 17–37. [CrossRef]
76. Keles, O.K.; Hakyemez, I.; Bagriyanik, M.; Kalenderli, O. A Study on Power System Retrofit for Cessna-172S Aircraft by Using Hydrogen Fuel Cell and Battery Hybrid. *IEEE Access* **2025**, *13*, 6089–6101. [CrossRef]
77. Doo, J. *Unsettled Issues Concerning the Opportunities and Challenges of eVTOL Applications During a Global Pandemic*; SAE International: Warrendale, PA, USA, 2020; Volume 2, p. 24.
78. Farajijalal, M.; Eslamiat, H.; Avineni, V.; Hettel, E.; Lindsay, C. Safety Systems for Emergency Landing of Civilian Unmanned Aerial Vehicles—A Comprehensive Review. *Drones* **2025**, *9*, 141. [CrossRef]
79. Zhang, Y.; Wang, J.; Yao, Z. Recent development of fuel cell core components and key materials: A review. *Energies* **2023**, *16*, 2099. [CrossRef]
80. Jamal, T.; Shafiullah, G.M.; Dawood, F.; Kaur, A.; Arif, M.T.; Pugazhendhi, R.; Elavarasan, R.M.; Ahmed, S.F. Fuelling the future: An in-depth review of recent trends, challenges and opportunities of hydrogen fuel cell for a sustainable hydrogen economy. *Energy Rep.* **2023**, *10*, 2103–2127. [CrossRef]

hydrogen

MDPI

Review

Fuel-Cell Thermal Management Strategies for Enhanced Performance: Review of Fuel-Cell Thermal Management in Proton-Exchange Membrane Fuel Cells (PEMFCs) and Solid-Oxide Fuel Cells (SOFCs)

Ibham Veza [1,2]

[1] Department of Mechanical Engineering, Faculty of Engineering, Universitas Bung Karno, Jalan Kimia No. 20. Menteng, Jakarta Pusat 10320, Indonesia; ibham@vezas.org

[2] Centre for Renewable Energy System Modeling and Policy Innovation, Aras Energy Consulting, Jalan HR Rasuna Said Kav. C-5, Setia Budi, Jakarta Selatan 12920, Indonesia

Abstract

Effective thermal management is crucial for optimizing the performance, efficiency, and durability of fuel-cell technologies, including proton-exchange membrane fuel cells (PEMFCs) and solid-oxide fuel cells (SOFCs). The operation of fuel cells involves complex heat generation mechanisms, primarily driven by electrochemical reactions, which can lead to significant energy loss as heat. This review examines the specific heat generation sources and challenges associated with different fuel-cell types, highlighting the critical importance of effective thermal management strategies. Key techniques for thermal regulation, including active and passive cooling systems, are examined in detail. Active cooling methods like liquid cooling and air cooling are effective in dissipating excess heat, while passive methods leverage advanced materials and optimized designs to enhance natural heat dissipation. Furthermore, innovative heat recovery systems are explored, demonstrating their potential to enhance overall energy efficiency by capturing and repurposing waste heat. The integration of machine learning techniques has arisen as a promising avenue for advancing temperature control in fuel cells. Reinforcement learning, deep learning algorithms, and support vector machines, along with artificial neural networks, are discussed in the context of their application in managing temperature dynamics and optimizing thermal performance. The review also emphasizes the significance of real-time monitoring, as well as adaptive control strategies to respond effectively to the dynamic operating conditions of fuel cells. Understanding and applying these thermal management strategies is essential for the successful commercialization of fuel cells across various sectors, ranging from automotive to stationary power generation. With the growing demand for clean energy solutions, progress in thermal management techniques will be crucial in improving the dependability and practicality of fuel-cell systems.

Keywords: fuel-cell thermal management; proton-exchange membrane fuel cells (PEMFCs); solid-oxide fuel cells (SOFCs); integrated thermal management system; fuel-cell hybrid cooling; fuel-cell heat recovery

1. Introduction: Fuel Cells

Fuel cells are innovative electrochemical mechanisms that change chemical energy directly into electrical energy via the reaction of H_2 with O_2, producing water as the only byproduct [1,2]. This clean and efficient process has gathered substantial attention recently, particularly in the context of renewable and sustainable energy solutions. Various kinds of fuel cells exist; every type comes with distinctive characteristics and applications, including proton-exchange membrane fuel cells (PEMFCs) [3–6], solid-oxide fuel cells (SOFCs) [7–10], molten carbonate fuel cells (MCFCs) [11–14], and phosphoric acid fuel cells (PAFCs) [15–18]. Figure 1 shows the curve characteristics of the PEMFC.

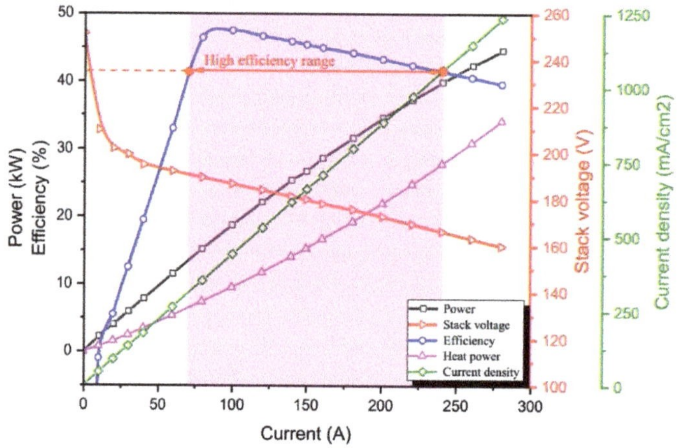

Figure 1. Curve characteristics of the PEMFC [19].

Proton-exchange membrane fuel cells (PEMFCs) are among the most extensively studied and utilized fuel cells, especially in the automotive industry [20–22]. They function at relatively low temperatures (typically 80 °C) and are characterized by their quick start-up times and high power density. Marghani [23] discusses the compatibility of PEMFCs for portable power sources because of their efficient conversion of hydrogen into electricity with minimal emissions. The simplicity of their design, combined with the availability of hydrogen as a fuel, makes PEMFCs a strong candidate for future energy systems. Despite significant progress, PEMFCs still face commercialization challenges, particularly related to durability and lifespan. Performance degradation over time results from factors such as electrode material degradation, catalyst loss, membrane mechanical damage, and fluctuating operating conditions, which ultimately affect efficiency, service life, and maintenance costs.

Solid-oxide fuel cells (SOFCs), in contrast, work at much higher temperatures (600 °C to 1000 °C) and utilize a solid-oxide or ceramic electrolyte. Wang et al. [24] highlight that SOFCs could employ various fuels, including natural gas, as well as biogas, which enhances their versatility. The high operating temperatures allow for greater efficiency in energy conversion, but they also pose challenges related to thermal management and material durability. Research into advanced materials and designs is ongoing to optimize SOFC performance while addressing these challenges.

Molten carbonate fuel cells (MCFCs) work at intermediate temperatures (around 600 °C to 700 °C) and utilize a molten carbonate electrolyte. These fuel cells are particularly effective for large-scale stationary applications, such as power plants. The possibility of using various fuels, including natural gas and hydrogen, makes MCFCs a flexible option for energy generation. Their thermal management strategies are crucial, as maintaining the appropriate operating temperature is essential for optimal performance and longevity.

Phosphoric acid fuel cells (PAFCs) are another type of fuel cell that functions at moderate temperatures (around 150 °C to 200 °C). They are famous for their robustness.

Their reliability makes them fit for stationary applications. Hossain et al. [25] note that PAFCs are used for numerous purposes, including backup power systems, as well as combined heat and power (CHP) systems, thanks to their capability to efficiently convert fuel into electricity while providing useful thermal energy.

The integration of machine learning techniques in fuel-cell technology is also gaining traction [26–32]. These techniques can enhance the control and optimization of fuel cells by analyzing data and predicting performance under various operating conditions. For instance, machine learning algorithms are used to monitor and manage temperature, pressure, and humidity within fuel cells, ensuring optimal operating conditions and improving overall efficiency [33]. Such a data-driven method enables more responsive and adaptive control strategies, which are critical for maximizing performance of fuel-cell systems in real-world applications.

In conclusion, fuel cells represent a potential technology for renewable and sustainable energy generation, with various types tailored for specific applications. The progress in fuel-cell research and development intends to enhance the efficiency, durability, and versatility of fuel cells, making them a vital element of future renewable energy systems. As machine learning and advanced materials continue to evolve, the potential for fuel cells to contribute to a low-carbon economy becomes increasingly significant.

Although the main focus of this review is PEMFCs and SOFCs, brief descriptions of MCFCs and PAFCs are included in the introduction to provide a complete overview of major fuel-cell types. This broader context clarifies the distinct operating ranges and thermal management requirements across technologies, and it helps explain why PEMFCs and SOFCs—representing low- and high-temperature extremes—were selected for a detailed comparative analysis.

This review provides a novel synthesis of current advancements in fuel-cell thermal management by bridging both proton-exchange membrane fuel cells (PEMFCs) and solid-oxide fuel cells (SOFCs), highlighting the unique challenges and strategies specific to each. The inclusion of innovative cooling systems, such as hybrid active–passive approaches, coupled with machine learning-driven adaptive control, marks a forward-thinking perspective on optimizing heat dissipation. The exploration of real-time monitoring techniques, combined with the potential of heat recovery systems to boost energy efficiency, underscores the need for dynamic, data-driven approaches in advancing fuel-cell technologies for broader purposes, ranging from the automotive sector to stationary power systems.

2. Fuel-Cell Thermal Management

Fuel cells, specifically proton-exchange membrane fuel cells (PEMFCs), are more and more recognized for their potential for numerous purposes, including automotive and stationary power generation. However, effective thermal management is of importance to optimize their performance and durability. The fuel cells' thermal management involves the regulation of temperature to ensure efficient operation, as excessive heat could result in performance degradation and decreased durability.

One primary drawback in fuel-cell thermal management is the heat generation during electrochemical reactions [34–36]. This heat must be effectively dissipated to maintain optimal operating temperatures. Various cooling methods have been explored, including liquid cooling, air cooling, and phase-change cooling strategies. Chen et al. [37] provide a comprehensive review of these cooling methods, highlighting their material properties and durability implications, which are essential for PEMFCs' reliability. Figure 2 illustrates a PEMFC and its cooling channels. Additionally, Zhang et al. [38] discuss the model predictive control application for open-cathode fuel-cell thermal management systems, emphasizing the need for advanced control strategies to enhance thermal performance.

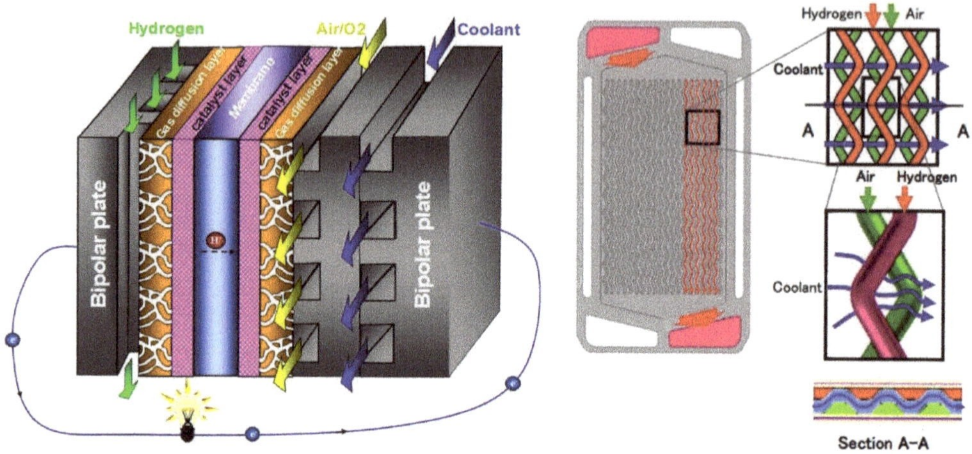

(a) (b)

Figure 2. (**a**) PEMFC with an embedded coolant channel system within the bipolar plates [39] and (**b**) coolant channels in an electric vehicle based on a fuel cell (FCEVs) [40].

Integrating waste heat recovery systems is another promising approach to improving whole-fuel-cell efficiency [41–43]. Nguyen and Shabani [42] explore the heat recovery opportunities in combined heating, cooling, and power applications, suggesting that such systems can significantly enhance energy utilization. Similarly, Sun et al. [44]. investigate the reuse of waste heat in fuel-cell vehicles, demonstrating that effective heat exchange can improve energy consumption and overall system efficiency. This aligns with the findings of Olabi et al. [45], who note that the waste heat use in fuel-cell systems has garnered considerable attention for its potential to increase energy conversion efficiency.

Moreover, the control and modeling of thermal management systems are critical for optimizing the performance of fuel-cell vehicles [35,46,47]. Xu et al. [19] present a model for vehicle-integrated thermal management systems, which allows for the better control of temperature dynamics in PEMFCs and is complemented by the work of Bai and He [48], who discuss system optimization for thermal management in a fuel-cell vehicle, emphasizing the importance of maintaining appropriate operating conditions to ensure reliability and performance.

Figure 3 illustrates the integration of fuel-cell systems at both the vehicle and energy network levels. The top schematic shows a vehicle PEMFC system where hydrogen from the storage tank and compressed air are supplied to the fuel-cell stack, producing electricity, heat, and water, with balance-of-plant components such as pumps, humidifiers, coolers, radiators, and purge valves ensuring proper thermal, water, and reactant management. The bottom schematic highlights how hydrogen can be produced via electrolysis using renewable sources such as solar and wind, stored, and later utilized in fuel cells to supply electricity back to the grid or vehicle, demonstrating hydrogen's role as an energy carrier that links intermittent renewables with end-use applications. Together, the diagrams emphasize that system scalability requires efficient balance-of-plant management, temperature control is critical to performance and durability, and techno-economic considerations remain central to achieving cost-effective deployment of large-scale hydrogen and fuel-cell systems.

In conclusion, effective thermal management is of importance for optimal fuel-cell performance. The integration of advanced cooling methods, waste heat recovery systems, and sophisticated control strategies can significantly improve PEMFCs' efficiency and lifespans. While previous reviews have focused on aspects such as flow channel design, two-phase flow simulation, and static performance evaluation, our paper provides a broader comparative perspective by covering both PEMFC and SOFC thermal management strategies, incorporating advanced techniques such as hybrid cooling and machine learning-based optimization approaches. Additionally, we have now included references and brief discussions on flow channel optimization and dynamic coolant control strategies to enhance the comprehensiveness of our review. As research continues to evolve in this area, the development of comprehensive thermal management systems will be vital for the broader adoption of fuel-cell technologies in various applications.

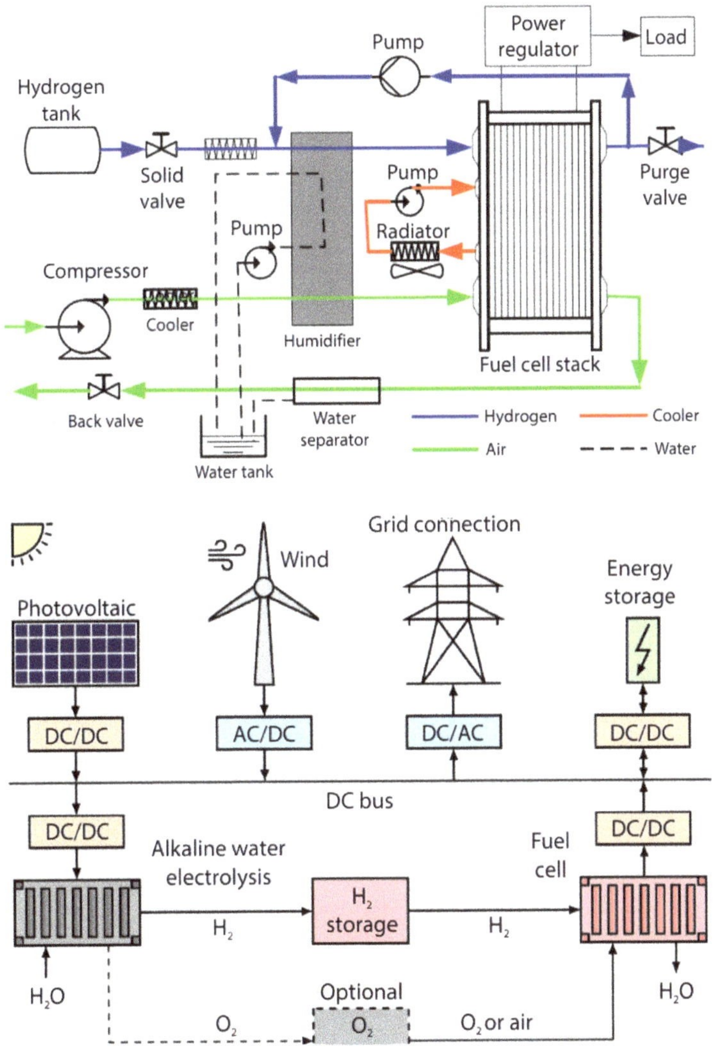

Figure 3. (**Top**) Vehicle fuel-cell system and (**bottom**) energy distribution within the stack system [48].

2.1. Heat Generation Process in Fuel Cells

2.1.1. Heat Generation Process in Proton-Exchange Membrane Fuel Cells

Heat generation in PEMFCs is a crucial element that influences fuel-cell efficiency and operational stability. The mechanisms of heat generation primarily stem from the electrochemical reactions taking place within a fuel cell and from various physical processes associated with the transport of reactants and products. Understanding these mechanisms is essential for optimizing fuel-cell design and improving thermal management strategies.

The primary source of heat in PEMFCs is the exothermic reaction that occurs during the conversion of the electrochemicals H_2 and O_2 into water and electricity. This reaction releases a significant amount of energy, which is primarily dissipated as heat. According to Murad [49], approximately 50% or up to 60% of the energy generation from reactions of these chemicals is lost as heat, highlighting the importance of effective thermal management to enhance overall efficiency. Furthermore, heat generation is exacerbated by the resistance encountered during proton transport via a membrane and electron transport via electrodes, which contributes to Joule heating [50].

In addition to electrochemical reactions, the flow dynamics of reactants, along with products within the fuel cell, also play a critical part in heat generation. The flow fields' design, as discussed by Atasay et al. [51], can significantly affect the reactants' distribution, heat removal, and water produced during operation. Inefficient flow field designs can lead to localized overheating, which can adversely affect the fuel-cell performance and durability. Hence, optimizing the flow-field structure is vital to enhance thermal management and minimize heat generation [52].

Moreover, the working conditions of PEMFCs, for instance, temperature and pressure, influence heat generation mechanisms. High-temperature PEMFCs (HT-PEMFCs) have been shown to exhibit improved kinetics and reduced heat generation due to enhanced reaction rates and better water management capabilities [53]. However, they also require careful thermal management to prevent overheating, as the reaction rates can lead to excessive heat if not properly controlled [54]. The balance between operating temperature and heat generation is crucial for maintaining optimal performance and preventing the thermal degradation of the membrane and other components. Figure 4 shows various membranes of proton exchange used in HT-PEMFCs, with all parameters' average values derived from an analysis of ten randomly selected articles [53].

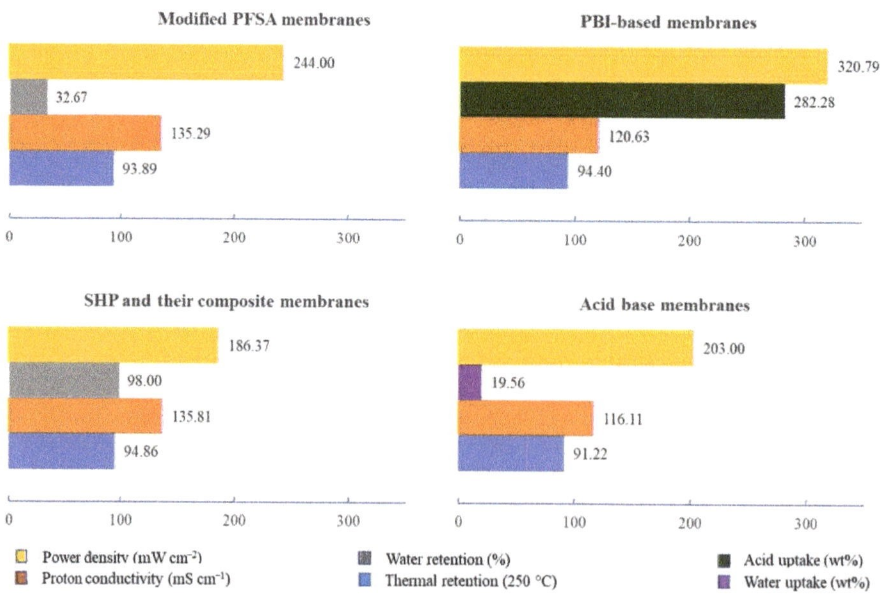

Figure 4. Various proton-exchange membranes used in HT-PEMFCs [53].

In conclusion, heat generation in PEMFCs arises from both electrochemical reactions and physical processes pertains to reactant transport. Effective thermal management strategies, including optimized flow field designs and the careful control of operating conditions, are essential to mitigate heat generation and improve PEMFCs' efficiency and durability. As research continues to advance in this field, a deeper understanding of these mechanisms will be vital for the development of next-generation fuel cells.

2.1.2. Heat Generation Process in Solid-Oxide Fuel Cells

Heat generation mechanisms in SOFCs are crucial for understanding their efficiency and operational stability [10,55]. SOFCs work at relatively high temperatures, normally between 600 °C and 1000 °C, which facilitates ionic conductivity of solid-oxide electrolyte. This enables efficient electrochemical reactions. The heat generated in SOFCs primarily arises from the exothermic reactions during the electrochemical conversion of fuel, as well as from various physical processes associated with reactant transport and resistance within the cell.

SOFCs are connected in series via interconnects to form a stack of fuel cells, allowing for higher efficiency at elevated voltages [56]. These interconnects, acting as bipolar plates, link individual cells while shielding the electrodes from environmental factors. Figure 5 displays the interconnected configuration within the stack of fuel cells, alongside the formation of native oxide scales at the interface.

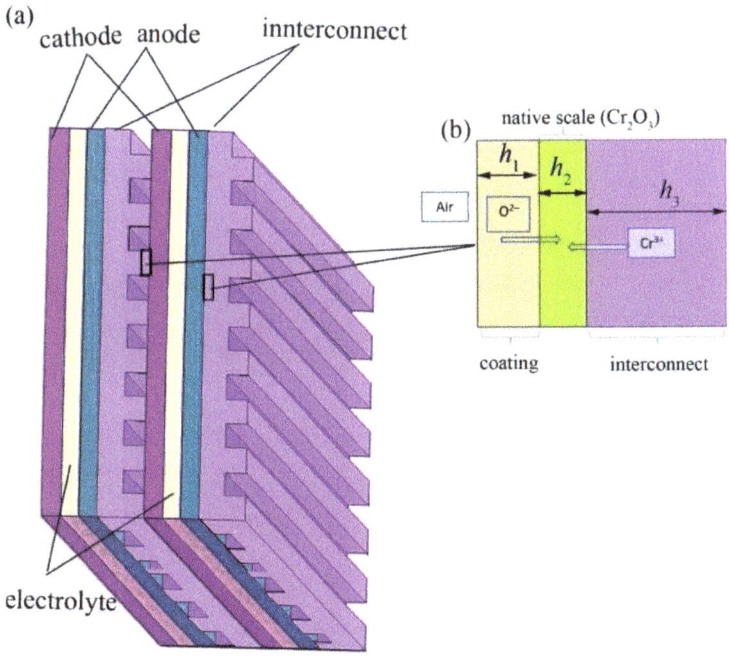

Figure 5. (**a**) Arrangement of interconnects in SOFCs. (**b**) Enlarged picture of the interconnect interface. The thicknesses are labeled as h_1 for the coating, h_2 for the native oxide scale, and h_3 for the interconnect [56].

The primary source of heat generation in SOFCs is the electrochemical oxidation of fuels, like H_2 or hydrocarbons, at the anode. This reaction releases energy, part of which is transformed into electrical energy, while the remainder is dispersed as heat. According to Xu et al. [57], the high operating temperatures not only enhance the reaction kinetics but also contribute to significant thermal energy generation due to the inherent inefficiencies in the electrochemical processes. The heat produced can be substantial, with estimates indicating that a considerable amount of energy from fuel vanishes as heat, rather than being converted into useful work [58].

In addition to the electrochemical reactions, thermal stresses and resistive heating also contribute to heat generation in SOFCs. As noted by Sai et al. [56], the structural integrity of the cell components, particularly the metallic interconnects, can be compromised because of thermal stresses due to the high operating temperatures. These stresses can lead to mechanical degradation, which not only influences the fuel-cell performance but also generates additional heat through friction and deformation. Furthermore, the resistance to ion and electron flow within the cell generates Joule heating, which adds to the overall thermal load [59].

The design and configuration of the SOFC also play a significant role in heat generation. For instance, the use of porous electrodes allows for better reactant distribution and heat dissipation, but an improper design can lead to localized overheating. Moreover, the interaction between the anode and the electrolyte can result in additional heat generation due to concentration gradients and reaction kinetics, as highlighted by Pirkandi et al. [60]. Therefore, optimizing the fuel-cell components' design is essential for managing heat generation effectively.

In conclusion, the generation of heat in solid-oxide fuel cells is a multifaceted process driven by electrochemical reactions, thermal stresses, and resistive heating. Understanding these mechanisms is vital for improving the efficiency and durability of SOFCs. As research progresses, advancements in materials and designs will be crucial for enhancing thermal management strategies, ultimately leading to higher efficiency and reliability of fuel cells. Table 1 summarizes the key points on thermal management of PEMFCs and SOFCs.

Table 1. Overview of the heat generation mechanisms, challenges, and thermal management strategies for PEMFCs and SOFCs.

Aspect	PEMFCs	SOFCs
Heat Generation Sources	• Electrochemical reactions converting hydrogen and oxygen into water. • Proton transport and electron flow create Joule heating.	• Electrochemical oxidation of fuels (e.g., hydrogen or hydrocarbons) at the anode. • High temperatures (600–1000 °C) enhance reaction kinetics but generate heat.
Energy Lost as Heat	• Fifty to sixty percent of energy generated is lost as heat.	• Significant thermal energy is produced due to inefficiencies in electrochemical processes.
Key Challenges	• Poor flow field design leads to localized overheating.	• Thermal stresses at high temperatures affect cell components (e.g., metallic interconnects) and generate heat.
Design Considerations	• Flow field design impacts the distribution of reactants and removal of heat.	• Porous electrode designs aid heat dissipation, but improper design can cause localized overheating.
Thermal Management Strategies	• Liquid cooling, air cooling, and phase-change cooling strategies.	• Managing concentration gradients and reaction kinetics through optimal anode-electrolyte interactions is essential.
Control Strategies	• Advanced control strategies like model predictive control for optimizing open-cathode fuel-cell systems.	• Efficient design of components (e.g., electrodes, interconnects) is critical to minimize resistive heating and stress.
Waste Heat Recovery	• Heat recovery systems can enhance energy utilization in combined heating, cooling, and power systems.	• Utilizing waste heat through system optimization can enhance overall system efficiency.

2.2. Significance of Thermal Management of Fuel Cells

2.2.1. Significance of Thermal Management of Proton-Exchange Membrane Fuel Cells

Proton-exchange membrane fuel cells' (PEMFCs') thermal management is of paramount importance because of its direct effect on the performance, efficiency, and longevity of these energy conversion devices. Effective thermal management ensures that the working temperature is maintained in the range of an optimal interval, which is crucial for maximizing electrochemical reactions while preventing the thermal degradation of the membrane and other components.

One of the primary reasons for the significance of thermal management in PEMFCs is the exothermic nature of the electrochemical reactions occurring within the cell. As highlighted by Wang et al. [54], the heat generated during these reactions can lead to increased temperatures that may exceed the fuel-cell material limits' components, particularly the proton-exchange membrane (PEM). High temperatures can cause membrane dehydration, which adversely affects proton conductivity and overall cell performance. Therefore, maintaining an appropriate thermal environment is essential to ensure that the membrane remains hydrated and functions efficiently.

Moreover, the operational efficiency of PEMFCs is closely linked to their thermal management strategies. Ji [61] emphasizes that high-temperature PEMFCs (HT-PEMFCs) exhibit enhanced reaction kinetics and improved tolerance to impurities, which can be attributed to effective thermal management. By optimizing the thermal conditions, the fuel-cell performance could be significantly enhanced, leading to higher energy conversion efficiencies. This is mostly relevant in usages in which rapid response times, as well as high power densities, are required, such as in automotive and portable power applications.

In addition to enhancing performance, effective thermal management also plays an essential part in prolonging the PEMFCs' lifespan. As noted by Hao et al. [62], the ability to manage heat effectively can mitigate thermal stresses that arise from temperature fluctuations during operation. These stresses could result in mechanical failures and cell components' degradation over time. By implementing robust thermal management systems, such as advanced cooling techniques or heat recovery systems, the durability of PEMFCs can be significantly enhanced, thereby reducing maintenance costs and improving reliability.

Furthermore, the integration of thermal management systems can facilitate better water management within the fuel cell. Feng et al. [63] discuss how effective thermal management can help control the water balance in PEMFCs, which is critical to maintain optimal hydration levels in a membrane. Proper water management not only supports proton conduction but also prevents issues such as flooding or drying out, which can severely impact performance. In line with the discussion on proper water management in PEMFC membranes, Figure 6a shows that water uptake decreases with increasing filler content at both 30 °C and 80 °C, while Figure 6b demonstrates a similar decreasing trend in swelling ratio, indicating that filler incorporation helps control hydration and dimensional stability of the membrane.

In conclusion, the significance of PEMFCs' thermal management cannot be overstated. It is essential for optimizing performance, enhancing efficiency, prolonging lifespans, and ensuring effective water management. At high current densities—necessary for large-scale commercialization—excessive liquid water accumulation can cause flooding, block gas diffusion, and accelerate performance degradation. Effective thermal management supports proper water removal and distribution, ensuring stable operation and sustained high-performance output of PEMFCs. As research continues to advance in this field, the development of innovative thermal management strategies will be critical for the successful commercialization and deployment of PEMFC technologies in various applications.

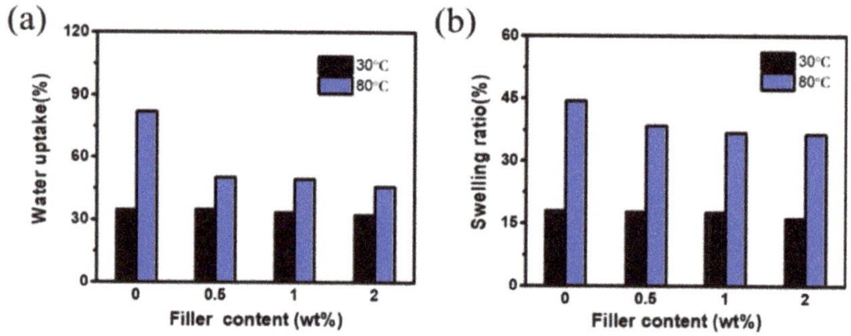

Figure 6. Sulfonated poly(arylene ether nitrile (SPEN) and SPEN/GO-Ca^{2+}-MWCNT [63].

2.2.2. Significance of Thermal Management of Solid-Oxide Fuel Cells

The importance of solid-oxide fuel cells' (SOFCs') thermal management is critical to ensure the optimal performance, efficiency, and longevity of these advanced energy conversion systems. SOFCs function at relatively high temperatures, normally from 600 °C to 1000 °C, which presents unique challenges and opportunities for thermal management strategies.

SOFCs consist of a solid ceramic electrolyte positioned between porous anode and cathode layers. Based on the type of species transported through the electrolyte, SOFCs can be classified into three categories: proton-conductive, O_2-ion-conductive, and mixed-ion-conductive. In O_2-ion-conductive SOFCs (Figure 7a), O_2 anions transport from the cathode to the anode, producing steam or CO_2 at the anode, which helps resist carbon deposition by promoting reforming reactions when fueled by hydrocarbons. Conversely, in proton-conductive SOFCs, only protons are transferred to the cathode for oxidation through oxygen (Figure 7b), increasing the risk of coking because of higher CO levels in the anode [64].

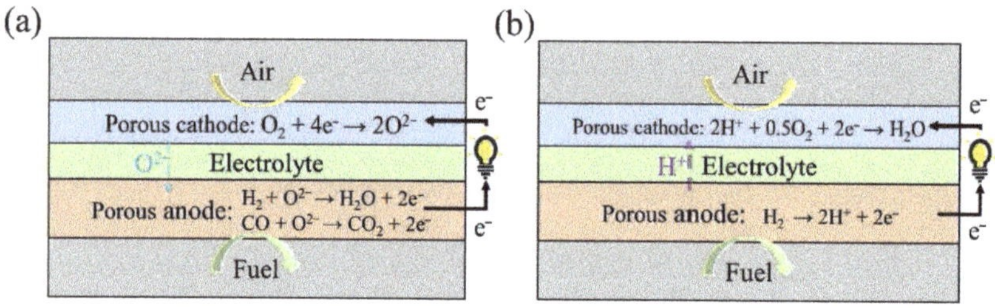

Figure 7. (**a**) O_2-ion-conductive SOFC and (**b**) proton-conductive SOFC [64].

One of the primary reasons why thermal management is crucial in SOFCs is the need to maintain the working temperature in a specific range to optimize electrochemical reactions. High operating temperatures enhance the solid electrolyte ionic conductivity, thereby improving a fuel cell's overall efficiency [64]. Nonetheless, excessive heat can lead to thermal stresses that may damage the cell components, including the cracking and delamination of the electrolyte and electrodes [65]; effective thermal management systems are essential to mitigate these risks and ensure stable operation.

Moreover, thermal management directly influences the efficiency of energy conversion in SOFCs. Maintaining optimal thermal conditions can significantly enhance the thermodynamic efficiency of the electrochemical reactions, allowing for better energy utilization. This is particularly important in applications where SOFCs are integrated into hybrid systems or combined heat and power (CHP) configurations since effective heat recovery can improve overall system efficiency [66]. By managing heat effectively, SOFCs can convert a larger portion of chemical energy in fuels into usable electrical energy, thus maximizing their performance.

In addition to performance and efficiency, thermal management is vital for the durability and reliability of SOFCs. Prolonged exposure to thermal cycling can result in the degradation of material and fuel-cell components' reduced lifespans [10]. Implementing robust thermal management strategies can help minimize the effects of thermal fatigue, thereby extending the operational life of SOFC systems. This is particularly important in commercial applications where reliability and maintenance costs are critical factors.

Furthermore, effective thermal management facilitates better water and fuel management within the SOFC. Temperature control can influence the fuel-cell water vapor content, which is vital for preserving optimal electrolyte hydration levels. Proper hydration is crucial for ensuring high ionic conductivity and preventing issues such as flooding or drying out, which can severely impact performance [67].

In conclusion, the significance of SOFCs' thermal management is multifaceted, encompassing aspects of performance optimization, efficiency enhancement, durability, and effective resource management. As research continues to advance, innovative thermal management solutions will be important for successful SOFC use in various applications, including distributed generation and renewable energy integration. Table 2 summarizes the importance of PEMFCs' and SOFCs' thermal management.

Table 2. Critical role of thermal management in ensuring the efficiency, performance, and longevity of both PEMFCs and SOFCs.

Aspect	PEMFCs	SOFCs
Key Importance	• Ensures optimal operating temperature for electrochemical reactions and membrane hydration.	• Maintains high operating temperatures (600–1000 °C) for ionic conductivity and efficient reactions.
Impact on Performance	• Prevents thermal degradation of PEM.	• Enhances electrochemical reaction efficiency by maintaining high temperatures.
	• Improves overall cell performance by optimizing reaction kinetics.	• Increases thermodynamic efficiency in energy conversion, especially in CHP systems.
Effect on Efficiency	• Effective heat management enhances energy conversion efficiency, particularly in high-power applications.	• Efficient heat management allows for better fuel-to-energy conversion, maximizing usable electrical energy.
Longevity and Durability	• Mitigates thermal stresses, reducing mechanical failures and component degradation.	• Reduces thermal cycling effects that can cause material degradation, extending system lifespan.
Water Management	• Facilitates better water management, ensuring membrane hydration and preventing flooding or drying out.	• Helps control water vapor content, ensuring proper hydration of the electrolyte and avoiding performance issues.
Advanced Strategies	• Integrates advanced cooling techniques, such as heat recovery systems, to enhance reliability and reduce maintenance costs.	• Implements robust thermal management systems to prevent damage to components like the electrolyte and electrodes.
Commercial Relevance	• Critical for high-performance, long-lifespan applications like automotive and portable power systems.	• Vital for reliable and efficient operation in hybrid systems, distributed generation, and renewable energy integration.

2.3. *Methods for Thermal Management of Fuel Cells*

Thermal management is a critical aspect of fuel-cell technology, especially for PEMFCs and SOFCs. Effective thermal management techniques are essential for maintaining optimal operating temperatures, enhancing efficiency, and prolonging fuel-cell lifespans. Various strategies are being developed to address the drawbacks related to heat generation and dissipation in these systems.

One prominent technique for thermal management in PEMFCs is the use of passive thermal management systems. According to Idi [68], these systems rely on the natural convection of air and the thermal properties of materials to dissipate heat without the need for mechanical components Idi (2023). This approach can be particularly beneficial in reducing complexity and improving reliability. Furthermore, phase-change materials (PCMs) have gained attention for their capability to absorb and release heat in phase transitions, thus offering a stable thermal environment for the fuel cell [69]. The integration of PCMs can help mitigate temperature fluctuations and improve overall thermal stability.

Active thermal management techniques are also widely employed in fuel cells. These include the use of liquid cooling systems, in which a coolant circulates through the fuel-cell stack to absorb excess heat. This method is effective in maintaining the desired operating temperature and preventing overheating, as highlighted by Kim and Yoo [70], who emphasize the importance of maintaining optimal temperature ranges for improved performance and longevity. Furthermore, advanced control strategies, such as adaptive control systems, have been developed to dynamically adjust the cooling rates based on real-time temperature measurements, ensuring efficient thermal management under varying load conditions [71].

Another innovative approach involves flow channel designs' optimization within the fuel cell. Research by Novotný et al. indicates that the shape, along with the distribution of flow channels, influences heat and mass transport within the cell [72]. By optimizing these designs, it is possible to enhance the dissipation of heat and improve the fuel-cell overall thermal management. Additionally, the use of serpentine flow patterns has been shown to facilitate better thermal and water management, addressing common issues such as flooding and dehydration [73].

In SOFCs, thermal management techniques often focus on the integration of heat exchangers and afterburners to recover waste heat and improve overall system efficiency. As discussed by Li [74], the design of these thermal management subsystems is crucial for maximizing energy utilization and ensuring stable operation. The implementation of heat recovery systems can significantly enhance the SOFCs' overall efficiency, making such fuel cells more viable for commercial purposes.

All in all, various techniques are used for fuel-cell thermal management, such as passive and active systems, flow channel optimization, and the integration of heat recovery technologies. These strategies are essential for maintaining optimal operating conditions, enhancing efficiency, and prolonging fuel-cell durability. As studies continue to advance, the development of innovative thermal management solutions will be vital, aiming at successful fuel-cell commercialization technologies in diverse applications.

2.3.1. Thermal Management of Proton-Exchange Membrane Fuel Cells (PEMFCs)

The thermal management of PEMFCs is critical for optimizing their performance, efficiency, and lifespan. The effective control of temperature within these systems ensures that electrochemical reactions take place at optimal rates while preventing damage to the components caused by excessive heat. Various techniques are being developed to address the thermal management drawbacks pertaining to PEMFCs.

One of the primary methods for thermal management in PEMFCs is the use of active cooling systems. These systems typically involve circulating a coolant, such as water or a specialized thermal fluid, through channels adjacent to the stack of fuel cells. This method helps dissipate excess heat generated during operation. According to Louis [75], maintaining the stack temperature within an optimal range is essential, as excessive temperatures can lead to catalyst deactivation and irreversible damage to the membrane. Active cooling systems can be designed to respond dynamically to load changes and temperature, ensuring that the fuel cell operates efficiently under varying conditions. Figure 8 shows the optimized controller exhibiting a smaller overshoot compared to the traditional controller, with a smoother and less volatile change curve. This contributes to more stable fuel-cell operation at the step load current. Specifically, Figure 8a illustrates the stack temperature response, where the optimized controller reaches the target temperature with less fluctuation and reduced oscillations compared to the conventional controller. Figure 8b presents the coolant outlet temperature response, demonstrating that the optimized controller maintains tighter regulation with smaller deviations, ensuring better thermal management and improved overall system stability.

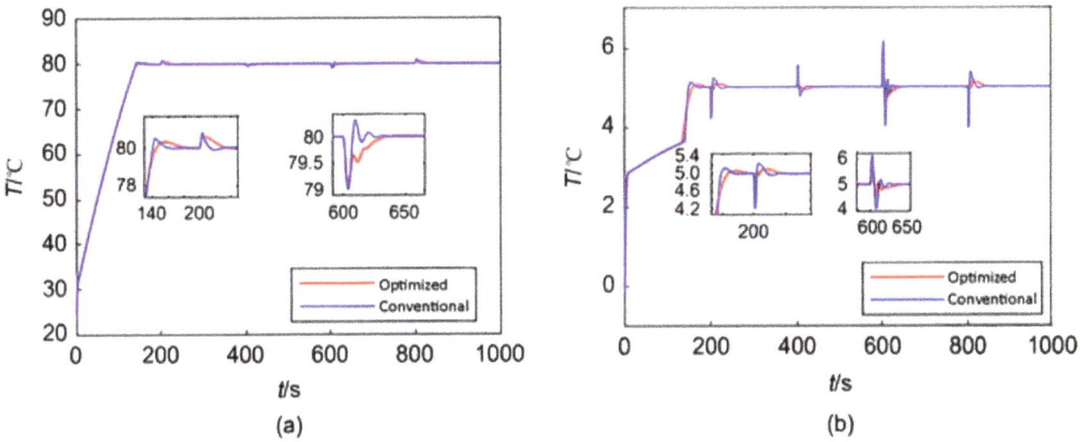

Figure 8. Temperature difference between stack and coolant at step load current [75].

Other than active cooling, passive thermal management techniques are also employed. These methods rely on the natural heat dissipation properties of materials and the fuel-cell design components. For instance, the use of materials with high thermal conductivity can facilitate heat transfer away from critical cell areas, thereby maintaining a more uniform distribution of temperature [76]. Moreover, the design of flow channels within the fuel cell can significantly impact thermal management. Research by Deng et al. [77] highlights that optimizing the design of a flow field could improve heat and mass transfer. This is important for maintaining the fuel-cell performance.

Water management is another critical aspect of thermal management in PEMFCs. The membrane of proton exchange requires a certain level of hydration to preserve its conductivity, as well as performance. However, excessive water accumulation can lead to flooding, which impairs gas transport and overall efficiency [78]. Techniques such as bypass configurations for membrane humidifiers have been developed to enhance water management and, consequently, thermal stability [76]. Effective water management not only supports the thermal balance but also enhances the fuel-cell electrochemical performance.

Furthermore, advancements in membrane materials have contributed to improved thermal management in PEMFCs. High-temperature PEMs (HT-PEMs) have gained interest because of their capability to function at elevated temperatures while maintaining performance [61]. These membranes facilitate better thermal and water management, allowing for more efficient operation and integration with fuel processing units. The development of new materials, such as phosphoric acid-doped polybenzimidazole membranes, has shown promise in enhancing both thermal stability and proton conductivity [79].

In conclusion, PEMFCs' thermal management is essential to optimize performance, ensuring longevity and facilitating effective water management. A mix of active and passive cooling methods, optimized flow channel designs, and advanced membrane materials plays an important part in keeping the operational PEMFCs' efficiency. As research continues to advance in this field, the development of novel thermal management approaches will be vital for the successful commercialization of fuel-cell technologies in various applications.

Active Cooling Methods

Thermal management is a critical aspect of proton-exchange membrane fuel cells (PEMFCs), particularly concerning the implementation of active cooling methods. Efficient thermal regulation guarantees that the fuel cell's temperature stays within the ideal range, which is vital for enhancing performance, efficiency, and durability. A variety of active cooling strategies have been introduced to tackle the issues linked to heat production in PEMFCs.

One of the most common active cooling methods is liquid cooling, in which a coolant, typically water or a water–glycol mixture, circulates via channels adjacent to the stack of fuel cells. This method effectively removes surplus heat that is generated during operation. According to Louis [75], maintaining the stack temperature within an optimal range is crucial, as excessive temperatures can lead to catalyst deactivation and irreversible damage to the membrane. Liquid cooling systems can be designed to dynamically adjust the flow rate of the coolant based on real-time temperature measurements, ensuring efficient thermal management under varying load conditions.

In addition to liquid cooling, air-cooling systems are also utilized in PEMFCs. These systems rely on forced air circulation to dissipate heat from the fuel-cell stack. Gan et al. [80] highlight that air-cooled systems are compact and require low energy consumption, thus suitable for usages in which weight and space are critical factors. The efficiency of air cooling can be enhanced by optimizing the design of the airflow paths and incorporating features such as heat exchangers to improve heat transfer rates.

Another innovative approach to active thermal management is the utilization of phase-change materials (PCMs). PCMs can absorb and release heat all through phase transitions, offering a stable thermal environment. Yan et al. [69] discuss the potential advantages of integrating PCMs into PEMFC systems, noting that they can help regulate temperature fluctuations and improve overall thermal stability. This method is particularly beneficial in applications with varying power demands, as it allows for the absorption of excess heat during peak loads and the release of stored heat during lower demand periods. It is important to note that PCMs can be deployed in both active and passive configurations. When integrated with forced fluid circulation or other auxiliary systems, PCMs form part of an active cooling approach, rapidly absorbing and redistributing heat. In contrast, when used without mechanical assistance, PCMs act as a passive system, relying solely on natural heat absorption and release during phase transitions. The inclusion of PCMs in both active and passive sections reflects these distinct operational contexts.

Moreover, advanced control strategies play a significant part in PEMFCs' thermal management. For instance, the implementation of multivariable control systems can optimize the cell cooling circuit, ensuring that temperature is maintained at an optimal level while minimizing energy consumption [81]. These control systems could adjust to variations in operational conditions, enhancing the reliability and efficiency of the thermal management process. Figure 9 depicts the illustration of a PEMFC stack-cooling system.

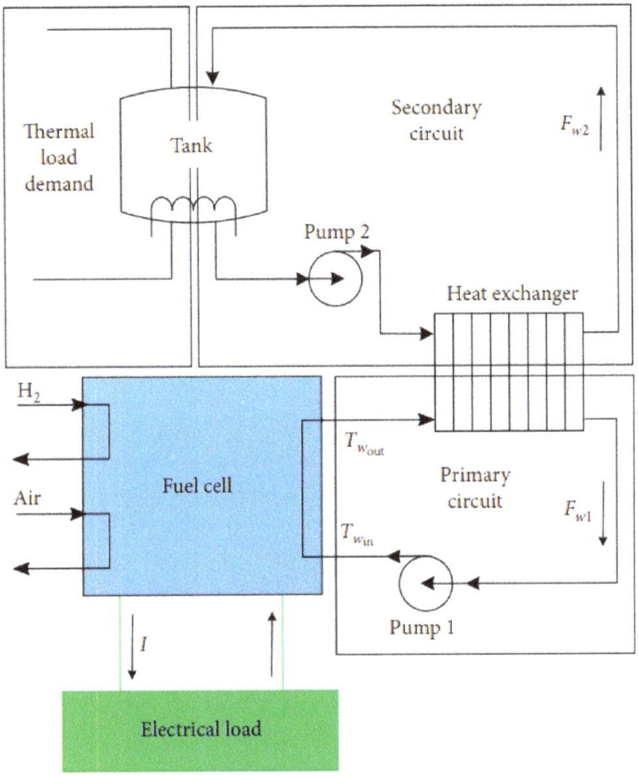

Figure 9. PEMFC stack cooling system for the PEMFC stack [81].

Furthermore, hybrid energy storage systems integrating supercapacitors with lithium-ion batteries have shown strong potential for improving power quality and mitigating the thermal load in renewable-integrated fuel-cell systems. Supercapacitors offer high power density and rapid charge/discharge capabilities, complementing the higher energy density but slower response of batteries. Experimental findings [82] demonstrate that such hybrid configurations can smooth renewable power fluctuations, maintain stable state-of-charge (SOC) profiles, and minimize peak thermal stress on fuel-cell stacks.

For example, in hybrid battery–supercapacitor systems, supercapacitors can achieve response times in the order of milliseconds and cycle lives exceeding one million cycles, while lithium-ion batteries typically provide 2000–5000 deep cycles with a higher energy capacity. Maintaining SOC in the optimal range (40–80%) through coordinated control reduces both battery degradation and fuel-cell thermal excursions.

In conclusion, active cooling methods are essential for effective PEMFCs' thermal management. Techniques like liquid cooling, air cooling, and the use of phase-change materials, along with advanced control strategies, contribute to maintaining optimal operating temperatures, thereby enhancing performance and prolonging the lifespan of PEMFCs. As research continues to advance in this area, the development of innovative thermal management solutions will be vital for the successful commercialization of fuel-cell technologies in various applications.

Passive Cooling Methods

PEMFCs' thermal management is important for maintaining optimal operating conditions, enhancing performance, and prolonging the fuel cells' lifespan. Among the various thermal management strategies, passive cooling methods have gained attention due to their simplicity, reliability, and energy efficiency. These methods utilize natural processes to dissipate heat without the need for external power sources, leading them to be desirable for usages in which energy conservation is critical.

One of the primary passive cooling techniques employed in PEMFCs is the utilization of phase-change materials (PCMs). PCMs can absorb and then release thermal energy at phase transitions, which assists in stabilizing the fuel-cell temperature. According to Wang et al. [83], integrating PCMs into the cooling channels of PEMFCs can significantly enhance thermal management by providing a buffer against temperature fluctuations. This method is particularly beneficial during periods of high thermal load, as the PCM can absorb excess heat and release it when the temperature drops, thereby maintaining a more uniform distribution of temperature in the fuel cell.

Another effective passive cooling strategy involves the use of enhanced thermal conductivity materials, such as nanofluids. Zakaria [84] discusses how incorporating nanofluids into the coolant can improve the thermal physical properties, thereby enhancing heat transfer efficiency in liquid-cooled PEMFCs. This approach allows for a reduction in the size of heat exchangers while maintaining effective cooling performance. The ability to utilize nanofluids not only improves thermal management but also contributes to the overall cell compactness.

Additionally, the fuel-cell components design plays an important part in passive cooling. The incorporation of heat sinks, as well as optimized flow channel geometries, can facilitate better heat dissipation. For instance, the arrangement of flow channels can be designed to maximize surface area exposure to the coolant, enhancing convective heat transfer [51]. This design consideration is vital for ensuring that heat is effectively removed from critical areas of the fuel cell, thus preventing overheating and maintaining optimal performance.

Moreover, natural convection can be harnessed as a passive cooling mechanism. In configurations where the fuel cell is oriented to allow for natural airflow, heat can be dissipated through convection without the need for mechanical fans or pumps. This method is particularly advantageous in applications where energy efficiency is paramount, as it reduces the whole system of energy consumption [85].

All in all, passive cooling methods are integral to the PEMFCs' thermal management. Techniques such as phase-change materials, enhanced thermal conductivity materials, optimized flow channel designs, and natural convection contribute to maintaining optimal operating temperatures, improving efficiency, and prolonging the lifespan of PEMFCs. As research continues to advance in this area, the development of innovative passive cooling strategies will be essential for successful fuel cell commercialization technologies in various applications.

Bipolar Plates for Proton-Exchange Membrane Fuel Cells

The thermal management of PEMFCs is important for optimizing fuel cell performance, as well as ensuring its longevity. Among the various elements involved in thermal management, bipolar plates play a significant role. These plates not only serve as current collectors but also facilitate heat dissipation and manage the flow of reactants and coolant within the fuel cell stack. The design and material selection of bipolar plates are therefore critical for effective thermal management.

Bipolar plates are typically made from materials that possess good electrical conductivity and thermal conductivity. Common materials include graphite and metal composites, which can efficiently conduct electricity while also allowing for effective heat transfer. The structural design of bipolar plates can significantly influence the thermal management capabilities of PEMFCs. Innovations such as tree-shaped fractal designs and tapered dual serpentine channels have been shown to enhance heat dissipation and promote uniform distribution of temperature across the membrane electrode assembly (MEA). This uniformity is essential for maintaining optimal operating conditions and preventing localized overheating, which can lead to performance degradation.

Moreover, the integration of cooling channels within the bipolar plates is a key aspect of thermal management. These channels allow for the circulation of coolant, which absorbs excess heat generated during the electrochemical reactions. Optimizing the flow paths within the bipolar plates can enhance heat transfer efficiency. By designing the cooling channels to maximize surface area and minimize flow resistance, the overall fuel-cell thermal management could be considerably enhanced.

In addition to their thermal management functions, bipolar plates also provide stack mechanical stability in the fuel cell. The ability of bipolar plates to withstand operational stresses while maintaining structural integrity is vital for the PEMFCs' durability. The materials' mechanical properties used in bipolar plates must be carefully considered to ensure that they can endure the thermal cycling and pressure variations that occur during operation. This durability is important for maintaining the fuel cell performance and reliability over time.

Furthermore, the surface treatment of bipolar plates can enhance their thermal management capabilities. Techniques such as coating the plates with materials that improve thermal conductivity or applying hydrophilic treatments can facilitate better heat transfer and water management within the fuel cell. Effective water management is critical in PEMFCs, as such a strategy helps uphold the membrane hydration while preventing flooding, which can adversely affect performance.

In conclusion, bipolar plates are integral to PEMFCs' thermal management. Their design, material selection, and cooling channel integration play a crucial part in optimizing heat dissipation and maintaining a uniform distribution of temperatures in a fuel-cell stack. As research continues to advance in this area, innovative designs and materials for bipolar plates will be important to improve the performance and durability of PEMFC technologies.

Heat Recovery Opportunities for Proton-Exchange Membrane Fuel Cells

Thermal management in PEMFCs is important to optimize the performance and ensure the longevity of the fuel cell system. One promising approach to enhancing thermal management is through heat recovery options, which allow for the efficient utilization of waste heat produced during electrochemical reactions. This not only improves the overall cell efficiency but also contributes to energy sustainability.

One primary method for PEMFCs' heat recovery involves the integration of heat exchangers. These devices can capture waste heat produced from the cell and then transfer it to other systems, such as water heating or space heating applications. According to Wang [86], the integration of heat exchangers could significantly enhance the overall system energy efficiency by utilizing the thermal energy that would otherwise be lost. This strategy is predominantly useful in applications in which both electricity and thermal energy are required, such as in CHP systems.

Another effective heat recovery strategy is using thermoelectric generators (TEGs). These devices transform temperature differences into electrical energy, allowing for waste heat recovery from PEMFC. As noted by Pourrahmani et al. [87], TEGs could be attached to a fuel cell to harness low-grade waste heat, thereby increasing the overall energy output. This method not only improves fuel-cell efficiency but also offers an additional source of power, which can be particularly advantageous in remote or off-grid applications.

Quantitatively, thermoelectric generators (TEGs) typically achieve thermal-to-electrical conversion efficiencies of 5–8% for low-grade waste heat (<200 °C) available from PEMFCs, and 12–15% for higher-grade heat (>500 °C) from SOFC exhaust streams. For example, in a PEMFC operating at ~70 °C with a 40–50 °C temperature gradient across the TEG module, each module can generate approximately 5–20 W of electrical power, depending on module design, thermal interface quality, and heat flux. These values underscore that, while TEGs cannot recover all waste heat, their ability to convert a fraction into useful electricity can improve system-level energy efficiency in both stationary and portable applications.

Furthermore, the design of the fuel-cell stack itself can facilitate heat recovery. For instance, optimizing the flow field configuration can enhance fuel cell heat transfer qualities. Research by Tolj et al. [88] indicates that specific flow field designs can improve heat distribution and reactants, thereby maximizing thermal energy utilization. This optimization could result in better thermal management and enhance the whole PEMFC performance. Four distinct geometries were created in Design Modeler, as shown in Figure 10. Model 1 serves as the baseline, calibrated to match an experimental PEM fuel cell, while Models 2, 3, and 4 represent various optimized configurations.

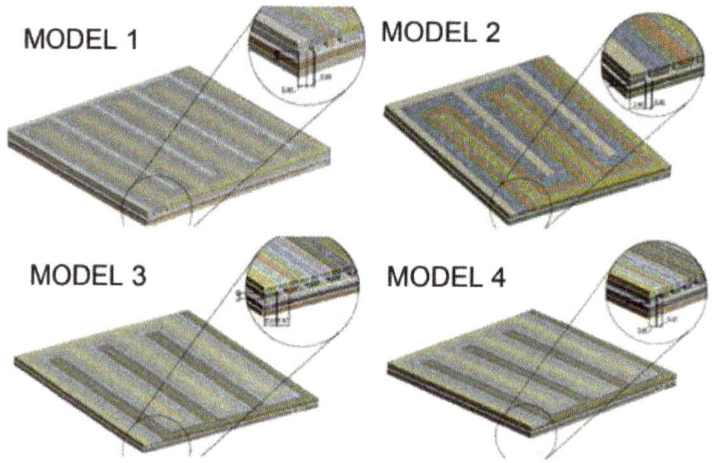

Figure 10. The computational domains for each of the models [89].

Additionally, the utilization of phase-change materials (PCMs) in conjunction with PEMFCs presents another avenue for heat recovery. PCMs can absorb excess heat during operation and release it when the temperature drops, helping maintain a stable thermal environment. Yan et al. [69] highlight the potential of PCMs to enhance thermal stability and improve the efficiency of PEMFCs by effectively managing the heat generated during electrochemical reactions. This approach can be particularly useful in applications with fluctuating power demands, as it allows for better temperature regulation.

In conclusion, the options of heat recovery for PEMFCs play a vital role in enhancing thermal management and improving overall system efficiency. The integration of heat exchangers, thermoelectric generators, optimized flow field designs, and phase-change materials can significantly contribute to the effective utilization of waste heat. As research continues to advance in this area, the development of innovative heat recovery strategies will be essential for the successful commercialization of PEMFC technologies in various applications. Table 3 summarizes the thermal management techniques in PEMFCs.

Table 3. Key techniques and strategies in PEMFCs thermal management.

Focus	Technique	Type	Description	Advantages
Thermal Management of PEMFCs	Active Cooling Systems	Active	Circulating coolant through channels to dissipate excess heat.	Prevents overheating, catalyst deactivation.
	Passive Cooling Systems	Passive	Using materials with high thermal conductivity and optimized designs to enhance natural heat dissipation.	Energy-efficient, improves reliability.
	Water Management Systems	Active	Humidifier bypass configurations to improve hydration and water distribution.	Enhances electrochemical performance and thermal balance.
	High-Temperature Proton Membranes	Advanced Materials	HT-PEMs maintain performance at elevated temperatures, improving thermal management.	Enhances thermal and water management, durable.
Active Cooling Methods	Liquid Cooling	Active	Water/glycol coolant circulates through stack channels.	Efficient heat removal, adaptable to load.
	Air Cooling	Active	Forced air dissipates heat from the fuel cell stack.	Compact, low energy consumption.
	PCMs	Active	Absorb/release heat during phase transitions to regulate temperature.	Stabilizes temperature during fluctuating loads.
Passive Cooling Methods	Phase-Change Materials (PCMs)	Passive	Stabilizes fuel-cell temperature by absorbing and releasing heat.	Reduces temperature fluctuations, improves stability.
	Nanofluids	Passive	Improves thermal conductivity in coolants, enhancing heat transfer.	Increases heat transfer efficiency, compact system.
	Natural Convection	Passive	Heat dissipation through natural airflow without mechanical assistance.	Energy-efficient, reduces power consumption.
Bipolar Plates	Bipolar Plate Design	Design Innovation	Tree-shaped fractal designs and serpentine channels for better heat dissipation and reactant flow.	Ensures uniform temperature, prevents overheating.
	Material Selection	Advanced Materials	Graphite and metal composites improve heat and electrical conductivity.	Enhances durability and performance.
	Surface Treatments	Design Innovation	Hydrophilic coatings enhance heat transfer and water management.	Improves thermal and water management.
Heat Recovery Options	Heat Exchangers	Heat Recovery	Transfers waste heat to other systems, such as space or water heating.	Improves overall energy efficiency in CHP systems.
	Thermoelectric Generators (TEGs)	Heat Recovery	Converts waste heat into electrical energy, enhancing energy output.	Harnesses low-grade waste heat, increases efficiency.
	Optimized Flow Field Designs	Design Innovation	Improves heat transfer by optimizing the flow of reactants and heat in the fuel cell stack.	Maximizes heat utilization, improves performance.
	PCMs for Heat Recovery	Heat Recovery	Absorbs excess heat and releases it during low-demand periods to maintain thermal balance.	Enhances efficiency and thermal stability.

2.3.2. Thermal Management of Solid-Oxide Fuel Cells (SOFCs)

Thermal management in SOFCs is a critical factor influencing their efficiency, performance, and longevity. SOFCs work at high temperatures, normally between 600 °C and 1000 °C, which allows for efficient electrochemical reactions but also presents challenges related to heat dissipation and thermal stress management. Effective thermal management strategies are essential to ensure optimal operating conditions and to avoid damage to the components of a fuel cell.

One of the primary methods for managing heat in SOFCs is through the design and integration of cooling systems. These systems can include active cooling methods, such as liquid cooling, where a coolant flows via channels next to the fuel cell stack to absorb surplus heat. According to Xu et al. [64], optimizing the cooling system design is crucial to maintain the temperature within an acceptable range, thus improving overall fuel-cell efficiency. The cooling channels should be cautiously designed to guarantee uniform heat distribution and prevent localized overheating, which can lead to material degradation.

In addition to active cooling, passive thermal management techniques are also employed in SOFCs. These methods leverage the natural heat dissipation properties of materials and the fuel cell design components. For instance, the use of materials with high thermal conductivity can facilitate heat transfer away from cell-critical areas, helping keep a stable operating temperature. The fuel cell structural design, including electrodes and electrolyte arrangement, can significantly impact thermal management. Sai et al. [56] emphasize that the geometry of the interconnects and electrodes plays an important part in managing thermal stresses and providing efficient heat dissipation.

Heat recovery systems are another important aspect of thermal management in SOFCs. By capturing waste heat generated during operation, these systems can improve overall energy efficiency. For example, integrating heat exchangers can allow for the recovery of thermal energy for use in other processes, such as heating water or providing space heating. This not only improves the overall SOFC efficiency but also promotes energy conservation, as well as sustainability.

Furthermore, the use of advanced materials in the construction of SOFC components can enhance thermal management. Research into new materials, such as advanced ceramics and metal alloys, is ongoing to improve thermal stability and reduce thermal expansion mismatches between different components. These advancements can help mitigate thermal stresses and improve fuel-cell durability, ensuring reliable and durable operation [89].

In conclusion, effective SOFC thermal management is important for optimizing performance, enhancing efficiency, and prolonging the lifespan of the system. The integration of active and passive cooling methods, the careful design of fuel cell components, and the implementation of heat recovery systems are all critical strategies in managing heat within SOFCs. As research continues to advance in this field, the development of innovative thermal management solutions will be vital for the successful commercialization of SOFC technologies in various applications.

Heat Recovery Opportunities for Solid-Oxide Fuel Cells

SOFCs' heat recovery is essential for improving the overall efficiency of these systems, particularly in applications where both electricity and thermal energy are required. SOFCs operate at high temperatures, normally between 600 °C and 1000 °C, which allows for significant amounts of waste heat to be generated during operation. Effectively capturing and utilizing this waste heat can lead to improved system performance and reduced energy costs.

One of the most common methods for heat recovery in SOFCs is the integration of combined heat and power (CHP) systems [90–93]. In a CHP configuration, waste heat produced by the SOFC is recovered and utilized for the purpose of heating applications, including water heating or space heating. Wang et al. [24] highlight that this integration could considerably enhance the system efficiency, achieving total efficiencies of up to 90% when both electrical and thermal outputs are considered. Electrolyzer efficiency is highly temperature-dependent; for example, PEM electrolyzers show improved kinetics at moderate temperatures (50–80 °C) but may suffer from membrane dehydration or catalyst degradation above this range. When coupled with fuel cells in renewable-powered systems, thermal management must balance the heat recovery benefits for electrolyzer preheating with the risk of overheating, ensuring stable hydrogen production rates and system longevity. The high-quality heat produced by SOFCs is particularly suitable for such applications, making CHP systems a viable option for maximizing energy utilization.

Another effective heat recovery strategy involves the use of heat exchangers. These devices can capture waste heat coming from SOFC exhaust and then transfer it to other processes, such as preheating feedwater or supplying heat to absorption chillers for cooling applications. However, Müller et al. [94] do not directly address the implementation of heat exchangers in SOFC systems, focusing instead on waste heat recovery for dehydrogenation processes.

Additionally, the design of the SOFC system can be optimized to facilitate heat recovery. For instance, Duong et al. [95] discuss the potential of integrating thermal storage systems with SOFCs to capture surplus heat during low-demand periods and release such excess heat during peak demand. This integration allows for better management of thermal energy and can help balance the supply and demand for heat in various applications. By utilizing thermal storage, SOFC systems can operate more flexibly and efficiently, adapting to varying energy needs.

Moreover, the use of advanced materials in the construction of SOFC components can enhance heat recovery capabilities. For example, the development of materials with improved thermal conductivity can facilitate better heat transfer within the system, allowing for more effective capture and utilization of waste heat. In conclusion, SOFCs' heat recovery is vital for enhancing overall system efficiency and sustainability. The integration of CHP systems and optimization of system design contribute to effective heat recovery strategies. As R&D in this field are continuing to progress, the implementation of innovative heat recovery solutions will be vital for the successful commercialization of SOFCs in various applications.

Possibilities to Decrease the Operating Temperature of Solid-Oxide Fuel Cells

The reduction in SOFCs' working temperatures is a significant area of research aimed at improving efficiency, reducing costs, and enhancing the longevity of these systems. Traditional SOFCs work at high temperatures, normally between 600 °C and 1000 °C, which can result in the degradation of material and increased thermal stresses. Therefore, various strategies have been and are being developed to reduce the working temperature, primarily focusing on advanced materials and innovative designs.

One promising approach to lowering the working temperature of SOFCs is the development of Intermediate Temperature SOFCs (IT-SOFCs). These cells typically work at temperatures ranging from 400 °C to 800 °C, significantly lower than conventional SOFCs. According to Mazlan et al. [96], the use of doped barium cerate ($BaCeO_3$) and barium zirconate ($BaZrO_3$) as electrolytes in IT-SOFCs has shown high proton conductivity and chemical stability, causing them to be suitable candidates for lower-temperature applications. This shift not only reduces the thermal stresses experienced by the cell components but also allows for the use of less expensive materials that can withstand lower operating temperatures.

From a thermal management perspective, reducing the operating temperature from the conventional ~1000 °C to intermediate (600–800 °C) or low-temperature regimes (<600 °C) decreases the overall heat rejection requirement, thereby lowering the size and complexity of heat exchangers and auxiliary cooling systems. The smaller thermal gradients between the active layers and the balance-of-plant components also reduce thermal stresses, which mitigates the risk of mechanical failure during start-up and shut-down cycles. Furthermore, lower peak temperatures improve insulation efficiency and allow more compact system designs, enhancing overall reliability and operational stability in both stationary and mobile applications.

Another effective strategy involves the optimization of electrode materials and structures. Zamudio-García et al. [97] discuss the use of tunable electrode architectures, which can enhance the electrochemical performance at lower temperatures. By adjusting the composition and microstructure of the electrodes, researchers can improve the reaction kinetics, thereby enabling efficient operation at reduced temperatures. This approach helps mitigate the sluggish electrochemical activity that often limits performance in traditional SOFCs.

Nanotechnology plays a critical part in decreasing the working temperatures of SOFCs. Lee et al. [98] highlight that the application of nanostructured materials can significantly enhance the performance of SOFCs while allowing for thinner electrolytes and smaller electrode particles. These advancements not only improve ionic conductivity but also reduce the thermal mass of the cell, facilitating quicker thermal responses and enabling operation at lower temperatures without sacrificing efficiency. Figure 11 depicts a thin-film SOFC, where the gas-tight electrolyte diameter columns increase with thickness. The white arrow designates the column diameter, which expands as the thickness gets bigger, resulting in residual compressive stress within the film.

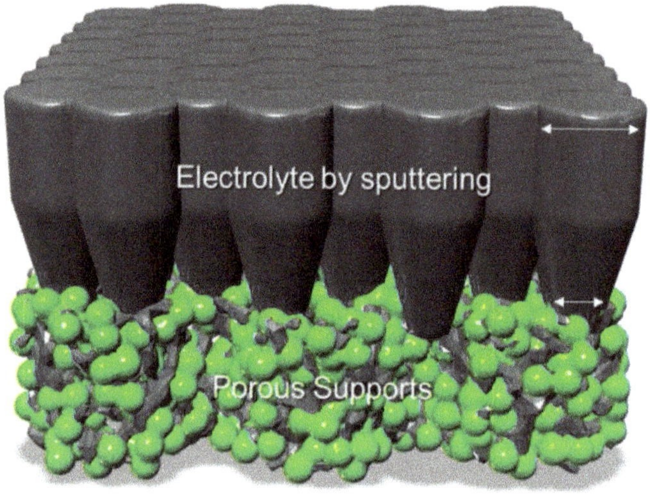

Figure 11. A thin-film SOFC where the gas-tight electrolyte diameter columns increase in thickness [98].

Furthermore, the integration of metal-supported SOFCs (MS-SOFCs) offers another pathway for reducing operating temperatures. These cells utilize a metal substrate that provides excellent thermal conductivity and mechanical strength, allowing for rapid thermal cycling and improved thermal management. Du et al. [99] note that the use of metal supports can enhance SOFCs' thermal shock resistance, allowing them to operate effectively under lower temperatures while maintaining performance. This innovation opens up new possibilities for applications in mobile and portable power applications.

Besides material advancements, the fuel cell design itself can be optimized to facilitate lower operating temperatures. For instance, the implementation of advanced thermal management strategies, such as improved heat recovery systems, can help maintain optimal thermal conditions within the fuel cell. By effectively capturing and utilizing waste heat, it is possible to lower the operational temperature while enhancing overall system efficiency.

In conclusion, lowering the SOFCs' working temperature is a multifaceted challenge that could be tackled through advanced materials' development, innovative electrode designs, and optimized system configurations. The transition to IT-SOFC, the application of nanotechnology, and the use of metal supports are all promising strategies that could improve the performance and longevity of SOFCs while enabling their broader application in various energy systems. Table 4 summarizes SOFCs' thermal management.

Table 4. Main strategies and considerations for thermal management, heat recovery, and temperature reduction in SOFCs.

Focus	Key Points
SOFCs Thermal Management	• SOFCs work at high temperatures (600 °C to 1000 °C), requiring effective thermal management. • Active cooling systems (e.g., liquid cooling) help manage heat and prevent localized overheating. • Passive methods include using materials with high thermal conductivity and optimizing component design. • Heat recovery systems capture waste heat for increased efficiency.
Heat Recovery Options for SOFCs	• CHP systems recover waste heat for heating applications, improving overall efficiency. • Heat exchangers capture and transfer waste heat to other processes. • Thermal storage systems balance heat supply and demand. • Advanced materials with improved thermal conductivity enhance heat recovery capabilities.
Opportunities to Decrease the Operating Temperature of SOFCs	• Intermediate Temperature SOFC (IT-SOFCs) function at 400 °C to 800 °C, reducing thermal stresses. • Optimization of electrode materials and structures improves performance at lower temperatures. • Nanotechnology enhances ionic conductivity and reduces thermal mass. • Metal-supported SOFCs offer improved thermal management and shock resistance. • System design improvements, such as advanced thermal management and heat recovery, contribute to lower operational temperatures.

2.3.3. Machine Learning to Regulate Fuel Cells

Machine learning has emerged as a powerful tool for controlling the fuel-cell temperature, especially in PEMFCs and SOFCs. The capability to accurately monitor and manage temperature is critical for optimizing fuel cell performance, enhancing efficiency, and prolonging fuel-cell lifespans. Numerous machine learning approaches are being explored to develop advanced control strategies that can adapt to the dynamic operating conditions of fuel cells.

One prominent technique is the application of reinforcement learning (RL) for temperature control in fuel cells. He et al. [100] discuss how RL can be utilized to develop adaptive control strategies that respond to the fuel cell's nonlinear and time-varying behaviors. Unlike traditional control methods, which may struggle to meet the dynamic optimization requirements, RL algorithms can learn from real-time data and adjust control actions accordingly. This adaptability allows for more effective temperature regulation, ensuring that the fuel cell operates within optimal thermal limits.

Another significant application of machine learning in temperature control is through the use of deep learning algorithms. Ma [101] explores the implementation of model-free, multiple-input, multiple-output control (MFMIMOC) centered upon deep reinforcement learning (DRL) for managing the temperature of electrolysis cells from a proton-exchange membrane. By leveraging the capabilities of DRL, the system can accurately simulate complex features and dynamics, leading to improved temperature control in hybrid fuel cell vehicles. This approach highlights the potential of machine learning to enhance the precision of temperature management systems.

Moreover, support vector machines (SVMs), along with artificial neural networks (ANNs), have been utilized to predict and control temperature dynamics in fuel cells. Chatterjee presents an SVM for hybrid electric vehicles powered with fuel cells, emphasizing the effectiveness of data-driven control systems in optimizing temperature management [102]. These machine learning models could examine historical data and recognize patterns, allowing proactive adjustments to maintain optimal operating temperatures.

Additionally, the integration of machine learning with traditional control methods can further enhance temperature management. Louis [75] discusses the use of linear active disturbance rejection control in conjunction with machine learning techniques for enhancing the PEMFCs temperature control. By combining the strengths of both approaches, it is possible to achieve more robust and reliable temperature regulation, which is essential for maintaining the performance and durability of fuel cells.

In conclusion, machine learning techniques offer innovative solutions for controlling the temperature of fuel cells. The application of reinforcement learning, deep learning algorithms, and hybrid control strategies could enhance the reliability and efficiency of temperature management systems in PEMFCs and SOFCs. As research continues to advance in this field, the integration of machine learning into fuel-cell technology has an essential part to play in optimizing performance and ensuring the successful deployment of fuel cells in various applications.

2.3.4. Thermal Management of Other Types of Fuel Cells

Thermal management is a fuel-cell-crucial element, affecting the efficiency, performance, and longevity of numerous fuel cells. While PEMFCs and SOFCs are commonly discussed, other fuel-cell types, such as molten carbonate fuel cells (MCFCs) and phosphoric acid fuel cells (PAFCs), also require effective thermal management strategies to optimize their operation.

Molten carbonate fuel cells (MCFCs) work at elevated temperatures, normally between 600 °C and 700 °C. Such high operating temperatures enable the use of carbonate electrolytes, which can facilitate electrochemical reactions. However, managing the thermal environment is essential to prevent overheating and ensure consistent performance. According to Muck [103], maintaining a uniform distribution of temperature in the MCFC is important, as temperature gradients could result in thermal stresses and the degradation of fuel cell components. Effective thermal management systems, such as heat exchangers and thermal storage solutions, can help regulate the temperature and improve overall efficiency.

Phosphoric acid fuel cells (PAFCs) also work at elevated temperatures, generally around 150 °C to 200 °C. The thermal management of PAFCs is vital for maintaining the optimal operating temperature, as insufficient heat can lead to reduced reaction rates and overall efficiency. Muck [103] emphasizes the importance of integrating advanced thermal management techniques, such as fiber sensing for spatially resolved temperature measurement, to monitor and control the temperature within PAFC systems. This real-time monitoring allows for proactive adjustments to the cooling systems, ensuring that the fuel cells work in the range of their optimal thermal interval.

Protonic ceramic fuel cells (PCFCs) are also gaining attention for their potential to operate at intermediate temperatures (400 °C to 800 °C). Shimada et al. [104] discuss the development of bi-layered structures in PCFCs, which can enhance thermal stability and performance. The ability to operate at lower temperatures compared to SOFCs while maintaining high efficiency is a significant advantage, and effective thermal management strategies are essential to optimize their performance.

Hybrid fuel cell systems, which combine different types of fuel cells, also present unique thermal management challenges. For example, integrating PEMFCs with SOFCs or other energy systems requires careful consideration of the thermal interactions between the components. Anselma et al. [71] propose a dynamic programming approach for optimizing hybrid fuel-cell thermal management, making sure that each component operates within its optimal temperature range. This integrated method could enhance the overall fuel cell efficiency as well as its reliability.

In conclusion, effective thermal management is critical for the performance and durability of various types of fuel cells, such as MCFCs, PAFCs, and PCFCs. The implementation of advanced thermal management methods, such as real-time monitoring, heat exchangers, and hybrid system optimization, can considerably improve the efficiency and reliability of such systems. As research continues to advance in this field, innovative thermal management solutions will be essential for the successful commercialization of diverse fuel-cell technologies. Table 5 summarizes the key points from Sections 2.3.3 and 2.3.4 regarding machine learning techniques for temperature control of fuel cells, along with the thermal management of various fuel-cell types:

Table 5. Main strategies, considerations, and research directions in using machine learning for temperature control and in the thermal management of various fuel-cell types.

Focus	Key Points
Machine Learning Methods to Regulate Fuel Cells Temperature	• Machine learning enhances temperature control in PEMFCs and SOFCs. • Reinforcement learning (RL) adapts to dynamic conditions for better temperature regulation. • Deep learning algorithms improve control accuracy in hybrid fuel cell vehicles. • SVM and ANN predict and manage temperature dynamics. • Integrating machine learning with traditional methods enhances robustness in temperature regulation.
Other Types of Fuel Cells: Thermal Management of	• Effective thermal management is crucial for various fuel-cell types: MCFCs, PAFCs, and PCFCs. • **MCFCs:** Function at 600 °C to 700 °C; uniform temperature distribution prevents thermal stresses. • **PAFCs:** Work at 150 °C to 200 °C; advanced techniques like fiber sensing enable real-time temperature control. • **PCFCs:** Work at 400 °C to 800 °C; bi-layered structures improve thermal stability. • **Hybrid Fuel Cell Systems:** Combining different types requires optimized thermal management to enhance efficiency.

3. Conclusions: Fuel-Cell Thermal Management

In conclusion, effective thermal management is an important component of ensuring optimal fuel cell performance, reliability, and the lifespan of various fuel-cell technologies, specifically PEMFCs and SOFCs. The complexity of heat generation within fuel cells, driven primarily by electrochemical reactions, necessitates a comprehensive approach to thermal management that addresses both heat generation and dissipation.

For PEMFCs, effective thermal strategies are critical in preventing overheating, which could result in proton-exchange membrane degradation. This degradation not only reduces fuel cell efficiency but also impacts its whole lifespan. In PEMFC systems, design and flow field optimization play important roles in distributing reactants and removing excess heat. The implementation of active cooling techniques, like liquid cooling and phase-change materials, has been shown to be effective in regulating temperature and enhancing fuel cell operation reliability. Moreover, the use of advanced materials that enhance thermal conductivity is crucial for improving heat dissipation, thereby maintaining optimal operating conditions.

Conversely, SOFCs, which work at significantly higher temperatures (600 °C to 1000 °C), face unique thermal management challenges. The high operating temperatures enhance ionic conductivity, which is essential for efficient electrochemical reactions; however, they also introduce the risk of thermal stresses that can lead to material degradation. To address these challenges, effective thermal management strategies must be implemented to maintain uniform temperature distribution within the cells. Advanced cooling techniques, such as liquid cooling systems and heat exchangers, can capture and manage waste heat effectively, thereby enhancing overall system efficiency.

Furthermore, the heat recovery systems' integration in both PEMFCs and SOFCs has the potential to significantly improve energy utilization by capturing waste heat to be used in other applications, for example, space heating or power generation. This not only improves the overall efficiency of fuel cell systems but also promotes the sustainability of energy use in various applications.

In addition to traditional thermal management techniques, the incorporation of machine learning approaches presents new opportunities for improving temperature control and thermal management in fuel cells. Algorithms of machine learning, including reinforcement learning and deep learning, can adapt to changing operating conditions, providing real-time adjustments to temperature management strategies. These advancements offer the potential for more robust and reliable fuel cell operation, ultimately improving performance and extending the lifespan of the systems.

Despite the promising advancements in thermal management strategies for fuel cells, several challenges remain that warrant further attention. The integration of advanced cooling methods and hybrid systems often increases system complexity, which can lead to higher installation and maintenance costs. Moreover, the scalability of these approaches from laboratory prototypes to large-scale commercial applications is not straightforward, as performance benefits observed at small scales may diminish under real-world conditions. Economic feasibility remains a critical hurdle, especially for systems incorporating novel materials or complex hybrid configurations.

Using estimation derived from reported cost ranges for hybrid supercapacitor–battery systems and advanced cooling technologies in comparable renewable–integrated fuel cell applications, incorporating these components can increase capital expenditure by approximately 8–15%. This estimate reflects the additional costs of power electronics, control integration, and specialized cooling hardware. However, these investments can yield lifecycle benefits—such as extended component life, reduced maintenance frequency, and improved operational efficiency—that offset the initial cost within a 10–15-year operational period. Furthermore, the integration of heat recovery in combined heat and power (CHP) configurations can enhance return on investment by monetizing thermal output in addition to electricity generation.

Scalability considerations include thermal uniformity in large fuel cell arrays, the modularity of hybrid storage units, and integration with grid or microgrid control systems. Ensuring efficient thermal management at scale may require distributed cooling loops, advanced heat exchangers, and control algorithms capable of coordinating multiple hybrid storage modules. Economic scaling can be supported with modular designs that share common components and standardized interfaces. Addressing these limitations through cost-effective designs, simplified control strategies, and scalable solutions will be crucial for the broader adoption and commercialization of fuel-cell technologies.

Recent reviews collectively show that effective fuel-cell commercialization depends on advances in thermal management, environmental adaptability, and integrated control strategies across both PEMFCs and SOFCs [105–109]. The ongoing research and development in fuel-cell thermal management will be critical for tackling challenges pertaining to heat generation and dissipation. By continuously optimizing thermal management strategies, the performance, efficiency, and lifespan of fuel-cell technologies can be enhanced, paving the way for their extensive adoption in numerous applications, from automotive to stationary power applications. As the energy landscape evolves toward cleaner and more sustainable solutions, the significance of effective fuel-cell thermal management will only grow, making it a key area of focus for future research and development efforts.

4. Future Research Direction: Fuel-Cell Thermal Management

The future of thermal management in fuel cells, particularly for PEMFCs and SOFCs, is ready to evolve significantly, driven by advancements in technology, materials, and computational techniques. As fuel-cell technologies become increasingly integral to the transition to clean energy systems, a concerted effort is needed to explore and refine thermal management strategies. Several key research directions can be identified to improve fuel-cell performance and reliability. Figure 12 shows the summary illustration of future research directions in fuel-cell thermal management.

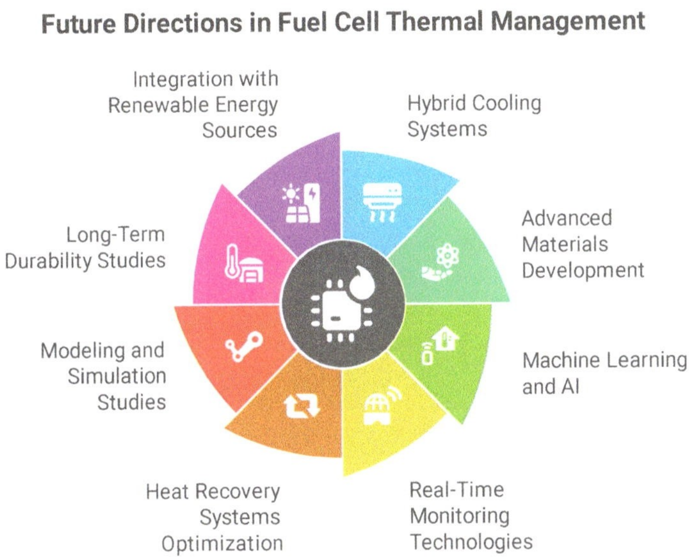

Figure 12. Future research direction in fuel-cell thermal management.

1. **Hybrid Cooling Systems**

One promising avenue for future research is the development of hybrid cooling systems that integrate both active and passive thermal management methods. Active cooling systems, such as liquid cooling, provide immediate temperature regulation, while passive systems leverage materials with high thermal conductivity and optimized designs to enhance natural heat dissipation. The synergistic combination of these approaches could lead to more efficient and reliable thermal management solutions. Research should focus on optimizing the design and control of these hybrid systems to improve their performance under varying operational conditions.

2. **Advanced Materials' Development**

The search for advanced materials that enhance thermal conductivity and stability is another critical area for research. The use of nanomaterials, composites, and innovative coatings can significantly improve heat transfer efficiency and reduce thermal stresses in fuel cells. Investigating the properties and applications of such materials in the context of fuel-cell thermal management will be essential for developing more effective systems. This includes exploring materials that can withstand the extreme conditions of SOFCs while maintaining high ionic conductivity.

3. **Machine Learning and Artificial Intelligence**

The integration of machine learning and artificial intelligence in thermal management represents a transformative opportunity. Future research should aim to leverage these technologies for the real-time monitoring and adaptive control of temperature in fuel cell systems. Machine learning algorithms can analyze vast amounts of data from fuel cell operations, identifying patterns and predicting performance under various conditions. By implementing adaptive control strategies based on this analysis, researchers can optimize thermal management systems to respond dynamically to changes in operating conditions, thus enhancing overall efficiency and reliability.

4. **Real-Time Monitoring Technologies**

The development of advanced sensing and monitoring technologies will play a vital role in improving thermal management. Integrating fiber-optic sensors, thermocouples, and infrared imaging techniques can facilitate real-time temperature measurement and distribution analysis within fuel cell systems. This data will be crucial to understanding thermal behaviors, identifying hotspots, and optimizing cooling strategies. Research should focus on developing compact and reliable sensing solutions that can be easily integrated into existing fuel cell designs.

5. **Heat Recovery Systems Optimization**

Enhancing heat recovery systems is essential for improving the overall efficiency of fuel-cell technologies. Future research should investigate innovative approaches to capture and utilize waste heat more effectively. This includes exploring advanced heat exchangers, thermal storage solutions, and integration with other energy systems. By optimizing the design and operation of heat recovery systems, researchers can significantly increase the overall energy conversion efficiency of fuel cells, making them more competitive with traditional energy sources.

6. **Modeling and Simulation Studies**

Advanced modeling and simulation studies will be critical for understanding the complex interactions within fuel cell systems. Future research should focus on developing comprehensive computational models that can accurately predict thermal behavior, efficiency, and performance under various conditions. These models will not only aid in the design of more effective thermal management strategies but also facilitate the optimization of fuel-cell systems as a whole. Collaborative efforts between researchers in computational fluid dynamics, materials science, and fuel-cell technology will be essential to achieve these goals.

7. **Long-Term Durability Studies**

Investigating the long-term durability of thermal management systems in fuel cells is crucial for ensuring reliable operation. Research should focus on understanding the effects of thermal cycling, material degradation, and operational stresses on fuel-cell performance over extended periods. This knowledge will inform the design of more robust thermal management solutions that can withstand the rigors of real-world applications. Long-term testing and evaluation will be necessary to assess the effectiveness of newly developed strategies and materials.

8. **Integration with Renewable Energy Sources**

As fuel cells are increasingly utilized in renewable energy applications, research should explore the integration of thermal management systems with renewable energy sources such as solar and wind. Understanding how these systems interact and the potential for synergistic benefits will be essential for developing efficient and sustainable energy solutions. This includes exploring the use of fuel cells in hybrid systems that combine multiple energy sources and storage solutions.

All in all, the future of thermal management in fuel cells presents numerous research opportunities that can considerably improve the performance, efficiency, and durability of fuel-cell systems. By exploring hybrid cooling systems, advanced materials, machine learning applications, real-time monitoring technologies, heat recovery optimization, modeling studies, durability assessments, renewable energy integration, and standardization efforts, researchers can pave the way for the successful commercialization of fuel-cell technologies. As the requirement for renewable and sustainable energy solutions is continuing to grow, these research directions will play a crucial part in advancing fuel-cell technology, as well as its applications in various sectors.

Funding: This research received no external funding.

Conflicts of Interest: The author declares no conflict of interest.

References

1. Si, F.; Liu, S.; Liang, Y.; Fu, X.-Z.; Zhang, J.; Luo, J.-L. Fuel cell reactors for the clean cogeneration of electrical energy and value-added chemicals. *Electrochem. Energy Rev.* **2022**, *5* (Suppl. S2), 25. [CrossRef]
2. Srivastava, R.K.; Boddula, R.; Pothu, R. Microbial fuel cells: Technologically advanced devices and approach for sustainable/renewable energy development. *Energy Convers. Manag. X* **2022**, *13*, 100160. [CrossRef]
3. Gao, X.; Chen, J.; Xu, R.; Zhen, Z.; Zeng, X.; Chen, X.; Cui, L. Research progress and prospect of the materials of bipolar plates for proton exchange membrane fuel cells (PEMFCs). *Int. J. Hydrogen Energy* **2024**, *50*, 711–743. [CrossRef]
4. Pourrahmani, H.; Yavarinasab, A.; Siavashi, M.; Matian, M.; Van Herle, J. Progress in the proton exchange membrane fuel cells (PEMFCs) water/thermal management: From theory to the current challenges and real-time fault diagnosis methods. *Energy Rev.* **2022**, *1*, 100002. [CrossRef]
5. Okonkwo, P.C.; Ige, O.O.; Barhoumi, E.M.; Uzoma, P.C.; Emori, W.; Benamor, A.; Abdullah, A.M. Platinum degradation mechanisms in proton exchange membrane fuel cell (PEMFC) system: A review. *Int. J. Hydrogen Energy* **2021**, *46*, 15850–15865. [CrossRef]
6. Zhao, J.; Tu, Z.; Chan, S.H. Carbon corrosion mechanism and mitigation strategies in a proton exchange membrane fuel cell (PEMFC): A review. *J. Power Sources* **2021**, *488*, 229434. [CrossRef]
7. Singh, M.; Zappa, D.; Comini, E. Solid oxide fuel cell: Decade of progress, future perspectives and challenges. *Int. J. Hydrogen Energy* **2021**, *46*, 27643–27674. [CrossRef]
8. Corigliano, O.; Pagnotta, L.; Fragiacomo, P. On the technology of solid oxide fuel cell (SOFC) energy systems for stationary power generation: A review. *Sustainability* **2022**, *14*, 15276. [CrossRef]
9. Peng, J.; Huang, J.; Wu, X.-L.; Xu, Y.-W.; Chen, H.; Li, X. Solid oxide fuel cell (SOFC) performance evaluation, fault diagnosis and health control: A review. *J. Power Sources* **2021**, *505*, 230058. [CrossRef]
10. Golkhatmi, S.Z.; Asghar, M.I.; Lund, P.D. A review on solid oxide fuel cell durability: Latest progress, mechanisms, and study tools. *Renew. Sustain. Energy Rev.* **2022**, *161*, 112339. [CrossRef]
11. Barckholtz, T.A.; Taylor, K.M.; Narayanan, S.; Jolly, S.; Ghezel-Ayagh, H. Molten carbonate fuel cells for simultaneous CO_2 capture, power generation, and H2 generation. *Appl. Energy* **2022**, *313*, 118553. [CrossRef]
12. Abdollahipour, A.; Sayyaadi, H. Thermal energy recovery of molten carbonate fuel cells by thermally regenerative electrochemical cycles. *Energy* **2021**, *227*, 120489. [CrossRef]
13. Ido, A.; Kawase, M. Development of a tubular molten carbonate direct carbon fuel cell and basic cell performance. *J. Power Sources* **2020**, *449*, 227483. [CrossRef]

14. Samanta, S.; Roy, D. Molten carbonate fuel cell integrated hybrid system for clean and efficient power generation. *Appl. Therm. Eng.* **2023**, *226*, 120294. [CrossRef]
15. Wilailak, S.; Yang, J.-H.; Heo, C.-G.; Kim, K.-S.; Bang, S.-K.; Seo, I.-H.; Zahid, U.; Lee, C.-J. Thermo-economic analysis of phosphoric acid fuel-cell (PAFC) integrated with organic ranking cycle (ORC). *Energy* **2021**, *220*, 119744. [CrossRef]
16. Oh, S.; Kim, T.; Kim, S.; Kang, S. Energetic, exergetic, economic, and exergoeconomic analysis of a phosphoric acid fuel cell-organic rankine cycle hybrid system. *Energy Convers. Manag.* **2023**, *284*, 116993. [CrossRef]
17. Park, C.; Jung, Y.; Lim, K.; Kim, B.; Kang, Y.; Ju, H. Analysis of a phosphoric acid fuel cell-based multi-energy hub system for heat, power, and hydrogen generation. *Appl. Therm. Eng.* **2021**, *189*, 116715. [CrossRef]
18. Guo, X.; Zhang, H.; Hu, Z.; Hou, S.; Ni, M.; Liao, T. Energetic, exergetic and ecological evaluations of a hybrid system based on a phosphoric acid fuel cell and an organic Rankine cycle. *Energy* **2021**, *217*, 119365. [CrossRef]
19. Xu, J.; Zhang, C.; Fan, R.; Bao, H.; Wang, Y.; Huang, S.; Chin, C.S.; Li, C. Modelling and Control of Vehicle Integrated Thermal Management System of PEM Fuel Cell Vehicle. *Energy* **2020**, *199*, 117495. [CrossRef]
20. Bielefeld, N.M.; Sørensen, R.D.; Jørgensen, M.; Kure, K.; Berning, T. A One-Dimensional Computational Model to Identify Operating Conditions and Cathode Flow Channel Dimensions for a Proton Exchange Membrane Fuel Cell. *Hydrogen* **2024**, *5*, 624–643. [CrossRef]
21. Alaswad, A.; Omran, A.; Sodre, J.R.; Wilberforce, T.; Pignatelli, G.; Dassisti, M.; Baroutaji, A.; Olabi, A.G. Technical and Commercial Challenges of Proton-Exchange Membrane (PEM) Fuel Cells. *Energies* **2021**, *14*, 144. [CrossRef]
22. You, S.-H.; Lee, W.; Jang, H.Y.; Kim, K.-S.; Baek, J.; Choe, G.; Ji, S.G.; Paidi, V.K.; Choi, C.H.; Back, S.; et al. Optimizing the Atomic Structure of Ruthenium Deposited on Pt/C Cathode Catalysts to Enhance Durability of Automotive Fuel Cell. *Appl. Catal. B Environ. Energy* **2024**, *359*, 124486. [CrossRef]
23. Marghani, A. Hydrogen Production and Applications: A Review. In *E3S Web of Conferences, International Conference on Energy and Green Computing (ICEGC'2023), Fez, Morocco, 23–24 November 2023*; EDP Sciences: Les Ulis, France, 2023; Volume 469, p. 00088.
24. Wang, Y.; Wehrle, L.; Banerjee, A.; Shi, Y.; Deutschmann, O. Analysis of a Biogas-Fed SOFC CHP System Based on Multi-Scale Hierarchical Modeling. *Renew. Energy* **2021**, *163*, 78–87. [CrossRef]
25. Hossain, S.S.; Ali, S.S.; Rushd, S.; Ayodele, B.V.; Cheng, C.K. Interaction effect of process parameters and Pd-electrocatalyst in formic acid electro-oxidation for fuel cell applications: Implementing supervised machine learning algorithms. *Int. J. Energy Res.* **2022**, *46*, 21583–21597. [CrossRef]
26. Wang, Y.; Seo, B.; Wang, B.; Zamel, N.; Jiao, K.; Adroher, X.C. Fundamentals, materials, and machine learning of polymer electrolyte membrane fuel cell technology. *Energy AI* **2020**, *1*, 100014. [CrossRef]
27. Ding, R.; Zhang, S.; Chen, Y.; Rui, Z.; Hua, K.; Wu, Y.; Li, X.; Duan, X.; Wang, X.; Li, J.; et al. Application of machine learning in optimizing proton exchange membrane fuel cells: A review. *Energy AI* **2022**, *9*, 100170. [CrossRef]
28. Raeesi, M.; Changizian, S.; Ahmadi, P.; Khoshnevisan, A. Performance analysis of a degraded PEM fuel cell stack for hydrogen passenger vehicles based on machine learning algorithms in real driving conditions. *Energy Convers. Manag.* **2021**, *248*, 114793. [CrossRef]
29. Deng, K.; Liu, Y.; Hai, D.; Peng, H.; Löwenstein, L.; Pischinger, S.; Hameyer, K. Deep reinforcement learning based energy management strategy of fuel cell hybrid railway vehicles considering fuel cell aging. *Energy Convers. Manag.* **2022**, *251*, 115030. [CrossRef]
30. Ming, W.; Sun, P.; Zhang, Z.; Qiu, W.; Du, J.; Li, X.; Zhang, Y.; Zhang, G.; Liu, K.; Wang, Y.; et al. A systematic review of machine learning methods applied to fuel cells in performance evaluation, durability prediction, and application monitoring. *Int. J. Hydrogen Energy* **2023**, *48*, 5197–5228. [CrossRef]
31. Tang, X.; Zhou, H.; Wang, F.; Wang, W.; Lin, X. Longevity-conscious energy management strategy of fuel cell hybrid electric Vehicle Based on deep reinforcement learning. *Energy* **2022**, *238*, 121593. [CrossRef]
32. Sun, H.; Fu, Z.; Tao, F.; Zhu, L.; Si, P. Data-driven reinforcement-learning-based hierarchical energy management strategy for fuel cell/battery/ultracapacitor hybrid electric vehicles. *J. Power Sources* **2020**, *455*, 227964. [CrossRef]
33. Su, D.; Zheng, J.; Ma, J.; Dong, Z.; Chen, Z.; Qin, Y. Application of Machine Learning in Fuel Cell Research. *Energies* **2023**, *16*, 4390. [CrossRef]
34. Rashidi, S.; Karimi, N.; Sunden, B.; Kim, K.C.; Olabi, A.G.; Mahian, O. Progress and challenges on the thermal management of electrochemical energy conversion and storage technologies: Fuel cells, electrolysers, and supercapacitors. *Prog. Energy Combust. Sci.* **2022**, *88*, 100966. [CrossRef]
35. Xu, J.; Zhang, C.; Wan, Z.; Chen, X.; Chan, S.H.; Tu, Z. Progress and perspectives of integrated thermal management systems in PEM fuel cell vehicles: A review. *Renew. Sustain. Energy Rev.* **2022**, *155*, 111908. [CrossRef]
36. Huang, Y.; Xiao, X.; Kang, H.; Lv, J.; Zeng, R.; Shen, J. Thermal management of polymer electrolyte membrane fuel cells: A critical review of heat transfer mechanisms, cooling approaches, and advanced cooling techniques analysis. *Energy Convers. Manag.* **2022**, *254*, 115221. [CrossRef]

37. Chen, Q.; Zhang, G.; Zhang, X.; Sun, C.; Jiao, K.; Wang, Y. Thermal Management of Polymer Electrolyte Membrane Fuel Cells: A Review of Cooling Methods, Material Properties, and Durability. *Appl. Energy* **2021**, *286*, 116496. [CrossRef]
38. Zhang, B.; Lin, F.; Zhang, C.; Liao, R.; Wang, Y.-X. Design and Implementation of Model Predictive Control for an Open-Cathode Fuel Cell Thermal Management System. *Renew. Energy* **2020**, *154*, 1014–1024. [CrossRef]
39. Wang, Y. Modeling of two-phase transport in the diffusion media of polymer electrolyte fuel cells. *J. Power Sources* **2008**, *185*, 261–271. [CrossRef]
40. Matsunaga, M.; Fukushima, T.; Ojima, K. Powertrain system of Honda FCX Clarity fuel cell vehicle. *World Electr. Veh. J.* **2009**, *3*, 820–829. [CrossRef]
41. Wilberforce, T.; Olabi, A.; Muhammad, I.; Alaswad, A.; Sayed, E.T.; Abo-Khalil, A.G.; Maghrabie, H.M.; Elsaid, K.; Abdelkareem, M.A. Recovery of waste heat from proton exchange membrane fuel cells–A review. *Int. J. Hydrogen Energy* **2024**, *52*, 933–972. [CrossRef]
42. Nguyen, H.Q.; Shabani, B. Proton Exchange Membrane Fuel Cells Heat Recovery Opportunities for Combined Heating/Cooling and Power Applications. *Energy Convers. Manag.* **2020**, *204*, 112328. [CrossRef]
43. Kwan, T.H.; Katsushi, F.; Shen, Y.; Yin, S.; Zhang, Y.; Kase, K.; Yao, Q. Comprehensive review of integrating fuel cells to other energy systems for enhanced performance and enabling polygeneration. *Renew. Sustain. Energy Rev.* **2020**, *128*, 109897. [CrossRef]
44. Sun, W.; Yi, F.; Hu, D.; Zhou, J. Research on Matching Design Method of Waste Heat Reuse System of Fuel Cell Vehicle Considering System Energy Consumption and Waste Heat Exchange Rate. *Int. J. Energy Res.* **2020**, *45*, 5470–5485. [CrossRef]
45. Olabi, A.; Wilberforce, T.; Sayed, E.T.; Elsaid, K.; Abdelkareem, M.A. Prospects of Fuel Cell Combined Heat and Power Systems. *Energies* **2020**, *13*, 4104. [CrossRef]
46. Xing, L.; Xiang, W.; Zhu, R.; Tu, Z. Modeling and thermal management of proton exchange membrane fuel cell for fuel cell/battery hybrid automotive vehicle. *Int. J. Hydrogen Energy* **2022**, *47*, 1888–1900. [CrossRef]
47. Kandidayeni, M.; Macias, A.; Boulon, L.; Kelouwani, S. Investigating the impact of ageing and thermal management of a fuel cell system on energy management strategies. *Appl. Energy* **2020**, *274*, 115293. [CrossRef]
48. Bai, W.; He, C. System Optimization of Thermal Management Performance of Fuel Cell System for Automobile. *Therm. Sci.* **2021**, *25 Pt B*, 2923–2931. [CrossRef]
49. Murad, A. Thermal Modeling and Performance Investigation of Proton Exchange Membrane (PEM) Fuel Cell. *Vfast Trans. Math.* **2023**, *11*, 112–125. [CrossRef]
50. Wang, Y.; Diaz, D.F.R.; Chen, K.S.; Wang, Z.; Adroher, X.C. Materials, Technological Status, and Fundamentals of PEM Fuel Cells–A Review. *Mater. Today* **2020**, *32*, 178–203. [CrossRef]
51. Atasay, N.; Atmanlı, A.; Yılmaz, N. Liquid Cooling Flow Field Design and Thermal Analysis of Proton Exchange Membrane Fuel Cells for Space Applications. *Int. J. Energy Res.* **2023**, *2023*, 1–16. [CrossRef]
52. Sun, M. Effect of Flow Field Structure and Operating Condition on the Distributions of Water and Phosphoric Acid in High Temperature Proton Exchange Membrane Fuel Cell (HT-PEMFC). *Ind. Eng. Chem. Res.* **2023**, *62*, 18720–18735. [CrossRef]
53. Haider, R.; Wen, Y.; Ma, Z.-F.; Wilkinson, D.P.; Zhang, L.; Yuan, X.; Song, S.; Zhang, J. High Temperature Proton Exchange Membrane Fuel Cells: Progress in Advanced Materials and Key Technologies. *Chem. Soc. Rev.* **2021**, *50*, 1138–1187. [CrossRef]
54. Wang, Y.; Sun, P.; Xia, Z.; Li, Z.; Ding, H.; Fan, Z.; Guo, H. Anchoring Highly Sulfonated Hyperbranched PBI Onto oPBI: Fast Proton Conduction with Low Leaching. *Acs Appl. Energy Mater.* **2022**, *5*, 10802–10814. [CrossRef]
55. Zeng, Z.; Qian, Y.; Zhang, Y.; Hao, C.; Dan, D.; Zhuge, W. A review of heat transfer and thermal management methods for temperature gradient reduction in solid oxide fuel cell (SOFC) stacks. *Appl. Energy* **2020**, *280*, 115899. [CrossRef]
56. Sai, D.P.; Sikarwar, B.S.; Singh, R.K. Estimation of Shearing and Peeling Thermal Stress of Spinel Coated Metallic Interconnects of Solid Oxide Fuel Cells. In Proceedings of the International Symposium on Fluids and Thermal Engineering (FLUTE 2021), Uttar Pradesh, India, 22 July 2021; Journal of Physics: Conference Series; IOP Publishing: Bristol, UK, 2022; Volume 2178, p. 012005.
57. Xu, Q.; Xia, L.; He, Q.; Guo, Z.; Ni, M. Thermo-Electrochemical Modelling of High Temperature Methanol-Fuelled Solid Oxide Fuel Cells. *Appl. Energy* **2021**, *291*, 116832. [CrossRef]
58. Zhang, B.; Wang, Y.; Zheng, J.; Liu, D. Study on Buildings CCHP System Based on SOFC. *Fluid Dyn. Mater. Process.* **2020**, *16*, 665–674. [CrossRef]
59. Ravi, R.; Kalib, N.S.; Muchtar, A.; Somalu, M.R. Potential of Integrating Solid Oxide Fuel Cell Based on Biomass in Power Generation in Malaysia: A Feasibility Study. In Proceedings of the 33rd Symposium of Malaysian Chemical Engineers (SOMChE 2022), Online, 8–9 August 2022; IOP Conference Series: Materials Science and Engineering; IOP Publishing: Bristol, UK, 2022; Volume 1257, p. 012018.
60. Pirkandi, J.; Maroufi, A.; Ommian, M. Exergy and Economic Investigation of Different Strategies of Hybrid Systems Consisting of Gas Turbine (GT) and Solid Oxide Fuel Cell (SOFC). *Int. J. Integr. Eng.* **2022**, *14*, 127–139. [CrossRef]
61. Ji, J. Guanidinium/Hydroxyl-Functionalized Polybenzimidazole for High-Temperature Proton Exchange Membrane Fuel Cell Applications. *ACS Appl. Energy Mater.* **2023**, *6*, 11754–11761. [CrossRef]

62. Hao, X.; Li, Z.; Xiao, M.; Han, D.; Huang, S.; Xi, G.; Wang, S.; Meng, Y. A Phosphonated Phenol-Formaldehyde-Based High-Temperature Proton Exchange Membrane with Intrinsic Protonic Conductors and Proton Transport Channels. *J. Mater. Chem. A* **2022**, *10*, 10916–10925. [CrossRef]
63. Feng, M.; Ma, Y.; Chang, J.; Lin, J.; Xu, Y.; Feng, Y.; Huang, Y.; Luo, J. Sulfonated Poly(arylene Ether Nitrile)-Based Composite Membranes Enhanced with Ca^{2+} Bridged Carbon Nanotube-Graphene Oxide Networks. *J. Inorg. Organomet. Polym. Mater.* **2022**, *32*, 2103–2112. [CrossRef]
64. Xu, Q.; Guo, Z.; Xia, L.; He, Q.; Li, Z.; Bello, I.T.; Zheng, K.; Ni, M. A Comprehensive Review of Solid Oxide Fuel Cells Operating on Various Promising Alternative Fuels. *Energy Convers. Manag.* **2022**, *253*, 115175. [CrossRef]
65. Hami, M.; Mahmoudimehr, J. Simulation-based Multiobjective Management of Transient Heating Process of Solid Oxide Fuel Cell. *Fuel Cells* **2023**, *23*, 188–201. [CrossRef]
66. Chen, T.-C.; Kumar, T.C.A.; Dwijendra, N.K.A.; Majdi, A.; Asary, A.R.; Iswanto, A.H.; Khan, I.; Madsen, D.Ø.; Alayi, R. Energy and Exergy Analysis of the Impact of Renewable Energy with Combined Solid Oxide Fuel Cell and Micro-Gas Turbine on Poly-Generation Smart-Grids. *Water* **2023**, *15*, 1069. [CrossRef]
67. Wu, X. Novel Hybrid Modeling and Analysis Method for Steam Reforming Solid Oxide Fuel Cell System Multifault Degradation Fusion Assessment. *Acs Omega* **2023**, *8*, 36876–36892. [CrossRef]
68. Idi, M.M.E. PEMFC Passive Thermal Management Systems: A Review. *Preprints* **2023**. [CrossRef]
69. Yan, X.; Peng, Y.; Shen, Y.; Shen, S.; Wei, G.; Yin, J.; Zhang, J. The Use of Phase-Change Cooling Strategy in Proton Exchange Membrane Fuel Cells: A Numerical Study. *Sci. China Technol. Sci.* **2021**, *64*, 2762–2770. [CrossRef]
70. Kim, B.M.; Yoo, S.J. Approximation-Based Adaptive Control of Constrained Uncertain Thermal Management Systems with Nonlinear Coolant Circuit Dynamics of PEMFCs. *IEEE Access* **2020**, *8*, 83483–83494. [CrossRef]
71. Anselma, P.G.; Luciani, S.; Tonoli, A. Dynamic Programming for Thermal Management of Automotive Fuel Cell Systems: Investigating Hydrogen Saving Potential. *IEEE Access* **2023**, *11*, 48080–48098. [CrossRef]
72. Novotný, J.; Nováková, L.; Čížek, R.; Kašpárek, M.; Machovská, I.; Dančová, P.; Novosád, J.; Pulec, J. Optimization of Air Mass Flow in a PEM Fuel Cell. In *EPJ Web of Conferences, EFM21–15th International Conference "Experimental Fluid Mechanics 2021", Liberec, Czech Republic, 23–26 November 2021*; EDP Sciences: Les Ulis, France, 2022; Volume 264, p. 01026.
73. Ahmed, A.; Mangi, F.; Kashif, M.; Chachar, F.; Ullah, Z. Parametric Analysis of a Serpentine Flow Pattern Proton Exchange Membrane Fuel Cell for Optimized Performance. *Int. J. Heat Technol.* **2020**, *38*, 69–76. [CrossRef]
74. Li, M. A Comprehensive Review of Thermal Management in Solid Oxide Fuel Cells: Focus on Burners, Heat Exchangers, and Strategies. *Energies* **2024**, *17*, 1005. [CrossRef]
75. Louis, K.B. Temperature Control of Proton Exchange Membrane Fuel Cell Based on Linear Active Disturbance Rejection Control. *J. Power Energy Eng.* **2024**, *12*, 1–23. [CrossRef]
76. Vu, H.N. Bypass Configurations of Membrane Humidifiers for Water Management in PEM Fuel Cells. *Energies* **2023**, *16*, 6986. [CrossRef]
77. Deng, Z.; Li, B.; Xing, S.; Zhao, C.; Wang, H. Experimental Investigation on the Anode Flow Field Design for an Air-Cooled Open-Cathode Proton Exchange Membrane Fuel Cell. *Membranes* **2022**, *12*, 1069. [CrossRef]
78. Wang, H.; Yang, G.; Li, S.; Shen, Q.; Li, Y.; Wang, R. Pore-Scale Modeling of Liquid Water Transport in Compressed Gas Diffusion Layer of Proton Exchange Membrane Fuel Cells Considering Fiber Anisotropy. *Membranes* **2023**, *13*, 559. [CrossRef]
79. Xu, Z. Porphyrin Helical Nanochannel-Assembled Polybenzimidazole Membranes Doped with Phosphoric Acid for Fuel Cells Operating in a Temperature Range of 25–200 °C. *Adv. Funct. Mater.* **2023**, *34*, 2310762. [CrossRef]
80. Gan, Z. Application Study of the Power Control System for Air-Cooled Proton Exchange Membrane Fuel Cell Based on Fuzzy Adaptive with Heuristic Adaptive Period Strategy. In Proceedings of the 2023 5th International Conference on Energy Systems and Electrical Power, Changsha, China, 19–21 May 2023; Journal of Physics: Conference Series; IOP Publishing: Bristol, UK, 2023; Volume 2584, p. 012031.
81. Pajares, A.; Blasco, X.; Herrero, J.M.; Simarro, R. Multivariable Controller Design for the Cooling System of a PEM Fuel Cell by Considering Nearly Optimal Solutions in a Multiobjective Optimization Approach. *Complexity* **2020**, *2020*, 8649428. [CrossRef]
82. Benavides, D.; Arévalo, P.; Tostado-Véliz, M.; Vera, D.; Escamez, A.; Aguado, J.A.; Jurado, F. An Experimental Study of Power Smoothing Methods to Reduce Renewable Sources Fluctuations Using Supercapacitors and Lithium-Ion Batteries. *Batteries* **2022**, *8*, 228. [CrossRef]
83. Wang, W.; Li, N.; Zhang, J.; Zhang, C.; Zhang, L.; Parale, V. Thermal Management Analysis of Proton Exchange Membrane Fuel Cell Filled with Phase Change Material in Cooling Channel. *Int. J. Energy Res.* **2023**, *2023*, 9077046. [CrossRef]
84. Zakaria, I.A. Performance of Hybrid Al2O3:SiO2 W:EG in PEM Fuel Cell Distributor Plate. *J. Mech. Eng.* **2023**, *20*, 335–354. [CrossRef]
85. Oleiwi, M.Q.; Mat, S.M.K.A.; Mohamed, M.F. Passive Cooling Strategies in the Hot Climate: A Review Study. *Arid. Int. J. Sci. Technol.* **2023**, *6*, 76–106. [CrossRef]

86. Wang, X. Study of a Proton Exchange Membrane Fuel Cell and Metal Hydride System Based on Double Spiral Structure Coupling. *Sustain. Energy Fuels* **2024**, *8*, 322–346. [CrossRef]
87. Pourrahmani, H.; Shakeri, H.; Herle, J.V. Thermoelectric Generator as the Waste Heat Recovery Unit of Proton Exchange Membrane Fuel Cell: A Numerical Study. *Energies* **2022**, *15*, 3018. [CrossRef]
88. Tolj, I.; Penga, Z.; Bosnic, P.; Radica, G. Proton Exchange Membrane Fuel Cell Flow Field Configuration: Modelling and Experimental Verification. *ECS Trans.* **2022**, *108*, 143–151. [CrossRef]
89. Feng, H.; Panthi, D.; Du, Y. Improving Current Collection of Tubular Solid Oxide Fuel Cells. *ECS Trans.* **2021**, *103*, 83–92. [CrossRef]
90. Costa, P.; Pinto, F.; André, R.N.; Marques, P. Integration of gasification and solid oxide fuel cells (SOFCs) for combined heat and power (CHP). *Processes* **2021**, *9*, 254. [CrossRef]
91. Alns, A.; Sleiti, A.K. Combined heat and power system based on Solid Oxide Fuel Cells for low energy commercial buildings in Qatar. *Sustain. Energy Technol. Assess.* **2021**, *48*, 101615. [CrossRef]
92. Iliev, I.K.; Filimonova, A.A.; Chichirov, A.A.; Chichirova, N.D.; Pechenkin, A.V.; Vinogradov, A.S. Theoretical and experimental studies of combined heat and power systems with SOFCs. *Energies* **2023**, *16*, 1898. [CrossRef]
93. Radenahmad, N.; Azad, A.T.; Saghir, M.; Taweekun, J.; Abu Bakar, M.S.; Reza, S.; Azad, A.K. A review on biomass derived syngas for SOFC based combined heat and power application. *Renew. Sustain. Energy Rev.* **2020**, *119*, 109560. [CrossRef]
94. Müller, K.; Skeledzic, T.; Wasserscheid, P. Strategies for Low-Temperature Liquid Organic Hydrogen Carrier Dehydrogenation. *Energy Fuels* **2021**, *35*, 10929–10936. [CrossRef]
95. Duong, P.A.; Ryu, B.; Jung, J.; Kang, H. Thermal Evaluation of a Novel Integrated System Based on Solid Oxide Fuel Cells and Combined Heat and Power Production Using Ammonia as Fuel. *Appl. Sci.* **2022**, *12*, 6287. [CrossRef]
96. Mazlan, N.W.; Osman, N.; Hassan, O.H.; Mohamed, Z. Lattice Expansion of BaCe0.54Zr0.36Y0.1O3-δ Ceramic Electrolyte. *Solid State Phenom.* **2020**, *307*, 149–153. [CrossRef]
97. Zamudio-García, J.; Porras-Vázquez, J.M.; Losilla, E.R.; Marrero-López, D. Tunable Electrode Architectures for La0.8Sr0.2Fe1-XTixO3-δ Based Symmetrical Solid Oxide Fuel Cells. *ECS Trans.* **2021**, *103*, 1601–1606. [CrossRef]
98. Lee, H.Y.; Lee, H.; Lee, Y.H. Effect of Sputtering Pressure on the Nanostructure and Residual Stress of Thin-Film YSZ Electrolyte. *Sustainability* **2022**, *14*, 9704. [CrossRef]
99. Du, P.; Wu, J.; Li, Z.; Wang, X.; Jia, L. Failure Mechanism and Optimization of Metal-Supported Solid Oxide Fuel Cells. *Materials* **2023**, *16*, 3978. [CrossRef] [PubMed]
100. He, L. Study on Waste Heat Recovery of Fuel-Cell Thermal Management System Based on Reinforcement Learning. *Energy Technol.* **2024**, *12*, 2400438. [CrossRef]
101. Ma, L. Research on Temperature Control of Proton Exchange Membrane Electrolysis Cell Based on MO-TD3. *IET Renew. Power Gener.* **2024**, *18*, 1597–1610. [CrossRef]
102. Chatterjee, D.; Biswas, P.K.; Sain, C.; Roy, A.; Minarul Islam, M.; Ustun, T.S. SVM classifier based energy management strategy for dual-source fuel cell hybrid electric vehicles. *J. Eng.* **2024**, *6*, e12392. [CrossRef]
103. Muck, N. Integrating Fiber Sensing for Spatially Resolved Temperature Measurement in Fuel Cells. *Energies* **2023**, *17*, 16. [CrossRef]
104. Shimada, H.; Yamaguchi, Y.; Ryuma, M.M.; Sumi, H.; Nomura, K.; Shin, W.; Mikami, Y.; Yamauchi, K.; Nakata, Y.; Kuroha, T.; et al. Protonic Ceramic Fuel Cell with Bi-Layered Structure of $BaZr_{0.1}Ce_{0.7}Y_{0.1}Yb_{0.1}O_{3-\delta}$ Functional Interlayer and $BaZr_{0.8}Yb_{0.2}O_{3-\delta}$ Electrolyte. *J. Electrochem. Soc.* **2021**, *168*, 124504. [CrossRef]
105. Su, H.; Xu, H.; Wang, L.; Liu, Z.; Xie, L. A review on thermal management strategy for liquid-cooling proton exchange membrane fuel cells: Temperature regulation and cold start. *Appl. Energy* **2025**, *393*, 126142. [CrossRef]
106. Cornet, M.; Tardy, E.; Poirot-Crouvezier, J.-P.; Bultel, Y. Impact of Coolant Operation on Performance and Heterogeneities in Large Proton Exchange Membrane Fuel Cells: A Review. *Energies* **2025**, *18*, 111. [CrossRef]
107. Zhao, C.; Fei Wang, F.; Wu, X. Analysis and review on air-cooled open cathode proton exchange membrane fuel cells: Bibliometric, environmental adaptation and prospect. *Renew. Sustain. Energy Rev.* **2024**, *197*, 114408. [CrossRef]
108. Yang, L.; Nik-Ghazali, N.-N.; Ali, M.A.H.; Chong, W.; Yang, Z.; Liu, H. A review on thermal management in proton exchange membrane fuel cells: Temperature distribution and control. *Renew. Sustain. Energy Rev.* **2023**, *187*, 113737. [CrossRef]
109. Lu, X.; Gang, W.; Tu, Z. Recent developments in control and integration of solid oxide fuel cells: From stack to system. *Renew. Sustain. Energy Rev.* **2025**, *223*, 116000. [CrossRef]

hydrogen

MDPI

Review

Hydrogen-Enabled Power Systems: Technologies' Options Overview and Effect on the Balance of Plant

Furat Dawood *, GM Shafiullah * and Martin Anda

School of Engineering and Energy, Murdoch University, Building 220, 90 South Street, Murdoch, WA 6150, Australia; m.anda@murdoch.edu.au
* Correspondence: furat.dawood@murdoch.edu.au (F.D.); gm.shafiullah@murdoch.edu.au (G.S.)

Abstract

Hydrogen-based Power Systems (H2PSs) are gaining accelerating momentum globally to reduce energy costs and dependency on fossil fuels. A H2PS typically comprises three main parts: hydrogen production, storage, and power generation, called packages. A review of the literature and Original Equipment Manufacturers (OEM) datasheets reveals that no single manufacturer supplies all H2PS components, posing significant challenges in system design, parts integration, and safety assurance. Additionally, both the literature and H2PS projects' database highlight a gap in a systematic hydrogen equipment and auxiliary sub-systems technology selection process, and how this selection affects the overall H2PS Balance of Plant (BoP). This study addresses that gap by providing a guideline for available technology options and their impact on the H2PS-BoP. The analysis compares packages and auxiliary sub-system technologies to support informed engineering decisions regarding technology and equipment selection. The study finds that each package's technology influences the selection criteria of the other packages and the associated BoP requirements. Furthermore, the choice of technologies across packages significantly affects overall system integrity and BoP. These interdependencies are illustrated using a cause-and-effect matrix. The study's significance lies in establishing a structured guideline for engineering design and operations, enhancing the accuracy of feasibility studies, and accelerating the global implementation of H2PS.

Keywords: balance of plant; hydrogen equipment packages; technology selection; H2PS design

1. Introduction

Excluding rare pure natural (geologic or white) hydrogen (H_2), other forms of hydrogen are an energy carrier, not an energy source; thus, it must be produced using energy to split it from hydrogen-containing materials such as water or hydrocarbons [1]. The inputs of energy, the production technology and the feedstock determine the carbon footprint or the colour code of the produced hydrogen, e.g., Green, Blue, Grey, Pink, etc. Although hydrogen technologies are over 100 years old in principle, recent efficiency improvements, capital costs decline [2], and awareness of global warming and climate change have accelerated the role of hydrogen in the economy, called the "Hydrogen Economy" [3]. Hence, the principle of utilising hydrogen to decarbonise the energy sectors is gaining momentum with tremendous support from Governments, academia, research centres, and industries [4,5]. Many countries have set a target to become zero-emission by 2050, where hydrogen plays a crucial role in achieving this target [2,5,6].

Using hydrogen to store Renewable Energy (RE), called Power to Gas [6], is a promising approach to decarbonise the electricity sector. Hydrogen is a superior energy carrier due to its high gravimetric energy density (141.9 MJ/kg), significantly surpassing liquefied petroleum (46–50 MJ/kg) and natural gas (50–55 MJ/kg) [7]. Therefore, Hydrogen-enabled Power Systems (H2PS) are gaining accelerating momentum globally [2,8] to reduce the cost of energy and dependency on fossil fuels [9,10]. Many remote, isolated, stand-alone microgrids worldwide with abundant RE resources have started evaluating the feasibility of 100% RE-H2PS, which has led to the installation of many trials and demonstrations of H2PS, and many more are underway [11–15]. Green hydrogen is the most proposed and used in H2PS development [15] for grid-support and off-grid power systems using various hydrogen equipment technologies [11,16]. A typical 100% RE-H2PS comprises an electrolyser, storage, and FC packages with all the necessary equipment for integration into the hosting or served power systems [9,11]. The operation's philosophy is to utilise excess RE to split water into hydrogen and oxygen, where hydrogen is stored, and oxygen is vented [9]. The stored hydrogen can be used to regenerate electricity via FCs when RE resources are insufficient [6,11,17]. Hydrogen Compressed Gas (H2CG) storage technology is the most used in the installed H2PS worldwide; however, other promising technologies like liquid hydrogen or material-based technologies are gaining momentum [11,18].

A scan of the worldwide H2PS projects' database and Original Equipment Manufacturers (OEM) products' datasheets revealed that no manufacturer produces all parts of the H2PS. A comprehensive literature review reveals no detailed Balance-of-Plant (BoP) guidelines, system integration requirements, and systems engineering design procedures from the BoP perspective. This presents challenges for feasibility studies, engineering design, system integration, and safe operations, as the system parts and technologies' selection determine the cost and complexity of the system [19].

This study focuses on the H2PS equipment's packages and technologies variation and their effect on the holistic system BoP to establish a guideline to support informed decision-making regarding the preliminary technologies' selection process and, therefore, the cost estimation. In addition, an H2PS-BoP guideline is essential to facilitate a systematic hydrogen-enabled systems' Engineering Design and Operations (EDO).

2. Method

The study was constructed at three levels of BoP: the hydrogen components at the OEM level, delivered packages, and the holistic H2PS from an EDO perspective, as illustrated in Figure 1.

The H2PS-BoP at level one discusses briefly (at high-level of detail) the hydrogen components, namely, Electrolysers and FC stacks, typical construction parts, and stacking methods at the manufacturer's facility, i.e., cell and stack fabrication. This level of detail is beyond the scope of the EDO processes, as hydrogen components (equipment) are typically designed and packaged by the OEM. However, these details are crucial for low-level maintenance, conducted by the OEM or the personnel they trained to do so. On the other hand, the storage equipment's high-level detailed BoP is discussed according to the storage technology used, as the components vary and differentiate according to the storage technology and the specific parameters, such as operation temperature, pressure, application, and required discharge duration.

The H2PS-BoP at level two discusses the hydrogen-delivered packages, including the equipment, essential auxiliary, and sub-systems. This level of BoP is crucial for the EDO and maintenance of H2PS. Therefore, from an EDO perspective, hydrogen equipment packages' typical components, such as functionality, interconnectivity, and interoperability, are discussed and listed for each part of the holistic H2PS. This study discusses the packages as an individual part of the holistic H2PS regardless of the assembly process and orientation. We conducted a comprehensive review of multiple OEM product specifications and datasheets [20–44], alongside selected knowledge-sharing reports from existing H2PS projects [3,11,13,14,19,45–57], to identify and categorise the distinct technology packages and solutions in use.

The holistic H2PS auxiliary and sub-systems are discussed at level three BoP. In addition, the external support systems for feedstock supply, cooling, control, safety, and waste management are discussed for both case-specific designed and pre-engineered-packaged H2PS.

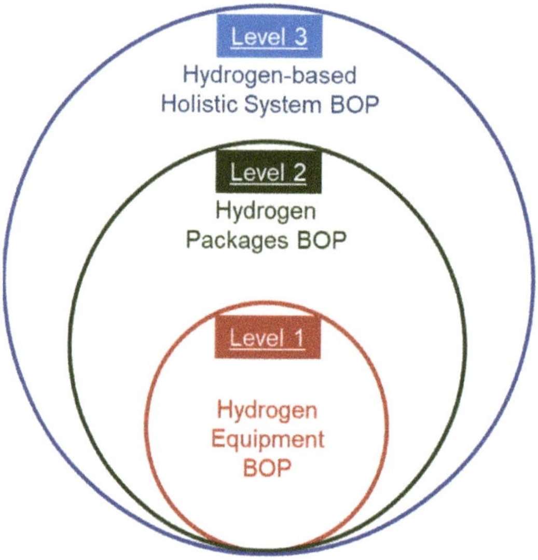

Figure 1. Typical H2PS: three levels Balance of Plant (BoP).

3. Water Electrolysis

Water electrolysis is the process of splitting water (H_2O) into hydrogen (H_2) and oxygen (O_2) by passing Direct Current (DC) through the electrolysis stack. The total reaction is explained in Equation (1) [58].

$$2\,H_2O\ (\text{liquid}) + \text{energy} \rightarrow 2\,H_2\ (\text{gas}) + O_2\ (\text{gas}) \tag{1}$$

where Energy = 237.2 kJ/mol (electricity) + 48.6 kJ/mol (heat) [59].

3.1. Electrolysis Technologies

Despite the many different types (technologies) of electrolysers, there are three well-known types of electrolysis technologies, namely, Alkaline Electrolyser (ALKE), Proton Exchange Membrane Electrolyser (PEME), and Solid Oxide Electrolyser (SOE) [60], as shown in Figure 2.

These electrolysers work under the same principle but in different ways [61]. The ALKE and PEME are the most used technologies in today's H2PS applications, while the SOE is gaining potential for Combined Heat and Power (CHP) applications, as the SOE operation temperature is very high compared to the other electrolysis technologies [62].

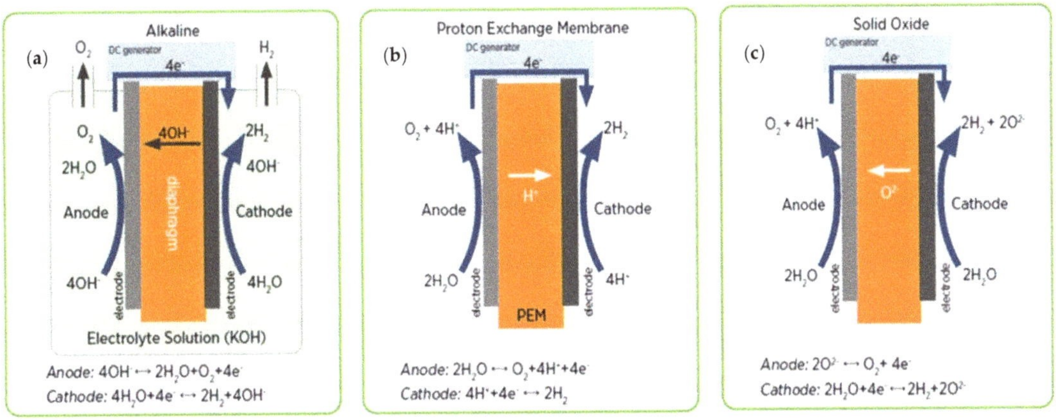

Figure 2. Different electrolysers work in different ways [7]: (**a**) Alkaline Electrolyser (ALKE), (**b**) Acidic, Polymer Electrolyte Membrane, or Proton Exchange Membrane Electrolyser (PEME), (**c**) Solid Oxide Electrolyser (SOE).

Electrolysers Emerging Technologies

A range of rapidly developing but not commercially ready water electrolysis technologies may require different BoP and engineering controls, such as the following:

1. Anion Exchange Membrane Electrolyser (AEME) possesses the advantages of both ALKE and PEM electrolysers technologies without their weak points [63].
2. Proton-Conducting Ceramic Electrolyser (PCCE) is another emerging high-temperature and comparably high-efficiency technology.
3. Unitized Regenerative Fuel Cell (URFC) is an electrolyser and FC in one stack arrangement, which can carry out the electrolysis of water in electrolysis mode and function in regenerative mode as a FC. The combination of both electrolysers and FCs in a single cell/stack requires special arrangements and engineering controls [64].
4. Capillary-fed Electrolysers (CFE) is a new category of electrolysis, introduced in 2022 by Hysata company in Wollongong, NSW, Australia. The concept was published in an article by Hodges A. et al. (2022) [65]. The efficiency mentioned in the literature is over 80% [65,66].

This study focuses on the most well-known electrolyser technologies and the most mentioned emerging technologies in the literature regarding H2PS. Some selected cell and stack specifications and properties of electrolysers are compared in Table 1.

Table 1. Types of electrolysers: cell and stack selected specifications and properties comparison.

Specifications	ALKE	PEME	SOE	AEME	PCCE
Electrode/Catalyst (O_2 side) (Anode)	Nickel coated. perforated stainless-steel [61]	Iridium oxide [61]	Perovskite-type (e.g., LSCF, LSM) [1] $Ni_{0.85}Co_{0.15}$-YSZ [61,67]	High surface area Nickel or NiFeCo alloys [61]	Perovskite-type (e.g., LSCF, LSM [63]
Electrode/Catalyst (H_2 side) (Cathode)	Nickel coated. perforated stainless-steel [61]	Platinum nanoparticles on carbon black [61]	Ni/YSZ [1] [61]	High surface area nickel [63]	Ni/YSZ, Ni-BZY/LSC, BCFYZ [63]
Electrolyte	Potassium Hydroxide (KOH) or potash lye 57/mol/L [61] Or NaOH 30 wt% [59]	PFSA membranes Nafion membrane [61]	oxide-ion conducting ceramics [63] Yttria-stabilized-Zirconia (YSZ)$Y_2O_3 + ZrO_2$ [1] [61]	DVB polymer support with KOH or $NaHCO_3$ 1 mol/L [63]	(Y,Yb)-Doped Ba(Ce,Zr)$O_{3-\delta}$ [63]
Separator/Diaphragms	ZrO_2 stabilised with PPS mesh [61]	Solid electrolyte (above) [61]	oxide-ion conducting ceramics electrolyte, as above [61]	Solid electrolyte (above) [63]	Solid electrolyte (above) [63]
Porous Transport Layer Anode	Nickel mesh (not always present) [63]	Platinum-coated sintered porous Titanium [61]	Coarse Nickel mesh or foam [63]	Nickel foam [63]	Coarse nickel mesh or foam [63]
Porous Transport Layer Cathode	Nickel mesh [61]	Sintered porous Titanium or Carbon cloth [61]	None	Nickel foam or carbon cloth [63]	None
Bipolar Plate Anode	Nickel-coated stainless steel [61]	Platinum-coated Titanium [61]	None	Nickel-coated or stainless steel [63]	None
Bipolar Plate Cathode	Nickel-coated stainless steel [61]	Gold-coated Titanium [61]	Cobalt-coated or stainless steel [61]	Nickel-coated or stainless steel [63]	Cobalt-coated [63]
Frames and sealing [1]	PSU, PTFE, EPDM [61]	PTFE, PSU, ETFE [61]	Ceramic glass [61]	PTFE, silicon [63]	Ceramic glass [63]
Ion Movement	hydroxyl (OH^-) [61]	Hydrogen Ion (H^+) [61]	Oxygen Ion (O^{2-}) [61]	hydroxyl (OH^-) [63]	Hydrogen Ion (H^+) [63]
Operating Temperature (°C)	70–90 [61]	50–80 [61] 30–80 [68]	700–850 [61] 800–1000 [63]	40–60 [63]	300–600 [63] Typically, 500
Cell Operating Pressure (bar)	1–30 [61] Up to 448 bar [60]	1–70, typically 30 [61] Min 11 bars cathode pressures to keep the H_2/O_2 mixture below 2% [68], Up to 700 bars [69,70]	<10 [61] Typically 1 bar [63]	<35 [61]	1 [63]
Electrical Efficiency (Stack) (kWh/Kg H_2)	50–78 [61] 47–66 [63]	50–83 [61] 47–66 [63]	45–55 [61] 35–50 [63]	51.5–66 [63]	None
Electrical efficiency (system) (kWh/Kg H_2)	50–78 [63]	50–83 [63]	40–50 [63]	57–69 [63]	None
Hydrogen Purity (%)	99.9–99.9998 [61,63]	99.9–99.9999 [61,63]	99.9 [63]	99.9–99.999 [63]	None
Nominal Cell Current Density (A/cm²)	0.2–0.8 [61,63]	1.6–2.3 [61] 3.7 @ 80 °C [68] 1–3 [63]	0.3–1 [63] Up to 3.6 @ 950 °C [68]	0.2–2 [63]	None

Table 1. *Cont.*

Specifications	ALKE	PEME	SOE	AEME	PCCE
Voltage range (Volts)	1.4–3 [63]	1.4–2.3 [63]	1–1.5 [63]	1.4–2 [63]	None
Load range (%)	15–100 [63]	5–130 [63]	30–125 [63]	5–100 [63]	None
Stack size	1 MW [63]	1–2 MW [63]	5 kW	2.5 kW [63]	None
Water feed	Cathode side [59]	Anode side [59]	Cathode side [59]	Cathode side [63]	Anode side [63]
Cold start to nom. Load (min)	<50 [63]	<20 [63]	>600 [63]	<20 [63]	None
Coupling with Variable RE (Solar, wind, and wave)	Awkward, i.e., long time to reach steady-state [63], performance deteriorates under part-load [59]	High dynamic range [63]	Awkward and require excess heat energy [63]	Awkward [63]	None
Lifetime (K hours)	60 [61]	50–60 [61] up to 80 [63]	20 and improving [61]	>5 [63]	None
Capital costs (small-scale stack)	800 to 1500 EUR/kW [60]	1400 to 2100 EUR/kW [60]	Unknown [60]	None	None
Capital costs (package) (USD/kW)	270 [63] 750 [71] 1000 EUR/kW [72] 571–1268 [73]	400 [63] 800 [71] 1800 EUR/kW [72] 385–268 [73]	>2000 [63] 677–2285 [73]	None	None
Capital costs (system) Minimum 10 MW (USD/kW)	500–1000 [63] 50 USD/kW additional for the BoP [73]	700–1400 [63] 50 USD/kW additional for the BoP [73]	Unknown [63] 50 USD/kW additional for the BoP [73]	None	None
State of Development	Mature/Marketed [61,63]	Mature/Marketed [61,63]	Developing [61]	None	None

[1] PFSA = Perfluorosulfonic acid; PTFE = Polytetrafluoroethylene; ETFE = Ethylene Tetrafluorethylene; PSF = poly (bisphenol-A sulfone); PSU = Polysulfone; YSZ = yttrium and scandium oxides; DVB = divinylbenzene; PPS = Polyphenylene sulphide; LSCF = $La_{0.58}Sr_{0.4}CO_{0.2}Fe_{0.8}O_{3-\delta}$; LSM = $(La_{1-x}Sr_x)1-yMnO_3$.

3.2. Electrolyser's Cell and Stack Arrangements

This section discusses typical electrolysis cells and stack arrangements (level 1) in high-level detail, as the cell and stack are not directly related to the EDO of H2PS. The designers typically use pre-assembled electrolyser stacks or packages, while the maintenance team may replace parts according to the OEM instructions and specifications. On the other hand, high-temperature electrolysers such as SOEs are not commercially available at this time and are not commonly used in H2PS, as they require high-quality excess heat energy [74]. Therefore, this study focuses on the AKLE and PEME cells and stacks arrangements.

The typical electrolysis cell consists of two electrodes, an anode and a cathode, plated with a catalyst and enclosed by an electrolyte as an ionic conductor (charge carrier) [61]. The two electrodes are separated by a membrane (separator) to pass a selected ion from one electrode side to the other side [62], as shown in Figure 3a. Typically, each electrode with the catalyst, electrolyte/membrane, and gas diffusion layer is assembled in what is known as the Membrane Electrode Assembly (MEA) [63], as illustrated in Figure 3b.

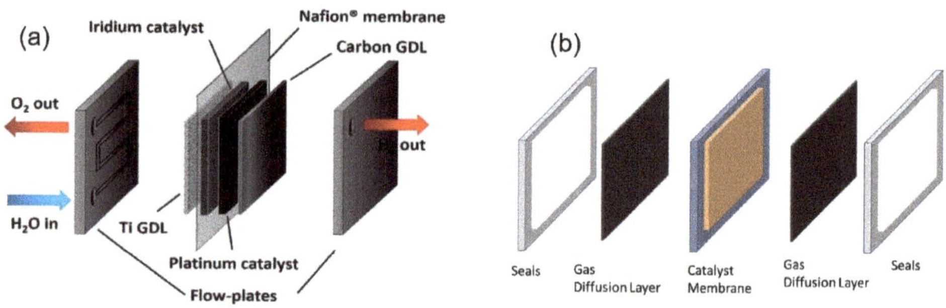

Figure 3. (**a**) General PEM electrolyser's single cell construction components, (**b**) Membrane Electrode Assembly (MEA). Pictures Source [7].

The design of the electrolysis cell complexity can vary significantly depending upon the technology type and application [63], but, in general, several essential components found in most electrolysis cells, as illustrated in Figure 3a,b, are as follows [63]:

1. The Membrane Electrode Assembly (MEA) consist of the following:
    a. Electrodes (anode and cathode);
    b. Catalyst (technology-specific);
    c. Membrane/Electrolyte (technology specific);
    d. Gas diffusion layer.
2. Flow plates.
3. Gaskets, clamping mechanism and seals.

The electrolyser's hydrogen production capacity (called size) increased by cells cascading, i.e., multiple cells connected in series to form an electrolyser stack, as illustrated in Figure 4 [63]. Typically, these cells are separated with spacers (insulating material between two opposite electrodes or bipolar plates), seals, frames (mechanical support), and end plates (to avoid leaks and collect fluids). However, each technology requires some specific stack arrangements, which also vary between manufacturers as compared in Table 1 above. The manufacturing cost factor is the cost of the materials used as catalysts and the mechanical construction, which depends on the electrolyser's technology and varies between manufacturers [61,63].

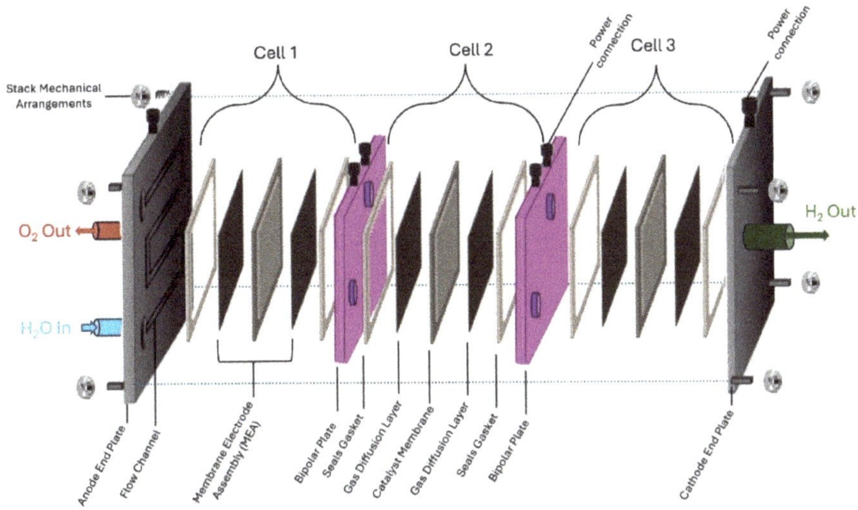

Figure 4. Basic cascading method of PEM electrolysis stack.

In general, several essential components can be found in most electrolysers' stacks [63,75]:

1. Feed stock (water) flow interconnection and recirculate water.
2. Bipolar plates interconnecting cells back to back (anode to cathode).
3. Anode and Cathode end plates at both ends of the stack.
4. Cooling plates (coolant flow).
5. Power connection arrangements.
6. Stack mechanical arrangements, such as the following:
    a. Bolts;
    b. Layers' stiffness;
    c. Machining gaskets;
    d. Seals;
    e. Clamping mechanism.

Figure 5 shows some examples of actual electrolyser stacks.

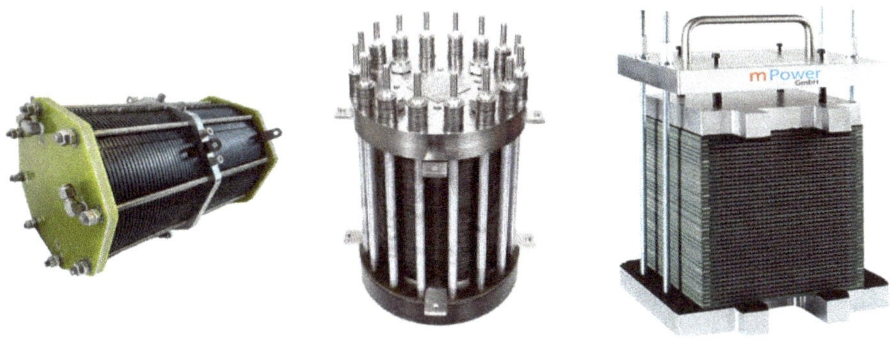

(**a**) ALKE stack (**b**) PEME stack (**c**) SOE stack

Figure 5. Examples of actual electrolyser stacks [7]: (**a**) ALKE, (**b**) PEME, and (**c**) SOE.

3.3. Electrolyser Package

The electrolyser stack requires support, auxiliary, and control systems to function, i.e., produce hydrogen [64,70]. These systems vary depending on the technology type and the scale. The electrolyser stack with all required sub-systems is referred to as an electrolyser package in this study. In general, the electrolyser packages' BoP (level 2), as illustrated in Figure 6 (a and b), comprises (depending on the stack technology and size) many supporting systems, such as (but not limited to) the following [60,63,76]:

1. Power supply rack (transformer and/or rectifier), Power Management System (PMS) and production control.
2. Water supply, including storage, treatment such as desalination and/or purification and de-ionisation, and feeding systems, such as water pumping.
3. Liquid electrolyte storage and circulating system (lye system) for the AKLE [76] or ion exchanger circulating system for the PEME.
4. Cooling system and/or heat exchangers.
5. Pressure control system.
6. Monitor, control, and Human interface, including remote control, communication, software, and PLC.
7. Hydrogen buffer tank.
8. Hydrogen purification system, i.e., impurities removal and deoxygenation to remove remnant oxygen.
9. Hydrogen dryers.
10. Ventilation system.
11. Venting valves and pipework.
12. Sensors and alarm system.
13. Interconnecting wires and pipelines.

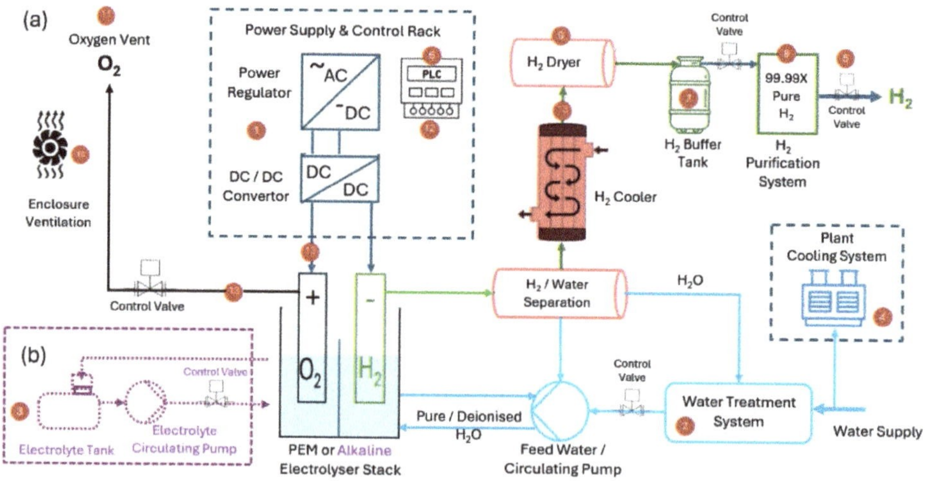

Figure 6. Typical BoP diagram of (**a**) Proton Exchange Membrane Electrolyser (PEME), (**b**) additional sub-system for Alkaline Electrolyser (ALKE).

Most electrolysers' packages supporting sub-systems are generally the same, while technology-specific systems vary [76]. The feedstock (water) system complexity and integration depend on the water source, i.e., whether the water source is desalinated, collects rainwater, or has potable water mains [77]. Typically, the water supply system is not considered a part of the electrolyser package, but purification (demineralisation and deionisation) and the feed pump are part of the package, even if some or all parts are located outside the main package enclosure [78]. The water purity depends on the electrolyser's technology but is typically required to be ultrapure and deionised feedstock [79]. The feedwater circulating system at the anode side of the stack in the PEME packages is essential to avoid the release of metal ions over time [76]. While liquid electrolyte storage tank and circulating system (typically pumps) can be found in the AKLE packages [80]. Also, the power supply rack may be located in the same enclosure or outside, depending on the electrolyser package scale. The power supply system varies depending on the power source, type, scale, and coupling with the hosting power system or grid connection requirements.

Typically, for small-to-medium-scale electrolysers, the stack and required support sub-systems, such as control and auxiliary systems, are delivered in one container (enclosure), as shown in Figure 7. However, some OEMs use more than one container or locate some sub-systems in other parts of the H2PS enclosures for safety or ease of access during operations and maintenance [26].

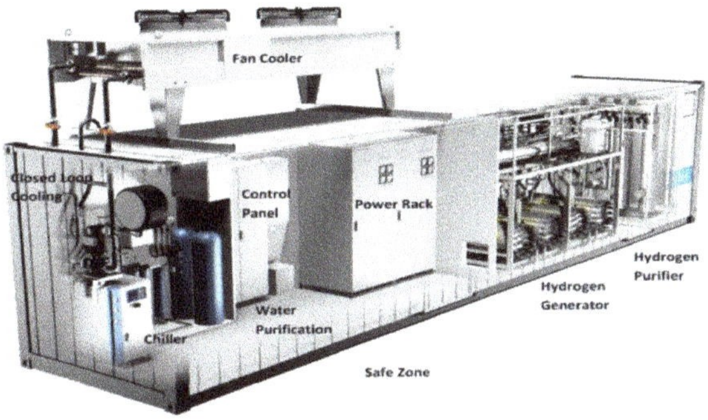

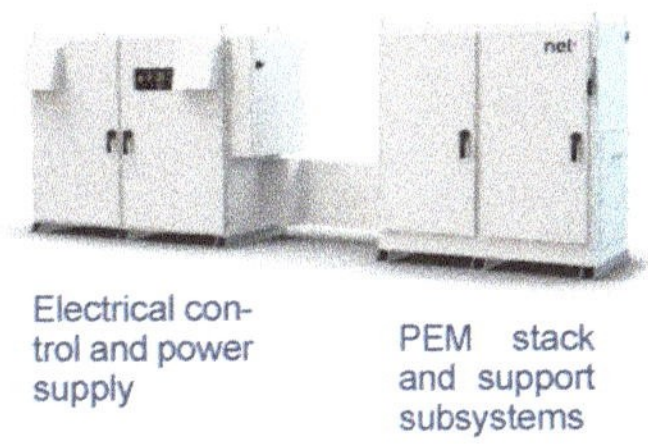

Figure 7. Examples of electrolyser's package BoP [7].

4. Fuel Cells

4.1. Fuel Cell Technologies

The Fuel Cell (FC) uses hydrogen as a fuel to generate electricity in the form of direct current (DC) by an electrochemical reaction; i.e., in a reverse way compared to electrolysers, where the anode of the electrolyser cell becomes the cathode of the FC, and the cathode becomes the anode. However, the directions of the migration of anions and cations with respect to current flow are unchanged [81].

The overall reaction is explained in Equation (2) [82] and as compared in Figure 8. Whilst hydrogen gas is the feedstock (fuel), oxygen is typically obtained from the air (~21% is oxygen).

$$2\,H_2\ (\text{gas}) + O_2\ (\text{gas}) \rightarrow 2\,H_2O\ (\text{liquid}) + \text{Electrical and Heat Energy} \quad (2)$$

The FCs were used for stationary and mobile electricity generation at a range of sizes. It is the heart of the H2PS to utilise the stored hydrogen for power regeneration, though hydrogen can be used as a fuel for internal combustion engines or hydrogen gas turbines to generate power [6]. The trend is to use FCs in the H2PS, as they are more efficient compared to hydrogen-burning generators [59]. However, if the generated heat is utilised in a combined heat and power (CHP) system, low-temperature FCs are considered less efficient. This study focuses on FCs for power regeneration as part of the holistic H2PS BoP.

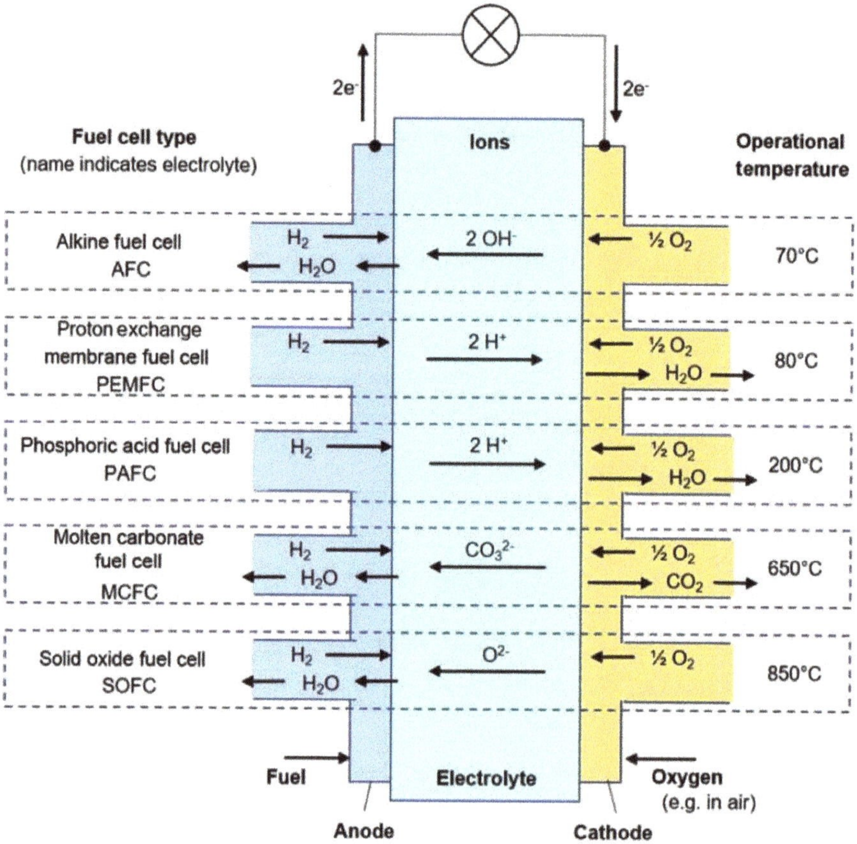

Figure 8. Fuel cell types with ions transported and typical operational temperatures [7].

There are five well-known types of hydrogen FCs, which are distinguished by their electrolytes and the reactions that take place on the electrodes, namely [81,83,84]:

1. Alkaline Electrolyte (AFC);
2. Proton-Exchange Membrane (PEMFC);
3. Phosphoric Acid PAFC;
4. Molten Carbonate (MCFC);
5. Solid Oxide (SOFC).

These technologies under the lens of this section are shown in Figure 8, and selected relevant specifications are compared in Table 2.

Table 2. Types of Fuel Cells: selected cell and stack specifications and properties comparison.

Specifications	AFC	PEMFC	PAFC	MCFC	SOFC
Electrode/Catalyst/Reaction (Anode)	Carbon electrodes Nickel coated. perforated stainless steel [81] $H_2 + 2(OH)- \rightarrow 2H_2O + 2e-$ [75,85]	$H_2 \rightarrow 2H+ + 2e-$ [85]	Platinum (Pt) alloys as the catalyst (0.10 mg Pt cm − 2) [81] $H_2 \rightarrow 2H+ + 2e-$ [85]	Ni–Cr or Ni–Al [81] $H_2 + CO_3 = \rightarrow H_2O + CO_2 + 2e-$ $CO + CO_3 = \rightarrow 2CO_2 + 2e-$ [85]	cermet of yttria-stabilised zirconia (YSZ) and Nickel [81] $H_2 + O= \rightarrow H_2O + 2e-$ $CO + O= \rightarrow CO_2 + 2e-$ $CH_4 + 4O= \rightarrow 2H_2O + CO_2 + 8e-$ [85]
Electrode/Catalyst/Reaction (Cathode)	Carbon electrodes, Silver-plated nickel screen [81] $½ O_2 + H_2O + 2e- \rightarrow 2(OH)-$ [85]	Platinum nanoparticles on carbon black [61] Pt and ruthenium (Ru) 0.2 mg Pt/cm^2 [81] $½ O_2 + 2H+ + 2e- \rightarrow H_2O$ [85]	Pt supported on carbon black Pt alloys as the catalyst (0.50 mg Pt cm^{-2}) [81] $½ O_2 + 2H+ + 2e- \rightarrow H_2O$ [85]	nickel oxide Lithiated NiO [81] $½ O_2 + CO_2 + 2e- \rightarrow CO_3=$ [85]	strontium-doped lanthanum manganite (LSM), $La_{1-X}Sr_XMnO_3$ [81] $½ O_2 + 2e- \rightarrow O=$ [85]
Electrolyte/Membrane	Liquid [86] Potassium Hydroxide (KOH), typically 33 wt.% [81]	Solid [86] sheet of electrolyte (Perfluorosulfonic acid) [87]. Solid or quasi-solid membrane. Nafion membrane [81]	Liquid [86] Inorganic acid, concentrated phosphoric acid (H_3PO_4) (100 wt.%) [81]	Solid [86] Eutectic mixtures of Li–Na carbonate (Li_2CO_3–Na_2CO_3) Recently: α-$LiAlO_2$ β-$LiAlO_2$ [81]	Solid [86] Zirconia doped with 8–10 mol.% yttria (YSZ) with a small amount of alumina [81]
Bipolar Plate	titanium, stainless steel, or nickel [88]	Graphite or stainless steel, titanium, aluminium, and several alloys [61,81]	multilayer porous graphitic carbon bonded on either side of a thin, non-porous carbon layer Ribbed bipolar plate [81]	thin sheets of stainless steel [81]	Tubular Design. Doped lanthanum chromite for the interconnect in HT-SOFCs [81]
Water or Steam Produced	Anode [81,89]	Cathode [81]	Cathode [81]	Anode [81]	Anode [81]
Ion Movement	hydroxyl (OH^-) [81]	Hydrogen Ion (H^+) [81]	Hydrogen Ion (H^+) [81]	CO_3^{2-} [81]	O^{2-} [81]
Operating Temperature (°C)	50–230 [81] 60–70 [86]	30 to 100+ [81] Up to 180 [90] Up to 200 [86]	~220 [81]	600–700 [81]	600–1000 [81]
Heat Quality	Poor but usable for hot water supply [81]	Poor but usable for hot water supply [81]	Usable typically, 150–180 °C [81]	Very good Suitable for CHP [87]	Very good Suitable for CHP [87]
H_2 Fuel Purity Requirements	Pure H_2, CO_2 poison the FC CO >10 ppm [81]	Pure H_2 and susceptible to poisoning by sulphur and CO [81], i.e., <10 ppm CO	H_2, (low S (H_2S and COS) < 50 ppm, poisoned by a few CO ppm, tolerant to CO_2) Or other fuel that will need refining or processing [81]	H_2 (No S) or Various hydrocarbon fuels O_2 and CO_2 to be supplied to the cathode to form CO_3 pure CO can be used as a fuel [81,83]	Impure H_2, (No S), or fuel that will need refining or processing. CO can serve as a fuels' S tolerance [81,83]
Current Density (A/cm2)	0.2–0.5 [81]	High densities [81] 0.6–2.0 [91]	0.2–0.4 [81]	Typically, 0.16, Pressurised stacks 0.5 [81]	0.2–1.0 [91] 1 at 1000 °C
Operating Pressure (bar)	Ambient–4 bar [91]	Ambient–5 bar [91]	Ambient–8 bar (typically ~1 to 3 bar) [91]	N/A	Ambient (atmospheric pressure) [91]
Stack Efficiency (%)	Up to 70 [81] Around 60 [92]	50–60 [92] 40 reformed fuel [87] 60, and (87 CHP) [86]	>50 (LHV H_2) [81] Over 80 [92] (80% CHP) [86]	60–80 [92] 50 [87] (80% CHP) [86]	>50% (LHV) [81] 60–80 [92] 60 [87]

Table 2. *Cont.*

Specifications	AFC	PEMFC	PAFC	MCFC	SOFC
Cell Voltage "open-circuit (no-load) voltage (OCV)" (Volts)	0.7–0.9 [88]	0.6 to 0.8 [93]	0.6–0.75 [81]	0.75–0.8 [81]	0.7–0.9 [91]
Typical Stack Size (kW)	1–100 [87]	1–100 [87] 1 MW [81]	5–400 [87]	300 modular, up to 3 MW [87]	1–2000 [87]
Cold start to Full Load	Quick start-up [87] Reservoir holding the electrolyte solution needs to be heated [81]	Quick start-up and load following [87]	Long start-up time [87]	Long start-up time [87]	Long start-up time and Limited number of shutdowns [87]
Lifetime (K hours)	>50 [88]	>40 [81]	40 [81]	40 [81]	40 [87]
Estimated Cost per kW (USD) [94]	500–1500	1200–3000	3000–5000	3000–6000	2000–5000
State of Development	Commercial [81]	Commercial [81]	Most Commercially Developed [81,85]	Mature/Marketed [81]	Commercial [81]

Fuel Cells Emerging Technologies

There are less well-known types of FCs, as they are still under development, such as Direct Borohydride (DBFC) [81], Direct Carbon (DCFC) [81], Biological (Microbial) MFC [81], and Reversible Fuel Cell [83]. These types of FCs are not used in the H2PS yet, while other technologies possess promising potential, such as Anion-Exchange Membrane Fuel Cells (AMFC) [81] and Unitized Regenerative Fuel Cells (URFC) [64].

4.2. Fuel Cell Package

This study focuses on the typical types of FCs found in existing and developing H2PS; therefore, it does not discuss Direct Methanol Fuel Cells (DMFCs) and other liquid-fed FCs because they do not use pure hydrogen directly to regenerate electricity. FCs fundamentally consist of similar construction parts (in general) of the electrolyser's stack construction and arrangement, i.e., level 1, as shown in Figures 3–5 [81,83,95,96]. However, there are some key differences in their BoP at levels 1 and 2 due to the fact that FCs work in opposite directions compared to electrolysers. Individual fuel cells are typically combined in series into a fuel cell stack to increase the single cell voltage from well below 1 V to a sufficient voltage suitable for applications [97]. The FC stacks' desired produced power and heat depends upon several factors, such as fuel cell type (technology), cell size, the operating temperature, and the pressure of the hydrogen gas supplied to the cell [83]. The FC's feedstock (fuel) is pure hydrogen gas, but each type of FC operates at a specific pressure and minimum accepted hydrogen purity [81].

Key Differences Between Electrolysers and Fuel Cell Stack Designs

There are some key differences between electrolysers and FCs stack designs and construction due to their different functions, such as the following [81,98]:

(1) Electrode materials.
(2) Electrolyte type.
(3) Operating conditions, such as operating heat and pressure.
(4) Gas management, i.e., modified, or different geometry.
(5) Auxiliary and sub-systems, e.g., air instead of water pumping system.
(6) Cell connections.

These differences are more related to the manufacturers than the H2PS designers, as shown in Tables 1 and 2. However, the designers should consider the equipment and supporting systems BoP for the FC package to function, i.e., regenerate electricity.

The FC stacks' BoP requires different auxiliary sub-systems compared to the electrolysers' packages [81]. In addition, the requirements for the FC package depend on the application; i.e., there are different auxiliary sub-systems for stationary power regeneration compared to the FC in mobility applications [81]. This study focuses on stationary power generation.

The essential auxiliary sub-systems within the FC package depend on the size (capacity) of the FC stack. Figure 9 illustrates an example of PEMFC system (package) integration and interconnections in a single line diagram. In general, the FC package comprises the following (but not limited to) auxiliary and sub-systems:

1. Hydrogen clean-up (purification) processors, e.g., carbon dioxide separators and sulphur removal, depend on the produced and/or stored hydrogen quality [17,81].
2. Air humidifier unit to prevent membrane dehydration [8,81,83,92].
3. Hydrogen pressure regulation (control) skid depends on the system's storage technology and operating pressure [99,100]. Typically, FCs operate in a low-pressure hydrogen inlet [101]; practically, it is between 8 and 16 bars [102].
4. Hydrogen recycling system for re-use of unconsumed hydrogen [81].

5. Air delivery units, i.e., air compressor or blower with a control unit [83,92,100].
6. Water management, i.e., water removal (prevent flooding), water purge (prevents freeze-out damage if the ambient temperature falls below zero °C), drainage and/or reclaim unit, typically a condenser and water pump [81].
7. Electrolyte circulating system for FCs using liquid electrolyte, i.e., AFC, typically a tank and pump with pipes [103].
8. Power-conditioning equipment, e.g., DC/DC converter and DC/AC inverter, to comply with the hosting grid connection requirements [81,83,92,100].
9. Intercooling system, which depends on the FC capacity, e.g., fans for air cooled or liquid coolant with heat exchange mechanism [90,92]. This required a third channel in the bipolar plate [90].
10. Overall control is called Fuel Cell Control Unit (FCCU) [90].
11. Human–Machine Interface (HMI) and PLC interface for remote control [104].
12. Power supply, typically battery for small-medium scale FCs, to provide power for controllers, valves, DC/DC converter, and pumps on start-up [81,100,104].
13. Electrical and Thermal insulation and protection [81]. This is essential for high-temperature FCs [105,106].
14. Enclosure ventilation system to prevent hydrogen leakage (if any) from accumulating within the closure [81].
15. Venting valves and pipework [100].
16. Sensors, safety systems [17], and alarm system [100].
17. Electrical interconnecting wires [81].
18. Pipework (hydrogen, air and water) [100].
19. Operating noise reduction mechanism [107].

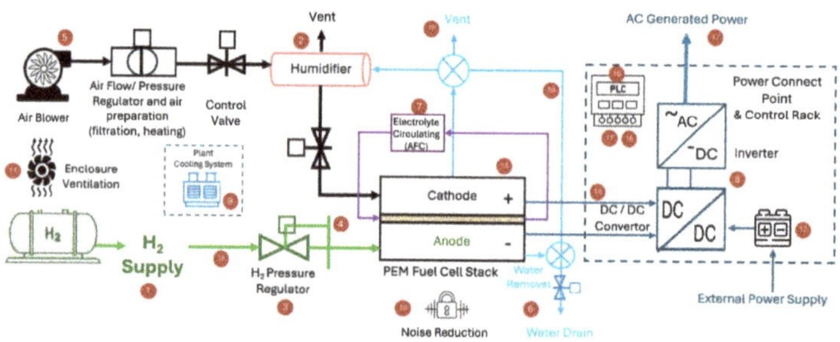

Figure 9. PEMFC package integration and interconnections single line diagram.

The FC integrated system or package can be enclosed in one or more containers. Figure 10 shows some examples of FC packages.

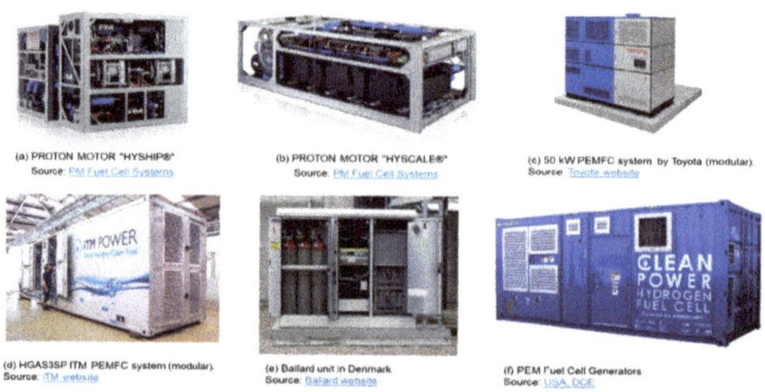

Figure 10. Examples of diverse types of FC packages [7].

5. Hydrogen Storage

5.1. Technologies

Hydrogen is gaseous at Standard (Room) Temperature and Pressure (STP) with a low volumetric energy density, i.e., a large footprint for low energy density compared to hydrocarbon fuels [17,81]. It is difficult to store in large quantities safely and cost-effectively at the STP. Therefore, different hydrogen storage technologies have been developed, and many other technologies are still being developed [15]. Nevertheless, hydrogen compression and cooling/heating are the main processes required for hydrogen storage technologies, regardless of the type of technology and storage containers [108]. Typically, the storage package comprises two parts: hydrogen process and storage vessels with all required auxiliary systems.

The manufacturing process, parts of the compressors, and thermal reactions within the storage vessel internally (at level 1) are out of the scope of the H2PS engineering design, though they can significantly change the operations and maintenance requirements. The system designers have no role in the level (1) BoP. Therefore, this study focuses on the storage package BoP at levels (2) and (3).

The choice of storage technology depends on land availability (footprint), safety, and cost-effectiveness. These different hydrogen storage technologies can be categorised into three main categories, namely, physical-based, material-based, and chemical-based (hydrogen carrier), as illustrated with some selected examples in Figure 11.

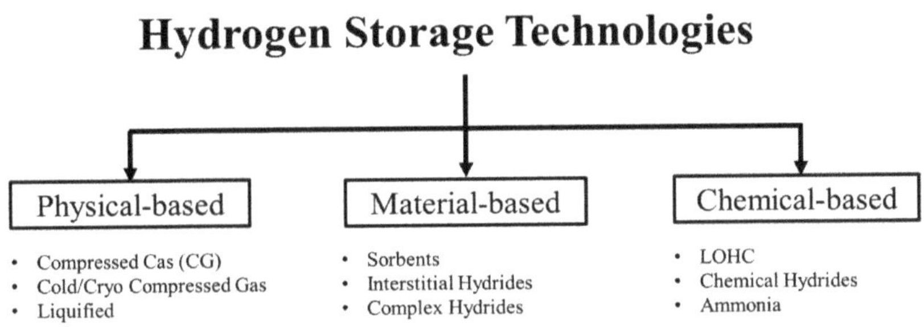

Figure 11. Hydrogen storage technologies (pathways).

5.2. Physical-Based Storage Technologies: BoP

Compression and cooling are the two main process requirements for all physical-based hydrogen storage, regardless of the type of technology and storage container shape, size, and make [17]. The physical-based technologies mean pure hydrogen is stored in its pure form at any matter phase without chemically reacting or binding with any other substance [108].

There are three well-known technologies to store pure hydrogen physically, which are defined by the pressure–temperature operating regimes, as follows [109]:

1. Hydrogen Compressed Gas (H2CG) operates at high pressures, as high as 70 MPa, and near ambient temperature.
2. Hydrogen Cryo-compressed Gas (H2CcG) typically operates at around and above 350 bars and temperatures less than -120 °C (150 K).
3. Liquid hydrogen (also called Cryogenic Liquid) operates at low pressures, i.e., typically less than 6 bars (<0.6 MPa) and low temperatures near the normal boiling point of the hydrogen, i.e., -253 °C (20 K).

Figure 12 shows the hydrogen pressure–temperature operating regimes, phase of matter and density at different temperatures and pressures.

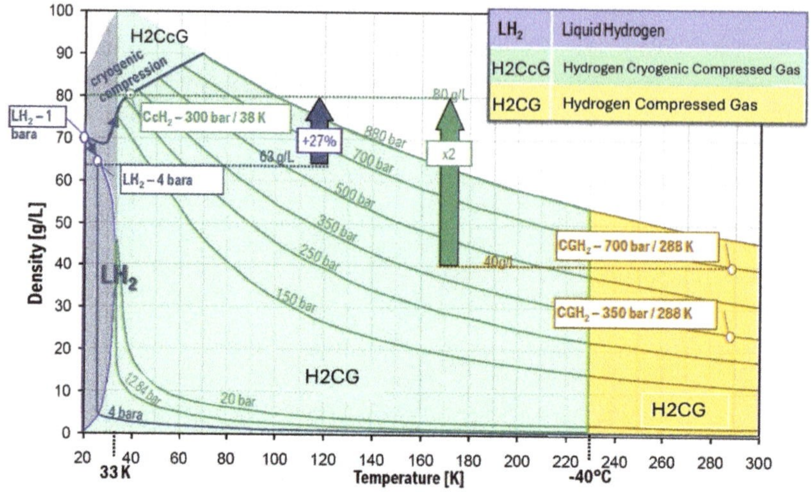

Figure 12. The hydrogen phases of matter and density at different temperatures and pressures. Edited from [7].

The BoP for the physical-based hydrogen storage technologies at level 2 (storage packages) depends on the compression type, cryogenic cooling system, storage container type, and operating parameters. This section discusses the common parts, equipment, and sub-systems for the well-known physical-based hydrogen storage technologies at level 2, i.e., packages.

The underground hydrogen storage in salt caverns, exhausted oil and gas fields or aquifers is acquiring attention as a physical-based pathway using H2-CG technologies [6]. However, it is geographically challenged; therefore, it is out of the scope of this study in the context of H2PS-BoP.

5.2.1. Compressed Gas Package

The Hydrogen Compressed Gas (H2CG) package comprises two main parts: the compressor unit and the storage vessel (typically low to 1000 bar pressure cylinders). Depending on the size of the H2PS (i.e., stored hydrogen mass), the compressor unit may come in one or two enclosures. In contrast, the multi-cylinder orientation arrangement determines the footprint of the vessel. The H2CG package BoP typically comprises (but is not limited to) the following:

1. The compressor unit (package) type depends on the required pressure, mass flow rate, and technology's techno-economic choice [110,111]. The compressor is the main part of H2CG technology, which operates as an intermediate system between the production and the storage vessel. Typically, the compressor unit is delivered from the OEM as a fully functioning unit packaged in one or more enclosures, which includes the following [25]:
    - (a) Power supply switchboard (depends on the compressor size).
    - (b) Bypass and pressure relief valves (manual or auto-actuated).
    - (c) Hydrogen and purge gas venting system.
    - (d) Control system and/or PLC unit (depending on the size of the system and its integration into the holistic H2PS requirements) for operations and remote control, if required.
    - (e) Safety systems, including hydrogen detectors, shut-off (stop) functions, and alarm systems, are typically automatic.

(f) Compressed air system and pipework if valves are air-actuated.
(g) Compressor cooling system for internal and compressed hydrogen outlet cooling. Cooling the outlet compressed hydrogen gas depends on the mass flow rate and the operating temperature of the storage vessel.
(h) Hydrogen pipelines.

Note: Using high-pressure electrolysers can eliminate the need for the compressor unit unless more compression is required.

2. Storage vessels are typically cylinders where the compressed hydrogen is stored within. The storage vessels system, in general, includes the following:
   (a) Storage vessel, where its type depends on the operation pressure, footprint, application, and cost-effectiveness.
   (b) Check and shut off (isolation) valves, safety raptures, inlet/outlet valves, and pipelines.
   (c) Ventilation system in the case of enclosed storage vessels.
   (d) Control system for operations and remote control (if required).
   (e) Safety systems, including leak detectors, isolation valves, and alarm systems.
   (f) Compressed air system and pipelines (if valves are air-actuated).
   (g) Venting system for hydrogen and purging gas.

5.2.2. Cryogenic Compressed Gas and Liquifying Package

Hydrogen gas can be compressed with cryogenic cooling to be less energy-intensive than liquified hydrogen and have more volumetric density than H2CG [17]. Hydrogen Cryogenic Compressed Gas (H2CcG) at high pressure, i.e., above 150 bars (typically 350 bars [109]) and temperature between the boiling temperature (−253 °C) and typically less than −124 °C [109], can have a higher density than liquid hydrogen, as shown in Figure 12 [17]. In addition, cryogenic operations at a low temperature, as low as −196.15 °C (77 K), maximise hydrogen absorption when an electrochemical compressor is used, resulting from the possibility to compress hydrogen to a high pressure of up to 700 bar in a single step [69]. Nonetheless, a notable advantage of H2CcG over Liquid Hydrogen (LH2) is that it requires less energy to produce, i.e., the theoretical compression to 200 bars and cooling work to −193 °C (80 K) is approximately 10 MJ/kg, which is similar to the work required to compress hydrogen to roughly 500 bars compared to 22–50 MJ/kg of the liquefaction work [109].

On the other hand, cryogenic (refrigeration) below −252.9 °C (usually rounded to −253 °C) [112], even with low pressure (as low as 6 bars), hydrogen becomes Liquid Hydrogen (LH2) [109], as shown in Figure 12.

In contrast, typical H2CcG or LH2 system BoP comprises the same main parts of the H2CG system, with the following differences:

(a) Cryogenic refrigerator is the heart of these technologies, typically one of the technologies, depending on the scale (hydrogen mass) and the system's cost-effectiveness [109,113].
(b) Cryogenic and/or liquid hydrogen containers (vessels), typically made of stainless steel and aluminium [114], are specifically designed to reduce heat leakage by thermal insulation. The H2CcG and LH2 vessels must be specifically designed to handle extremely low temperatures [115]. Specifically, LH2 tanks are expected to be built at larger (capacity) units compared to H2-CG tanks, as they do not handle high pressure [116]. Additionally, in LH2 plants, liquid hydrogen boiloff management system is crucial to reduce the heat transfer (losses) from the surroundings when hydrogen is stored for a long time [17,116].

(c) Vaporisers are used to vaporise LH2 when hydrogen is needed [17]. Vaporisers use heat to convert liquid hydrogen back into its gaseous state. Several types of vaporisers exist, including ambient air vaporisers, steam-heated vaporisers, electric vaporisers, and microwave-induced plasma [117].

(d) A hydrogen boiloff venting system is crucial, and safety concerns must be addressed as vented cold hydrogen is denser than ambient air and may fall back to the plant grounds [118].

(e) Cryogenic pumps transfer LH2 from the vessel to the vaporiser and/or the buffer tanks [103].

Cryogenic-suitable materials of any type in contact with LH2 or H2CcG, such as package's sub-systems, auxiliary, valves, and pipelines, shall be embrittlement-resistant and capable of withstanding extremely low temperatures, i.e., sufficiently flexible to provide for the effect of expansion and contraction due to temperature changes [110,114,119].

5.3. Material-Based Storage Technologies: BoP

Material-based hydrogen storage technologies have the potential for high volumetric density at lower pressure (close to ambient conditions), significantly improving safety [120]. Materials chemistry, structure, and properties have attracted the interest of many leading researchers as a promising, safe, and high-density technology to store hydrogen [121]. The choice of the material to store hydrogen depends on the required storage capacity, operating conditions, safety, and cost-effectiveness. Thermal management is at the heart of material-based technologies that control the hydrogen loading and release rates at desired levels [122]; some examples of hydrogen storage materials are shown in Figure 13.

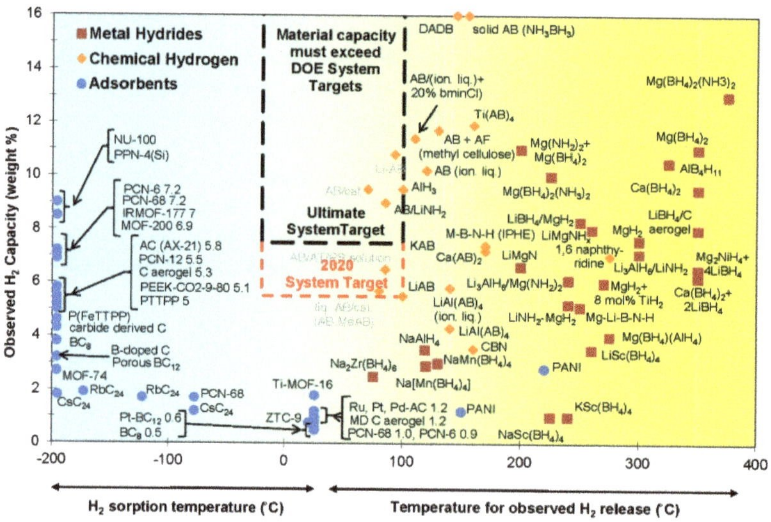

Figure 13. Plot of hydrogen storage materials observed capacity (weight %) at sorption and release temperatures [7].

The BoP in the context of a material-based hydrogen storage package in use within the H2PS is significantly different compared to physical-based storage pathways [122]. These technologies work on low pressure, which may eliminate the need for a compressor unit. Also, the hydrogen uptake and control of release are managed via a thermal management system [123]. A typical material-based storage package consists of the following:

1. A thermal management system controls the hydrogen uptake (hydrogenation) and the release at the required flow rate [92]. The uptake reaction is exothermic; i.e., heat has to be removed while heat is required to enable the endothermic decomposition of the hydride (dehydrogenation) [123]. Therefore, the released hydrogen must be cooled and compressed to the FC inlet-specific operation pressure [124]. The required release heat, typically, can be obtained from an external source or utilise the heat releases from the regenerating package, i.e., the FC [123]. The external heat energy can be obtained from burning a portion of the released hydrogen to control the release flow rate [124], which adds complexity to the package. The H2PS-associated battery may be used for the startup of the release process. Also, the package includes a cooling and/or cryogenic refrigerator for the intake process and control [125].
2. The material-specific storage vessel type and specific design depend on the type of material in use, storage size, operating pressure, and temperature range.
3. Typically, the package requires moisture sensors in addition to the H_2 and O_2 sensors [124].
4. A hazardous waste-materials disposal management system depends on the type of storage material and the system operation conditions [85].
5. A low-pressure compressor might be required (depending on the storage system H_2 release pressure) to pressure the released H_2 to the FC inlet operating pressure [124].
6. Power supply switchboard (depends on the compressor size and the thermal management technology).
7. Pressure relief valves (typically auto-actuated).
8. Hydrogen and purge gas (typically nitrogen) venting system.
9. Control system and/or PLC unit (depending on the system's size and integration requirements into the H2PS) for operations and remote control (if required).
10. Safety systems include a shut-off (stop) function and an alarm system, typically automatic.
11. Compressed air system and pipelines (if valves are air-actuated).
12. Compressor cooling system for internal cooling and the released H_2 cooling after compression.

5.4. Chemical-Based (Hydrogen Carriers): BoP

Hydrogen storage by chemically binding with other materials is a promising pathway to efficiently and cost-effectively ease the H_2 storage difficulties [124]. However, this pathway is impractical for H2PS when hydrogen is produced on-site, as converting hydrogen to synthetic hydrocarbon fuels, LOHC, or non-organic hydrogen carriers adds significant complexity to the system [104]. Therefore, this H_2 storage pathway is outside the scope of this study.

6. Holistic H2PS Systems' BoP

The literature review reveals many research studies, industry knowledge-sharing reports, and OEM datasheets related to the H2PS parts and equipment. However, it is rare to find publications related to the BoP of a holistic H2PS. Therefore, this study depends on the authors' experience, personal communications, and publicly shared information from the implemented H2PS worldwide [11,13,14,45–57,126,127].

Typically, H2PSs comprise three main components or packages: hydrogen production, storage, and power regeneration. Each package relies on additional auxiliary systems and sub-systems to operate effectively. These packages are integrated and interconnected to make a holistic H2PS. The holistic H2PS must meet the integration requirements of the hosting power system or the system it serves, both at the input and output. Figure 14 visually represents how the typical packages in an H2PS at level 3 (the holistic system) are integrated with the supporting auxiliary and sub-systems.

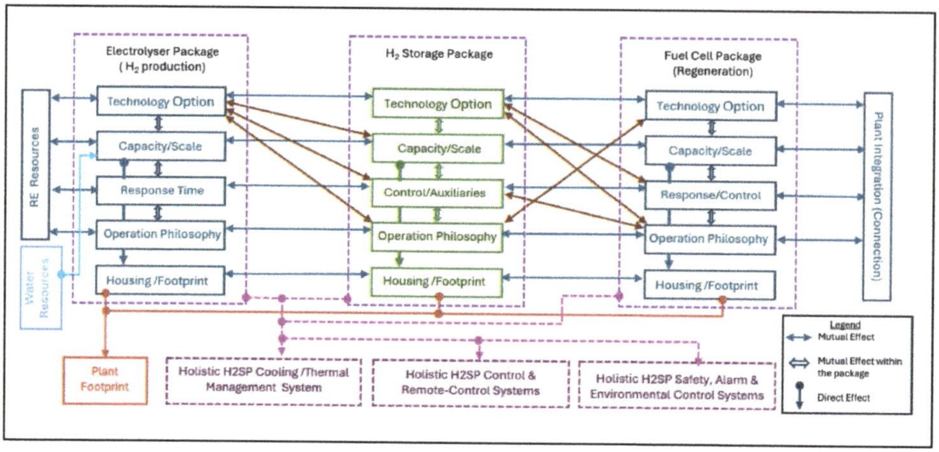

Figure 14. The hydrogen-based power systems (H2PS) three main parts (packages), namely, electrolyser(s), storage, and power regeneration fuel cell(s) with supporting and auxiliary sub-systems.

The hydrogen packages are typically fully functional devices, incorporating various supporting auxiliary, sub-systems, and peripherals from original equipment manufacturers (OEMs). These peripherals can include control, sensors, safety devices, and essential equipment for package functionality. The peripherals included may differ based on the intended use of the package, the type of technology employed, and the processes followed by the OEM company. The previous sections have covered the balance of plant (BoP) for each package at levels 1 and 2, considering the equipment type and technology used. The subsequent sub-sections will discuss the BoP of the holistic H2PS at level 3 from an EDO perspective.

6.1. Non-Hydrogen Auxiliary and Sub-Systems

Typical H2PS, as shown in Figure 14, requires many non-hydrogen systems to function effectively and operate safely. The minimum (i.e., but not limited to) requirements are as follows:

1. Renewable Energy (RE) power generation plants, such as solar PV, wind turbines, oceanic energy, etc., depend on the area's RE resources availability, whilst the plant size is related to the scale of the H2PS.
2. Water supply, storage, and purification systems and related pipework, whereby complexity depends on the H2PS scale and water sources, i.e., seawater, underground, or potable water.
3. The cooling system and its related pipework depend on the H2PS scale and packages' technologies, i.e., the cooling requirements.
4. The compressed air system depends on the H2PS scale and the pneumatic devices in use.
5. Non-product stream treatment and disposal include brine water, coolant, lubricants, and any other waste related to specific technologies in use. The treatment and disposal process shall follow environmental codes and regulations at all levels.
6. Point of connection switchboard and electrical connections to and from the H2PS I/O terminals, i.e., cables, trays, converters, inverters, and protection systems. The

electrical equipment shall be compatible with an explosive atmosphere requirement, depending on the safe distance from the hydrogen environment.

7. Electrical buffer system to manage the load rejection and shedding, as well as the start-ups and shutdowns of the H2PS equipment. These buffers are typically batteries and/or supercapacitor banks. The buffer system's size depends on the scale of the H2PS and the hosting or served power system size and orientation.
8. Fire detection and suppression systems associated with the holistic H2PS.
9. Civil work and safeguards include concrete foundations, footpaths, heavy machinery access, barriers, and fences. Note: Civil works are out of the scope of this study.

The non-hydrogen systems are typically designed to follow the control philosophy and align with the holistic H2PS control system integration requirements and operational communications protocols, such as connections to plant PLC.

6.2. Interconnections and Interoperability

The holistic H2PS typically acts as an energy storage system consisting of three main packages and many other sub-systems amalgamated to function as one device, like a battery bank functionality. The holistic system packages and subsystems' interconnections and interoperability are the main challenges for system engineering designers. These challenges depend on the package's technology type, which determines the type of equipment connecting any two packages and the integration into the holistic H2PS. The packages' mutual interdependence is discussed in the following sections.

6.2.1. Production-Storage Packages' Interconnection

The interconnection between the production (electrolyser(s)) and the storage packages depends on the electrolyser outlet pressure and flow rate (production rate) and the specific inlet operation pressure and control philosophy of the storage package. The required interconnections and interoperability equipment can be summarised (but are not limited to) as follows:

(1) Hydrogen transfer pipework with proper mechanical support and safety valves.
(2) The hydrogen buffer (bladder) tank depends on the storage package inlet operation pressure and control philosophy.
(3) Control, sensors, and communication commands (PLCs) wiring with proper mechanical support.
(4) The cooling system that is typically part of the holistic H2PS cooling system but can be part of one or both supplied packages.
(5) A heat recovery system for CHP can be used in thermal management when material-based or cryogenic storage technology is used.
(6) A battery bank or buffer unit can be included in the electrolyser package or the holistic H2PS.

6.2.2. Storage-Regenerative (FC) Packages' Interconnection

The integration and interconnection of the storage and the regenerative (FC) packages depend on the storage outlet pressure and the FC package-specific inlet operation pressure and operations philosophy. The required interconnections and interoperability equipment can be summarised (but are not limited to) as follows:

(1) Hydrogen transfer pipework with proper mechanical support and safety valves.
(2) Hydrogen pressure reduction skid when high-pressure storage technology is in use or a low-pressure compressor when using low-pressure material-based or LH2 storage technology. The pressure regulation depends on the FC package inlet operating pressure and the control philosophy.

(3) Control, sensors, and communication commands (PLCs) wiring with proper mechanical support.
(4) The cooling system that is typically part of the holistic H2PS cooling system but can be part of one or both supplied packages.
(5) Heat recovery system for CHP or to control the hydrogen release flow rate when using material-based or LH2 storage technology.
(6) Reclaimed water and transferred from the FC to the water supply system, which depends on the FC technology, operating temperature, and water scarcity.
(7) The DC/DC converter unit can be included in the FC package or associated with the external DC/AC inverter, which is the gateway to the point of connection.
(8) The auxiliary power supply, typically a battery bank, can be included in the FC package or the holistic H2PS. Also, it could be an external power supply.

6.3. *Off-the-Shelf Engineered-Packaged Systems*

The literature reveals that many companies are offering pre-engineered assembly from different OEM companies and packaged as off-the-shelf products (H2PS), which are ready to be installed with fewer engineering design requirements. These systems may be provided in one or more enclosures with all the necessary operations and safety measures [128,129]. However, these systems may require external sub-systems and/or power regulation or inversion to match the site point of connection requirements. Typically, these systems comprise the same three main packages and the sub-systems of the case-specific-designed H2SP [128]. In addition, some OEM companies offer case-specific custom-built systems that may require the system engineers or the owner's engineer contractors' involvement [129]. Figure 15 shows two selected examples of engineered H2PS.

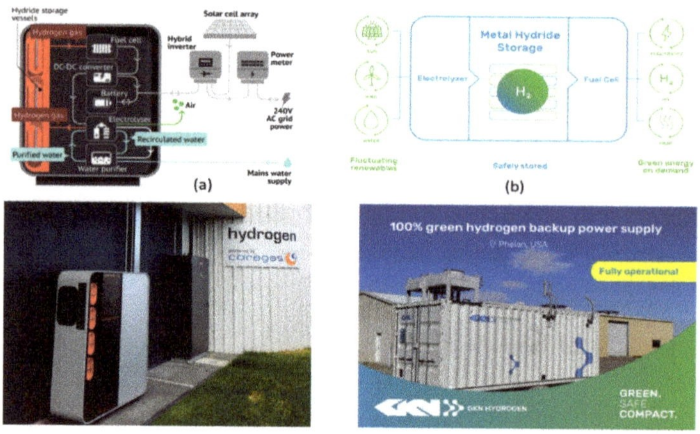

Figure 15. Examples of Pre-engineered H2PS [7]: (**a**) 40 kWh capacity (modulable) H2PS by LAVO, (**b**) small-medium scale H2PS by GKN.

6.4. *Technologies' Selection: Cause and Effect*

The hydrogen equipment technologies at a high level are typically decided during the preliminary stages, i.e., in the feasibility study stage. This requires informed decisions on the hydrogen equipment specifications, interconnection, and integration requirements. This study presents engineering design guidelines for the technologies' selection and the mutual effect of the technologies' options on holistic H2PS BoP integration and functionality, i.e., at level 3. The mutual impact is presented as a cause-and-effect matrix, as shown in Table 3.

Table 3. Technology selection and specifications' parameters cause and interdependent effect matrix.

Technology Selection Interdependent Cause & Effect Matrix Legend "√" = Can Have an Effect		Effect on the H2PS Packages																										
	Selected Technology (Cause) / **Effect**	**Grid-Connected or Stand-Alone**	**Site Location/Constraints**	**New or Replacing Existing PS**	**Stand-Alone or Part of a System**	**RESS and/or Grid Stabilising**	**Electrolyser Technology**	**Storage Technology**	**FC Technology**	**Type of RE Generation**	**H2PS Supporting BESS**	**Operating Pressure**	**Operating Temperature**	**Cooling System**	**Water Treatment System**	**Hydrogen Purity**	**Hydrogen Compression**	**Hydrogen Pressure Reduction**	**Thermal Management System**	**Buffer Unit**	**Holistic System Response Time**	**Cryogenic System**	**Point of Connection**	**Holistic H2PS Safe Operations**	**Hazardous and Exclusion Zones**	**Venting and Ventilation**	**Holistic H2PS Footprint**	**Environmental Controls**
Electrolyser(s)	Technology Option		√			√		√		√	√	√	√	√		√	√		√	√	√			√	√	√		√
	Production Capacity/Size in kW	√	√		√	√	√	√		√	√	√	√	√		√	√		√	√	√	√	√	√	√	√		√
	Response Time (Ramp Up/Down)	√			√	√	√			√	√																	
	Control/Operation Philosophy	√		√	√	√	√	√		√	√			√			√		√	√			√	√	√	√		
	Housing (Outdoor/Containerised)		√				√							√					√				√	√	√	√	√	
Storage	Technology Option		√			√	√	√	√			√	√	√		√	√	√	√	√	√	√		√	√	√	√	√
	Storage (H_2 Mass) Capacity	√			√	√		√		√	√	√		√			√		√			√		√	√	√	√	√
	Control/Operation Philosophy	√			√	√	√	√	√			√	√	√			√	√	√		√	√		√		√		
	Housing (Outdoor/Containerised)		√					√						√					√			√		√	√	√	√	√
Fuel Cell (FC)	Technology Option	√				√			√		√	√		√	√	√	√	√	√	√	√		√	√	√	√	√	√
	Regenerating Capacity/Size in kW		√		√	√		√	√		√	√	√	√	√		√	√	√	√	√		√	√	√	√	√	√
	Load Following/Steady Generating	√		√	√	√		√	√		√			√	√		√	√	√	√	√			√		√		
	Response Time (Ramp Up/Down)	√		√	√	√		√	√		√						√	√	√	√	√							
	Control/Operation Philosophy	√	√	√	√	√	√	√	√	√	√			√	√		√	√	√	√						√		
	Housing (Outdoor/Containerised)		√						√															√	√	√	√	
	Water Reclaiming		√											√	√				√									√

7. Conclusions and Way Forward

This study aimed to fill the literature gap concerning the Hydrogen-enabled Power Systems (H2PS) equipment, devices, auxiliary, and sub-systems Balance of Plant (BoP). The H2PS-BoP was researched regarding publicly available information such as manufacturers' specifications and datasheets, the literature, authors' experience, and personal communications across three critical levels. The first (level 1) encompasses the manufacturing (construction) BoP of hydrogen cells and stacks with their related devices regarding different technologies. This level was discussed briefly as it is not directly related to the engineering design of the H2PS projects. The packages (level 2) comprising the hydrogen stacks and devices, along with their auxiliary and sub-systems delivered as ready-to-use packages, were thoroughly discussed. The holistic H2PS (level 3) addressing the interconnection and interoperability BoP requirements were discussed from an engineering design perspective. This comprehensive study revealed that many common auxiliary and sub-systems are typically included in the OEM-delivered packages. However, many others vary depending on the choice of technology, which significantly affects other packages and the holistic H2PS BoP. The variation in technology choices' effect on other packages was presented as a cause-and-effect matrix. This matrix demonstrated the expected variations in the BoP at the packages' level 2 and the holistic H2PS level 3.

The findings of this study establish engineering design and operations guidance and accelerate the feasibility studies and implementations of H2PS worldwide. The list of equipment, devices, and sub-systems required to operate H2PS in this study introduced a database of existing and developing technologies, OEM packages, and system integration requirements. Nevertheless, expanding this study to other hydrogen systems like fuelling stations, mobility, transport, and export, and hydrogen for green industries, is vital to advance the Hydrogen Economy.

Funding: This research received no external funding.

Conflicts of Interest: The authors declare that they have no known competing financial interests or personal relationships that could have appeared to influence the work reported in this paper.

Abbreviations

The following abbreviations are used in this study:

H2PS	Hydrogen-based Power Systems
OEM	Original Equipment Manufacturer
BoP	Balance of Plant
RE	Renewable Energy
H2CG	Hydrogen Compressed Gas
EDO	Engineering Design and Operations
DC	Direct Current
ALKE	Alkaline Electrolyser
PEME	Proton Exchange Membrane Electrolyser
SOE	Solid Oxide Electrolyser
CHP	Combined Heat and Power
AEME	Anion Exchange Membrane Electrolyser
PCCE	Proton-Conducting Ceramic Electrolyser
URFC	Unitized Regenerative Fuel Cell
CFE	Capillary-fed Electrolysers
MEA	Membrane Electrode Assembly
PMS	Power Management System
FC	Fuel Cell

PAFC	Phosphoric Acid Fuel Cell
MCFC	Molten Carbonate Fuel Cell
DBFC	Direct Borohydride Fuel Cell
MFC	Microbial Fuel Cell
DMFC	Direct Methanol Fuel Cells
HMI	Human–Machine Interface
PLC	Programmable Logic Controller
FCCU	Fuel Cell Control Unit
STP	Standard Temperature and Pressure
H2CcG	Hydrogen Cryo-compressed Gas
LH2	Liquid Hydrogen
LOHC	Liquid Organic Hydrogen Carrier

References

1. Dawood, F.; Anda, M.; Shafiullah, G.M. Hydrogen production for energy: An overview. *Int. J. Hydrogen Energy* **2020**, *45*, 3847–3869. [CrossRef]
2. IRENA (International Renewable Energy Agency). *Geopolitics of the Energy Transformation the Hydrogen Factor*; International Renewable Energy Agency: Abu Dhabi, United Arab Emirates, 2022.
3. Lund, C.; Islam, F.D.K.; Shafiullah, G.M.; Anda, M.; Bahri, P. *Hybrid Solar Pv-Battery-Hyrdorgen System for Renewable Energy Standalone Microgrid Development*; Murdoch University: Perth, WA, Australia, 2023.
4. Al Ghafri, S.Z.S.; Revell, C.; Di Lorenzo, M.; Xiao, G.; Buckley, C.E.; May, E.F.; Johns, M. Techno-economic and environmental assessment of LNG export for hydrogen production. *Int. J. Hydrogen Energy* **2023**, *48*, 8343–8369. [CrossRef]
5. Salehi, F.; Abbassi, R.; Asadnia, M.; Chan, B.; Chen, L. Overview of safety practices in sustainable hydrogen economy—An Australian perspective. *Int. J. Hydrogen Energy* **2022**, *47*, 34689–34703. [CrossRef]
6. Srinivasan, V.; Delaval, B.; Dollman, R.; Towns, A.; Charnock, S.; Palfreyman, D.; Hayward, J.; Graham, P.; Foster, J.; Reedman, L.; et al. *Renewable Energy Storage Roadmap*; CSIRO: Canberra, Australia, 2023.
7. Dawood, F. Hydrogen-Based Power Systems Evolution and Applications: An Engineering Design and Operation Advisory Model. Ph.D. Thesis, Murdoch University, Perth, WA, Australia, 2023.
8. Colpan, C.O. *Fuel Cell and Hydrogen Technologies in Aviation*; Springer: Cham, Switzerland, 2022.
9. Furat, D.; Shafiullah, G.M.; Martin, A. Stand-Alone Microgrid with 100% Renewable Energy: A Case Study with Hybrid Solar PV-Battery-Hydrogen. *Sustainability* **2020**, *12*, 2047. [CrossRef]
10. Arup Australia Pty Ltd. *Australian Hydrogen Hubs Study*; COAG Energy Council Hydrogen Working Group: Sydney, Australia, 2019. Available online: https://www.dcceew.gov.au/sites/default/files/documents/nhs-australian-hydrogen-hubs-study-report-2019.pdf (accessed on 11 June 2023).
11. CSIRO. HyResource. Available online: https://research.csiro.au/hyresource/ (accessed on 11 June 2023).
12. AHC. Regulations and Standards. Available online: https://h2council.com.au/policy-regulation/standards (accessed on 11 June 2023).
13. CSIRO. Hyresearch. Available online: https://research.csiro.au/hyresource/projects/rd/ (accessed on 11 June 2023).
14. Power, H. *Horizon Power Denham Hydrogen Demonstration*; ARENA: Canberra, Australia, 2020. Available online: https://arena.gov.au/assets/2023/04/horizon-power-denham-hydrogen-demonstration-lessons-learnt-1-2.pdf (accessed on 11 June 2023).
15. CSIRO. Hyresource-Projects. Available online: https://research.csiro.au/hyresource/projects/ (accessed on 11 June 2023).
16. The Hydrogen Strategy Group. *Hydrogen for Australia's Future*; The Hydrogen Strategy Group: Canberra, Australia, 2018. Available online: https://www.chiefscientist.gov.au/sites/default/files/HydrogenCOAGWhitePaper_WEB.pdf (accessed on 11 June 2023).
17. Kotchourko, A.; Jordan, T. *Hydrogen Safety for Energy Applications—Engineering Design, Risk Assessment, and Codes and Standards*; Elsevier: Amsterdam, The Netherlands, 2022.
18. Shet, S.P.; Shanmuga Priya, S.; Sudhakar, K.; Tahir, M. A review on current trends in potential use of metal-organic framework for hydrogen storage. *Int. J. Hydrogen Energy* **2021**, *46*, 11782–11803. [CrossRef]
19. Dawood, F.; Shafiullah, G.M.; Anda, M. 100% Renewable Energy: A Stand-alone Hybrid Solar PV-Hydrogen-Battery Power Systems for Homeland Communities in Regional Western Australia. In *Transition Towards a Carbon Free Future: Selected Papers from the World Renewable Energy Congress (WREC) 2023*; Sayigh, A., Ed.; Springer Nature: Cham, Switzerland, 2024; pp. 39–55.
20. Sunfire-GmbH. Sunfire-Hylink SOEC. Available online: https://backend.sunfire.de/wp-content/uploads/2024/10/Sunfire_Fact-Sheet_SOEC_EN_digital.pdf (accessed on 11 June 2023).

21. Sunfire-GmbH. Sunfire-Hylink Alkaline. Available online: https://backend.sunfire.de/wp-content/uploads/2024/10/Sunfire_Fact-Sheet_AEL_EN-digital.pdf (accessed on 11 June 2023).
22. Seiemens-Energy. Siemens Silyser 200 and 300. Available online: https://www.siemens-energy.com/global/en/offerings/renewable-energy/hydrogen-solutions.html (accessed on 11 June 2023).
23. Proton-Motors-Power-Systems-PLC. Fuel Cell Power Systems. Available online: https://www.protonmotor-powersystems.com/en/ (accessed on 11 June 2023).
24. PowerCell-Group. Hydrogen Fuel Cells—P Stack and V Stack. Available online: https://powercellgroup.com/fuel-cell-stacks/ (accessed on 11 June 2023).
25. PDC. PDC Machines-Diaphragm Hydrogen Compressors. Available online: https://www.pdcmachines.com/ (accessed on 11 June 2023).
26. nel-hydrogen. Hydrogen Production. Available online: https://nelhydrogen.com/podcasts/ (accessed on 11 June 2023).
27. Nedstack. PEM Fuel Cell Solutions. Available online: https://nedstack.com/en (accessed on 11 June 2023).
28. McPhy. McLyzer. Available online: https://mcphy.com/en/equipment-services/electrolyzers/?cn-reloaded=1 (accessed on 11 June 2023).
29. MYNT First Element, Hydrogen-Electric Power Generators- H2-50. Available online: https://myntgroup.com.au/purple-hydrogen/h2-50/ (accessed on 11 June 2023).
30. LAVO Hydrogen. Industrial Clean Energy Storage. Available online: https://www.lavo.com.au/lavo-hydrogen (accessed on 11 June 2023).
31. ITM-Power. PEM Electrolysers. Available online: https://itm-power.com/ (accessed on 11 June 2023).
32. Intelligent-Energy. PEM Fuel Cells. Available online: https://www.intelligent-energy.com/ (accessed on 11 June 2023).
33. Intelligent-Energy. Cylinder Options for UAVs. Available online: https://www.intelligent-energy.com/ (accessed on 11 June 2023).
34. Hydrogen-Pro. High Pressure Alkaline Electrolysers. Available online: https://hydrogen-pro.com/solutions/ (accessed on 11 June 2023).
35. HYDROGENICS. Alkaline Water Electrolysis. Available online: https://www.accelerazero.com/ (accessed on 11 June 2023).
36. Green-Hydrogen-Systems. HyProvide™ A-Series. Available online: https://greenhydrogen.dk/wp-content/uploads/2021/02/A-Series-brochure-120421.pdf (accessed on 11 June 2023).
37. Fuel-Cell-Energy. SOFC SureSource Series. Available online: https://www.fuelcellenergy.com/ (accessed on 11 June 2023).
38. ErreDue-s.p.a. Mercury Hydrogen Generators. Available online: https://www.erreduegas.it/wp-content/uploads/brochure_mercury_2019.pdf (accessed on 11 June 2023).
39. ELUM-Energy. Product Catalogue. Available online: https://elum-energy.com/en/products-2/ (accessed on 11 June 2023).
40. Doosan-FC. PureCell® Model 400. Available online: https://www.doosanfuelcell.com/en/prod/prod-0101/ (accessed on 11 June 2023).
41. DOOSAN. Hydrogen Tank. Available online: https://www.doosanfuelcell.com/en/serv-0201/ (accessed on 11 June 2023).
42. Blue-World. Stationary System for Power Generation. Available online: https://www.blue.world/wp-content/uploads/2023/04/202304_Stationary-system_data-sheet_ENG-1.pdf (accessed on 11 June 2023).
43. Bloom-energy. Bloom Energy Server. Available online: https://www.bloomenergy.com/technology/ (accessed on 11 June 2023).
44. AREVA-H2Gen. Hydrogen Quality from PEM Electrolyzers. Available online: https://hydraite.eu/wp-content/uploads/2018/07/20180307_HYDRAITE_ws_AH2GEN-presentation_diff.pdf (accessed on 11 June 2023).
45. Sunfire-GmbH. The RWE Pilot Project. Available online: https://www.sunfire.de/en/news/detail/rwe-realizes-electrolysis-project-with-sunfire (accessed on 11 June 2023).
46. Siemens-Energy. Hydrogen Power Plants. Available online: https://www.siemens-energy.com/global/en/offerings/power-generation/power-plants/hydrogen-power-plants.html (accessed on 11 June 2023).
47. Salzgitter-AG. Green Industrial Hydrogen. Available online: https://salcos.salzgitter-ag.com/en/grinhy-20.html (accessed on 11 June 2023).
48. P2X-Solutions. 20 MW Electrolyzer Plant. Available online: https://www.sunfire.de/en/news/detail/rwe-realizes-electrolysis-project-with-sunfire (accessed on 11 June 2023).
49. Oil-India-Limited. India's First Pure Green Hydrogen Plant Commissioned in Jorhat. Available online: https://pib.gov.in/PressReleasePage.aspx?PRID=1818482 (accessed on 11 June 2023).
50. Hyflexpower-Consortium. First Tests for Power-to-Hydrogen-to-Power HYFLEXPOWER. Available online: https://press.siemens-energy.com/global/en/pressrelease/first-tests-power-hydrogen-power-hyflexpower-demonstrator-successfully-completed (accessed on 11 June 2023).
51. Hydrogen-Europe. Demonstration of 4MW Pressurized Alkaline Electrolyser for Grid Balancing Services (Demo4Grid). Available online: https://www.demo4grid.eu/ (accessed on 11 June 2023).

52. Gonernment-of-South-Australia. Off-Grid Projects. Available online: https://www.energymining.sa.gov.au/industry/modern-energy/off-grid-energy/off-grid-projects (accessed on 11 June 2023).
53. CSIRO. Hydrogen Map. Available online: https://www.csiro.au/en/maps/hydrogen-projects (accessed on 11 June 2023).
54. Australian-Government. Australia a Global First with Dual-Fuel Hydrogen Power Plant. Available online: https://www.austrade.gov.au/news/success-stories/australia-a-global-first-with-dual-fuel-hydrogen-power-plant (accessed on 11 June 2023).
55. ARENA. Knowledge Bank. Available online: https://arena.gov.au/knowledge-bank/?technology=hydrogen&page=3 (accessed on 11 June 2023).
56. IEA, International Energy Agency. *The Future of Hydrogen; Seizing Today's Opportunities*; IEA: Paris, France, 2019.
57. Goldmeer, J. *Power To Gas: Hydrogen For Power Generation*; GE Power: Schenectady, NY, USA, 2019.
58. Chisholm, G.; Kitson, P.J.; Kirkaldy, N.D.; Bloor, L.G.; Cronin, L. 3D printed flow plates for the electrolysis of water: An economic and adaptable approach to device manufacture. *Energy Environ. Sci.* **2014**, *7*, 3026–3032. [CrossRef]
59. Mohammadi, A.; Mehrpooya, M. A comprehensive review on coupling different types of electrolyzer to renewable energy sources. *Energy* **2018**, *158*, 632–655. [CrossRef]
60. Rizwan, M.; Alstad, V.; Jäschke, J. Design considerations for industrial water electrolyzer plants. *Int. J. Hydrogen Energy* **2021**, *46*, 37120–37136. [CrossRef]
61. IRENA (International Renewable Energy Agency). *Green Hydrogen Cost Reduction: Scale up Electrolysers to Meet the 1.5 °C Climate Goal*; International Renewable Energy Agency: Abu Dhabi, United Arab Emirates, 2020.
62. Silva, Y.S.K.D. Design of an Alkaline Electrolysis Stack. Master's Thesis, University of Agder, Kristiansand, Norway, 2017.
63. Chatenet, M.; Pollet, B.G.; Dekel, D.R.; Dionigi, F.; Deseure, J.; Millet, P.; Braatz, R.D.; Bazant, M.Z.; Eikerling, M.; Staffell, I.; et al. Water electrolysis: From textbook knowledge to the latest scientific strategies and industrial developments. *Chem. Soc. Rev.* **2022**, *51*, 4583–4762. [CrossRef]
64. Gayen, P.; Saha, S.; Liu, X.; Sharma, K.; Ramani, V.K. High-performance AEM unitized regenerative fuel cell using Pt-pyrochlore as bifunctional oxygen electrocatalyst. *Proc. Natl. Acad. Sci. USA* **2021**, *118*, e2107205118. [CrossRef]
65. Hodges, A.; Hoang, A.L.; Tsekouras, G.; Wagner, K.; Chong-Yong, L.; Swiegers, G.F.; Wallace, G.G. A high-performance capillary-fed electrolysis cell promises more cost-competitive renewable hydrogen. *Nat. Commun.* **2022**, *13*, 1304. [CrossRef]
66. Swiegers, G.; Hoang, A.L.; Owen, R.E.; Tsekouras, G.; Brett, D.J.L. Bubble Detection on the Cathode and Anode of a High-Performing Capillary-Fed Water Electrolysis Cell. *Sustain. Energy Fuels* **2023**, *7*, 4450–4460.
67. Kida, K.; Nicharee, W.; Chaianansutcharit, S.; Sato, K. Electrochemical performance and stability of $Ni_{1-x}Co_x$-based cermet anode for direct methane-fuelled solid oxide fuel cells. *MATEC Web Conf.* **2017**, *130*, 03005. [CrossRef]
68. Trinke, P.; Bensmann, B.; Hanke-Rauschenbach, R. Current density effect on hydrogen permeation in PEM water electrolyzers. *Int. J. Hydrogen Energy* **2017**, *42*, 14355–14366. [CrossRef]
69. Salehmin, M.N.I.; Husaini, T.; Goh, J.; Sulong, A.B. High-pressure PEM water electrolyser: A review on challenges and mitigation strategies towards green and low-cost hydrogen production. *Energy Convers. Manag.* **2022**, *268*, 115985. [CrossRef]
70. Hancke, R.; Holm, T.; Ulleberg, Ø. The case for high-pressure PEM water electrolysis. *Energy Convers. Manag.* **2022**, *261*, 115642. [CrossRef]
71. Reksten, A.H.; Thomassen, M.S.; Møller-Holst, S.; Sundseth, K. Projecting the future cost of PEM and alkaline water electrolysers; a CAPEX model including electrolyser plant size and technology development. *Int. J. Hydrogen Energy* **2022**, *47*, 38106–38113. [CrossRef]
72. Proost, J. State-of-the art CAPEX data for water electrolysers, and their impact on renewable hydrogen price settings. *Int. J. Hydrogen Energy* **2019**, *44*, 4406–4413. [CrossRef]
73. Christensen, A. *Assessment of Hydrogen Production Costs from Electrolysis: United States and Europe*; International Council on Clean Transportation: Washington, DC, USA, 2020.
74. Tang, E.; Wood, T.; Brown, C.; Casteel, M.; Pastula, M.; Richards, M.; Petri, R. *Solid Oxide Based Electrolysis and Stack Technology with Ultra-High Electrolysis Current Density (>3 A/cm2) and Efficiency*; Department of Energy (DOE): Washington, DC, USA, 2018.
75. Jamal, T.; Shafiullah, G.M.; Dawood, F.; Kaur, A.; Arif, M.T.; Pugazhendhi, R.; Elavarasan, R.M.; Ahmed, S.F. Fuelling the future: An in-depth review of recent trends, challenges and opportunities of hydrogen fuel cell for a sustainable hydrogen economy. *Energy Rep.* **2023**, *10*, 2103–2127. [CrossRef]
76. SynerHy. Technical Articles, Balance of Plant (BoP) of an Electrolyser. Available online: https://synerhy.com/en/category/technical-articles/ (accessed on 11 June 2023).
77. Dokhani, S.; Assadi, M.; Pollet, B.G. Techno-economic assessment of hydrogen production from seawater. *Int. J. Hydrogen Energy* **2022**, *48*, 9592–9608. [CrossRef]
78. Mancera, J.J.C.; Manzano, F.S.; Andújar, J.M.; Vivas, F.J.; Calderón, A.J. An Optimized Balance of Plant for a Medium-Size PEM Electrolyzer: Design, Control and Physical Implementation. *Electronics* **2020**, *9*, 871. [CrossRef]

79. Madsen, H.T. White Paper: Water Treatment for Green Hydrogen, Eurowater. Available online: https://www.eurowater.com/Admin/Public/DWSDownload.aspx?File=%2fFiles%2fFiles%2feurowater%2fCountry%2fInternational%2fLeaflets%2fWhite-paper_water-treatment-for-hydrogen_EUROWATER.pdf (accessed on 11 June 2023).
80. Marini, S.; Salvi, P.; Nelli, P.; Pesenti, R.; Villa, M.; Berrettoni, M.; Zangari, G.; Kiros, Y. Advanced alkaline water electrolysis. *Electrochim. Acta* **2012**, *82*, 384–391. [CrossRef]
81. Dicks, A.L.; Rand, D.A.J. *Fuel Cell Systems Explained*; John Wiley & Sons Ltd: Hoboken, NJ, USA, 2018; 460p.
82. Shaygan, M.; Ehyaei, M.A.; Ahmadi, A.; Assad, M.E.H.; Silveira, J.L. Energy, exergy, advanced exergy and economic analyses of hybrid polymer electrolyte membrane (PEM) fuel cell and photovoltaic cells to produce hydrogen and electricity. *J. Clean. Prod.* **2019**, *234*, 1082–1093. [CrossRef]
83. DOE. Fuel Cell Systems. Available online: https://www.energy.gov/eere/fuelcells/fuel-cell-systems#:~:text=The%20fuel%20cell%20stack%20is,is%20insufficient%20for%20most%20applications (accessed on 11 June 2023).
84. Mayyas, A.; Mann, M. Emerging Manufacturing Technologies for Fuel Cells and Electrolyzers. *Procedia Manuf.* **2019**, *33*, 508–515. [CrossRef]
85. Hydrogen, D. The Hydrogen Tools Portal. Available online: https://h2tools.org/ (accessed on 11 June 2023).
86. GenCell. Comparing Fuel Cell Technologies. Available online: https://www.gencellenergy.com/news/comparing-fuel-cell-technologies/ (accessed on 11 June 2023).
87. DOE. Comparison of Fuel Cell Technologies. Available online: https://www.energy.gov/eere/fuelcells/comparison-fuel-cell-technologies (accessed on 11 June 2023).
88. Ferriday, T.B.; Middleton, P.H. Alkaline fuel cell technology—A review. *Int. J. Hydrogen Energy* **2021**, *46*, 18489–18510. [CrossRef]
89. GÜLzow, E.; Schulze, M. 3—Alkaline fuel cells. In *Materials for Fuel Cells*; Gasik, M., Ed.; Woodhead Publishing: Cambridge, UK, 2008; pp. 64–100.
90. Lehmann, J.T.J. (Ed.) *Hydrogen and Fuel Cell Technologies and Market Perspectives*; Springer: Berlin/Heidelberg, Germany, 2016.
91. O'Hayre, R.; Cha, S.-W.; Colella, W.; Prinz, F.B. *Fuel Cell Fundamentals*; John Wiley & Sons Ltd: Hoboken, NJ, USA, 2016; pp. 117–166.
92. Yue, M.; Lambert, H.; Pahon, E.; Roche, R.; Jemei, S.; Hissel, D. Hydrogen energy systems: A critical review of technologies, applications, trends and challenges. *Renew. Sustain. Energy Rev.* **2021**, *146*, 111180. [CrossRef]
93. Daud, W.R.W.; Rosli, R.E.; Majlan, E.H.; Hamid, S.A.A.; Mohamed, R.; Husaini, T. PEM fuel cell system control: A review. *Renew. Energy* **2017**, *113*, 620–638. [CrossRef]
94. IEA. *Global Hydrogen Review 2023*; IEA: Paris, France, 2023.
95. Li, X. *Principles of Fuel Cells*, 1st ed.; CRC Press: Boca Raton, FL, USA, 2005.
96. Wang, H.; Li, H.; Yuan, X.-Z. (Eds.) *PEM Fuel Cell Durability Handbook*; CRC Press: Boca Raton, FL, USA, 2011.
97. Zhang, G.; Bao, Z.; Xie, B.; Wang, Y.; Jiao, K. Three-dimensional multi-phase simulation of PEM fuel cell considering the full morphology of metal foam flow field. *Int. J. Hydrogen Energy* **2021**, *46*, 2978–2989. [CrossRef]
98. Barbir, F. Fuel Cell Stack Design Principles with Some Design Concepts of Micro-Mini Fuel Cells. In *Mini-Micro Fuel Cells*; NATO Science for Peace and Security Series C: Environmental Security; Kakaç, S., Pramuanjaroenkij, A., Vasiliev, L., Eds.; Springer: Dordrecht, The Netherlands, 2008. [CrossRef]
99. Peng, F.; Mao, B.; Li, L.; Shang, Z. Development of online systematic condition assessment architecture for integrated PEMFC systems based on data-driven random matrix analysis. *Int. J. Hydrogen Energy* **2020**, *45*, 27675–27693. [CrossRef]
100. DOE (Battelle Memorial Institute). *Manufacturing Cost Analysis of PEM Fuel Cell Systems for 5- and 10-kW Backup Power Applications*; Battelle Memorial Institute: Washington, DC, USA, 2016.
101. DOE. *Fuel Cell Handbook*, 7th ed.; U.S. Department of Energy: Washington, DC, USA, 2004.
102. Bloom Energy Corporation. ES-5700 Energy Serve ES 5700 DataSheet. 2015. Available online: https://www.documentcloud.org/documents/2106918-bloom-energy-datasheet-es-5700/ (accessed on 11 June 2023).
103. Sasaki, K.; Yamabe, J.; Li, H.-W.; Ogura, T.; Hayashi, A.; Lyth, S.M. (Eds.) *Hydrogen Energy Engineering: A Japanese Perspective*; Springer: Tokyo, Japan, 2016.
104. IRENA (International Renewable Energy Agency). *World Energy Transitions Outlook 2022: 1.5 °C Pathway*; International Renewable Energy Agency: Abu Dhabi, United Arab Emirates, 2022.
105. Nechache, A.; Hody, S. Alternative and innovative solid oxide electrolysis cell materials: A short review. *Renew. Sustain. Energy Rev.* **2021**, *149*, 111322. [CrossRef]
106. Laurencin, J.; Kane, D.; Delette, G.; Deseure, J.; Lefebvre-Joud, F. Modelling of solid oxide steam electrolyser: Impact of the operating conditions on hydrogen production. *J. Power Sources* **2011**, *196*, 2080–2093. [CrossRef]
107. Tawalbeh, M.; Alarab, S.; Al-Othman, A.; Javed, R.M. The Operating Parameters, Structural Composition, and Fuel Sustainability Aspects of PEM Fuel Cells: A Mini Review. *Fuels* **2022**, *3*, 449–474. [CrossRef]

108. Coppola, C.M.; Tolbatov, I.; Tranca, I.C.; Coletti, C.; Marrone, A.; Storchi, L.; Profio, P.D.; Re, N.; Kazandjian, M.V.; Pellecchia, A.; et al. A database approach for materials selection for hydrogen storage in aerospace technology. *Rend. Fis. Acc. Lincei* **2019**, *30*, 287–296. [CrossRef]
109. Stetson, N.T.; McWhorter, S.; Ahn, C.C. 1—Introduction to hydrogen storage. In *Compendium of Hydrogen Energy*; Gupta, R.B., Basile, A., Veziroğlu, T.N., Eds.; Woodhead Publishing: Cambridge, UK, 2016; pp. 3–25.
110. Basile, A.; Dalena, F.; Tong, J.; Veziroglu, T.N. *Hydrogen Production, Separation and Purification for Energy*; Institution of Engineering & Technology: Stevenage, UK, 2017.
111. Sdanghi, G.; Maranzana, G.; Celzard, A.; Fierro, V. Review of the current technologies and performances of hydrogen compression for stationary and automotive applications. *Renew. Sustain. Energy Rev.* **2019**, *102*, 150–170. [CrossRef]
112. Viswanathan, B. Chapter 10—Hydrogen Storage. In *Energy Sources*; Viswanathan, B., Ed.; Elsevier: Amsterdam, The Netherlands, 2017; pp. 185–212.
113. Zhang, S.; Feng, Y.; Zhang, D.; Jiang, Y.; Qin, J.; Bao, W. Parametric numerical analysis of regenerative cooling in hydrogen fueled scramjet engines. *Int. J. Hydrogen Energy* **2016**, *41*, 10942–10960. [CrossRef]
114. Aziz, M. Liquid Hydrogen: A Review on Liquefaction, Storage, Transportation, and Safety. *Energies* **2021**, *14*, 5917. [CrossRef]
115. Tan, X.F.; Kim, M.; Yasuda, K.; Nogita, K. Strategies to enhance hydrogen storage performances in bulk Mg-based hydrides. *J. Mater. Sci. Technol.* **2023**, *153*, 139–158. [CrossRef]
116. Berstad, D.; Gardarsdottir, S.; Roussanaly, S.; Voldsund, M.; Ishimoto, Y.; Nekså, P. Liquid hydrogen as prospective energy carrier: A brief review and discussion of underlying assumptions applied in value chain analysis. *Renew. Sustain. Energy Rev.* **2022**, *154*, 111772. [CrossRef]
117. Chehade, G.; Lytle, S.; Ishaq, H.; Dincer, I. Hydrogen production by microwave based plasma dissociation of water. *Fuel* **2020**, *264*, 116831. [CrossRef]
118. Hansen, O.R. Liquid hydrogen releases show dense gas behavior. *Int. J. Hydrogen Energy* **2020**, *45*, 1343–1358. [CrossRef]
119. Rivkin, C.; Burgess, R.; Buttner, W. *Hydrogen Technologies Safety Guide*; National Renewable Energy Laboratory (NREL): Golden, CO, USA, 2015.
120. Buckley, C.E.; Chen, P.; van Hassel, B.A.; Hirscher, M. Hydrogen-based Energy Storage (IEA-HIA Task 32). *Appl. Phys. A* **2016**, *122*, 141. [CrossRef]
121. Hirscher, M.; Yartys, V.A.; Baricco, M.; Bellosta von Colbe, J.; Blanchard, D.; Bowman, R.C.; Broom, D.P.; Buckley, C.E.; Chang, F.; Chen, P.; et al. Materials for hydrogen-based energy storage—Past, recent progress and future outlook. *J. Alloys Compd.* **2020**, *827*, 153548. [CrossRef]
122. Nguyen, H.Q.; Shabani, B. Review of metal hydride hydrogen storage thermal management for use in the fuel cell systems. *Int. J. Hydrogen Energy* **2021**, *46*, 31699–31726. [CrossRef]
123. Tong, L.; Yuan, C.; Yang, T.; Yuan, Y.; Chahine, R.; Xiao, J. Thermal management of metal hydride hydrogen storage tank coupled with proton exchange membrane fuel cells. *Case Stud. Therm. Eng.* **2023**, *43*, 102812. [CrossRef]
124. Adametz, P.; Müller, K.; Arlt, W. Energetic evaluation of hydrogen storage in metal hydrides. *Int. J. Energy Res.* **2016**, *40*, 1820–1831. [CrossRef]
125. DOE. Materials-Based Hydrogen Storage. Available online: https://www.energy.gov/eere/fuelcells/materials-based-hydrogen-storage (accessed on 11 June 2023).
126. European-Commission. In Focus: Renewable Hydrogen to Decarbonise the EU's Energy System. Available online: https://commission.europa.eu/news/focus-renewable-hydrogen-decarbonise-eus-energy-system-2022-11-15-0_en (accessed on 11 June 2023).
127. Australian-Government. Australia's First Large Scale Hydrogen Plant to Be Built in Pilbara. Available online: https://www.energy.gov.au/news-media/news/australias-first-large-scale-hydrogen-plant-be-built-pilbara (accessed on 11 June 2023).
128. LAVO Website. Available online: https://www.lavo.com.au/ (accessed on 11 June 2023).
129. GreenHy2 Technology, Solid State and Low Pressure Hydrogen Storage. Available online: https://www.greenhy2.com.au/solid-state-hydrogen-storage (accessed on 11 June 2023).

MDPI AG
Grosspeteranlage 5
4052 Basel
Switzerland
Tel.: +41 61 683 77 34

Hydrogen Editorial Office
E-mail: hydrogen@mdpi.com
www.mdpi.com/journal/hydrogen

www.ingramcontent.com/pod-product-compliance
Lightning Source LLC
Chambersburg PA
CBHW061813210926
51850CB00006B/693